T0317592

Grain Boundary Engineering in Ceramics

—From Grain Boundary Phenomena
to Grain Boundary Quantum Structures

Related titles published by The American Ceramic Society:

The Magic of Ceramics
By David W. Richerson
©2000, ISBN 1-57498-050-5

Advances in Ceramic Matrix Composites V (Ceramic Transactions Volume 103)
Edited by Narottam P. Bansal, J.P. Singh, and Ersan Ustundag
©2000, ISBN 1-57498-089-0

Ceramic Innovations in the 20th Century
Edited by John B. Wachtman Jr.
©1999, ISBN 1-57498-093-9

Ceramic Material Systems with Composite Structures: Towards Optimum Interface Control (Ceramic Transactions Volume 99)
Edited by Nobuo Takeda, Laurel M. Sheppard, Jun-ichi Kon
©1998, ISBN 1-57498-065-3

Innovative Processing and Synthesis of Advanced Ceramics, Glasses, and Composites (Ceramic Transactions Volume 85)
Edited by Narottam P. Bansal and J.P. Singh
©1998, ISBN 1-57498-030-0

Mass and Charge Transport in Ceramics (Ceramic Transactions Volume 71)
Edited by Kunihito Koumoto, Laurel M. Sheppard, and Hideaki Matsubara
©1996, ISBN 1-57498-018-1

Materials Processing and Design (Ceramic Transactions Volume 44)
Edited by Koichi Niihara, Kozo Ishizaki, and Mitsui Isotani
©1994, ISBN 0-944904-78-5

For information on ordering titles published by The American Ceramic Society, or to request a publications catalog, please contact our Customer Service Department at 614-794-5890 (phone), 614-794-5892 (fax),<customersrvc@acers.org> (e-mail), or write to Customer Service Department, 735 Ceramic Place, Westerville, OH 43081, USA.

Visit our on-line book catalog at <www.ceramics.org>.

Ceramic Transactions
Volume 118

Grain Boundary Engineering in Ceramics

—From Grain Boundary Phenomena to Grain Boundary Quantum Structures

Proceedings of the Grain Boundary Engineering in Ceramics—From Grain Boundary Phenomena to Grain Boundary Quantum Structures Japan Fine Ceramics Center Workshop, March 15–17, 2000, in Nagoya, Japan.

Edited by

Taketo Sakuma
University of Tokyo

Laurel M. Sheppard
Lash Publications International

Yuichi Ikuhara
University of Tokyo

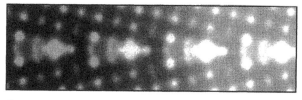

Published by
The American Ceramic Society
735 Ceramic Place
Westerville, Ohio 43081

Proceedings of the Grain Boundary Engineering in Ceramics—From Grain Boundary Phenomena to Grain Boundary Quantum Structures Japan Fine Ceramics Center Workshop, March 15–17, 2000, in Nagoya, Japan.

Copyright 2000, The American Ceramic Society. All rights reserved.

Statements of fact and opinion are the responsibility of the authors alone and do not imply an opinion on the part of the officers, staff, or members of The American Ceramic Society. The American Ceramic Society assumes no responsibility for the statements and opinions advanced by the contributors to its publications or by the speakers at its programs. Registered names and trademarks, etc., used in this publication, even without specific indication thereof, are not to be considered unprotected by the law.

No part of this book may be reproduced, stored in a retrieval system, or transmitted in any form or by any means, electronic, mechanical, photocopying, microfilming, recording, or otherwise, without written permission from the publisher.

Authorization to photocopy for internal or personal use beyond the limits of Sections 107 and 108 of the U.S. Copyright Law is granted by the American Ceramic Society, provided that the appropriate fee is paid directly to the Copyright Clearance Center, Inc., 222 Rosewood Drive, Danvers, MA 01923 USA, www.copyright.com. Prior to photocopying items for educational classroom use, please contact Copyright Clearance Center, Inc.

This consent does not extend to copying items for general distribution or for advertising or promotional purposes or to republishing items in whole or in part in any work in any format.

Please direct republication or special copying permission requests to Copyright Clearance Center, Inc., 222 Rosewood Drive, Danvers, MA 01923 USA 978-750-8400; www.copyright.com.

Cover photo is courtesy of the Japan Fine Ceramic Center.

Library of Congress Cataloging-in-Publication Data
A CIP record for this book is available from the Library of Congress.

For information on ordering titles published by The American Ceramic Society, or to request a publications catalog, please call 614-794-5890.

4 3 2 1–03 02 01 00

ISSN 1042-1122
ISBN 1-57498-115-3

Contents

Diffusion and Transformation

Electronic Ceramics

Structural Ceramics

Intergranular Film

Hereto Interface and Others

Preface

This volume contains the selected papers presented at the 10th international workshop sponsored by the Japan Fine Ceramics Center, "Grain Boundary Engineering in Ceramics—From Grain Boundary Phenomena to Grain Boundary Quantum Structures"—which was held on March 15–17, 2000, in Nagoya, Japan. This workshop is held every two years on timely topics of ceramic materials. Past workshops discussed mass and charge transport, as well as composite structures.

Since most advanced ceramics are fabricated by sintering raw powders, grain boundaries play a key role in sintering behavior and affect various properties. Therefore, grain boundary-related subjects were often discussed in past JFCC workshops, as well as at this meeting. However, this meeting discussed grain boundaries in ceramics not only from conventional approaches, but also from novel approaches, including microscopic analysis of local atomic bonding or even quantum structures of grain boundaries. This type of microscopic analysis will enable us to design new ceramic materials in the near future. Recent data has shown that various properties in ceramics can be controlled by the presence of small dopant ions in grain boundaries, which supports this prediction.

Fortunately, a number of top-level scientists came to join this meeting from around the world, making the workshop an exciting event with many lively discussions. The high-quality papers compiled in this proceedings are evidence that leading experts attended. The editors and the Japan Fine Ceramics Center hope that this volume will be a guide to designing new ceramic materials in the future.

This workshop was subsidized by the Japan Keirin Association with promotion funds from KEIRIN RACE.

Taketo Sakuma

Laurel M. Sheppard

Yuichi Ikuhara

KEIRIN

JFCC International Workshop on Fine Ceramics 2000 Organizing Committee

Board Chairman

Masaaki Ohashi (JFCC)

Board Sub-Chairman

Emeritus Prof. Hiroaki Yanagida (JFCC)

Chairman

Prof. Taketo Sakuma (University of Tokyo)

Members

Tsukasa Hirayama (JFCC)

Yuichi Ikuhara (JFCC/University of Tokyo)

Jun-ichi Kon (JFCC)

Michiko Kusunoki (JFCC)

Hideaki Matsubara (JFCC)

Yasusi Matsuo (NGK/NTK)

Hiroaki Sakai (NGK)

Isao Tanaka (Kyoto University)

Yoshio Ukyo (Toyota Central R&D Lab, Inc.)

Yoshiyuki Yasutomi (JFCC)

Secretariat:

Mikio Ishikawa (JFCC)

Mari Yamada (JFCC)

Speakers

T. Sakuma

L.M. Sheppard

Y. Ikuhara

J.M. Albuquerque

P.F. Becher

R.M. Cannon

C.B. Carter

Y.M. Chiang

W.Y. Ching

S.D. de la Torre

F. Dogan

Y. Enomoto

F. Ernst

C.A.J. Fisher

K. Hayashi

T. Hirayama

N. Hirosaki

Y. Inagaki

K. Kaneko

H. Kawasaki

V.V. Kiseliev

S. Kitaoka

M. Kohyama

M. Kusunoki

J. Li

X.L. Ma

H. Matsubara

K. Matsunaga

M. Mitomo

N. Mizutani

R. Monzen

H. Murotani

F. Oba

S. Ogata

C.W. Park

xiv

G. Pezzotti

M. Rühle

T. Saito

S. Sakaguchi

J. Shibata

N. Shibata

D.J. Srolovitz

Y. Sugawara

A.P. Sutton

T. Suzuki

Y. Suzuki

Y. Takigawa

I. Tanaka

K. Tanaka

S. Tanaka

H. Tsubakino

K. Tsurata

Y. Ukyo

Y. Waku

Tadao Watanabe

T. Watanabe

K. Watari

D.S. Wilkinson

R. Xie

T. Yamamoto

J. Yang

Y. Yasutomi

D.Y. Yoon

H. Yoshida

M. Yoshiya

T. Yukiko

G.-D. Zhan

J. Zhu

General, Theory, and Simulation

Grain boundaries in metals:
Current understanding and some future directions

A P Sutton

Department of Materials, Oxford University, OX1 3PH, UK.

Abstract

After a survey of the history of the thinking about grain boundaries in metals over the past 50 years, the need for simple analytic models to provide understanding is stressed. The links between grain boundary geometry and energy are now well understood, and the reasons for the failures of almost all early geometric criteria for low interfacial energy are known. The influence of periodicity, in one or two dimensions in the boundary plane, on grain boundary energy has been explained rigorously. The influences of the boundary plane and the boundary expansion are revealed clearly through a simple model that enables 4 of the 5 dimensions of the geometrical parameter space of grain boundaries to be explored in far greater detail than will ever be possible by computer simulation. The boundary plane and expansion are shown to affect directly the cleavage energy rather than the boundary energy itself. These results have been encapsulated in a set of simple (and approximate) formulae for the energies of all grain boundaries in metals. They also form the basis of a classification of interfaces that is consistent with many observed grain boundary properties. With increasing temperature the free energies of boundaries are shown to fall within a smaller range. This is brought about by the dependence of the excess vibrational entropy on the boundary expansion, and by the diminishing influence of periodicity on grain boundary energy at higher temperatures.

In the future more work is likely to be done on the influence of grain boundaries on the properties of polycrystals. Watanabe's concept of the grain boundary character distribution is quantified. Functional forms for distributions of grain boundary misorientations and plane normals, that are suitable for fitting to experimental data, are derived. These distributions enable quantitative comparisons between grain boundary character distributions in polycrystals to be made. This will enable quantitative assessments to be made of the success of thermo-mechanical processing treatments designed to achieve particular grain boundary character distributions.

To the extent authorized under the laws of the United States of America, all copyright interests in this publication are the property of The American Ceramic Society. Any duplication, reproduction, or republication of this publication or any part thereof, without the express written consent of The American Ceramic Society or fee paid to the Copyright Clearance Center, is prohibited.

1. Introduction
Grain boundaries in metals have attracted interest throughout the second half of the 20th century because of their influence on mechanical properties and on microstructural evolution in metallic alloys. The widespread belief that a knowledge of the structure of interfaces enables one to modify their properties in a predictive and controlled manner, has led to sustained efforts in observing and modelling the atomic structures of grain boundaries. In the past 10-20 years we have seen atomic resolution imaging of low index tilt grain boundaries and computer simulations of a wider range of grain boundaries. Density functional theory has enabled the structures of certain grain boundaries to be predicted from first principles, i.e. the Schrödinger equation for the solid containing the grain boundary has been solved within the (excellent) local density approximation. The trend in recent years has been to use ever more sophisticated experimental and theoretical tools to determine atomic structures in ever greater detail and accuracy. In this paper I shall argue that the ability of such work to meet the original objective of understanding the relationships between structures and properties of grain boundaries is limited: the geometrical parameter space of grain boundaries is 5 dimensional and far too large to sample with such methods. I shall argue that simple analytic models, some of which can be solved with very low technology (a pencil and paper), are needed to meet this objective. The very detailed experimental and theoretical work on individual interfaces provides valuable checks on these simple models. By sampling more of the five dimensional geometrical parameter space associated with grain boundaries these simple models have enabled a classification of interfaces to be introduced, which relates directly to a wide range of observable properties. This is one of the main themes of ref.[1].

In section 4 the nature of the 5 dimensional parameter space of grain boundaries is considered in greater depth. Individual boundaries are represented by points in this space, but how can a distribution of boundaries in the space (as occurs in a polycrystal) be defined? Specific answers to this question are put forward in section 4.2. The section also includes a brief discussion of the relation between such distributions and properties of polycrystals.

The paper is intended to be tutorial in nature. For a deeper understanding of the subject the reader is referred to [1], although sections 2.4, 3.3 and 4.2 contain new material that does not appear in [1].

1.1 Brief history, spanning 50 years
Geometry and crystallography have played a central role in much of the thinking about grain boundary structure over the past 50 years. Several early experiments established that grain boundaries differ and have properties (and hence structures) that vary with their geometrical parameters. When one surveys the kinetic, mechanical and electrical properties of grain boundaries one is indeed struck by the diversity of their properties. Solute segregation, self-diffusion, mobility during migration, the source/sink efficiency for point defects, sliding, slip transmission, the propensity to cleavage, the resistance to electron current flow, and the extent to which they act as weak links in high T_c superconductors, are all examples of phenomena that occur at grain boundaries to markedly varying degrees from one boundary to another in the same material [1].

In an elemental metal, like Cu or Fe, there are 5 macroscopic geometrical degrees of freedom associated with a grain boundary. They may be specified in a variety of equivalent ways. Historically, the most popular has been to specify the orientation relationship, i.e. the rotation matrix, relating the two crystals (3 degrees of freedom), and the boundary plane in one crystal (2 degrees of freedom). The boundary plane in the other crystal is then automatically specified. Another specification involves the concept of the mean boundary plane, which is particularly useful for analysing experiments on tilt boundaries [2]. One can also view the creation of a boundary as the bringing together of two free surfaces (4 degrees of freedom) followed by a rotation about their common normal (1 degree of freedom). This last specification is useful for analysing cleavage energies theoretically.

The coincidence site lattice (CSL) was introduced to explain why certain grain boundaries were observed to have 'special' properties, that is, sufficiently different properties from other boundaries nearby in the geometrical parameter space. In general, when the lattices of both crystals are allowed to interpenetrate there are no common sites, except possibly at the origin. But at certain orientation relationships, i.e. certain rotation axes and angles, a three dimensional lattice of common or 'coincident' sites appears. A grain boundary can be constructed by selecting a plane in the interpenetrating lattices and discarding one crystal lattice on one side of the plane and the other crystal lattice on the other side. If the boundary plane is a rational plane of the CSL then the boundary will be periodic, with a periodicity determined by the corresponding 2D section of the CSL. The ratio of the volume of the unit cell of the CSL to the volume of the unit cell of the crystal lattice is designated Σ. There are still some who believe that grain boundary properties are correlated with Σ. Here are three arguments to encourage them to abandon that belief:

- On physical grounds it is unreasonable to expect the properties of an interface to vary discontinuously with geometrical parameters such as the rotation angle and axis. But Σ does change discontinuously with rotation angle and axis.
- In a given coincidence system there is an infinite number of grain boundaries, which are obtained by choosing the cut plane on an infinite number of possible planes in the CSL. All these boundaries are associated with the same value of Σ, and yet it is unreasonable physically to expect them all to have similar properties. For example, faceting would not arise if all boundaries in the same coincidence systems had the same energy.
- Some values of Σ are not uniquely associated with an orientation relation, e.g. there are two distinct CSLs with $\Sigma=13$, two with $\Sigma=17$, two with $\Sigma=19$, two with $\Sigma=21$ and so on. There is no reason to expect the boundaries from different CSLs, but sharing the same value of Σ to have similar properties.

The concept of the CSL arose out of studies of grain boundaries in cubic systems. But in non-cubic systems the occurrence of CSLs is much rarer. Some have tried to overcome the problem by allowing some flexibility in the crystal lattice parameters. But the requirement of *three*-dimensional periodicity has never been justified by any physical argument. All one can truthfully say about the CSL is that its existence leads to boundaries with periodic structures. Sound physical arguments *have* been given for the significance of periodicity in the boundary plane, and we shall review them in section 2.2. But, it is *not* necessarily true that a boundary that is periodic in 1 or 2 dimensions is a member of a *three* dimensional CSL. In conclusion, we cannot attribute any physical significance to whether or not a boundary is associated with a 3 dimensional CSL, and certainly not to Σ.

The O-lattice was introduced [3] to deal with some of the perceived limitations of the CSL. An O-point is a generalisation of a coincidence of crystal lattice sites to a coincidence of points occupying equivalent positions within unit cells of the two crystal lattices. Unlike the CSL, the O-lattice does vary continuously with misorientation parameters. But, like the CSL, the O-lattice has no real physical significance, although it can be mapped directly onto the formal description of interfaces in terms of dislocations developed by Frank and Bilby [4].

The most comprehensive and rigorous treatment of the crystallography of interfaces has been given by Pond and coworkers [5]. Their treatment, which has become known as 'bicrystallography', is based on a systematic analysis of the symmetry of a bicrystal. Starting with the interpenetrating lattices of the two adjoining crystals one goes through a sequence of steps to create a relaxed bicrystal. At each step the symmetry is lowered and a set of variants is produced.

The variants are related by the symmetry elements that are lost or 'suppressed' in the lowering of the symmetry. These symmetry elements enable line defects that could exist at an interface separating energetically degenerate domains to be enumerated systematically. In this way one arrives at a complete classification of interfacial line defects that conserve the symmetry of an interface.

For example, before the sectioning of the interpenetrating lattices to create a bicrystal we can translate each crystal lattice separately by any of its own translation vectors. Such a composite operation will not change the interpenetrating lattices, and two domains of the interface related by such a composite operation will be identical, and separated by a dislocation with a Burgers vector equal to the sum of the two crystal lattice translation vectors. If the two crystal lattice translation vectors almost cancel then the Burgers vector will be small. The set of all possible sums of lattice translation vectors from both crystal lattices is more commonly known as the DSC lattice. Pond and Bollmann [6] gave the description 'displacements that are symmetry conserving' to DSC. In this analysis the only significance of the existence of a CSL is that the DSC lattice has a smallest lattice vector and it is periodic. Otherwise, in the absence of a CSL, there is no smallest DSC lattice vector, and the DSC lattice is quasiperiodic [7].

The analysis of symmetry and interfacial defects by Pond and coworkers is based on the principle of symmetry compensation: 'if symmetry is destroyed at one structural level it arises and is preserved at another' [8]. The mathematical basis of this geometrical principle is Lagrange's theorem in group theory. The sense in which symmetry is preserved in the context of interfaces is through the existence of variants and interfacial defects that separate them. Bicrystallography has found widespread applications in both homophase and heterophase interfaces, often explaining the existence and nature of observed line defects that could not be explained by other means.

Finally in this section let us consider the concept of long-range order at a grain boundary. We begin by asking the question of whether a flat grain boundary in a pure metal can ever be considered to be amorphous or 'liquid-like'. All the available evidence indicates that grain boundaries in pure metals are no more than a few atoms wide. That is less than the distance over which modulations in the density of a liquid metal decay at a solid-liquid metal interface [9]. Therefore, the boundary region is always strongly influenced in a direction normal to the boundary plane by the order in the adjoining crystals. It is inaccurate to describe the boundary region as liquid-like or amorphous because this ignores the ordering influence normal to the boundary plane of the adjoining crystals. The question of the degree of order parallel to the boundary plane is much more difficult to treat in general terms. But it is clear that at some boundaries it is likely to be dependent on temperature, and the reasons for this will be discussed in section 3.3. At very low temperatures even general boundaries are ordered parallel to the boundary plane in the sense that the pair correlation function does not decay. This is a consequence of the quasiperiodic order at such boundaries, where we are using the word 'quasiperiodic' in the mathematical sense of quasiperiodic functions [10], and not in the perhaps more widely used sense of something that lacks periodicity but which does not look amorphous!

1.2 The need for simple models.
Suppose we span the whole 5 dimensional geometrical parameter space with a grid at 2° intervals. There would be approximately 6×10^9 grid points. This is the number of grain boundaries one would have to observe experimentally or model by computer simulation if one were to sample the whole five dimensional space at 2° intervals. Modern high resolution electron microscopes can provide atomic resolution images of low index tilt boundaries, i.e. <110> and <001> tilt boundaries in f.c.c. crystals. The set of all <110> and <001> tilt boundaries separated by 2° intervals consists of 10,125 boundaries, which is just 2×10^{-6} of all boundaries in the 5 dimensional parameter space spanned by 2° intervals. The sum total of all boundaries observed by

high resolution electron microscopy (HREM) or modelled by computer simulation is of the order of a few thousand (at most) i.e. of order 10^{-6} of all possible boundaries in the 5D parameter space spanned by 2° intervals. This is the very limited extent to which the results of all computer simulations and HREM experiments to date are representative of the structures and energies of all possible boundaries in a material. The fraction becomes even smaller when we admit other variables such as temperature and impurity content.

Simple models that predict the variations and trends of grain boundary energies in a physically transparent and defensible manner are needed to explore much more of this 5D parameter space than is possible through experiments and computer simulations. However, experiments and detailed simulations are needed to check the results of the simple models.

2. Geometrical parameters and grain boundary energy

2.1 Introduction
There have been several attempts to identify geometrical parameters that may be used to pick out relatively low energy interfaces in the 5D parameter space. In section 1.1 we gave 3 reasons why a small value of Σ fails as a criterion for low grain boundary energy. There have been other criteria, most of which have been abject failures when compared with experimental results [11]. But it must be true that grain boundary structure and energy are dependent on the 5 macroscopic degrees of freedom. The difficulty is to identify the geometrical parameters that can be related in a *sound physical manner* to energy. This challenge has been met in the past 10 years, and it has formed the basis of a classification of interfaces [12] that rationalises a huge amount of experimental data on grain boundary properties.

2.2 The role of periodicity in the boundary plane
Ever since the discovery of the CSL, periodicity in the boundary plane has been associated with relatively low energy in the minds of many researchers. To my knowledge Fletcher and Adamson [13] were the first to present a theory that explained why there is an association between periodicity and energy. Like all good ideas it has been rediscovered many times, and it continued to be rediscovered in the 1990s. The outline given here is based on the detailed treatment given in [14].

Consider the energy of interaction between two semi-infinite crystals meeting at a common interface, in which there is 1 or 2 dimensional periodicity. Regardless of the description of atomic interactions, or of the atomic relaxation at the interface (provided it does not destroy the periodicity), it is always possible to express the *interaction energy* between the 2 semi-infinite crystals as a Fourier expansion:

$$\Xi(\tau,e) = \sum_{\mathbf{G}^c} v\,(\mathbf{G}^c,e)\ e^{i\mathbf{G}^c\cdot\tau} \qquad (1)$$

The expansion at the boundary (i.e. normal to the boundary plane) is denoted by e. Each crystal presents a lattice plane at the boundary. A 2D reciprocal lattice is constructed for each of these direct lattice planes. The vectors $\mathbf{G}^c$ are common members of both of these 2D reciprocal lattices. The vector τ is the rigid-body translation of one semi-infinite crystal, relative to the other, parallel to the boundary plane. The 1 or 2 dimensional direct lattice that is reciprocal to the lattice of common reciprocal lattice vectors $\mathbf{G}^c$ has lattice vectors $\mathbf{X}$. The Wigner-Seitz cell formed by the set of vectors $\{\mathbf{X}\}$ is called the <u>cell of non-identical displacements</u> or <u>c.n.i.d.</u> Any vector τ lying outside the c.n.i.d. is equivalent to a vector within it. Let the area of the c.n.i.d. be A_c. The Fourier coefficients, $v(G^c, e)$, depend on the relaxation and the nature of the atomic interactions and they are given by:

$$v(G^c, e) = \frac{1}{A_c} \int_{A_c} \Xi(\tau, e)\ e^{-iG^c \cdot \tau} d^2\tau \qquad (2)$$

where the integral is taken over the c.n.i.d.

Equations (1) and (2) are already quite revealing. The existence of periodicity at the interface leads immediately to the conclusion that the boundary energy is a periodic function of the rigid body displacement, τ. But the periodicity of the energy is *not* necessarily the periodicity of the boundary structure, but the periodicity of the c.n.i.d. How can this possibly be? Consider the $\Sigma=5$ (001) twist boundary in an f.c.c. crystal. The boundary is two dimensionally periodic with primitive repeat vectors 1/2[310] and 1/2[3-10]. But for this boundary it is well known that the energy is a periodic function of the rigid body displacement τ within the Wigner-Seitz cell of the DSC lattice, defined by 1/10[310] and 1/10[3-10]. This Wigner-Seitz cell is the c.n.i.d. in this case. More generally, *the c.n.i.d. is defined by the Wigner-Seitz cell of the DSC lattice vectors in the boundary plane that do not result in a relocation of the boundary plane.* In some cases these DSC lattice vectors are the repeat vectors of the boundary structure. For example at the $\Sigma=5(310)$ symmetric tilt boundary the repeat vectors of the boundary structure are [001] and 1/2[3-10]. These vectors are also the smallest DSC lattice vectors that preserve the location of the boundary plane. They also define the c.n.i.d.

Notice that eqns.(1) and (2) do not rely on the existence of a three dimensional CSL: they rely only on the existence of 1 or 2 dimensional periodicity *in the boundary plane*.

So far, all we have established is the geometrical structure of the interaction energy $\Xi(\tau, e)$. To make further progress we have to insert a physical model for the atomic interactions. Following [14] we consider a pair interaction model, in which the energy of interaction between any two atoms is $\phi(x, z)$, where x is parallel to the boundary and z is their separation normal to the boundary. Consider the energy, E, of interaction, per unit area, between two unrelaxed lattice planes, one from each crystal, parallel to the boundary plane. We assume that each lattice site has an atom sitting at it, and that there are no other atoms in the basis. In ref.[14] the interaction energy is shown to be

$$E(\tau, z) = \frac{1}{A_c^b A_c^w} \sum_{G^c} f(G^c, z)\ e^{iG^c \cdot \tau} \ . \qquad (3)$$

A_c^b and A_c^w are the areas of the 2D repeat cells of the crystal lattice planes parallel to the boundary. The sum is taken over the same set of common reciprocal lattice vectors as in eqn.(1) above. The rigid body displacement, τ, is again defined uniquely within the c.n.i.d. The separation of the two lattice planes is z. The function $f(q, z)$ is the two dimensional Fourier transform of the pair potential:

$$f(q, z) = \int \phi(x, z)\ e^{-iq \cdot x} d^2 x \ , \qquad (4)$$

where the integral extends over all 2D space parallel to the boundary plane.

Again we see that the interaction energy is a periodic function of τ, with the periodicity defined by the c.n.i.d. If there are no matching reciprocal lattice vectors then only $G^c = 0$ contributes to the sum in eqn.(3), and the interaction energy is independent of τ, as should be the case for a quasiperiodic interface. But when periodicity is present the interaction energy oscillates with τ, with an amplitude determined by the Fourier transform $f(\mathbf{q},z)$.

For a Lennard-Jones potential of the form

$$V(r) = \varepsilon \left(\left(\frac{r_0}{r} \right)^{12} - 2\left(\frac{r_0}{r} \right)^6 \right) , \qquad (5)$$

which has a minimum at $r=r_0$ of depth ε, the 2D Fourier transform, $f(\mathbf{q},z)$, decays exponentially with $|\mathbf{q}|z$ at large values of $|\mathbf{q}|z$ (see ref.[14]). This indicates that

as the wavelength of the modulation of the boundary energy with rigid body displacement, τ, tends to zero, the amplitude of the modulation tends to zero exponentially.

Only those boundaries with relatively large c.n.i.d.s, and hence small $|\mathbf{G}^c|$, will show significant variations of the boundary energy with rigid body displacement τ.

There is considerable experimental evidence in support of the above analysis. Sutton and Balluffi [11] reviewed the results of rotating ball experiments in metals and ionic crystals where small single crystal balls are placed on a single crystal substrate and allowed to rotate during high temperature annealing to low energy misorientations. While the boundary plane is fixed on the substrate side, it is completely free on the ball side. For any given plane on the ball side there is an infinite number of possible misorientations, because the ball may rotate about an axis normal to the substrate without changing the boundary plane. It was found that for each plane selected on the ball side, the observed misorientations always corresponded to those with the largest c.n.i.d. In some cases not only the largest c.n.i.d. misorientation was seen but also the two or three next largest. The rationalisation for this observation is that the largest c.n.i.d. (hence smallest $|\mathbf{G}^c|$) on a given boundary plane is associated with the largest variation in the boundary energy as a function of τ. Thus the boundary can attain the lowest energy state on a given plane at these orientations by adopting a suitable value of τ. In a few cases, balls also became trapped at smaller local minima where the c.n.i.d. was relatively large, but not the largest available on a given plane. As discussed in ref.[15] there is also considerable support for the above analysis from computer simulations.

The energy lowering that occurs when there are matching $\mathbf{G}$ vectors in the boundary plane is the driving force for the introduction of screw dislocations at twist boundaries with misorientations that deviate slightly from an orientation where there are exactly matching $\mathbf{G}$ vectors. However, it is clear that the energy lowering decreases exponentially with the size of $|\mathbf{G}|$, and unless the periodic cell of the nearby lower energy twist boundary is relatively small, this driving force will not be large. Nevertheless, screw dislocations have been observed at manufactured (001) twist boundaries up to $\Sigma=29$ [16], where the smallest $\mathbf{G}^c$ vectors are $2\pi/a<730>$. Similarly, boundaries in which $\mathbf{G}^c$ vectors exist along only one direction in the interface have also been found experimentally to have relatively low energy [17] . These boundaries are among those that have been called 'plane matching' or 'coincident axial direction' boundaries. In this case the driving force for the introduction of the observed dislocations at nearby boundaries is the energy gain from introducing patches where relatively small common $\mathbf{G}$ vectors exist locally along one direction in the boundary plane. It is noteworthy that when there is just one direction of matching $\mathbf{G}$ vectors the value of Σ is infinite. This demonstrates the usefulness of our analysis, which does not depend on the existence of a 3 dimensional CSL. We have here the physical basis for many of the experimentally observed modes of relaxation at interfaces leading to localisation of particular arrays of dislocations, and the near coincidence model (in 2 dimensions) for incommensurate interfaces.

2.3 The role of the boundary plane and the boundary expansion

In the previous section we saw that the significance of periodicity parallel to the boundary plane is that the boundary may lower its energy through a rigid body displacement, τ, parallel to the boundary plane. Irrespective of the presence of periodicity, the boundary may also undergo an expansion, e, normal to the boundary plane. The expansion is an important mode of relaxation at all grain boundaries. To isolate its contribution to the boundary relaxation we consider a conceptually important limit: the incommensurate limit [18]. Consider the set of boundaries obtained by rotations about the normal of a boundary with a fixed plane in the two crystals. If the boundary plane is of the same crystallographic form in both crystals then we are considering a set of pure twist boundaries. But if the crystallographic forms of the boundary plane in the two crystals differ, such as {411} in one crystal and {110} in the other, then we are considering a set of mixed tilt and twist boundaries. The incommensurate limit is obtained at the misorientation at which the boundary energy is a maximum. At this twist misorientation the boundary structure is least affected by the presence of possible periodicities in the boundary plane. Not only is there no periodicity in this boundary but it is also the most remote from all boundaries, sharing the same plane, at which periodicities may exist. We address the following question: *how does the boundary energy in the incommensurate limit depend on the boundary plane?* To answer this question we will have to explore 4 of the 5 dimensions of the geometrical parameter space of the boundary energy, since the boundary plane normal in each crystal is associated with 2 degrees of freedom.

I devised a simple *analytic* model, using a Lennard-Jones potential, to treat all boundaries in the incommensurate limit, *throughout the 4 dimensional parameter space* [14]. The only form of relaxation was the boundary expansion, and I assumed that the expansion was taken up entirely between the first atomic planes of both crystals on either side of the geometrical boundary plane. The model is described in detail in [14]. Here we summarise its main conclusions. But first it is stressed that the model calculates the cleavage energy not the grain boundary energy itself. The cleavage energy is the energy required to separate the grain boundary reversibly into two free surfaces[1]. The main results may be summarised as follows:

In the incommensurate limit the larger the average spacing, <d>, of planes parallel to the boundary the larger the cleavage energy and the smaller the boundary expansion. Also, the larger the boundary expansion the smaller the cleavage energy.

For a given value of <d> there is a range of possible interplanar spacings, parallel to and on either side of the boundary, between <d> in both crystals, and 2<d> in one crystal and zero in the other. The model predicts that, at a given value of <d>, the more unlike the interplanar spacings the smaller the expansion and the greater the cleavage energy. This is a more subtle

---

[1]It is interesting from a theoretical point of view that the cleavage energy appears, in this analysis at least, to have a greater fundamental significance that the boundary energy itself. This is because the cleavage energy is the (negative of the) total interaction energy between the grains. When this interaction energy is zero we have two free surfaces. It is sometimes helpful to think of the grain boundary energy as the two surface energies minus the cleavage energy, which corresponds to the creation of the grain boundary by bringing two free surfaces together and allowing them to interact fully. Conceptually this makes it clear that the boundary energy is determined by the surface energies and their interaction energy.

Grain Boundary Engineering in Ceramics

result which presumably reflects the anharmonicity of the atomic interactions in the Lennard-Jones model.

The trends predicted by the simple analytic model have been confirmed [14] by full atomistic relaxations with the same Lennard-Jones pair potential. However, the trends *must* eventually break down for good reasons. The trends relate the cleavage energy and the expansion to the average interplanar spacing of a boundary in the incommensurate limit. Consider a boundary plane that is only very slightly inclined to a rational boundary plane at which <d> is finite and equal to d_o. This slightly deviated boundary plane will be irrational, and its value of <d> will be zero. The simple model would therefore predict a discontinuous decrease in the cleavage energy, and a discontinuous increase in the expansion, relative to the values for the rational boundary. This implausible result stems from the lack of atomic relaxation (other than the expansion) in the simple model. In reality the slightly deviated boundary will have steps, or edge dislocations, or both, to localise the deviation of the boundary plane into small defective regions, so that most of the deviated boundary will assume a local value of <d> equal to that of the rational boundary.

The insight that the simple model brings is very useful despite the limitations of the trends at small interplanar spacings. First, *the model predicts a monotonic increase of the cleavage energy, not the boundary energy, with increasing <d> of boundaries in the incommensurate limit.* Second, *the model predicts a monotonic decrease of the cleavage energy, not a monotonic increase of the boundary energy, with increasing expansion of boundaries in the incommensurate limit.* Variations in the boundary energy with <d> include variations of the <u>free surface energies</u>. This is a partial explanation for why Sutton and Balluffi [11] found that the criterion of maximising <d> to minimise the boundary energy fails frequently. The other reason is that periodicity can affect the boundary energy away from the incommensurate limit. These two reasons also explain why there is so much scatter in plots of the boundary energy against expansion.

As the model is analytic it can be unpicked to expose the physical reasons for the trends it predicts. In the unrelaxed state the separation of planes at the boundary is <d>. Owing to the misorientation there will be atomic overlap even at boundaries with the maximum possible value of <d>. This atomic overlap gives rise to strong repulsive forces which force the grains apart. As the boundary expands *the repulsive forces do work against the attractive interactions across the boundary plane*, and both the repulsive *and* attractive forces are weakened. The attractive forces become weaker with increasing expansion owing to the anharmonicity of atomic interactions. The equilibrium expansion is determined by a balance of these weakened attractive and repulsive forces. The greater the initial overlap, the stronger the initial repulsive forces, and the greater the expansion required to relieve the overlap. The cleavage energy decreases with increasing expansion because the residual attractive forces across the boundary, against which further work has to be done to cleave the bicrystal, are weakened owing to the anharmonic nature of atomic interactions. It is clear that the cleavage energy and the expansion are as much determined by attractive interactions as by repulsive interactions. Short-range repulsion is common to all classes of materials since it arises from the exclusion principle and electrostatic repulsion. But attractive interactions have different physical origins in metallic, covalent and ionic systems, and one cannot ignore these differences if one is to study grain boundaries in these different systems. In the literature one sometimes reads the bold assertion that the physics of interfaces is dominated by short-range repulsion through the exclusion principle. It is no more dominant than either partner in a good marriage.

By considering boundaries in the incommensurate limit we have isolated the roles of the interplanar spacing and the boundary periodicity. A boundary displaying one or two dimensional periodicity will have lower boundary energy, and hence larger cleavage energy, than the boundary on the same plane in the incommensurate limit. The extent of the energy lowering will depend on the size of the smallest common $\mathbf{G^c}$ vector, and it will become exponentially small as $|\mathbf{G^c}|$ increases. The presence of $\mathbf{G^c} \neq 0$ terms in the boundary energy will also affect the boundary expansion.

2.4 Classification of interfaces

In this section we will pull together the results of sections 2.2 and 2.3 to classify interfaces in a way that is most useful for understanding the relationships between their structures and properties. At the same time we will capture the results for the grain boundary energy in a sequence of empirical formulae which may be useful in applications where one wants to know *roughly* how the energy of any grain boundary depends on its macroscopic degrees of freedom. The formulae offered here are 'quick and dirty' reproductions of the trends described in sections 2.2 and 2.3; their accuracy should not be relied on too heavily. I have written them down because several researchers have indicated to me that there is a need for such formulae.

Many grain boundary studies have focused attention on periodic structures that are amenable to high resolution electron microscopy or computer simulations. Our starting point for the classification of interfaces [12] is the other extreme. In a given material the simplest possible boundaries are those at local energy *maxima* with respect to at least one macroscopic degree of freedom. These boundaries are called general boundaries. Their energies are independent of the rigid body displacement τ because there are no common $\mathbf{G}$ vectors in the boundary plane, or $|\mathbf{G^c}|$ is so large that its influence on the boundary energy is negligible (see below). All general boundaries for which <d> is less than some $<d>_c$ have more or less the same (minimum possible) cleavage energy. If we assume (reasonably) that the energies of the corresponding free surfaces are also more or less constant, then the grain boundary energy is given by:

$$ \frac{\sigma_1}{\sigma_{max}} = 1 \qquad \text{where } <d> << <d>_c \qquad (6) $$

where σ_{max} is the maximum possible grain boundary energy in the material. Computer simulations [14] indicate that $<d>_c$ is about $<d>_{max}/5$, where $<d>_{max}$ is the maximum possible value of <d> in the material.

Consider the case where $<d> > <d>_c$, but there are still no significant periodicities in the boundary plane. The cleavage energy increases approximately quadratically with <d> [14]. But there may also be a reduction, $-|\Delta\sigma_s|$, in the corresponding free surface energies; a reduction because the surfaces are lower index on average. In general, the orientation dependence of $\Delta\sigma_s$ does not appear to be representable as a dependence only on d, even with $<d> > <d>_c$. Unlike the boundary cleavage energy, there is no good physical reason for believing that the orientation dependence of $\Delta\sigma_s$ may be expressed as a function of the interplanar spacing. The boundary energy becomes:

$$ \frac{\sigma_1}{\sigma_{max}} = 1 - \frac{\Delta\sigma_d}{\sigma_{max}} \qquad \text{provided } <d> > <d>_c, $$

$$ \text{where} \qquad \frac{\Delta\sigma_d}{\sigma_{max}} = \frac{|\Delta\sigma_s|}{\sigma_{max}} + \frac{|\Delta\sigma_{cl}|}{\sigma_{max}} \times \frac{<d>^2}{<d>_{max}^2} \qquad (7) $$

The maximum possible value of $<d>$ in the crystal structure is denoted by $<d>_{max}$. $|\Delta\sigma_{cl}|$ is the variation of the cleavage energy among all boundaries in the incommensurate limit. Computer simulations [14] in metals indicate that $|\Delta\sigma_{cl}|/\sigma_{max}$ is about 1/2, but it may be regarded as an adjustable parameter in eqn.(7). The surface energy term is more difficult. We need a mathematical representation of the orientation dependence of the surface energy that is used in the Wulff construction for the equilibrium shape of a single crystal. It is unwise simply to ignore $\Delta\sigma_s$ because there are two surfaces contributing to it!

Boundaries with $<d> > <d>_c$ are among those classified as <u>singular</u>. A singular interface is one for which the energy is at a local minimum with respect to at least one macroscopic degree of freedom [12]. The lowering of the boundary energy, stemming from the relatively large value of $<d>$, leads to localisation of regions where the boundary plane deviates as a result of small changes in the boundary parameters. For example, such a boundary will support localised steps if the boundary inclination changes. Boundaries that undergo relaxations to introduce patches of nearby singular interfaces are classified as <u>vicinal</u> [12].

Finally, a further reduction of the boundary energy to less than σ_2 is possible if the boundary structure displays one or two dimensional periodicity. Thus,

$$\frac{\sigma_3}{\sigma_{max}} = 1 - \frac{\Delta\sigma_d}{\sigma_{max}}\left(1 + \exp\left(-\frac{\lambda}{c}\right)\right) \qquad (8)$$

where c is the largest length of the c.n.i.d. and λ is a parameter with the dimensions of length. Computer simulations [14] indicate that λ is of the order of the crystal lattice parameter. The largest length of the c.n.i.d. is equal to the largest primitive DSC vector *in the boundary plane*. The exp($-\lambda$/c) form is the asymptotic form derived in [14] and referred to in section 2.2 above. Eqn.(8) applies to the most general case of mixed tilt and twist boundaries where the perfect crystal is not a configuration that is generated by rotations about the boundary normal. The perfect crystal is accessed by rotations about the normal of pure twist boundaries, and in that case eqn.(8) may need to be slightly modified to give $\sigma_3 = 0$ at the perfect crystal orientation.

Boundaries in which the periodicity is such that exp($-\lambda$/c) is non-negligible are further examples of <u>singular</u> interfaces. As noted in section 2.2 such boundaries localise deviations of the twist angle to maximise the area of the boundary that retains locally the structure of the periodic configuration. Boundaries sufficiently close to twist orientations containing common **G** vectors that localised screw dislocations are formed, are further examples of <u>vicinal</u> interfaces.

The classification [12] of all interfaces as general, singular and vicinal enables variations in many boundary properties to be understood. First of all it is clear that most grain boundaries have $<d> < <d>_c$ and they are therefore either general or vicinal. Singular interfaces are relatively rare in the 5 dimensional parameter space. *If all boundaries are equally probable* then it is reasonable to neglect the small populations of singular interfaces and their associated vicinal boundaries, and concentrate on the properties of general boundaries. Since general boundaries are characterised by eqn.(6) a first approximation is to say that all boundaries have the same properties. A distribution in which all boundaries are equally probable is most likely to be produced under non-equilibrium conditions. But as soon as the microstructure evolves in such a way as to reduce the overall grain boundary energy, the lower energy grain boundaries are more likely to appear. In that case singular and vicinal boundaries will play a larger role in the properties of the material.

As an example of the usefulness of the classification let us consider the propensity to segregation of large misfitting solute atoms, such as sulphur atoms in nickel or iron. To lower their elastic strain energy the solute atoms will segregate preferentially to boundaries offering larger free volume. These will be boundaries with the largest expansions normal to the boundary plane - the general boundaries. Thus the energies of general boundaries will be lowered more through segregation than those of singular and vicinal boundaries, and this may result in a change of the Wulff form. As another example consider grain boundary migration. Since singular and vicinal grain boundaries can support localised defects such as steps and dislocations with Burgers vectors parallel to the interface, these boundaries may migrate by mechanisms involving the motion of such defects. On the other hand general boundaries cannot support such localised defects and they can move only by mechanisms involving rearrangements of single atoms or small groups of atoms at the interface.

3. The influence of temperature

3.1 Introduction
As the temperature is raised we may expect grain boundaries to disorder in various ways to produce structures of higher entropy. The boundary may undergo a roughening or melting transition or a phase change. The thermodynamics of interfaces and phase changes at interfaces are discussed in chapters 5 and 6 of [1]. Here we consider just two points which indicate, in different ways, that as temperature is increased, the range of free energies of grain boundaries decreases. There is experimental evidence for this conclusion from the rotating ball experiments, where the numbers of relatively low energy boundaries are found to decrease with temperature [19].

In section 3.2 we derive, following [20], a simple relation between the excess vibrational entropy of the boundary and the boundary expansion. We also show that the temperature dependence of the excess vibrational entropy is directly proportional to the excess thermal expansivity. Then, in section 3.3, we consider the influence of temperature on the lowering of the boundary energy arising from periodicity in the boundary structure. We show that increasing temperature diminishes the size of the reduction of the boundary energy due to periodicity.

3.2 Vibrational entropy and the expansion of grain boundaries
The free energy of an interface is an excess quantity. That is to say, it is that part of the free energy of the system which is attributable to the interface, and it is determined by comparing the free energy of the bicrystal with the free energy of a reference single crystal. Since the free energy of the interface is obtained by comparing the free energies of two entire systems - a bicrystal and a reference single crystal - one may be tempted to conclude that an atom by atom breakdown of the free energy of the interface is impossible. After all, what is the meaning of 'the free energy of an atom', when the free energy is a property of the system as a whole?

There is an analogy with the electronic structure of the interface that is instructive. In general, electronic eigenstates extend throughout the bicrystal and one may again conclude that an atom by atom breakdown of the electronic structure of the bicrystal is impossible. However, every atom participates to some degree in each eigenstate because each eigenstate can be viewed as a linear combination of atomic states. This line of thought leads to the concept of the local density of electronic states, which is a measure of the extent to which an atom participates in each eigenstate of the entire system. Mathematically the local density of states is a projection of the global density of states onto the atomic states at a particular site. The variation of the local atomic environment at a grain boundary is reflected in variations of the local densities of states.

The normal vibratory modes of a bicrystal are linear combinations of displacements of atoms throughout the entire system. One can define a local density of states for vibratory modes in exactly the same way as for electronic modes. The local density of vibratory modes measures the extent to which an atom participates (through its displacement) in each normal mode of the entire system. The vibrational contribution to the free energy of the entire system is determined by the normal modes of the entire system. By projecting the global density of vibratory states of the entire system onto individual atomic sites one can obtain the atom by atom breakdown of the free energy of the system that we thought was impossible. Mathematically, the projection operation is achieved through the use of the local density of states. The dependence of the local density of states on the local atomic environment then enables us to examine in detail which atoms contribute most to the excess free energy of the interface. This is the basic idea underlying ref.[21] (which is described in section 3.9 of [1]).

When we speak, rather loosely, of 'the free energy of an atom' let us be clear about what we mean. First, there is the potential energy that we associate with the atom arising from the interatomic forces. Second, there is the contribution that may be attributed to the atom, through the local density of states, to the vibrational free energy of the entire system.

The equilibrium position of the atom is determined by minimisation of the free energy of the entire system with respect to its coordinates. The equilibrium position is its position averaged over a long period of time compared with the periods of the vibratory modes. If the interatomic forces were harmonic then the equilibrium position of the atom would not change as the temperature is raised; its amplitudes of vibration would increase but its mean position would not change. Real interatomic forces are always anharmonic, and as the temperature is raised the vibrational free energy is lowered by driving the mean position of the atom into softer environments, where the vibration amplitudes are greater and the vibrational entropy is larger. This tendency is counter acted by the potential energy, and the final position of the atom is determined by a balance of "forces" arising from both the potential energy and the vibrational free energy. Here we have the origin of thermal expansion in perfect crystals. If we use harmonic lattice theory to write down expressions for the vibrational free energy, but allow the mean positions of atoms to be displaced so as to minimise the free energy of the system at each temperature, we are using the quasiharmonic approximation. Since the vibrational free energy is derived, through the theory of lattice dynamics, from the interatomic forces in the system, we may think of the effective atomic interactions as being temperature dependent [21].

The excess vibrational entropy of an atom in the grain boundary may be expressed approximately as [20]:

$$\Delta S = 3k_B \ln(\theta_b / \theta_{gb}) , \qquad (9)$$

where θ_{gb} and θ_b are local Debye temperatures at the grain boundary and bulk sites respectively, and k_B is Boltzmann's constant. Eqn.(9) assumes that the temperature is higher than both Debye temperatures, so that all normal modes are occupied. The ratio of the Debye temperatures is related to the stiffnesses of the two sites:

$$\frac{\theta_b}{\theta_{gb}} = \sqrt{\frac{\nabla_b^2 E_p}{\nabla_{gb}^2 E_p}} , \qquad (10)$$

where $\nabla_{gb}^2 E_p$ and $\nabla_b^2 E_p$ are the curvatures of the potential energy at grain boundary and bulk sites respectively. Thus, the softer the atomic environment in the grain boundary, the smaller the value of $\nabla_{gb}^2 E_p$, and the larger the vibrational entropy of the interface. When an atom is at a hydrostatically compressed site $\nabla_{gb}^2 E_p$ is relatively large. The atom is unable to vibrate so much because it is more restricted by its neighbours. Conversely, if an atom is in a relatively large hole then $\nabla_{gb}{}^2 E_p$ is smaller, and the atom vibrates with relatively large amplitudes. Consequently, we expect that a greater free volume at an interface will give rise to a larger excess vibrational entropy. That is, *the excess vibrational entropy should increase with the boundary expansion normal to the boundary plane.* We now present a simple model that supports this statement.

Consider a model [20] of a grain boundary in which all atoms within a slab of thickness h, centred on the boundary, have a Debye temperature θ_{gb} and all atoms outside the slab have a Debye temperature θ_b. Computer simulations indicate that h is about 6Å in pure metals [21]. Expanding $\ln(\theta_{gb})$ about $\ln(\theta_b)$ to first order in the boundary expansion, e, we have

$$\ln(\theta_{gb}) = \ln(\theta_b) + \frac{d \ln \theta}{d \ln V} \frac{d \ln V}{de} e \qquad (11)$$

where V is the volume of the slab. Since $d\ln\theta/d\ln V = -\gamma$, which is the Grüneisen constant, and $d\ln V/de = 1/h$, we have

$$\Delta S = \frac{3k_B \gamma e}{h} \qquad \text{per atom.} \qquad (12)$$

We have assumed that the Grüneisen constants in the boundary slab and in the bulk are the same. Computer simulations [21] indicate that they are the same to within about 10%. To express ΔS as the excess entropy per unit area we multiply by the volume of the slab per unit area, which is h, and divide by the atomic volume in the slab, which is Ω_{gb}:

$$\Delta S = \frac{3k_B \gamma}{\Omega_{gb}} e \qquad \text{per unit area.} \qquad (13)$$

For $e = 0.5\text{Å}$, $\Omega_{gb} = 2 \times 10^{-29} m^3$ and $\gamma = 3$ we obtain $\Delta S \approx 0.3 mJ/m^2/K$, which agrees with typical experimental measurements and computer simulations of grain boundary excess entropies [20]. Note that eqn.(13) predicts that the excess entropy increases linearly with the boundary expansion, as we anticipated above.

The greater degree of anharmonicity at grain boundaries, compared with the bulk, leads to larger local thermal expansivities. We expect the thermal expansivity parallel to the boundary to be the same as the thermal expansivity in the bulk because otherwise long-range compatibility stresses would be generated. This expectation is borne out by experimental measurements [22]. But there is no such restriction on the thermal expansivity, α_n, normal to the boundary plane of a bicrystal. If the boundary expansion is e^0 at a temperature $T = T^0$ then the (excess) boundary expansion becomes

$$e = e^0 + (\alpha_n - \alpha)(T - T^0) h \qquad (14)$$

where α is the thermal expansivity of the bulk and h is the thickness of the grain boundary slab. Experimental measurements and computer simulations for twist boundaries indicate that α_n in Au is about $40 \times 10^{-6} K^{-1}$, compared with $\alpha = 10 \times 10^{-6} K^{-1}$. The thermal contribution to the boundary expansion over a 1000K temperature range is thus about 0.2Å. If the vibrational contribution to the excess entropy of the boundary is ΔS^0 at $T = T^0$, then the excess entropy at temperature T is given by substituting eqn.(14) into eqn.(13):

Grain Boundary Engineering in Ceramics

$$\Delta S = \Delta S_0 + \frac{3k_B \gamma (\alpha_n - \alpha)(T - T^0)h}{\Omega_{gb}} \qquad \text{per unit area} \qquad (15)$$

The significant point about eqn.(15) is that the temperature dependence of the excess vibrational entropy of the boundary is directly proportional to the excess thermal expansivity of the boundary.

In conclusion, we have shown that the excess vibrational entropy of a boundary increases with the boundary expansion. The thermal contribution to the boundary expansion is the origin of the temperature dependence of the vibrational excess entropy. Boundaries associated with larger expansions, i.e. general boundaries, will undergo larger reductions in their free energies with increasing temperature, than singular and vicinal boundaries. Thus the free energies of all boundaries in the material will become more alike as the temperature is increased.

3.3 The diminishing influence of periodicity with increasing temperature

In section 2.2 we pointed out that the significance of periodicity in the boundary plane is to enable the boundary energy to be lowered through a rigid body displacement, τ, parallel to the boundary. At low temperatures the extent of the energy lowering that can be achieved is determined by the size of the c.n.i.d.: the smaller the c.n.i.d. the smaller the energy lowering.

At a finite temperature the analysis of section 2.2 has to be modified. The phase factor $\exp(i\mathbf{G}^c . \tau)$ in eqns.(1) and (3) is multiplied by a Debye-Waller factor, which is approximately $\exp(-k_B T |\mathbf{G}^c|^2 / \nabla_{gb}{}^2 E_p)$. The contributions to the boundary energy from the larger $|\mathbf{G}^c|$ terms are therefore severely reduced by thermal vibrations in the boundary plane. As the temperature is increased the lowering of the energy that arises from periodicity in the boundary structure is reduced in magnitude. For example, for $\nabla_{gb}{}^2 E_p = 10 eV/Å^2$ and $T = 600K$ the Debye-Waller factor is equal to e^{-1} when $|\mathbf{G}^c| \approx 14 Å^{-1}$, corresponding to an in-plane DSC vector of size $2\pi/14 \approx 0.4 Å$. In Au (001) twist boundaries those boundaries with $\Sigma \approx 100$ and larger have in-plane DSC vectors of 0.4Å or less.

Since the lowering of the energy due to periodicity is reduced in magnitude at higher temperatures the energies of certain singular boundaries (i.e. those with periodic structures) will tend towards those of general boundaries at elevated temperatures. At the same time the cores of intrinsic dislocations, with Burgers vectors *parallel* to the boundary plane, in related vicinal boundaries are expected to broaden. There is some experimental evidence for such behaviour [23], although the broadening does appear to stop short of complete delocalisation of the intrinsic dislocation cores.

4. Future developments

4.1 Introduction

After all the work that has been done over the past 50 years on understanding individual grain boundaries we should try to apply this knowledge to the properties of polycrystals. This is an enormously difficult and challenging task, and one that is likely to occupy at least the first few years of this millennium.

Suppose we knew how some target grain boundary property varied with the 5 macroscopic degrees of freedom. Suppose further that for every boundary in the polycrystal we knew its area and its 5 degrees of freedom. Could we calculate the property for the whole polycrystal? The answer to this question depends very much on the property concerned. For example, suppose the property is the interfacial free energy per unit volume in the polycrystal.

This is an important quantity, for example in assessing the stability of the polycrystal to grain growth. To a good approximation we could evaluate the total interfacial free energy by summing the energies of the individual boundaries in the polycrystal.

But suppose the property were the self-diffusivity of the material at a temperature where mass transport is dominated by grain boundary diffusion. Now, in addition to evaluating the diffusivity of each boundary, we would have to consider the boundary connectivity so as to balance the fluxes at grain boundary junctions. In this case the spatial distribution of boundaries is just as significant as the distribution of boundary types.

Consider the fracture toughness of a polycrystal. Certain (low cleavage energy) grain boundaries may be much more susceptible to intergranular fracture than others. The fracture toughness of the material is determined by the spatial distribution of these weak boundaries, the orientations of their boundary planes to the stress axis, and the path that the crack will follow between these weak interfaces. In this case the target property of the polycrystal is strongly influenced by the spatial distribution of a (possibly small) subset of all the boundaries in the material. Another example of such a property is the super-current in a polycrystalline high T_c material, where many boundaries may be weak links. In that case the current seeks out the path of least resistance which takes it across boundaries where the supercurrents are strongly coupled.

Over the coming years there is likely to be much more work done on the properties of polycrystals, not least because they have greater industrial significance than bicrystals! But there are considerable conceptual problems to be overcome if any real progress is to be made. Again, suppose that we have characterised the 5 macroscopic degrees of freedom of every boundary in a polycrystal. The number of boundaries as a function of the five degrees of freedom has been called a "grain boundary character distribution" by Watanabe [24]. We need a way of comparing *quantitatively* the character distributions for two polycrystals, for example to gauge the success of a thermo-mechanical processing treatment to modify the grain boundary properties of the material. Some first steps in this direction have been taken [25] and are outlined in section 4.2. But I am not aware of any attempts to combine a grain boundary character distribution with the spatial distribution of grain boundaries, which will include information about the grain boundary connectivity. That is such a daunting problem one wonders whether a direct assault is even feasible. An alternative approach is to borrow some recent ideas [26] from the closely related field of composites. There one uses variational arguments to find upper and lower bounds on certain properties of the composite, such as its elastic and dielectric properties. But one still needs to input some features of the character distribution into such variational arguments, notably their lowest order moments.

4.2 Grain boundary character distributions
There are several equivalent ways in which one may choose to specify the 5 macroscopic degrees of freedom [27], and one is free to select the most convenient. Let the 5 macroscopic degrees of freedom be v_1 ,.... v_5 . Then

$$dW = T(v_1, v_2, v_3, v_4, v_5)\, dv_1\, dv_2\, dv_3\, dv_4\, dv_5 \qquad (16)$$

is the fraction of all boundaries in the "volume element" $dv_1\, dv_2\, dv_3\, dv_4\, dv_5$ at $(v_1 ,v_2 ,v_3 ,v_4 ,v_5)$. Equation (16) defines the character distribution $T(v_1 ,v_2 ,v_3 ,v_4 ,v_5)$. It is normalised to facilitate subsequent comparisons between different distributions:

$$\int T\, (v_1, v_2, v_3, v_4, v_5)\, dv_1\, dv_2\, dv_3\, dv_4\, dv_5 = 1 \qquad (17)$$

There exist experimental techniques to measure the orientations of the crystal axes within a large number of grains of a polycrystalline material in a straightforward manner. For example, electron back-scattering patterns in the scanning electron microscope have been used in a fully automated manner [28] to collect large amounts of data on crystal lattice orientations. From these orientational data one can readily evaluate the boundary misorientations. The measurement of grain boundary planes is less straightforward using such techniques because one sees only the trace of the boundary plane at the surface of the sample; the inclination of the boundary plane to the surface is unknown. Nevertheless it remains true that the boundary plane normal (in either crystal) is needed for a full specification of the 5 macroscopic degrees of freedom.

In view of the generation of crystal orientational data experimentally it is useful to choose the *misorientation*, **R**, of a boundary as 3 of the 5 degrees of freedom. We may then choose the *boundary plane* normal, $\underline{n}$, in either crystal for the remaining two degrees of freedom. Thus, $T = T(\mathbf{R},\underline{n})$. This is not a convenient form if **R** is represented by a rotation *matrix*. It is much more convenient to use other representations of the misorientation, notably the unit *quaternion* representation [25]. In that case, a misorientation is represented by the quaternion $q = (q_1, q_2, q_3, q_4)$ where $q_1 = \rho_1 \sin\omega/2$, $q_2 = \rho_2 \sin\omega/2$, $q_3 = \rho_3 \sin\omega/2$ and $q_4 = \cos\omega/2$, and $\underline{\rho} = (\rho_1, \rho_2, \rho_3)$ is a unit vector representing the rotation axis relative to a reference frame defined by the crystal axes in either crystal, and ω is the misorientation angle. It follows that $q_1^2 + q_2^2 + q_3^2 + q_4^2 = 1$, which is why there are only 3 degrees of freedom associated with the quaternion. Quaternion representations of rotations have very useful properties. For example, consider two rotations, A and B, represented by the quaternions $q^{(A)}$ and $q^{(B)}$, or the rotation matrices $\mathbf{R}^{(A)}$ and $\mathbf{R}^{(B)}$, respectively. If we want the angle of rotation, α, required to get from A to B we form the matrix $\mathbf{R}^{(B)}\mathbf{R}^{(A)-1}$ and deduce the rotation angle from the trace of this matrix. By contrast, the corresponding recipe using quaternions is a direct and explicit formula:

$$\cos(\alpha/2) = \cos[\omega^{(A)}/2]\cos[\omega^{(B)}/2] + \underline{\rho}^{(A)} \cdot \underline{\rho}^{(B)} \sin[\omega^{(A)}/2]\sin[\omega^{(B)}/2]. \tag{18}$$

The right hand side of (18) is the inner ('dot') product of the quaternions $q^{(A)}$ and $q^{(B)}$ regarded as four-dimensional vectors, $\cos(\alpha/2) = q^{(A)}.q^{(B)}$, but note that only *half* of α appears in the cosine on the left-hand side. Moreover, the average of a set of misorientations represented by matrices is non-trivial to deduce, but the average misorientation is found immediately using the quaternion representation:

$$m = \frac{1}{N}\sum_{i=1}^{N} q^{(i)} \tag{19}$$

Let $F(q)$ be the distribution of boundary misorientations represented by quaternions. If we represent the misorientation axis $\underline{\rho}$ in usual spherical coordinates it becomes $\rho = (\sin\theta \cos\phi, \sin\theta \sin\phi, \cos\theta)$. Then, the components of the quaternion representation of the misorientation are:

$$
\begin{aligned}
q_1 &= \sin\theta \cos\phi \sin\omega/2 \\
q_2 &= \sin\theta \sin\phi \sin\omega/2 \\
q_3 &= \cos\theta \sin\omega/2 \\
q_4 &= \cos\omega/2 ,
\end{aligned}
\tag{20}
$$

where $0 \le \theta \le \pi$, $0 \le \phi \le 2\pi$ and $0 \le \omega \le 2\pi$. The misorientation is now characterised by the angles θ and ϕ specifying the misorientation axis and ω specifying the misorientation angle. Then, $dF = F(q)d^3q$

is the number of boundaries within the volume element, d^3q, of quaternion space at the misorientation q. The volume element is readily evaluated using eqns.(20):

$$d^3q = \frac{1}{2} \sin\theta \sin^2(\omega/2)\, d\theta\, d\phi\, d\omega. \tag{21}$$

How do we fit a continuous misorientation distribution to experimental data consisting of a discrete set of N misorientations $q^{(1)}$, $q^{(2)}$, ... , $q^{(N)}$? To answer this question let us first define the n'th moment of the distribution as follows:

$$\mu^{(n)} = \int F(q)\, (q.m)^n\, d^3q \tag{22}$$

where m is the quaternion representing the average misorientation, given by eqn.(19). The term $q.m = \cos(\alpha/2)$, where α is the rotation angle between the misorientations q and m, and it is given by:

$$\cos(\alpha/2) = \cos(\omega/2)\cos(\omega_m/2) + (\rho_m.\rho)\sin(\omega/2)\sin(\omega_m/2), \tag{23}$$

where ρ_m and ω_m are the axis and angle of the average misorientation, m. It is convenient to choose the z-axis along ρ_m. Then, $\rho_m.\rho = \cos\theta$, and

$$\cos(\alpha/2) = \cos(\omega/2)\cos(\omega_m/2) + \cos\theta\sin(\omega/2)\sin(\omega_m/2). \tag{24}$$

Thus, $\mu^{(n)} = <\cos^n(\alpha/2)>$, which may be evaluated directly from the experimental data. For example,

$$\mu^{(2)} = \frac{1}{N} \sum_{j=1}^{N} (q^{(i)}.m)^2 . \tag{25}$$

Suppose we evaluate the first M moments of the N experimental data points, i.e. $\mu^{(1)}$, $\mu^{(2)}$, ... , $\mu^{(M)}$. Then, maximum entropy arguments [25] enable F(q) to be determined rigorously:

$$F(q) = \frac{1}{Z} \exp\left[-\left(\sum_{n=1}^{M} \lambda_n (q.m)^n\right)\right], \tag{26}$$

where λ_n, n=1,2,...M, are Lagrange multipliers and Z is defined by

$$Z = \int \exp-\left(\sum_{n=1}^{M} \lambda_n (q.m)^n\right) d^3q . \tag{27}$$

The Lagrange multipliers may be found by maximising the entropy, S, of the distribution with respect to λ_1, λ_2, ... , λ_M, where

Grain Boundary Engineering in Ceramics

$$S = \ln Z + \sum_{n=1}^{M} \lambda_n \mu^{(n)} . \tag{28}$$

Maximisation of S with respect to the M Lagrange multipliers guarantees that the first M moments of the distribution are reproduced exactly.

We are now able to answer quantitatively the question of how alike or unlike two distributions of grain boundary misorientations are. Suppose we have two distributions $F_1(q)$ and $F_2(q)$. Then the overlap, C, between these distributions may be defined as follows:

$$C^2 = \frac{< F_1 | F_2 >^2}{< F_1 | F_1 >< F_2 | F_2 >} \tag{29}$$

where

$$< F_1 | F_2 > = \int F_i F_j d^3 q \tag{30}$$

C is a number lying between 0 and 1. When C=0 the distributions have no overlap at all. When C=1 the distributions are identical. Equation (29) is simply the degree of correlation between $F_1(q)$ and $F_2(q)$.

Having defined the distribution of boundary misorientations in eqn.(26) let us return to the character distribution, which we wrote as $T=T(q,\underline{n})$. We recall that each boundary is characterised by the misorientation q and the boundary normal $\underline{n}$. Although the boundary normal is obviously the same direction in both crystals, it generally has different components when it is expressed in the two coordinate systems. When we come to find moments of the distribution of boundary normals we will get different answers depending on which of the two crystal coordinate systems we choose. It is better, therefore, to use the mean boundary plane $\underline{N}$ instead of $\underline{n}$. The mean boundary plane is defined in the median lattice [27]. The boundary plane normals $\underline{n}$ and $\underline{n}'$ in the two crystals are related to the mean boundary plane as follows:

$$\underline{n} = \underline{N} - \underline{N} \times \rho^R$$

$$\underline{n}' = \underline{N} + \underline{N} \times \rho^R \tag{31}$$

where ρ^R is the Rodrigues vector representation of the misorientation, $\rho^R = \rho \tan(\omega/2)$. Equation (31) is derived on p.22 of ref.[1].

A grain boundary with normals $\underline{n}$ and $\underline{n}'$ has a mean boundary plane $\underline{N} = \underline{n} + \underline{n}'$. The adjoining crystal lattices are generated by applying equal and opposite rotations of magnitude $\omega/2$ about the axis ρ to a crystal lattice. Conversely, if we imagine a process in which the boundary misorientation shrinks to zero, through the crystals rotating by equal and opposite amounts about the axis ρ, then we generate a single crystal lattice, which is the median lattice, and the boundary plane becomes a plane in that lattice with the same normal as the mean boundary plane. In this way all grain boundaries may be related back to planes of a unique single crystal lattice. For example, all symmetric tilt boundaries in cubic crystals have {110} or {100} mean boundary planes.

The mean boundary plane normal $\underline{N}$ may always be expressed as a unit vector using spherical polar coordinates as follows:

$$N_1 = \sin\theta_N \cos\phi_N$$
$$N_2 = \sin\theta_N \sin\phi_N$$
$$N_3 = \cos\theta_N \tag{32}$$

where $0 \le \theta_N \le \pi$ and $0 \le \phi_N \le 2\pi$ are not to be confused with the variables θ and ϕ we have used to define the rotation axis. Thus, the 5 independent variables of the character distribution may all be specified as angles, i.e. θ, ϕ, ω, θ_N, ϕ_N, and the 'volume element' in this space is $(1/2)\sin\theta$ $\sin^2(\omega/2) \sin\theta_N \, d\theta \, d\phi \, d\omega \, d\theta_N \, d\phi_N$.

Just as we have defined a distribution $F(q)$ for the boundary misorientations we may also define a distribution $f(\underline{N})$ for the mean boundary plane normals. Then, $df = f(\underline{N}) \sin\theta_N \, d\theta_N \, d\phi_N$ is the fraction of mean boundary plane normals passing through the surface area element $\sin\theta_N \, d\theta_N \, d\phi_N$ at (θ_N, ϕ_N) on the unit sphere. The distribution $f(\underline{N})$ should also be normalised: $\int f(\underline{N}) \sin\theta_N \, d\theta_N \, d\phi_N = 1$.

Since q and $\underline{N}$ are independent variables, we may consider the distributions $F(q)$ and $f(\underline{N})$ separately by fitting their moments $<\cos^p\alpha/2>$ and $<\cos^p\theta_N>$ respectively, where p is an integer. The character distribution may then be expressed in a separated form:

$$T(q,\underline{N}) = F(q)f(\underline{N}). \tag{33}$$

The mean boundary plane normal distribution $f(\underline{N})$ is derived by the same maximum entropy arguments as were used to derive $F(q)$:

$$f(\underline{N}) = \frac{1}{Z_N}\exp\left[-\left(\sum_{n=1}^{K}\beta_n \cos^n\theta_N\right)\right], \tag{34}$$

where β_n, n=1,2,...K, are Lagrange multipliers and Z_N is defined by

$$Z_N = 2\pi\int_0^\pi \exp-\left(\sum_{n=1}^{K}\beta_n \cos^n\theta_N\right)\sin\theta_N d\theta_N \tag{35}$$

The Lagrange multipliers may be found by maximising the entropy, S_N, of the distribution with respect to β_1, β_2, ... , β_K, where

$$S_N = \ln Z_N + \sum_{n=1}^{K}\beta_n <\cos^n\theta_N>. \tag{36}$$

In conclusion, we have shown how grain boundary character distributions may be defined quantitatively. A particularly convenient choice of the 5 macroscopic degrees of freedom is the boundary misorientation and the mean boundary plane normal. The character distribution then factorises into two distributions, one for the misorientation and the other for the mean boundary plane normal. Experimental data have already been collected for boundary misorientations, and these data could be used to generate distributions of misorientations for quantitative assessments and comparisons. The experimental determination of grain boundary plane normals is more difficult, but no less important. Clearly, there is much work to be done on data collection and fitting of character distributions.

Grain Boundary Engineering in Ceramics

References

[1] A. P. Sutton and R. W. Balluffi, *Interfaces in crystalline materials*, Oxford University Press: Oxford and New York (1995).

[2] A.P. Sutton and V. Vitek, Phil. Trans. R. Soc. Lond. A309, 37 (1983).

[3] W. Bollmann, *Crystal defects and crystalline interfaces*, Springer: Berlin (1970).

[4] Section 2.6 of ref.[1].

[5] Section 1.5 of ref.[1].

[6] R. C. Pond and W. Bollmann, Phil. Trans. R. Soc. Lond. A292, 449 (1979).

[7] Section 1.3.1 of ref.[1].

[8] A. V. Shubnikov and V. A. Koptsik, *Symmetry in science and art*, Plenum: New York (1977).

[9] A. D. J. Haymet and D. W. Oxtoby, J. Chem. Phys., 74, 2559 (1981).

[10] Section 1.9 of ref.[1].

[11] A.P. Sutton and R.W. Balluffi, Acta Metall., 35, 2177 (1987).

[12] Section 4.3.1.5 of ref.[1].

[13] N. H. Fletcher and P. L. Adamson, Phil. Mag. 14, 99 (1966).

[14] A.P. Sutton, Phil. Mag. A, 63, 793 (1991); A.P. Sutton, Prog. Mat. Sci., 36, 167-202 (1992); Section 4.3 of ref.[1]

[15] p.257-258 of ref.[1]

[16] S. E. Babcock and R. W. Balluffi, Phil. Mag. A 55, 643 (1987).

[17] P. H. Pumphrey, Scr. Metall., 6 ,107 (1972); R. Schindler, J. E. Clemans and R. W. Balluffi, Phys. Stat. Sol. (a), 56, 749 (1979).

[18] D. Wolf and S. R. Phillpot, Mater. Sci. Eng. A, 107, 3 (1989).

[19] U. Erb and H. Gleiter, Scr. Metall., 13, 61 (1979).

[20] Section 4.3.1.10 of ref.[1].

[21] A.P. Sutton, Phil. Mag A., 60, 147 (1989).

[22] M. R. Fitzsimmons, E. Burkel and S. L. Sass, *Phys. Rev. Lett.*, 61, 2237 (1988).

[23] T. E. Hsieh and R. W. Balluffi, Acta Metall., 37, 1637 (1989).

[24] T. Watanabe, Res. Mechanica, 11, 47 (1984).

[25] A. P. Sutton, Phil. Mag. Letts., 74, 389 (1996).

[26] J. R. Willis, *Variational and related methods for the overall properties of composites*, Advances in Applied Mechanics, 21, ed. C. S. Yih, Academic Press: New York, p.1-78 (1981).

[27] Section 1.4 of ref.[1].

[28] B.L. Adams, Mat. Sci. Forum, 207-208, 13 (1996).

RECENT PROGRESS IN THE ELECTRONIC STRUCTURE THEORY OF COMPLEX CERAMICS[#]

Wai-Yim Ching*, Shang-Di Mo, Lizhi Ouyang and Young-Nian Xu
University of Missouri-Kansas City, Kansas City Missouri, 64110 USA

ABSTRACT

Recent progress in the electronic structure theory of complex ceramics is reported. The basic tool used is the density-functional-theory based electronic structure method, the first-principles orthorgnalized linear combination of atomic orbitals (OLCAO) method. Reported results include the implementation of a geometry optimization scheme without direct force calculation, inclusion of the core-hole relaxation effect in the ELNES calculation, and supercell calculation of impurity states in laser crystals. All these new developments can be directly applied to the study of grain boundary structures in ceramics.

INTRODUCTION

In recent years, the density functional theory (DFT) in its local density approximation (LDA) has been extremely successful in predicting the structures and properties of many different materials with increasing accuracy [1]. With the unprecedented advance in computing technology within the last two decades, the LDA-DFT becomes a driving force in modern materials research. The first-principles orthogonalized linear combination of atomic orbitals (OLCAO) method is a DFT-based method that has been effectively applied to the study of many fundamental properties of complex ceramics and their microstructures [2]. In this article, we summarize some of the most recent advances in the computational methodology that would be particularly helpful for studying the properties of grain boundaries (GB) in ceramics. First, the implementation of an efficient geometry optimization scheme without direct force calculation enables us to explore the structure of unknown compounds, such as the new cubic phase of spinel nitrides. The method can be used to accurately model the GB structures. Second, the electron-loss near-edge spectroscopy (ELNES), in conjunction with high-resolution electron microscopy, is a potent tool for GB characterization. The OLCAO method has been used for ELNES calculations of some crystals and their GBs with only limited success [3,4]. It is now shown that a realistic inclusion of the core-hole effect in the calculation can lead to superior agreement between theoretical and experimental spectra. Finally, impurity segregation at the

----

Work supported by U.S. DOE grant NO. DE-FG02-84DR45170.

* Member, American Ceramic Society, (e-mail: chingw@umkc.edu).

To the extent authorized under the laws of the United States of America, all copyright interests in this publication are the property of The American Ceramic Society. Any duplication, reproduction, or republication of this publication or any part thereof, without the express written consent of The American Ceramic Society or fee paid to the Copyright Clearance Center, is prohibited.

GB is a well-recognized fact. Its electronic structure and bonding are important and can be studied by large supercell calculations. We demonstrated that such calculations can be done by reporting the first calculations of the $(Cr^{4+} + Ca^{2+})$ and $(Cr^{4+} + Mg^{2+})$ centers in the YAG $(Y_3Al_5O_{12})$ crystal. It is envisioned that by applying these newly developed techniques to the realistically modeled GB structures, a much higher level of understanding at the fundamental level can be achieved that may facilitate GB engineering.

GEOMETRY OPTIMIZATION AND APPLICATIONS

It is now well established that *ab-initio* total energy (TE) and force calculations based on DFT can predict the crystal structures and properties of materials with virtually no experimental input. However, for a relatively complex crystal with many internal parameters, such an endeavor is still a time-consuming task if the full geometry of the crystal is to be optimized. One drawback of the OLCAO method is the difficulty of obtaining the forces on each atom directly, as the basis functions in the OLCAO method are nuclear-position-dependent. There is a significant Pulay correction [5] to the Hellmann-Feynman forces, which is difficult to obtain with the local orbital methods. This makes full optimization of a complex crystal a very difficult, if not impossible task.

To circumvent the difficulty of direct force calculation in the OLCAO method, we have introduced a simple scheme of evaluating the force by the finite difference of the TE. Let E_T $(a,b,c,\alpha,\beta,\gamma; x_1, x_2, x_3 \ldots)$ be the TE of a general crystal with lattice parameters specified by $a,b,c,\alpha,\beta,\gamma$, and internal parameters by $x_1, x_2, x_3 \ldots$etc. The energy gradient for each parameter P_i is obtained by the finite difference approach according to:

$$\frac{\partial E_T}{\partial P_i} \approx \frac{E_T(P_i + \Delta P_i) - E_T(P_i - \Delta P_i)}{2 \times \Delta P_i} \tag{1}$$

The TE of the crystal is minimized in the multi-parameter space and the geometry fully optimized using the first-order-gradients-only algorithms such as the conjugate gradient. We have integrated the OLCAO method with the General Lattice Utility Program (GULP) for geometry optimization [6]. The key element of the above simple scheme is that the TE calculation must be sufficiently fast and efficient. For each energy gradient, a minimum of two full self-consistent field (SCF) calculations is needed. For each cycle of parameter refinement, 2 x N_p SCF calculations are necessary where N_p is the total number of parameters to be optimized. It generally takes 10 to 20 cycles for the TE of a complex crystal to fully converge.

The above scheme was used to optimize the geometry optimization of the four experimentally reported crystalline phases of Si_3N_4 [7]. They are ß-Si_3N_4 (space group P6$_3$/m or P6$_3$), α-Si_3N_4 (space group P3$_1$/c), and the cubic spinel phase c-Si_3N_4 (space group Fd$_3$/m). Past investigations on Si_3N_4 were limited to equilibrium lattice constants and volumes without relaxing the internal parameters [8,9]. The exact structure of the β-phase was a subject of considerable debate for some time [10,11], but the difference between the P6$_3$/m structure with a mirror symmetry and that of the P6$_3$ structure with no mirror symmetry was generally ignored in the past theoretical investigations. The discovery of the cubic spinel phase c-Si_3N_4 under high pressure and temperature is truly significant [12] since it demonstrates group IV nitrides can be formed in the cubic structure. The lattice constant and the internal parameter of the c-Si_3N_4 obtained using the geometry optimization scheme were found to be in close agreement with experiments [13]. We have since predicted the crystal structure of many other cubic spinel nitrides. Electronic structure calculations show that they have many interesting physical properties and potential applications [14].

Table I lists the results of geometry optimization and the calculated values of the bulk modulus for the four Si_3N_4 crystals. Also listed are the number of atoms N_A in the unit cell, the total number of parameters N_P relaxed, and the number of k-points N_K in the irreducible portion of the Brillouin zone used. As can be seen, the agreements between the calculated and the measured lattice constants are excellent. The average deviations are less than 0.5%. The differences in the internal parameters are also satisfactory but were listed elsewhere [7]. In general, the agreement is better for the Si parameters than for the N parameters. The bulk modulus B_0 of each crystal is obtained by fitting the energy vs. volume data to the Murnaghan equation of state [15]. In each calculation with a specific volume expansion or contraction, the c/a ratio is fixed at the equilibrium value while all internal parameters are relaxed. We believe the present calculation provides the most accurate and consistent B_0 values for the Si_3N_4 phases since they were obtained by using a single method with all parameters relaxed and with sufficient number of k points. From Table I, it is seen that β-Si_3N_4 (P6$_3$) is lower in energy than β-Si_3N_4 (P6$_3$/m) by 0.238 eV per formula unit. On the other hand, α-Si_3N_4 (p3$_1$) is higher than β-Si_3N_4 (p6$_3$) by 0.455 eV per formula unit. This is a much smaller difference than the value (1.65 eV/molecule) reported before with no optimization for the internal parameter [9]. More detailed discussion can be found in ref. 7.

To apply this general scheme to GB modeling or defect structure relaxation in ceramics, the demand for computational power will be significantly increased, since each atom in the model will involve three parameters corresponding to three Cartesian coordinates. Techniques for speeding up the relaxation process are currently under development.

Table I. Calculated and measured (in parenthesis) crystal parameters of Si_3N_4 phases. B_0, bulk modulus, E_o, total energy per formula unit (excluding core electrons).

Space Group	β-Si_3N_4 [P6$_3$/m]	β-Si_3N_4 [P6$_3$]	α-Si_3N_4 [P3$_1$/c]	c-Si_3N_4 [Fd$_3$/m]
N_A	14	14	28	14
N_P	6	8	16	2
N_K	32	32	16	84
a (Å)	7.622 (7.607)	7.623 (7.595)	7.792(7.766)	7.837 (7.80)
x	-	-	-	0.3844(.3875)
c (Å)	2.910 (2.911)	2.901 (2.902)	5.614 (5.615)	-
c/a	2.619 (2.613)	2.627 (2.617)	0.720 (0.723)	-
B_0 (Gpa)	274	278	257	280
E_0 (eV)	-1415.088	-1415.326	-1414.871	-1411.795

ELNES CALCULATIONS

Electron energy-loss spectroscopy and the related energy-loss near-edge structure (ELNES) observed in the inner-shell ionization process is a powerful technique for material characterizations [16]. ELNES edges obtained by using analytical transmission electron microscopy show a rich variety of structures, which can be correlated to the structure and composition of the material under study. This is the so-called "fingerprints" approach by looking for common features of the same ELNES edge in different compounds and inferring their electronic or structural property [17].

To properly interpret the experimental ELNES edges, especially for complex microstructures such as grain boundaries, a theoretical calculation with sufficient predictive power is extremely helpful. In the band structure approach, the site-projected and orbital-resolved local density of states (LDOS) is widely used to interpret the ELNES spectra [3,4,18]. However, the agreement between the calculated and measured spectra is not always satisfactory. A good agreement in one edge in one material is not accompanied by the equally good agreement of the other edges in the same compound, or the same edge in different compounds. This lack of agreement casts some doubt in the full understanding of the ELNES spectra of complex ceramic structures.

The main reason for poor agreement is the neglect of the core-hole relaxation effect. In the expression for the ELNES intensity I(E), which is proportional to the differential scattering cross section [16], the final state is the one with an electron in the conduction band (CB) and a hole in the atomic core. The electron and hole can interact strongly, thereby modifying the final state to be drastically different from that of the ground state calculation. Strictly speaking, the core-hole effect is a two-particle interaction and a correct treatment requires separate calculations of the initial and final states. A rigorous scheme to include the core-hole relaxation effect has been implemented using the OLCAO method [18]. The initial and the final states for the transition matrix elements in the ELNES intensity are calculated separately. Self-consistent calculations are performed on supercells in which the core states of the excited atom are retained in the basis set and those of other atoms are eliminated by the orthogonalization process. The final state is obtained by removing one electron from the core of the target atom, and placing it in the lowest CB. The interaction between the excited electron and the hole left is fully accounted for by self-consistent iterations.

We have applied this new scheme to calculate the ELNES spectra of MgO, α-Al_2O_3, and $MgAl_2O_4$. These crystals have the same cation in different coordinations and local environments, hence very different ELNES spectra. Furthermore, rather complete experimental data are available for comparison. Large supercells are necessary to obtain all the fine details in the spectra. Detailed tests lead us to use supercells of 128, 120 and 112 atoms for MgO, α-Al_2O_3 and $MgAl_2O_4$ respectively. To achieve high resolution, 27 k-points in the reduced Brillouin zone of the supercell were used for the k-space integration.

The eleven ELNES edges in the three crystals together with the available experimental data [19-22] are presented in Fig. 1. The energy scale is set at the top of the valence band. It is gratifying to see that all the fine details of the peak positions and relative intensities are reproduced by the calculation. This is particularly impressive when one realizes that the edges of the same atom in different crystals are very different. The so-called spike structure in Mg-K at 1310.2 eV has never been successfully reproduced in the past. The Mg-K and the Mg-$L_{2,3}$ edges in spinel $MgAl_2O_4$ are the first-time theoretical results. No experimental data for the Al-K edge in $MgAl_2O_4$ has been reported. Mg-K data in $MgAl_2O_4$ [ref. 22] is from XANES, which has a better resolution. The only discrepancy in these comparisons appears to be that the experimental peak at 80.5 eV in Al-$L_{2,3}$ of $MgAl_2O_4$ is much more pronounced than the same peak in the theoretical curve. This can be explained by the fact that the $MgAl_2O_4$ crystal used in the experiment was a natural mineral sample which contains a certain fraction of inverse spinel [22]. In the inverse spinel, Al ions occupy both the octahedral (Al_{oct}) and the tetrahedral (Al_{tet}) sites while in normal spinel, Al ions occupy only the Al_{oct} sites. There is strong evidence that Al_{tet} has a stronger peak at a higher energy than Al_{oct} [18].

Grain Boundary Engineering in Ceramics

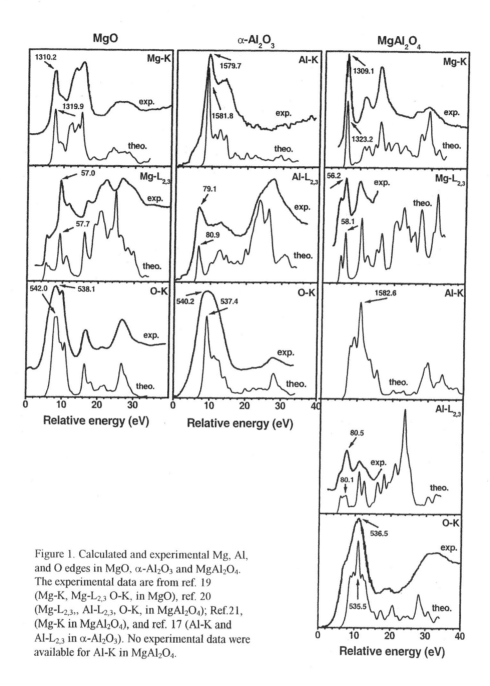

Figure 1. Calculated and experimental Mg, Al, and O edges in MgO, α-Al₂O₃ and MgAl₂O₄. The experimental data are from ref. 19 (Mg-K, Mg-L$_{2,3}$ O-K, in MgO), ref. 20 (Mg-L$_{2,3}$,, Al-L$_{2,3}$, O-K, in MgAl₂O₄); Ref.21, (Mg-K in MgAl₂O₄), and ref. 17 (Al-K and Al-L$_{2,3}$ in α-Al₂O₃). No experimental data were available for Al-K in MgAl₂O₄.

The OLCAO method has been used to study the point defects and impurities in α-Al_2O_3 using a supercell [23,24]. For studying impurities in the GBs, it is envisioned that the structural model for the GB will be much larger, containing at least several hundreds of atoms. The OLCAO method, with its economic use of the localized basis set, can meet this challenge. Here we present some examples by showing the result of calculations of Cr ions with co-doping elements in the laser crystal YAG. Recently, there has been an increased interest in studying the energy levels of 3d or 4f ions in laser crystals and the associated absorption and emission spectra. Cr^{4+}:YAG is used as a passive Q-switch material and as a source for near-infra-red tunable lasers. To fully understand the mechanism and to improve the laser performance, it is important to study the fundamental energy levels of Cr ions in YAG based on realistic calculations [25].

YAG has two Al sites, Al_{oct} (16a) and Al_{tet} (24d), and one site each for Y (24c) and O (96h). Al_{oct} is octahedrally coordinated and Al_{tet} is tetrahedrally coordinated. A Cr substituting Al_{oct}(Al_{tet}) will be in the Cr^{3+} (Cr^{4+}) state. For Cr^{4+}, it was envisioned that a simultaneous substitution of Y by Ca^{2+} or Al_{oct} by Mg is necessary for charge compensation. This co-doping scenario introduces sufficient complication that the traditional method of using a free ion in the crystal field approach [26] can no longer be used for the energy level analysis. One-electron calculation using an LDA-DFT method and a large supercell offers a viable alternative. The complexity of the problem is similar to that will be encountered in the case of impurities at the GB of ceramics. Electronic structure calculations are carried out using the OLCAO method with (Cr^{4+} + Ca^{2+}) and (Cr^{4+} + Mg^{2+}) in YAG [25,27]. The cubic cell of YAG crystal with 160 atoms is large enough to serve as a 'supercell' such that impurity-impurity interaction is minimized. All calculations are self-consistent with no adjustable parameters.

Figure 2 shows the defect levels in the band gap of the YAG crystal (labeled as A, B, C and D). A should be identified as the doubly occupied $^3A_2$ state in the traditional analysis of a d^2 ion in the elongated tetrahedral environment with D_{2d} symmetry [26]. The highest defect level D in both cases is almost doubly degenerate, but is split by the presence of the co-doping element. The Cr^{4+} levels in the gap are very similar for the Ca and Mg cases. However, it is also noticed that charge transfer states (CTS) appear near the top of the valence band. The CTS originate from the O atoms, which are the nearest neighbors of the substituted Ca or Mg. It can be seen that the CTS in the Mg case are higher than that of the Ca case, mainly because Mg substitutes an Al_{oct} site which has a shorter bond length. This may explain different degrees of laser crystal degradation for Ca or Mg doped Cr^{4+}:YAG crystal [28]. It is also shown that excited state absorption (ESA) may play a role in the laser degradation in such crystals. These and other results are reported elsewhere [27]. Here we only demonstrate that impurities in complex GB models of ceramics can be studied using a similar approach. More detailed analysis will require the consideration of the atomic relaxation due to impurity substitution and the presence of other defects in the crystal.

CONCLUSION

It is shown that tremendous progress has been made in the electronic structure theory using the first-principles OLCAO method. The method is ideal for complex ceramics and their microstructures. Combining with experimental investigations, ab-initio theoretical calculation with sufficinet predictive power can be an effective tool for complex materials characterization and for understanding the micrscopic interactions that lead to specific structures and properties.

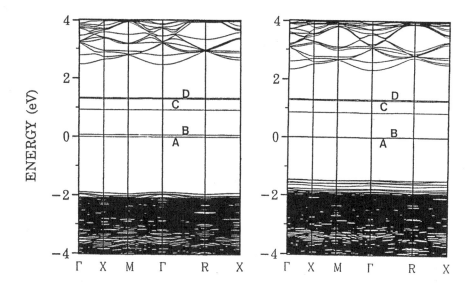

Figure 2. Supercell band structures of calculated impurity levels of $(Cr^{4+} + Ca^{2+})$ in YAG (left panel) and $(Cr^{4+} + Mg^{2+})$ in YAG (right panel).

REFERENCES

1. W. Kohn, "Nobel Lecture: Electronic Structure of Matter-wave Functions and Density Functionals," Rev. of Mod. Phys. **71**, 1253 (1998).
2. W. Y. Ching, "Theoretical Studies on the Electronic Structures of Ceramics," J. Am. Ceram. Soc. **73**, 3135 (1990).
3. Shang-Di Mo, W.Y. Ching , M. Chelshim and G. Duscher, "Electronic Structure of Grain Boundary Model of $SrTiO_3$," Phys. Rev. **B60**, 2416(1999).
4. Shang-Di Mo, W.Y. Ching, and R.H. French, "Electronic Structure of a Near Σ-11 a-axis Tilt Grain Boundary in α-Al_2O_3," J. of Amer. Ceram. Soc. **79**, 627(1996).
5. P. Pulay, "Ab-initio Calculation of Force Constant and Equilibrium Geometries in Polyatomic Molecules," Mol. Phys. **17**, 197 (1969).
6. J.D. Gale, "GULP: A Computer Program for the Symmetry-adapted Simulation of Solids," J. Chem.Soc. Faraday Trans. **93**, 629 (1997).
7. W.Y. Ching, L. Ouyang, and J. Gale, "Full ab-initio Geometry Optimization of All Known Crystalline Phases of Si_3N_4," Phys. Rev. **B61**, 8696 (2000).
8. A.Y. Liu and M.L. Cohen, " Structural Properties and Electronic Structure of Low-

compresibility Materials: ß-Si$_3$N$_4$ and Hypothetical ß-C$_3$N$_4$," Phys. Rev. **B41**, 10727 (1994).

9. W.Y. Ching, Yong-Nian Xu, J. D. Gale and M. Rühle, "Ab-initio Total Energy Calculation of α- and β-phases of Silicon Nitride and the Derivation of Effective Inter-atomic Pair Potentials with Application to Lattice Dynamics," J. Amer. Ceram. Soc. **81**, 3189 (1998).

10. R. Grün, "The Crystal Structure of β-Si$_3$N$_4$: Structural and Stability Considerations Between α- and β-Si$_3$N$_4$", Acta Crystallog. **B 35**, 800 (1979).

11. P. Goodman and O'Keeffe, "The space Group of ß-Si$_3$N$_4$," Acta Cryst. **B36**, 2891 (1980).

12. A. Zerr, G. Miehe, G. Serghiou, M. Schwarz, E. Kroke, R. Riedel, H. Fueß, P. Kroll, and R. Boehler, "Synthesis of Cubic Silicon Nitride," Nature **400**, 340 (1999).

13. Shang-Di Mo, L. Ouyang, W.Y. Ching, and I. Tanaka, Y. Koyama, and R. Riedel, "Interesting Physical Properties of the New Spinel Phase of Si$_3$N$_4$ and C$_3$N$_4$," Phys. Rev. Lett. **83**, 5046 (1999).

14. W. Y. Ching, Isao Tanak, S.-D. Mo, Lizhi Ouyang, and M. Yoshiya, "Prediction of the New Spinel Phase of Ti$_3$N$_4$ and SiTi$_2$N$_4$, and Metal-Insulator Transition," Phys Rev. **B61**, 10609 (2000).

15. F.D. Murnaghan, "The Compressibilty of the Media under Extreme Pressure," Proc. Natl. Acad. Sci., U.S.A. **30**, 244 (1944).

16. R. F. Egerton, *Electron Energy Loss Spectroscopy in the Electron Microscope* (Plenum, New York, 1986).

17. R. Bryson, H. Sauer, and W. Fugel, "Electron Energy Loss Near-Edge Structure as an Analytical Tool – The Study of Minerals," in *Transmission Electron Energy Loss Spectroscopy in Materials Science*, edited by M.M. Disko, C.C. Ahn and B. Fuitz, P131 (TMS, Warrendale, PA, 1992).

18. M. Gülgün, W.Y. Ching , Y.-N. Xu and M. Rühle, "Electron States in YAG probed by Electron Loss Near Edge Spectroscopy and Ab-initio Calculations," Philo. Mag. **B79**, 921(1999).

19. Shang-Di Mo and W.Y. Ching, "Ab-inito Calculation of Core-hole Effect in the Electron Energy-loss Near Edge Spectra.", submitted to Phys. Rev. B.

20. T. Lindner, H. Sauer, W. Engel and K. Kambe, "Near-edge Structure in Electron-energy-loss Spectra of MgO," Phys. Rev. **B33**, 22(1986).

21. J. Bruley, M.-W. Tseng and D.B. Williams, "Spectrum-line Profile Analysis of a Magnesium Alumina Spinel/ Sapphire Interface," Microsc. Microanal. Microstr. **6**, 1 (1995).

22. D. Li, M. -S. Peng, T. Murata, "Coordination and Local Structure of Magnesium in Silicate Minerals and Glasses, Mg K-edge XANES Study," Cana. Miner. **37**, 199(1999).

23. Yong-Nian Xu, Zhong-Quan Gu, Xue-Fue Zhong and W.Y. Ching, "Ab-initio Calculation for the Neutral and Charged O Vacancy States in Sapphire," Phys. Rev. **B56**, 7277 (1997).

24. W.Y. Ching, Y.-N. Xu and M. Rühle, "Ab-initio Calculation of Y substitutional impurities in α-Al$_2$O$_3$," J. Am. Ceram. Soc. **80**, 3199 (1997).

25. W.Y. Ching, Yong-Nian Xu and B. Brikeen, "Ab-initio Calculation of Excited State Absorption of Cr^{4+} in Y$_3$Al$_5$O$_{12}$," Appl. Phys. Lett. **74**, 3755 (1999).

26. R.C. Powell, *Physics of Solid-State Laser Materials*, AIP Press (New York, 1997).

27. W.Y. Ching, Y.-N. Xu, and B. Brickeen, "Photoconductivity and Excited State Absorption in Cr4+:YAG," submitted to J. Appl. Phys.

28. I. T. Sorokina, S. Naumov, and E. Sorokin, E. Winter, and A.V. Shestakov, "Directly diod-pumped Tunable Continuous-wave room-temperature Cr4+:YAG Laser", Opticas Lett., **24** [22], 1578 (1999).

TIGHT-BINDING MOLECULAR DYNAMICS OF CERAMIC NANOCRYSTALS USING PARALLEL PC CLUSTER

Kenji Tsuruta, Hiroo Totsuji, and Chieko Totsuji
Department of Electrical and Electronic Engineering, Okayama University,
3-1-1 Tsushima-naka, Okayama 700-8530, JAPAN
Email: tsuruta@elec.okayama-u.ac.jp
URL: http://www.mat.elec.okayama-u.ac.jp

ABSTRACT

Tight-binding molecular dynamics (TBMD) simulations are performed to study atomistic/electronic structures and electronic transport in nanocrystalline silicon-carbide ceramic (nc-SiC). The simulations are based on an $O(N)$ algorithm (the Fermi-operator expansion method) for calculating electronic contributions in the energy and forces. The code has been fully parallelized on our Pentium-based parallel machines.

In a sintering simulation of aligned (no tilt or twist) SiC nanocrystals at $T \approx 730°C$, we find that a neck is formed promptly without formation of defects. Analyses reveal that unsaturated bonds exist only in grain surfaces accompanying the gap states. In the case of tilted (<122>) nanocrystals, surface structures formed before sintering affect significantly the grain-boundary formation. An electrical conductivity of the sintered (aligned) nc-SiC system is estimated via the Kubo-Greenwood formula with tight-binding representation. Effects of surface states on the conductivity is analyzed

INTRODUCTION

Silicon carbide (SiC) has been attracting much interest as an enabling material for new electronic applications (high-power, high-frequency, high-temperature electronic devices, gas-sensing devices, and irradiation detectors) [1]. Silicon or gallium-arsenic devices cannot effectively work for these applications, especially in harsh environments. The attractiveness of SiC for these applications is mainly

To the extent authorized under the laws of the United States of America, all copyright interests in this publication are the property of The American Ceramic Society. Any duplication, reproduction, or republication of this publication or any part thereof, without the express written consent of The American Ceramic Society or fee paid to the Copyright Clearance Center, is prohibited.

due to its unique properties including high barrier for electric breakdown, chemical stability, and high thermal conductivity. However, the brittleness of SiC is a major drawback in reliability and in tailoring the materials for specific applications. Nanocrystalline SiC (nc-SiC) synthesized by consolidating nano-particles is therefore a very promising material because it has shown [2] enhanced mechanical stability compared to coarse-grained counterpart. Computer-aided optimization of nanocrystalline materials will be very important for reducing the costs required for development and fabrication of this material. The semi-empirical tight-binding molecular dynamics method [3] is a powerful tool to investigate theoretically both atomistic and electronic processes in synthesizing nanocrystalline materials and their effects on the transport properties.

In this paper, we report on tight-binding molecular-dynamics simulations for sintering of SiC nanocrystals at a temperature of 730℃. The Fermi-operator expansion method (FOEM) [3,4] is employed to calculate efficiently the electronic part of the energy and forces, and it has been run on our eight-node parallel PC cluster. Using the parallel TBMD, we investigate the processes of neck formation between aligned nanocrystals, and between tilted nanocrystals. We also investigate effects of surface/grain boundary structures on electronic transport through Kubo-Greenwood analysis of dc conductivity. Results indicate that the gap states appeared due to surface structures in a sintered nc-SiC may contribute to an increase in conductivity.

TIGHT-BINDING MODEL FOR SILICON CARBIDE SYSTEM

For a system consisting of N atoms, the tight-binding total energy model is defined as

$$E_{tot} = E_{kin} + E_{bs} + E_{rep} , \qquad (1)$$

where E_{kin} stands for the kinetic energy, E_{bs} is the band-structure energy, and E_{rep} represents the repulsive term that takes into account the core-core interactions and neglected contributions in E_{bs} to the true electronic energy, such as a correction for double counting of electron-electron interactions. In semi-empirical TB methods, the repulsive energy is modeled by the sum of short-range 2-body interaction. The band-structure energy is calculated by diagonalizing the effective one-electron Hamiltonian matrix, $\mathbf{H}$: Each off-diagonal element of the Hamiltonian involves interactions between valence electrons within the two-center hopping approximation. The electronic contribution to interatomic forces, i.e., derivatives of E_{bs} with respect to the atomic coordinates, can be obtained through the Hellmann-Feynman (HF) theorem. The TB Hamiltonian ($\mathbf{H}$) plays an essential role in TBMD simulations.

Among a number of TB calculations for SiC systems [5], we have chosen Mercer's parametrization of the TB Hamiltonian[6], which is based on the sp^3 orthogonal basis. It includes "environment-dependent" contributions to the onsite energies through intra-atomic terms. These terms give important contributions to variation of the charge transfer between Si and C atoms, especially for inhomogeneous systems such as nc-SiC. The original parametrization by Mercer, however, involves some discrepancies with experimental data such as the lattice constant and the interfacial energies. Minor modifications of the parameters have therefore been made so that magnitudes of these discrepancies are reduced. Table I summarizes comparisons of the present TB model with other works.

Table I Comparison of the physical quantities in the present TB model with other works

Physical quantities	Present	Other work
lattice constant [Å]	4.36	4.36 [1]
cohesive energy [eV]	6.53	6.34 [1]
bulk modulus [GPa]	234	224 [1]
grain-boundary energy[2] [J/m^2]	1.75	1.27 [2]

1) Experimental values referred to in Ref. [7]
2) $\{122\}\Sigma=9$ non-polar grain boundary energy in Ref. [5]

O(N) ALGORITHM AND PARALLELIZATION STRATEGY

Diagonalization of the TB Hamiltonian matrix, involved in conventional TB methods, takes the computational cost proportional to N^3, where N is the number of atoms in the simulation cell. This becomes overwhelming when the system size of simulation exceeds a few hundred atoms, which is too small for realistic simulations of nanostructured materials. Several new methods have been proposed recently to resolve this problem. Among them we adopt the Fermi-operator expansion method, proposed originally by Goedecker and Colombo [4]. This algorithm is based on moment expansion of a pseudo-density matrix (the Fermi operator) of the TB Hamiltonian by Chebyshev polynomials. Combining an appropriate truncation at a finite order of the expansion and a truncation in the multiplication between each element of the matrix at a physical cut-off distance, the computational complexity in evaluating the band-structure energy and the HF forces is reduced to that proportional to N.

In addition, this algorithm is suitable for parallel computing, since the truncation at the physical distance between atomic sites can make the data structures localized. We employ the spatial domain-decomposition technique for calculating matrix-matrix multiplication. In this algorithm, data transmission between neighboring nodes is performed only for near-boundary atoms and matrix elements because of the spatially localized nature of the density matrix.

Figure 1 shows results of parallel performance test for the systems of bulk Si crystal by the present algorithm using our parallel PC cluster. The parallel machine consists of Dual Pentium III-based personal computers connected by the Myrinet switch. The MPI is used for parallelization of the code. In Fig. 1, it is shown that the total CPU time per MD step is nearly constant for the system size up to 8,000 atoms, when the numbers of atoms per node are kept constant. Thus, the present strategy for parallelization is a highly scalable algorithm on a distributed parallel computer.

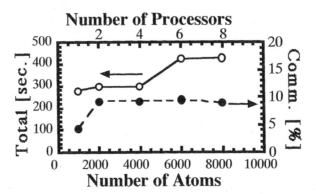

Figure 1: Parallel performance of the present algorithm on 8-node parallel PC cluster: Total CPU time per MD step and communication overhead. The systems calculated are Si crystals and the numbers of Si atoms per node are fixed to be 1,000. The TB parameters for Si crystal are taken from Ref. [8].

SINTERING SIMULATIONS OF SILICON-CARBIDE NANOCRYSTALS

Understanding dynamics of neck formation at high temperatures is very important because it may determine not only mechanical behavior but also carrier transport properties of sintered samples. We investigate effects of surface relaxation (or surface reconstruction) and of grain-boundary orientation on the neck formation processes between β-SiC nanocrystals by the TBMD simulation.

Figure 2 shows xy projections of the initial and final configurations of 'aligned' (no tilt or twist) nanocrystals: The nanocrystals are initially prepared by removing a cylindrical system with radius 14Å from β-crystalline SiC bulk. The total number of atoms in the simulation box is 460 (230 Si atoms and 230 C atoms) and the initial size of the simulation box is 32.05Å x 30.82Å x 7.55Å. (The periodic-boundary condition is used for all the simulations in this study).

Before the sintering simulation begins, the system is heated gradually to room temperature and subsequently to 730°C. At each temperature, the system was thermalized for several hundred femto seconds. After this preparation procedure, the system is compressed gradually along y direction ([11$\bar{2}$]) by reducing the box length in the y direction. Figure 2(b) for the final configuration shows that even after compression at $T \approx 730$°C the structure at the neck is crystal-like while surface reconstruction occurs on cluster surfaces. Analyses of effective charges (Mulliken charges) and local density of states indicate that unsaturated bonds or bond distortions exist only on the free surfaces, not in the grain boundary.

We have also performed TBMD simulation of nanocrystals with a tilt angle. Figure 3(a) shows the initial configuration of 816-atom system. Initially, the grain at the center of the MD cell and the surrounding grains are tilted to each

Grain Boundary Engineering in Ceramics

other, so that it corresponds to a coincidence grain boundary ({122}Σ=9) found typically in experiments [1,5]. After the same thermal process as in the previous case (the aligned nanocrystals), the system was compressed along the x and y direction at $T \approx 730°C$. Figure 3(b) shows the configuration after compression. Contrary to the previous case, the figure indicates that disordered structure is formed in the grain boundaries.

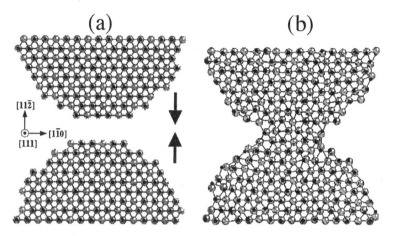

Figure 2: Neck formation of SiC 'aligned' nanocrystals: (a) initial configuration, (b) configuration after compression at $T=730°C$. Large and small spheres represent Si atom and C atom, respectively.

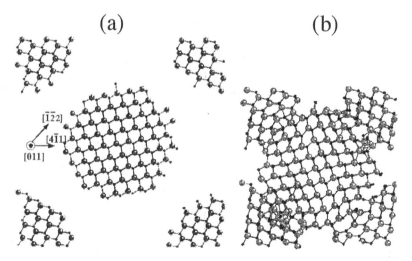

Figure 3: Same as Fig. 2, but with a tilt angle between nanocrystals.

ELECTRICAL CONDUCTIVITY

We have investigated the electronic transport properties in the sintered nc-SiC through the Kubo-Greenwood formula of dc electrical conductivity. We used the recursion method, derived by Roche and Mayou [9], for an efficient calculation of dc conductivity of disordered systems. Their prescription is based on a moment expansion of the Heisenberg operator with orthogonal polynomials, and it involves essentially the same procedure as the Chebyshev expansion of the density matrix in our TBMD simulation.

Figure 4 shows dc conductivity across the grain boundary in the sintered nc-SiC system, shown in Fig. 2(b), as a function of electronic Fermi energy. We find an increase of the conductivity in the band gap, which is absent in the perfect crystal. Analyses indicate the main contribution to this increase is from electrons at atomic sites on the free surface. Since the structures in the grain boundary and inside the grains are found to be regular (no defects), as described in the previous section, the possible channels of electron transport are probably only along the free surfaces.

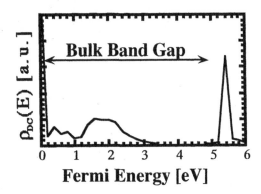

Figure 4: DC electrical conductivity across grain boundary formed after sintering between aligned nc-SiC [Fig.2(b)]. The origin of the Fermi energy is defined at the top of valence band.

CONCLUSIONS

We have shown the results of large-scale TBMD simulations of SiC nanocrystals. The $O(N)$ algorithm has been implemented for efficient computation of the band-structure energy and the Hellmann-Feynman forces on atoms. The code has been fully parallelized on PC-based parallel machines, and the algorithm has been shown to be scalable. We have found in the sintering simulations that the neck formation at 730°C occurs very quickly in the regular grain boundary, whereas the surface structure formed before compression affects significantly the tilt grain-

Grain Boundary Engineering in Ceramics

boundary formation. Preliminary results of the dc electrical conductivity in the sintered nc-SiC system have shown an enhancement of the conductivity in the gap region. This is presumably due to channels of electronic transport along grain surfaces.

ACKNOWLEDGMENTS

This work was supported partially by the Grants-in-Aid for Scientific Research (B) 08458109 and (B) 11480110 from the Ministry of Education, Science, Sports, and Culture of Japan.

REFERENCES

1. E.g.,"Special Issue on Silicon Carbide Electronic Devices", *IEEE Trans. Elec. Dev.* **46** [3], 441-619 (1999); S. Somiya and Y. Inomata, *Silicon Carbide Ceramics - 1: Fundamental and Solid Reaction*, Elsevier Applied Science, London and New York, 1991.
2. S. Komarneni, J. C. Paker, and H. J. Wollenberger(Eds.), Materials Research Society Symposium Proceedings Vol. 457, *Nanophase and Nanocomposite Materials II*, Materials Research Society, Warrendale, PA, 1997; K. Tsuruta, A. Omeltchenko, A. Nakano, R. K. Kalia, P. Vashishta, "Structure, mechanical properties, and dynamic fracture in nanophase silicon nitride via parallel molecular dynamics", *ibid* Vol. 457, 205-210 (1997).
3. P. Turchi, A. Gonis, L. Colombo, Materials Research Society Symposium Proceedings Vol. 491, *Tight-Binding Approach to Computational Materials Science*, Materials Research Society, Warrendale, PA, 1998.
4. S. Goedecker and L. Colombo, "Efficient linear scaling algorithm for tight-binding molecular dynamics", *Phys. Rev. Lett.* **73**[1], 122-125 (1994)
5. M. Kohyama, S. Kose, M. Kinoshita, R. Yamamoto, "The self-consistent tight-binding method: Application to silicon and silicon carbide", *J. Phys. Condens. Matter* **2**, 7791-7808 (1990); J. Robertoson, "The electronic and atomic structure of hydrogenated amorphous Si-C alloys", *Phil. Mag. B* **66**[5], 615-638 (1992); D. Sanchez-Portal, E. Artacho, J. M. Soler., "Projection of plane-wave calculations into atomic orbitals", *Solid State Comm.* **95**[10], 685-690 (1995)
6. J. L. Mercer, "Tight-binding models for compounds: Application to SiC", *Phys. Rev. B* **54**[7], 4650-4659 (1996).
7. K. J. Chang and M. L. Cohen, "Ab initio pseudopotential study of structural and high-pressure properties of SiC", *Phys. Rev. B* **35**[15], 8196-8201 (1987).
8. L. Goodwin, A. J. Skinner, and D. G. Pettifor, "Generating transferable tight-binding parameters: application to silicon", *Europhys. Lett.* **9**[7], 701-706 (1989).
9. S. Roche and D. Mayou, "Conductivity of quasiperiodic systems: A numerical study", *Phys. Rev. Lett.* **79**[13], 2518-2521 (1997).

FIRST-PRINCIPLES STUDY OF CERAMIC INTERFACES: SiC GRAIN BOUNDARIES AND SiC/METAL INTERFACES

Masanori Kohyama and John Hoekstra*
Osaka National Research Institute, AIST,
1-8-31, Midorigaoka, Ikeda,
Osaka 563-8577, Japan

ABSTRACT

SiC grain boundaries and SiC/metal interfaces have been studied using the *ab initio* pseudopotential method based on the density-functional theory. The stable configuration of the non-polar interface of the $\{122\}\Sigma=9$ tilt boundary has been obtained. The tensile strength and fracture of this interface have been clarified through the behaviour of electrons and atoms by using the *ab initio* tensile test. The Si-terminated and C-terminated SiC(001)/Al and SiC(001)/Ti interfaces have been studied in order to clarify the effects of the surface species and metal species. The four kinds of interfaces have quite different features in regard to atomic configurations, bonding nature, adhesive energy and Schottky-barrier heights.

INTRODUCTION

It is of great importance to investigate ceramic interfaces such as grain boundaries and ceramic/metal interfaces. Grain boundaries dominate various properties of ceramics. It is crucial to fabricate ceramic/metal interfaces with desirable properties for practical applications of ceramics. It is desirable to understand such ceramic interfaces at the atomic and electronic scales. Currently, it is possible to perform theoretical calculations of structure and properties of ceramic interfaces by virtue of the development of the density-functional theory (DFT) [1] and novel computational schemes such as the first-principles molecular-dynamics (FPMD) method [2]. We present our recent first-principles calculations of SiC grain boundaries [3] and SiC/metal interfaces [4].

For SiC grain boundaries, we deal with the $\{122\}\Sigma=9$ boundary in cubic SiC, which is a typical coincidence tilt boundary in chemical vapor-deposited SiC observed by high-resolution transmission electron microscopy (HRTEM) [5]. We examine the tensile strength and fracture of this boundary by applying an *ab initio* tensile test. For SiC/metal interfaces, we deal with Si-terminated and C-terminated interfaces of the SiC(001)/Al and SiC(001)/Ti systems. We investigate the effects of the surface species and metal species on the atomic configuration, bonding nature, bond adhesion and Schottky-barrier height (SBH).

---

*Present Address: Dept. of Materials, Oxford University, OX1 3PH, UK

To the extent authorized under the laws of the United States of America, all copyright interests in this publication are the property of The American Ceramic Society. Any duplication, reproduction, or republication of this publication or any part thereof, without the express written consent of The American Ceramic Society or fee paid to the Copyright Clearance Center, is prohibited.

THEORETICAL METHOD

The ground-state properties of solids can be dealt with accurately by the DFT [1], which settles the many-body problem of electrons. The local density approximation [6] or generalized gradient approximation have to be used in the execution of the DFT, although this can reproduce a variety of properties within errors of a few percent of experimental values. The plane-wave pseudopotential method [7] can provide the electronic structure, total energy, atomic forces, and stress tensor practically based on the DFT. This method deals with only valence electrons and ions apparently with frozen core electrons, which is a good approximation for usual ceramics. Calculations of interfaces with some two-dimensional periodicity such as coincidence boundaries or coherent interfaces are performed using the supercell technique. Currently, computations of large supercells can be executed using the techniques of the FPMD method [2], where the electronic ground state is obtained efficiently through novel iterative schemes such as conjugate-gradient (CG) methods [8,9], and the number of plane-wave basis is reduced by using soft or optimized pseudopotentials [10]. We use the CG method by Teter *et al.* [8] for a SiC boundary and that by Bylander *et al.* [9] for SiC/metal systems with the Troullier-Martins-type pseudopotentials [10]. A plane-wave cutoff energy of 816 eV is used for a SiC boundary, and that of 680 eV is used for SiC/metal systems.

SiC GRAIN BOUNDARY

Stable Configuration

The {122}Σ=9 boundary is constructed by rotating one grain by about 38.9° around the $\langle 011 \rangle$ axis. Two polar and one non-polar interfaces can be constructed by inverting the polarity of each grain. We constructed atomic models for these interfaces with the same bonding network as the same boundary in Si [11]. All the atoms are four-fold coordinated, and two sets of five-membered and seven-membered rings constitute one period. The non-polar interface contains C–C and Si–Si wrong bonds, although the polar interfaces contain either C–C or Si–Si wrong bonds. In this study, we examine the non-polar interface using a 64-atom supercell. We use two special **k** points per irreducible eighth of the Brillouine zone.

In the relaxed configuration, all the interfacial bonds are well reconstructed. Bond lengths and bond charges of the interfacial C–C and Si–Si wrong bonds are rather similar to those in bulk diamond and Si. Bond-length and bond-angle distortions of the other Si–C bonds are rather small, which range from –2.9% to +2.9% and from –22.4° to +27.9°, respectively. The boundary energy is 1.27 Jm^{-2}, which is very much smaller than the two surfaces. This interface has been observed by HRTEM [5]. The agreement between the calculated and observed configurations is rather good.

Ab Initio Tensile Test

In the *ab initio* tensile test [12], uniaxial tensile strain is introduced into the stable configuration. First, the supercell is stretched in a small increment in the direction normal to the interface, and the atomic positions are changed by uniform scaling. Second, all the atoms are relaxed through iterative electronic structure calculations until all the atomic forces are less than 0.18 eV/Å. Third, the total energy and stress tensor are calculated. This cycle is iterated until the interfaces are broken. In this procedure, the most stable configuration is attained for each tensile strain through the behavior of electrons and atoms. This procedure corresponds to a real slow tensile test at T=0K.

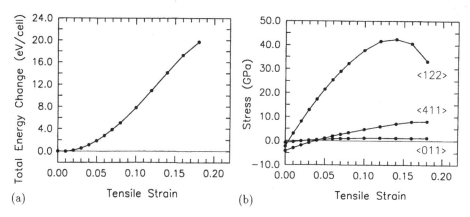

Figure 1. (a) Energy increase and (b) stress–strain curve in the tensile test of the non-polar interface of the $\Sigma=9$ boundary in SiC.

Figure 1 shows the energy increase and the stress–strain curve. Stresses along the $\langle 411 \rangle$ and $\langle 011 \rangle$ directions parallel to the interface are also generated, because the cell sizes along these directions are fixed. Strictly speaking, the cell sizes along these directions should be adjusted in each step so as to remove such stresses. However, this effect is not so serious.

The Young's modulus of the interface is 519 GPa from the curve in Fig. 1(b). This value is comparable to the bulk SiC value along the $\langle 111 \rangle$ direction, 558 GPa, by a similar first-principles calculation [13]. The maximum tensile stress of the interface is about 42 GPa at the strain of 14%. This is about 80% of the theoretical strength of bulk SiC along the $\langle 111 \rangle$ direction, 50.8 GPa [13], which is in good agreement with the experimental value of 53.4 GPa for a SiC nanorod [14]. The strength of this interface is very large because all the interfacial bonds are well reconstructed. This strength is very much larger than the observed macroscopic strength of SiC ceramics, which is usually less than 1 GPa. In most ceramics, fracture starts from pre-existing cracks. The growth and propagation of cracks occurs by the stress concentration under lower macroscopic stresses. The calculated strength of the interface should correspond to the necessary local tensile stress at the crack tip for the crack propagation along the interface. It does not seem that the present interface acts as an initial point of fracture, or necessarily acts as a preferential fracture path.

The energy in Fig. 1(a) corresponds to the fracture energy. The energy-derived fracture toughness K_C can be estimated by the empirical formula, $K_C=(2\gamma E)^{1/2}$ [15]. 2γ is the fracture energy of one interface per unit area and E is the Young's modulus. If we use an energy value twice as large as the value at 18% in Fig. 1(a) because the energy is still increasing, K_C is $2.4\,\mathrm{MPa\,m^{1/2}}$. This value is also comparable to the bulk SiC value, about $3\,\mathrm{MPa\,m^{1/2}}$ [16].

As shown in Fig. 2, the back Si–C bond of the C–C wrong bond, indicated by 'a', is broken first. Then the two interfacial Si–C bonds, indicated by 'c' and 'd', are broken. Finally, the

Si–Si wrong bond is broken. The reason why the back Si–C bond of the C–C bond is broken first is due to the local stress concentration caused by the special large strength and short bond length of the C–C bond like a diamond bond.

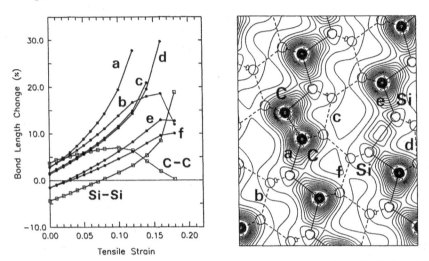

Figure 2. Bond-length changes in the tensile test. In a left pannel, Si–C, C–C and Si–Si bond-length changes are shown against the bulk bond lengths of SiC, diamond and Si, respectively. The bonds are specified in a right pannel, where one period of the stable configuration with no strain is shown with the contours of the valence electron density from $0.015 \, \text{a.u.}^{-3}$ to $0.295 \, \text{a.u.}^{-3}$ in spacing of $0.020 \, \text{a.u.}^{-3}$. In this pannel, circles indicate the atomic positions. Solid lines between atoms indicate the bonds on the same (011) plane on which the electron density is plotted. Broken lines are the bonds on different planes.

Figure 3. Breaking of the back Si–C bond of the C–C bond. Arrows indicate the contours of the same value of valence density at the back bond. Solid or dashed lines indicate bonds of which the stretching is less than 30%.

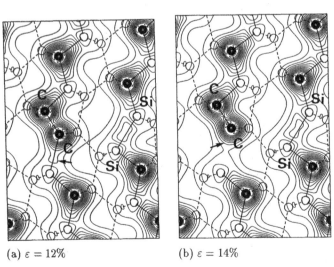

(a) $\varepsilon = 12\%$ (b) $\varepsilon = 14\%$

Grain Boundary Engineering in Ceramics

Figure 3 shows the breaking of the back Si–C bond of the C–C bond. At the tensile strain of 10%, the bond stretchings of the C–C bond and the back Si–C bond are +7.0% and +19.4% against the bond lengths of bulk diamond and SiC, respectively. At the strain of 12%, the back bond is suddenly stretched into +27.8%, while the C–C bond shrinks into +6.4%. The C–C bond shrinks much more hereafter. Thus the back Si–C bond has been almost broken at this point. At the strain of 14%, the stretching of the back Si–C bond is +46.3%, and the bond charge clearly disappears as shown in Fig. 3(b). About the breaking of the back Si–C bond and the two interfacial Si–C bonds, we have found the critical Si–C bond stretching. Once the bond stretching exceeds about 20%, the Si–C bond cannot sustain the stress, and is rapidly stretched and broken. The bond charge clearly disappears when the stretching exceeds about 30%.

SiC/METAL INTERFACES

Supercell

Each supercell contains a slab of 9 SiC(001) atomic layers, where both surfaces are terminated by the same species. Two sets of five Al or Ti(001) layers are stacked on both surfaces. Fcc Ti is dealt with similarly to Al, although hcp Ti is the most stable. Two free metal surfaces are separated by a vacuum region of about 15 a.u. in the supercell, which ensures stable interlayer distances without any constraint. Coherent interfaces with ideal (1×1) periodicity are dealt with, where metal layers are expanded along the interface by about 8% for Al and about 6% for Ti. About the rigid-body translation parallel to the interface, we examine four special translations corresponding to the energy extrema. Ten special **k** points per irreducible part are used.

Configurations and Adhesive Energy

Figures 4 and 5 show the stable configurations and electron density distributions of the SiC/Al interface. For the C-terminated interface, the interfacial C atoms are three-fold coordinated. The bond length and bond charge of the C–Al bond are rather similar to the C–Si bond. The charge density is pulled from the Al atom toward the C atom, which generates a charge depletion region near the interfacial Al atom as shown in Fig. 5(a). For the Si-terminated interface, the interfacial Si atoms are four-fold coordinated. The Si–Al bond length is much larger than the C–Al bond length. The interfacial charge density is rather broadly distributed. As listed in Table I, the C–Al bond with both covalent and ionic characters is twice as strong as the Si–Al bond with more metallic characters.

Figures 6 and 7 show the results of the SiC/Ti interface. In both configurations, the interfacial C or Si atoms are four-fold coordinated. The Si-terminated interface has features rather similar to that of the SiC/Al interface. The Si–Ti bond length is similar to the Si–Al bond length, and the interfacial charge density is rather broadely distributed, which indicates a metallic character. Thus, the bond adhesion is relatively small in Table I.

The C-terminated interface has peculiar features quite different from the other interfaces. The C–Ti interlayer distance is very small. The interfacial charge distribution clearly reveals the p–d σ covalent bonds between C $2p$ and Ti $3d$ orbitals, similar to bulk TiC. The humps of the $3d$ electron density at the interfacial Ti atom directed to neighboring C atoms are typical of the e_g-type hybridization in bulk TiC [17]. The local zigzag chain of the interfacial Ti and C atoms is common to the local configuration of bulk TiC, although the C–Ti bond length is a little less than that in bulk TiC. There also exists partial electron transfer from the interfacial Ti atom toward the C atom as shown in Fig. 7(a). The strong C–Ti bond with both covalent and ionic characters causes very large adhesive energy in Table I.

Figure 4. Stable configurations of the SiC(001)/Al interface. (a) $(1\bar{1}0)$ cross section of the C-terminated interface, and (b) (110) cross section of the Si-terminated interface. Contours of the valence electron density are plotted from 0.01 a.u.$^{-3}$ to 0.29 a.u.$^{-3}$ in spacing of 0.02 a.u.$^{-3}$.

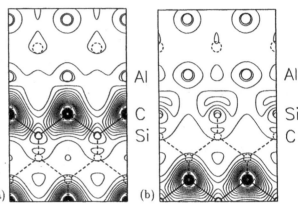

Figure 5. Valence density averaged on each (001) plane in the supercell plotted along the ⟨001⟩ axis. Asterisks indicate atomic positions. (a) The C-terminated interface and (b) the Si-terminated interface.

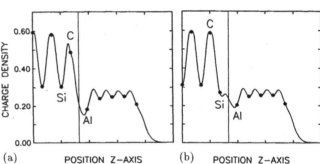

Table I. Calculated adhesive energy E_{ad} and p-type SBH of the SiC(001)/Al and SiC(001)/Ti interfaces. E_{ad} is the energy gain per unit area in forming one interface from two relaxed (1×1) surfaces. SHB is the difference between the Fermi level and the valence-band top of the bulk SiC region in the supercell by the analysis of the local densities of states.

SiC(001)/Al interface	E_{ad}	SBH
C-terminated	6.42 Jm^{-2}	0.08 eV
Si-terminated	3.74 Jm^{-2}	0.85 eV
SiC(001)/Ti interface		
C-terminated	8.74 Jm^{-2}	0.22 eV
Si-terminated	2.52 Jm^{-2}	0.50 eV

The interfacial C–Ti layers can be regarded as some type of TiC compound layers quite different from bulk SiC and bulk Ti. As shown in Fig. 7(a), the interlayer distance at the back Ti–Ti bond is increased by about 21% than the averaged value, which greatly reduces the interlayer charge density. Thus, the plane between the 1st and 2nd Ti layers should be regarded as a new interface between TiC and Ti. It can be said that the C-terminated SiC(001) surface can react with Ti atoms even at low temperature, and that TiC compound layers can be formed spontaneously. This point is in good agreement with experiments [18].

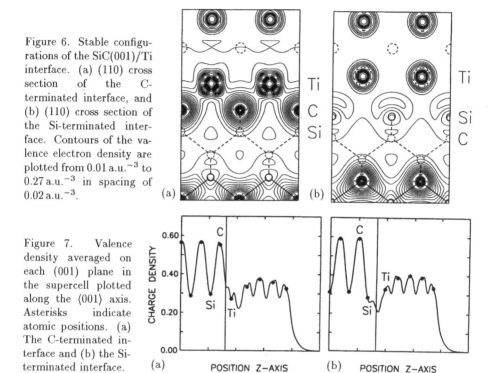

Figure 6. Stable configurations of the SiC(001)/Ti interface. (a) (110) cross section of the C-terminated interface, and (b) (110) cross section of the Si-terminated interface. Contours of the valence electron density are plotted from 0.01 a.u.$^{-3}$ to 0.27 a.u.$^{-3}$ in spacing of 0.02 a.u.$^{-3}$.

Figure 7. Valence density averaged on each (001) plane in the supercell plotted along the $\langle 001 \rangle$ axis. Asterisks indicate atomic positions. (a) The C-terminated interface and (b) the Si-terminated interface.

Electronic Structure and Schottky-Barrier Heights

The local densities of states (LDOS's) of the interface regions of the Si-terminated interfaces clearly show the metallic characters of the Si-Al and Si-Ti bonds, which is consistent with the charge distributions. The LDOS's of the C-terminated interfaces also show the covalent interactions between the C and metal atoms. As listed in Table I, the SBH can be determined by the LDOS's. The SBH is one of the most important electronic properties of ceramic/metal interfaces. The SBH of the Si-terminated SiC(001)/Al interface is in good agreement with the experimental value, 0.9 eV [19]. It is clear that the C-terminated interface has smaller p-type SBH for each system. This can be explained by the interface dipole caused by the interfacial charge transfer from the interfacial Al or Ti layer to the C layer as shown in Figs. 5(a) and 7(a). The present dependence of the SBH on the interface structure is contrary to previous models for the SBH [20] which deny such dependence. However, this dependence is consistent with experiments and *ab initio* calculations of semiconductor/metal interfaces without interfacial defects [21]. We think that the following two factors should dominate the SBH if the effects of interfacial defects can·be ignored. The first is the absolute relation between the band structure of two materials. The second is the interface dipole, which should greatly affect the connection of the two bands. This is determined by the interfacial charge distribution, and should seriously depend on the interface structure.

CONCLUSION

A coincidence tilt boundary in SiC has large tensile strength because of the reconstruction, although the wrong bonds have large effects. The properties of SiC/metal interfaces greatly depend on the surface species and metal species. *Ab initio* calculations of ceramic interfaces can provide useful information for the interface design. *Ab initio* tensile or shear [22] tests should be promising tools to investigate the basic mechanical properties of interfaces.

REFERENCES

[1] P. Hohenberg and W. Kohn, *Physical Review,* **136**, B864 (1964); W. Kohn and J. L. Sham, *ibid.,* **140**, A1133 (1965).
[2] R. Car and M. Parrinello, *Physical Review Letters,* **55**, 2471 (1985); M. C. Payne *et al., Reviews of Modern Physics,* **64**, 1045 (1992).
[3] M. Kohyama, *Philosophical Magazine Letters,* **79**, 659 (1999).
[4] J. Hoekstra and M. Kohyama, *Physical Review B,* **57**, 2334 (1998); M. Kohyama and J. Hoekstra, *ibid.,* **61**, 2672 (2000).
[5] K. Tanaka and M. Kohyama, p.581 in *Electron Microscopy 1998, Vol. II,* ed. by H. A. C. Benavides and M. J. Yacamán (Institute of Physics, 1998).
[6] J. P. Perdew and A. Zunger, *Physical Review B,* **23**, 5048 (1981).
[7] W. E. Pickett, *Computer Physics Report,* **9**, 115 (1989).
[8] M. P. Teter, M. C. Payne, and D. C. Allan, *Physical Review B,* **40**, 12255 (1989).
[9] D. M. Bylander, L. Kleinman, and S. Lee, *Physical Review B,* **42**, 1394 (1990).
[10] N. Troullier and J. L. Martins, *Physical Review B,* **43**, 1993 (1991).
[11] M. Kohyama *et al., Journal of Physics: Condensed Matter,* **2**, 7809 (1990); **3**, 7555 (1991); M. Kohyama, *Materials Chemistry and Physics,* **50**, 159 (1997).
[12] V. B. Deyirmenjian *et al., Physical Review B,* **52**, 15191 (1995).
[13] W. Li and T. Wang, *Physical Review B,* **59**, 3993 (1999).
[14] E. W. Wong *et al., Science,* **277**, 1971 (1997).
[15] A. A. Griffith, *Philosophical Transactions of the Royal Society of London,* **A221**, 163 (1920); G. R. Irwin, *Journal of Applied Mechanics,* **24**, 361 (1957).
[16] J. L. Henshall and C. A. Brookes, *Journal of Materials Science Letters,* **4**, 783 (1985).
[17] P. Blaha *et al., Physical Review B,* **31**, 2316 (1985).
[18] S. Hasegawa *et al., Surface Science,* **206**, L851 (1988).
[19] V. M. Bermudez, *Journal of Applied Physics,* **63**, 4951 (1988).
[20] A. P. Sutton and R. W. Balluffi, *Interfaces in Crystalline Materials* (Oxford, 1995), §11.2.
[21] R. T. Tung, *Physical Review Letters,* **52**, 461 (1984); G. P. Das *et al., ibid.,* **63**, 1168 (1989); H. Fujitani and S. Asano, *Physical Review B,* **42**, 1696 (1990).
[22] C. Molteni *et al., Physical Review Letters,* **76**, 1284 (1996).

MOLECULAR DYNAMICS SIMULATIONS OF SURFACES AND GRAIN BOUNDARIES IN YTTRIA-STABILIZED ZIRCONIA

Craig A. J. Fisher and Hideaki Matsubara
Japan Fine Ceramics Center
2-4-1 Mutsuno, Atsuta-ku, Nagoya 456-8587
Tel: +81-52-871-3500   Fax: +81-52-871-3599   E-mail: fisher@jfcc.or.jp

ABSTRACT

Molecular dynamics (MD) simulations of four surfaces and four grain boundaries in 8 mol% yttria-stabilized zirconia (YSZ) are presented. Two-body Buckingham potentials were used to describe short-range interatomic forces. Oxide ion conductivity was found to vary with orientation and structure of the interfaces.

INTRODUCTION

Fully-stabilized zirconia is an industrially important ceramic used in oxygen gas sensors and solid oxide fuel cells (SOFCs) because of its exceptionally rapid ionic conductivity and good mechanical properties.[1] Several studies have shown that interfaces in zirconia act as microstructural barriers to oxygen diffusion; the more interfaces, the greater the resistance of the material compared to a single crystal.[2,3] Finer grained ceramics, however, typically have higher strengths, so increasing the number of grain boundaries improves the mechanical properties. Optimizing the microstructure, and particularly interfaces, is therefore necessary to simultaneously improve the electrical and mechanical properties of this material.

The high temperature cubic phase of zirconia is stabilized to lower temperatures by addition of aliovalent cations such as Ca^{2+}, Mg^{2+} and Y^{3+}. The oxygen vacancies that form to maintain charge balance are responsible for stabilized zirconia's rapid ionic conductivity.[1] Both experiment[1] and simulation[4] have shown that there is a maximum in conductivity for any given dopant. Although scandia-doped zirconia displays the highest conductivity of the zirconia alloys, yttria-stabilized zirconia (YSZ) is most commonly used because of its lower cost. The maximum in conductivity for YSZ is found at about 8 to 9 mol% Y_2O_3.

To the extent authorized under the laws of the United States of America, all copyright interests in this publication are the property of The American Ceramic Society. Any duplication, reproduction, or republication of this publication or any part thereof, without the express written consent of The American Ceramic Society or fee paid to the Copyright Clearance Center, is prohibited.

In addition to studies of single crystals,[4-8] simulations of interfaces in tetragonal and cubic zirconia have been reported, although these were all by the energy-minimization technique for structures at 0 K.[9-13] Ours is the first study to model both surfaces and grain boundaries in YSZ at elevated temperatures *via* MD calculations, and the first to report their effect on the oxide ion conductivity.

SIMULATION METHOD

Potential Model

The MD method involves the solution of Newton's equations of motion for a well-defined system of particles by finite difference methods over a large number of time-steps, Δt. The interatomic forces are calculated from potential energy functions, which in the case of purely ionic solids consist of an electrostatic component combined with a short-range term to take into account Pauli repulsion and van der Waal's forces due to electron cloud overlap. A common expression for the short-range term is the Buckingham potential used here:

$$\phi(r_{ij}) = A\exp(-\frac{r_{ij}}{\rho}) - \frac{C}{r_{ij}} \tag{1}$$

where A, ρ and C are parameters particular to interactions between two ions i and j separated by distance r_{ij}. In this study, we used the potential parameters of Lewis and Catlow,[14] as these successfully reproduce the lattice parameters and cell geometry of cubic YSZ to within a couple of percent. All simulations were performed using the MOLDY code.[15]

Initial Configurations

Four different surfaces were constructed by truncating single crystals of ZrO_2 at two positions parallel to either the (111), (110), (112) or (310) planes. The surfaces were at least 60 Å apart in both directions to ensure there was no interaction between them. Surface systems contained around 1500 atoms.

Initial grain boundary configurations were constructed according to CSL theory[16] by taking two ZrO_2 single crystals and tilting or twisting one of them by a given angle, θ, relative to the other until a proportion $(1/\Sigma)$ of the atomic sites of both crystals coincided. Each grain boundary was placed at the center of a simulation box containing around 2000 atoms and at least 70 Å in length. Application of 3D periodic boundary conditions created a second (equivalent) grain boundary at the box edges parallel with, but in the reverse direction to, the first boundary.

The four grain boundaries investigated were the $\Sigma 5a$ (310)/[001], $\Sigma 13$ (320)/[001], $\Sigma 5b$ $(2\bar{1}1)$/[011] symmetrical tilt boundaries and $\Sigma 3$ (111) $\theta = 60°$ symmetrical twist boundary. The $\Sigma 3$ boundary is equivalent to a $(1\bar{1}1)$/[011] tilt boundary, owing to the symmetry of the cubic fluorite structure, but we choose to distinguish it from the other three grain boundaries because of this extra symmetry.

Systems of YSZ were created by randomly substituting the number of yttrium ions that gave a composition closest to 8 mol% Y_2O_3 (usually within ± 0.1 mol%). The corresponding number of oxygen ions were then removed at random to create oxygen vacancies and maintain charge balance.

Simulation Conditions

The simulation temperature for grain boundaries was 1000°C, and for surfaces, 1500°C. Systems were equilibrated (*i.e.* "thermalized") at the appropriate temperature for at least 10,000 time steps before commencing the simulation run proper. In the case of grain boundary systems, the simulation box was allowed to expand/contract in the direction perpendicular to the boundary plane to allow for restructuring of the boundary during equilibration. Once the average potential energy reached a constant value, the average length of the simulation box was calculated, and this was used as the box length in the simulation run. Both surface and grain boundary simulations were performed in the microcanonical ensemble for 60,000 time steps, giving a total simulation time of 120 ps.

Simulation Analysis

Particle positions were periodically recorded during the course of each run to allow calculation of transport properties, in particular the mean square displacement (*msd*). The *msd* of a species numbering N particles is defined as:

$$msd = \frac{1}{NN_t} \sum_{n=1}^{N} \sum_{t_0}^{N_t} |r_n(t_o + t) - r_n(t_o)|^2 \tag{2}$$

where N_t is the number of initial times, t_0, averaged over, and $r_i(t)$ is the position of ion i at time t. The *msd* gives a measure of the diffusion undergone by a particular species during the course of the simulation, since according to the Einstein equation, the gradient of a plot of *msd* versus time is proportional to the diffusion coefficient, D_i.

For a cubic crystal, the conductivity, σ, can be calculated using:

$$\sigma_i = \frac{f(Z_i e)^2 ND_i}{k_B T} \qquad (4)$$

where Z_i is the charge of species i, e is the unit of electronic charge, k_B is Boltzmann's constant, T is temperature, and f is the Haven ratio, which we take to be 0.7815.[7]

RESULTS & DISCUSSION

Surfaces

The {110}, {112} and {310} planes of the cubic fluorite structure are Type I faces, each containing stoichiometric numbers of cations and anions, so that there is no dipole moment perpendicular to these planar surfaces.[17] The {111} surface, however, is a Type II face, in which alternating layers contain only anions or cations, but for a finite crystal thickness the layers are symmetric about the central layer and contain a stoichiometric number of ions, so that again there is no net dipole perpendicular to the surface if the crystal is cut in such a way that the stiochiometric crystal layer groups remain intact. Type III faces, which have a net dipole moment perpendicular to the surface, are not studied here as they are intrinsically unstable in the pure material.[17]

The {111} surface has the highest density of ions, and undergoes almost no changes during equilibration. Table I shows that this surface has the lowest energy at 1500°C, in agreement with previous static lattice simulations,[9,10] as well as the experimental observation that this surface dominates the morphology of cubic zirconia single crystals. The {110}, {112} and {310} surfaces, on the other hand, all undergo varying degrees of reconstruction. In the case of the {110} surface, oxygen ions extend a small way out of the surface above their normal positions, in agreement with calculations by Mackrodt.[10]

Table I: Calculated Surface Energies for 8YSZ at 1773 K.

Surface	Surface Energy (J/m^2)
{111}	1.2
{110}	2.2
{112}	2.3
{310}	3.2

Grain Boundary Engineering in Ceramics

As shown in Fig. 1, the {112} and {310} surfaces undergo substantial reconstruction compared to a simple truncation of the crystal. In both cases, oxygen atoms move out of the initial surface plane to form an extra layer, thereby "shielding" the otherwise exposed cations at the surface.

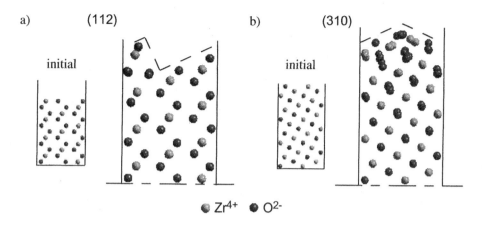

Fig. 1. Time-averaged positions of ions at surfaces a) (112) and b) (310) of ZrO$_2$.

Msds were calculated for bulk (b) and surface (s) regions of each system, and plotted against time (Fig. 2). The {111} surface supports faster oxygen diffusion in the top two surface layers than in the bulk, even when the surface region contained a slightly higher yttrium content than the bulk. The third and fourth layers, which we include as part of the bulk *msd* here, actually have a lower oxygen diffusion rate than the innermost crystal, which is why the slope of the (111) bulk component is lower than that of the single crystal.

The three other surfaces, however, showed lower diffusion rates in the surface region than the bulk. Interestingly, while the {111}, {110} and {112} surfaces displayed negligible cation diffusion, even at the surface, the cations in the {310} surface region were found to diffuse, albeit much more slowly than the anions.

Grain Boundaries

Upon relaxation, the Σ5a and Σ5b tilt boundaries underwent shifts from the initial CSL configuration equivalent to introducing an extra layer of vacancies at the boundary plane. This resulted in these two boundaries having the highest excess volumes and lowest atomic densities of the four boundaries studied (Table II).

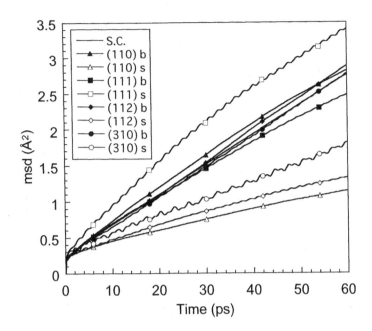

Fig. 2. Surface (s) and bulk (b) components of oxygen *msd*s for various surface planes compared with a single crystal (S.C.) of 8 YSZ at 1773 K.

Table II: Calculated Properties of Grain Boundaries.

Boundary	Misorientation Angle (°)	Excess Volume (%)	Formation Energy (J/m²)	Binding Energy (J/m²)
Σ5a tilt	36.9	1.1	2.2	4.3
Σ13 tilt	67.4	0.8	1.7	-
Σ5b tilt	109.5	1.0	1.9	-
Σ3 twist	60.0	0.3	1.2	1.2

The plot of overall system *msd*s versus time in Fig. 3 reveals that the three tilt boundaries significantly degrade the ionic conductivity. The Σ3 twist boundary, however, has a slightly higher conductivity than a single crystal (S.C.), which can be correlated with the small excess volume and narrow boundary width. However, the low binding energy (the energy released on forming the boundary from two surfaces per unit area) suggests that the Σ3 boundary is more easily fractured than the other grain boundaries, which may be detrimental to the mechanical performance of the material.

　　　　　　　　　　　　　Grain Boundary Engineering in Ceramics

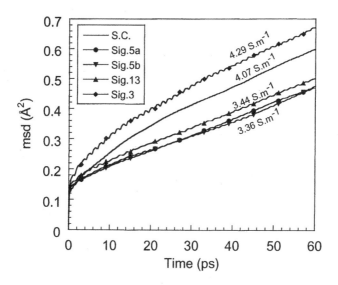

Fig. 3. Overall oxygen *msd*s for grain boundary systems in 8YSZ compared with a single crystal (S.C.) at 1000°C.

CONCLUSIONS

Our MD simulations of surfaces and grain boundaries in 8 mol% YSZ show that, on a microscopic level, the oxygen ion diffusion varies with interface structure and crystal orientation. Of the four surfaces examined, only the lowest energy, highest density surface, {111}, displayed the same diffusion rate as in the bulk crystal, while the other three surfaces all had lower conductivities. Similarly, the three high energy, high excess volume grain boundaries all decreased the overall anion conductivity, while the Σ3 twist boundary supported a slightly higher conductivity due to the fortuitous matching of cation sublattices between the two crystals constituting the grain boundary.

REFERENCES

[1] R. Stevens, *Zirconia and Zirconia Ceramics*, 2nd ed, Magnesium Elektron, Twickenham, 1986.
[2] M. J. Verkerk, B. J. Middelhuis and A. J. Burggraaf, "Effect of Grain Boundaries on the Conductivity of High-Purity ZrO$_2$-Y$_2$O$_3$ Ceramics," *Solid State Ionics*, **6**, 159-70 (1982).

[3] M. Aoki, Y.-M. Chiang, I. Kosacki, L. J.-R. Lee, H. Tuller and Y. Liu, "Solute Segregation and Grain-Boundary Impedance in High-Purity Stabilized Zirconia," *J. Am. Ceram. Soc.*, **79** [5] 1169-80 (1996).

[4] X. Li and B. Hafskjold, "Molecular Dynamics Simulations of Yttrium-stabilized Zirconia," *J. Phys.: Condens Matter*, 7, 1255-74 (1995).

[5] F. Shimojo, T. Okabe, F. Tachibana, M. Kobayashi and H. Okazaki, "Molecular Dynamics Studies of Yttria Stabilized Zirconia, I. Structure and Oxygen Diffusion," *J. Phys. Soc. Jpn*, **61** [8] 2848-57 (1992).

[6] F. Shimojo and H. Okazaki, "Molecular Dynamics Studies of Yttria Stabilized Zirconia. II. Microscopic Mechanism of Oxygen Diffusion," *J. Phys. Soc. Jpn*, **61** [11] 4106-18 (1992).

[7] H. W. Brinkman, W. J. Briels and H. Verweij, "Molecular Dynamics Simulations of Yttria-stabilized Zirconia," *Chem. Phys. Lett.*, **247** 386-90 (1995).

[8] K. Suzuki, M. Kubo, Y. Oumi, R. Miura, H. Takaba, A. Fahmi, A. Chatterjee, K. Teraishi and A. Miyamoto, "Molecular Dynamics Simulation of Enhanced Oxygen Ion Diffusion in Strained Yttra-stabilized Zirconia," *Appl. Phys. Lett.*, **73** [11] 1502-4 (1998).

[9] D. Bingham, P. W. Tasker and A. N. Cormack, "Simulated Grain-Boundary Structures and Ionic Conductivity in Tetragonal Zirconia," *Phil. Mag.* **A 60** [1] 1-14 (1989).

[10] W. C. Mackrodt, "Atomistic Simulation of the Surfaces of Oxides," *J. Chem. Soc., Fara. Trans. 2*, **85** [5] 541-54 (1989).

[11] G. Balducci, J. Kaspar, P. Fornasiero, M. Graziani, M. S. Islam and J. D. Gale, "Computer Simulation Studies of Bulk Reduction and Oxygen Migration in CeO_2-ZrO_2 Solid Solutions," *J. Phys. Chem. B*, **101**, 1750-3 (1997).

[12] G. Balducci, J. Kaspar, P. Fornasiero, M. Graziani and M. S. Islam, "Surface and Reduction Energetics of the CeO_2-ZrO_2 Catalysts," *J. Phys. Chem. B*, **102** [3] 557-61 (1998).

[13] A. Christensen and E. A. Carter, "First-Principles Study of the Surfaces of Zirconia," *Phys. Rev. B*, **58** [12] 8050-64 (1998).

[14] G. V. Lewis and C. R. A. Catlow, "Potential Models for Ionic Oxides," *J. Phys. C: Solid State Phys.*, **18**, 1149-1161 (1985).

[15] K. D. Refson, "Moldy: A Portable Molecular Dynamics Simulation Program for Serial and Parallel Computers," *Comp. Phys. Comm.*, **126** [3] 309-28 (2000).

[16] H. F. Fischmeister, "Structure and Properties of High-Angle Grain Boundaries," *J. de Physique*, **46** [C4] 3-23 (1985).

[17] P. W. Tasker, "The Stability of Ionic Crystals," *J. Phys. C*, **12**, 4977 (1979).

ATOMISTIC COMPUTER SIMULATIONS OF AMORPHOUS SILICON NITRIDE BASED CERAMICS

Katsuyuki Matsunaga and Hideaki Matsubara
Japan Fine Ceramics Center
2-4-1 Mutsuno Atsuta-ku, Nagoya 457-8587 JAPAN
E-mail: matunaga@jfcc.or.jp

ABSTRACT

Molecular dynamics simulations were carried out for amorphous silicon nitride (Si-N) and borosilicon nitride ceramics (Si-B-N). Short-range atomic arrangement and self-diffusion behavior in these materials were investigated. In amorphous Si-B-N, boron atoms were bonded to three nitrogen atoms. It was found that the self-diffusion constants of atoms in Si-B-N were much decreased, as compared to those in amorphous Si-N. This indicates that boron atoms play a role in reducing atomic diffusivities in Si-B-N at high temperatures, which may explain the observed thermal stability of amorphous Si-B-N due to the presence of boron atoms.

INTRODUCTION

In recent years, amorphous covalent Si_3N_4-based ceramics have received much interest because of their high-temperature properties. The materials can be synthesized by pyrolysis of polymer precursors, and are candidates for novel applications, such as ceramic fibers and coatings, due to their excellent thermal stability and mechanical properties at high temperatures.

In this class of materials, the amorphous state of boron-doped silicon nitride (Si-B-N) has excellent thermal stability. Baldus et al. [1] reported that polymer-derived $Si_3B_3N_7$ preserves its amorphous state up to around 1700°C without crystallization into Si_3N_4 and BN. The thermal stability of amorphous $Si_3B_3N_7$ is much higher than that of amorphous Si-N where the amorphous state is maintained up to around 1000°C. Riedel et al. [2] also found that the amorphous state of Si-B-C-N is maintained up to 2000°C.

To the extent authorized under the laws of the United States of America, all copyright interests in this publication are the property of The American Ceramic Society. Any duplication, reproduction, or republication of this publication or any part thereof, without the express written consent of The American Ceramic Society or fee paid to the Copyright Clearance Center, is prohibited.

As stated above, the doping of boron or carbon into the amorphous state plays an important role in improving the thermal stability of amorphous silicon nitride. In order to consider the enhanced thermal stability of amorphous silicon nitride due to boron, it is important to study atomic structures of the amorphous network and their related properties. Thus, we have employed molecular dynamics (MD) simulations for the study of boron-doped amorphous silicon nitride. Special attention has been paid to short-range atomic arrangements and atomic transport properties in the amorphous state. In particular, the self-diffusion behavior of atoms in the amorphous state at high temperatures may be closely related to the atomic rearrangement of the amorphous network toward crystallization. Effects of boron on the thermal stability are discussed in terms of atomic diffusivities.

COMPUTATIONAL PROCEDURE

For MD simulation, we used Tersoff potentials to describe interactions between atoms in the present simulated systems. Details of the formalism of Tersoff potentials have been given elsewhere.[3] The Tersoff potentials used in this study include three-body bonding interactions of Si-N, Si-Si and B-N pairs, which originate from the covalent nature of bonding. Available potential parameters were obtained for Si_3N_4[4] and for BN[5] which were demonstrated to successfully describe structural and elastic properties of crystalline and amorphous Si_3N_4 and cubic BN.[5,6]

MD simulations were performed using the MASPHYC program (Fujitsu). Systems of 896 atoms for amorphous Si_3N_4 and of 832 atoms for amorphous $Si_3B_3N_7$ with three-dimensional periodic boundary condition were treated in the constant temperature and volume ensemble. The temperature was controlled by the scaling of particle velocities. At the beginning of simulation runs, the constituent atoms were randomly distributed in a cubic simulation box, where the density was 2.5 g/cm³. The initial temperature was kept at 7727°C for 3.0 ps, in order to agitate atoms in a simulation cell and eliminate the effect of the initial atomic arrangement. The system was then cooled to 1127°C for 1.0 ps, and equilibrated at 1127°C for more than 5.0 ps. Properties of simulated systems were obtained from time averaged data over the last 50.0 ps at 1127°C. The equations of motion were integrated using a time interval of 0.2 fs for all runs.

RESULTS AND DISCUSSION

Short-range atomic arrangement in amorphous Si-B-N

Figure 1 shows the typical example of the atomic structure of amorphous $Si_3B_3N_7$

obtained by MD at 1127°C. In the amorphous network, Si and B atoms are bonded mainly to N atoms, and yet some Si-Si bonding pairs are found. In contrast, N atoms have bonding to both Si and B atoms. These bonding pairs are homogeneously distributed through the random amorphous network.

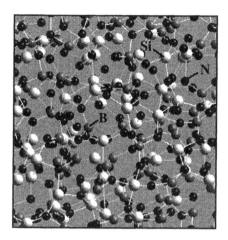

Figure 1. Snapshot of the atomic structure of amorphous $Si_3B_3N_7$ generated by MD.

We also examined the coordination numbers of atoms in the MD-simulated amorphous $Si_3B_3N_7$. We found that the coordination number of B to N was found to be 3.32 and the average bond length of a B-N pair was 1.49 Å. This indicates that boron mainly forms a three-fold structural unit of BN_3 as found in crystalline h-BN, although there are some four-fold coordinated BN_4 units in the simulated system. Such a structural unit around boron was experimentally observed in Si-B-N or Si-B-C-N by Baldus et al.[1], so that the characteristic of the amorphous structure of Si-B-N can be described by the present MD simulation.

Atomic mobility in Si-B-N

As shown above, the amorphous states of Si-N and Si-B-N have a homogeneous distribution of constituent atoms. The homogeneity of atomic distribution in the amorphous state was also observed experimentally.[1] In contrast, the amorphous states of Si-N and Si-B-N crystallize into thermodynamically stable phases of Si_3N_4 and BN by annealing at elevated temperatures. The amorphous network will rearrange itself so as to give rise to phase separation into Si_3N_4-rich and BN-rich regions before crystallization. Such rearrangement of the amorphous network requires atomic diffusion of Si, B and N through the amorphous state. Thus, we examined the self-diffusion behavior in amorphous Si_3N_4 and $Si_3B_3N_4$ by MD.

Figure 2 displays trajectories of nitrogen in amorphous Si_3N_4 (Fig. 2(a)) and $Si_3B_3N_7$ (Fig. 2(b)) obtained by MD at 1127°C. The upper figures in Fig. 2(a) and (b) indicate the atomic positions of nitrogen in the simulation cells. It can be seen from Fig. 2(a) that each N atom shows a broad trajectory. This means that N atoms not only make significant thermal vibration, but also are able to diffuse through the amorphous state. While the trajectory of each nitrogen in $Si_3B_3N_7$ (Fig. 2(b)) is restricted to a smaller region than that in Si_3N_4. The atomic mobility of nitrogen is much reduced by the incorporation of boron into amorphous silicon nitride.

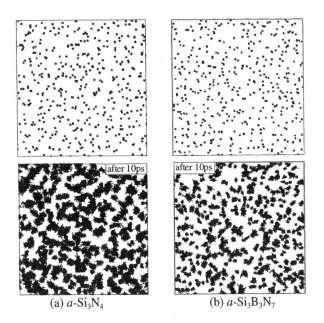

(a) a-Si_3N_4 (b) a-$Si_3B_3N_7$

Figure 2. Trajectories of nitrogen during 10.0 ps in (a) amorphous Si_3N_4 and (b) $Si_3B_3N_7$ at 1127°C.

Atomic diffusivities can be examined by analysis of mean square displacement (*msd*) curves. Figure 3 shows the *msd* curves of constituent atoms in amorphous Si_3N_4 and $Si_3B_3N_7$ at 1127°C. We can see that the *msd* curves of Si and N in amorphous Si_3N_4 increase rapidly with time, indicating that atomic diffusion of Si

                     Grain Boundary Engineering in Ceramics

and N takes place through the amorphous state.

The *msd* profiles are strongly influenced by addition of boron atoms. The amplitudes of the *msd* curves for amorphous $Si_3B_3N_7$ are much smaller than those for amorphous Si_3N_4. The *msd* curves for $Si_3B_3N_7$ increase with time in a similar way to those in Si_3N_4. However, the slopes of the *msd* curves exhibit smaller values, as compared to those in Si_3N_4. These results mean that the motions of Si and N atoms in $Si_3B_3N_7$ are significantly restricted by addition of boron.

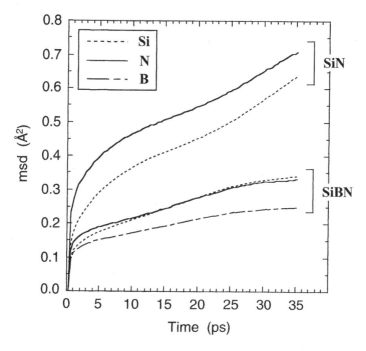

Figure 3. Mean square displacement (*msd*) curves of amorphous Si_3N_4 and $Si_3B_3N_7$ at 1127°C.

The decrease of atomic mobility in amorphous $Si_3B_3N_7$ can be attributed to the presence of B-N linkages in the amorphous network. According to experimental data for BN and SiN molecules, the bond energy of B-N (474 kJ/mol) is higher than that of Si-N (371 kJ/mol).[7] The strongly bonded B-N linkages in $Si_3B_3N_7$ become rigid nodes in the amorphous network. When rigid linkages are homogeneously distributed, the atomic rearrangement throughout the amorphous state can be inhibited even at high temperatures. This may explain the observed thermal stability of the amorphous state of Si-B-N.

CONCLUSIONS

We performed MD simulation of amorphous Si_3N_4 and $Si_3B_3N_7$ using the Tersoff potentials. We examined their thermal stability in terms of the local atomic arrangement and atomic diffusivity in the amorphous network. In amorphous $Si_3B_3N_7$, B atoms are bonded to three N atoms. It was found that the self-diffusion constants of constituent atoms in $Si_3B_3N_7$ were much smaller than those in amorphous Si_3N_4. Our simulation clearly shows that the presence of B-N linkages in Si-B-N can reduce the atomic mobility in the amorphous state. This may explain the enhanced thermal stability of boron-doped amorphous Si_3N_4, which has been reported experimentally.

References

[1] H.-P. Baldus, M. Jansen and O. Wagner, " New Materials in The System Si-(N,C)-B and Their Characterization," *Key Eng. Mater.* **89-91** 75-80 (1994).

[2] R. Riedel, A. Kienzle, W. Dressler, L. Ruwisch, J. Bill and F. Aldinger, "A Silicoboron Carbonitride Ceramic Stable to 2000 °C," *Nature* **382** 796-798 (1996).

[3] J. Tersoff, "New Empirical Approach for The Structure and Energy of Covalent Systems," *Phys. Rev. B* **37** [12] 6991-7000 (1988).

[4] P. M. Kroll, "Computer Simulations and X-Ray Absorption Near Edge Structure of Silicon Nitride and Silicon Carbonitride Ceramics," Ph. D. Thesis, Technische Hochschule Darmstadt (1996).

[5] K. Matsunaga, C. Fisher and H. Matsubara, "Tersoff Potential Parameters for Simulating Cubic Boron Carbonitrides," *Jpn. J. Appl. Phys.* **39** [1] L48-L51 (2000).

[6] K. Matsunaga, Y. Iwamoto, C. Fisher and H. Matsubara, "Molecular Dynamics Study of Atomic Structures in Amorphous Si-C-N Ceramics," *J. Ceram. Soc. Jpn*, **107** [11] 1025-1031 (1999).

[7] M. W. Chase, Jr., C. A. Davis, J. R. Downey, Jr., D. J. Frurip, R. A. McDonald and A. N. Syverud, JANAF Thermochemical Tables, 3rd ed., J. Phys. Chem. Ref. Data, Suppl. 1, **14** pp. 247 and pp. 1540 (1985).

AB INITIO CALCULATIONS OF 3C-SiC(111)/Ti POLAR INTERFACES

Shingo Tanaka (SWING) and Masanori Kohyama
Department of Material Physics, Osaka National Research Institute,
Agency of Industrial Science and Technology,
1-8-31, Midorigaoka, Ikeda, Osaka 563-8577, Japan

ABSTRACT

Ab initio pseudopotential calculations of the 3C-SiC(111)/Ti polar interface (equivalent to 6H-SiC(0001)/Ti interface) have been performed, and are compared with our previous results of the 3C-SiC(001)/Ti interface [Phys. Rev. B **61** [4] 2672-79 (2000)]. The C-terminated and Si-terminated interfaces have interesting features, different from the 3C-SiC(001)/Ti interface. The C-Ti bond at the C-terminated interface has strong p-d covalent interactions between C-$2p$ and Ti-$3d$ orbitals, similar to the 3C-SiC(001)/Ti interface. On the other hand, the Si-Ti bond at the Si-terminated interface also has covalent interactions such as dimer, different from the metallic bond at the Si-terminated interface of 3C-SiC(001)/Ti. This result appears that the stable Ti layers grow on the Si-terminated 3C-SiC(111) (or 6H-SiC(0001)) surface. The adhesive energy value of the C-terminated interface is larger than that of the Si-terminated interface; however, the difference in their values is small compared with the 3C-SiC(001)/Ti interface. The value of the Si-terminated interface is about two times half larger than that in the 3C-SiC(001)/Ti, while the value of the C-terminated interface is slightly small. Therefore, the present Si-terminated interface is rather strong.

INTRODUCTION

SiC is a very important material for high-temperature, high-speed, and high-power electronic and optelectronic devices, as well as high-temperature structural ceramics. SiC/metal interfaces are essential to develop such devices. To fabricate SiC/metal interfaces with desirable electronic, mechani-

To the extent authorized under the laws of the United States of America, all copyright interests in this publication are the property of The American Ceramic Society. Any duplication, reproduction, or republication of this publication or any part thereof, without the express written consent of The American Ceramic Society or fee paid to the Copyright Clearance Center, is prohibited.

cal and thermal properties is crucial for electronic and structural applications. Numerous experiments have been performed for the SiC/Ti system, because Ti or Ti-containing alloys are often used for such applications[1-5].

On the theoretical side, *ab initio* calculations of the atomic and electronic structure of the SiC(111)/TiC interface[6] has been performed using the full-potential linear-muffin-tin orbital (FP-LMTO) method based on the density functional theory (DFT) with local density approximation (LDA)[7]. However, it is of great importance to clarify the nature of direct interfaces between SiC and Ti before generating compound layers in order to understand the interface reactions. To understand different reactivity of C and Si atoms at SiC surfaces with Ti, semi-empirical molecular-orbital calculations[8] and our first-principles pseudopotential calculations, 3C-SiC(001)/Al[9] and 3C-SiC(001)/Ti[10], were performed. In our calculations, the C-terminated and Si-terminated interfaces have quite different features.

In this paper, we perform ab initio calculations of the 3C-SiC(111)/Ti polar interface for the first time. We deal with both C-terminated and Si-terminated (111) interfaces in order to examine C-Ti and Si-Ti interactions, respectively. We use *ab initio* pseudopotential method, and obtain stable atomic configurations at zero temperature, adhesive energies and electronic properties of the two interfaces. In addition, the present results are compared with our preceding calculations[10].

From studies using recent electron microscopy techniques and theoretical calculations[9-17], it can be generally said that the bonding nature and adhesion of ceramic/metal interfaces are dominated by the following two factors. The first is the bonding nature of respective ceramics and metals and their combination[14-16]. The second is the problem of polar interfaces, termination atoms, or interface stoichiometry. It has been observed in *ab initio* calculations of the MgO/Cu[17] and $Al_2O_3(0001)$/Nb interfaces[16], where polar and non-polar MgO surfaces or O-terminated and Al-terminated Al_2O_3 surfaces generate interfaces with quite different features, respectively. For the 3C-SiC(001)/Al interface[9] and 3C-SiC(001)/Ti interfaces[10], we have also found that the C-terminated and Si-terminated interfaces have quite different features including atomic configurations, adhesion and bonding nature. The same problem is examined for the present interface.

THEORETICAL METHOD

The details of the computational scheme are given in previous papers[9,10]. Total energies, stable conflgurations and electronic properties are given in the framework of the *ab initio* pseudopotential method based on the DFT-LDA[7,18]. The electronic ground state is efficiently obtained using the conjugate-gradient technique[19] for metallic systems with an effective mixing scheme[20] preventing the charge sloshing instability[21]. Stable configurations are obtained through relaxation according to Hellmann-Feynman forces. We use the TM-type optimized pseudopotentials[22]. The separable form[23] is used with the local p component for Si and C and with the local s component for Ti. The plane-wave cut-off energy of 30 Ry is used.

Each supercell contains a slab of 14 3C-SiC(111) atomic layers, where both surfaces are terminated by the same species and the stacking-faults are introduced in the center of the slab. Two sets of four fcc-Ti(111) layers are stacked on both surfaces. Although hcp-Ti is the most stable at low temperature, the reason why Ti atoms are dealt with the fcc structure is that the experimental result of the transmission electron microscopy (TEM) at the 6H-SiC(0001)/Ti interface[24] shows fcc-structure in the Ti thin-layer. In our preliminary calculation for bulk Ti, the total energy difference between the fcc and the hcp structure is quite small (< 10 mRy). Two free metal surfaces are separated by a vacuum region of about 18 a.u. in the supercell, which ensures stable interlayer distances without any constraint. All the configurations have special symmetry of the point group D_{3h} and the space group is P6m2 (#182). The 16 special k-points per irreducible Brillouine zone are used. For fractional occupanceis, the Gaussian broadening scheme[25] are used.

RESULTS AND DISCUSSION

The Si-terminated interface

Figure 1(a) shows the stable atomic configuration and valence charge distribution on the [011] cross-sections. At the interface, the Ti atom is on top of the Si atom. The interlayer distance between the Si and Ti layers, equal to Si-Ti bond length, is 4.76 a.u. This is larger than that of the C-terminated interface mentioned below. The density has a peak above the interface Si atom and d electrons are distributed only near Ti atoms. The d electrons of Ti atom at the interface reveal somewhat polarized distribution differently from other Ti atoms.

From Fig. 1(a), the atomic configuration and electron distribution recover the bulk features quickly at the back Si-C and back Ti-Ti bonds. The features of the charge distribution are different from those of the 3C-SiC(001)/Ti interface. The charge density of Ti atom at the interface has four humps and the peak value of the valence density above the Si atom is rather large. It seems that the interfacial Si-Ti bond has covalent interactions as a dimer and is stronger than the metallic bond at the Si-terminated 3C-SiC(001)/Ti interface.

Figure 2(a) shows the local density of state (LDOS) of the Si-terminated 3C-SiC(111)/Ti interface. The LDOS is calculated for each region between succesive (111) layers of the supercell[9,10]. The LDOS's at the back C-Si bond and the back Ti-Ti bond recover the bulk features, which is consistent with the features of charge distribution and atomic configuration.

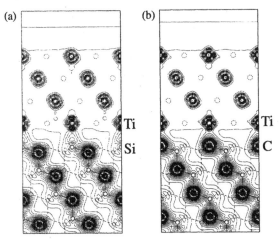

Fig.1 Relaxed configulation and valence charge density of (a) Si-terminated and (b) C-terminated 3C-SiC(111)/Ti interface. Atoms and Si-C bonds are represented by circles and straight lines. Broken circles and lines indicate those not located on the same plane.

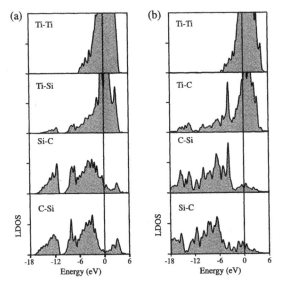

Fig.2 Local density of states of (a) Si-terminated and (b) C-terminated 3C-SiC(111)/Ti interface. The energy zero is the Fermi level.

Grain Boundary Engineering in Ceramics

The C-terminated interface

Figure 1(b) shows the stable atomic configuration and valence charge distribution of the C-terminated interface. The interlayer distance between the C and Ti layers, the C-Ti bond length, is 4.34 a.u. These are smaller than those of the Si-terminated interface, 4.76 a.u. However, the C-Ti bond length is much larger than that of the 3C-SiC(001)/Ti interface. It is derived from the difference of the atomic configuration, because the Ti atoms are fixed on top of the interface C atoms while the interface Ti atoms are coordinated between the interface C atoms at the 3C-SiC(001)/Ti like bulk TiC. Concerning the charge distribution, it is clear that the *d* electrons of Ti atoms at the interface have special hybridization quite different from *d* electrons of other Ti atoms. As the Si-terminated interface, the charge density of the Ti atoms has four humps. It can be said that the C-Ti bonds with strong *p-d* covalent interactions are formed at the interface, similar to the 3C-SiC(001)/Ti interface.

Figure 2(b) shows the LDOS's of the C-terminated SiC/Ti interface. The LDOS at the interface also shows the strong covalent interactions between the C and Ti atoms. However, the LDOS's at the back C-Si bond is much smaller below the Fermi level. The cause of reduction is not clear. The first possibility is the effect of stacking-faults. If the distance between stacking-faults and nearly interface atoms is not enough, it is possible to occur the charge sloshing. The second possibility is the consistency of the cut-off energy. The cut-off energy for the LDOS calculation, 30 Ry, is not so large. Thus, more larger cut-off energy is neccesary in the calculation.

Bond adhesion

The adhesive energies of the interfaces have been obtained by comparing the total energies of relaxed interface systems with those of the relaxed (1x1) surfaces of pure material. Results are listed in Table I with those of the 3C-SiC(001)/Ti interface. For the Si-terminated interface, the adhesive energy is 6.42 Jm^{-2} per one interface. It is 2.5 times larger than that of the Si-terminated 3C-SiC(001)/Ti interface, 2.52Jm^{-2}. This is consistent with the experimental results that the Si-terminated interface tended to form in 6H-SiC(0001)/Ti[4,24] equivalent to 3C-SiC(111)/Ti. For the C-terminated interface, the adhesive energy is 7.83 Jm^{-2}. This value is slightly smaller than that of the C-terminated 3C-SiC(001)/Ti interface, 8.74 Jm^{-2}. If the interface Ti atoms are the same site as that in the 3C-SiC(001)/Ti, it seems that the value is rather large. The present calculation of the interfaces shows that both surface atoms can react with Ti atoms even at zero temperature. In any case, the strong interactions between the surface atoms and Ti shown in the present calculations should dominate interface reactions in many applications.

Table I. Calculated bond adhesion of the Si-terminated and
C-terminated 3C-SiC/Ti interfaces.

	3C-SiC(111)/Ti	3C-SiC(001)/Ti[10]
Si-terminated	6.42 Jm^{-2}	2.52 Jm^{-2}
C-terminated	7.83 Jm^{-2}	8.74 Jm^{-2}

CONCLUSION

Ab initio pseudopotential calculations of the 3C-SiC(111)/Ti polar interface have been performed, and are compared with our previous results of the 3C-SiC(001)/Ti interface [Phys. Rev. B **61** [4] 2672-79 (2000)]. The C-terminated and Si-terminated interfaces have interesting features, different from the SiC(001)/Ti interface. The Si-Ti bond at the Si-terminated interface has covalent interactions as a dimer and is stronger than the bond at the Si-terminated 3C-SiC(001)/Ti interface. On the other hand, the C-Ti bond at the C-terminated interface has strong *p-d* covalent interactions between C-2*p* and Ti-3*d* orbitals, similar to the 3C-SiC(001)/Ti interface. The stable distance between the interfacial layers at the Si-terminated interface is larger than that of the C-terminated interface. The adhesive energy value of the C-terminated interface is larger than that of Si-terminated interface; however, the difference between the values is much smaller than that of the 3C-SiC(001)/Ti interface. The value of Si-terminated interface is about two times larger than that of the 3C-SiC(001)/Ti, while the value of the C-terminated interface is similar. These points indicate that the present Si-terminated interface is strong, compared with the metallic bonding nature of the 3C-SiC(001)/Ti interface.

ACKNOWLEDGMENTS

The present study was supported by the Science and Technology Agency of Japan as the project "*Frontier Ceramics*". The calculations were done by "*Promoted Research Projects for High Performance Computing*" using the high-performace supercomputing system, HITACHI-SR8000 and IBM RS6000-SP at Tsukuba Advanced Computing Center (TACC) of the Agency of the Industrial Science and Technology of Japan.

REFERENCES

[1] T. Yano, H. Suematsu, and T. Iseki, "High-resolution electron microscopy of a SiC/SiC joint brazed by a Ag-Cu-Ti alloy", J. Mater. Sci. **23** [9] 3362-6 (1988).
[2] C. Iwamoto and S. -I. Tanaka, "Grain-boundary character of titanium carbide produced by the reaction between titanium-containing molten alloy and silicon carbide" Phil. Mag. A **78** [4] 835-44 (1998).

[3] J. J. Bellina, Jr. and M.V. Zeller, "*Novel Refractory Semiconductors*"; pp. 265 in MRS Symposium Proceedings Vol. 97. Edited by D. Emin, T. Aselage, and C. Wood. Materials Research Society, Pittsburgh, 1987.

[4] L. M. Porter, R. F. Davis, J. S. Bow, M. J. Kim, R. W. Carpenter, and R. C. Glass, "Chemistry, microstructure, and electrical properties at interfaces between thin films of titanium and alpha (6H) silicon carbide (0001)", J. Mater. Res. **10** [3] 668-79 (1995).

[5] F. R. Chien, S. R. Nutt, J. M. Carulli, Jr., N. Buchan, C. P. Beetz, Jr., and W. S. Yoo, "Heteroepitaxial growth of beta -SiC films on TiC substrates: interface structures and defects", J. Mater. Res. **9** [8] 2086-95 (1994).

[6] S. N. Rashkeev, W. R. L. Lambrecht, and B. Segall, "Electronic structure, Schottky barrier, and optical spectra of the SiC/TiC 111 interface", Phys. Rev. B **55** [24] 16472-86 (1997).

[7] P. Hohenberg and W. Kohn, " Inhomogeneous electron gas", Phys. Rev. **136**, B864-71 (1964); W. Kohn and J. L. Sharn, "Self-consisten equations including exchange and correlation effects", Phys. Rev. **140** A1133-38 (1965).

[8] A. B. Anderson and Ch. Ravimohan, "Bonding of alpha -SiC basal planes to close-packed Ti, Cu, and Pt surfaces: molecular-orbital theory", Phys. Rev. B **38** [2] 974-7 (1988); S. P. Mehandru and A. B. Anderson, "Adhesion and bonding of polar and nonpolar SiC surfaces to Ti(0001)", Surf. Sci. **245**, 333-44 (1991).

[9] J. Hoekstra and M. Kohyama, "Ab initio calculations of the beta -SiC(001)/Al interface", Phys. Rev. B **57** [4] 2334-41 (1998).

[10] M. Kohyama and J. Hoekstra, "Ab initio calculations of the beta -SiC(001)/Ti interface", Phys. Rev. B **61** [4] 2672-79 (2000).

[11] *"Metal-Ceramic Interfaces"*; all chapter. Edited by M. Ruehle, A. G. Evans, M. F. Ashby, and J. P. Hirth. Pergamon, Oxford, 1989.

[12] F. S. Ohuchi and M. Kohyama, "Electronic structure and chemical reactions at metal-alumina and metal-aluminum nitride interfaces", J. Am. Ceram. Soc. **74** [6] 1163-87 (1991).

[13] M. W. Finnis, "The theory of metal-ceramic interfaces", J. Phys. Condens. Matter **8** [32] 5811-36 (1996).

[14] J.R. Smith, T. Hong and D. J. Srolovitz, "Metal-ceramic adhesion and the Harris functional", Phys. Rev. Lett. **72** [25] 4021-4 (1994).

[15] P. W. Tasker and A. M. Stoneham, "The stabilization of oxide and oxide-metal interfaces by defects and impurities", J. Chimie Phys. **84** [2] 149-55 (1987); D. M. Duffy, J. H. Harding, and A. M. Stoneham, "Atomistic modelling of metal-oxide interfaces with image interactions", Phil. Mag. A **67** [4] 865-82 (1993).

[16] C. Kruse, M. W. Finnis, J. S. Lin, M. C. Payne, V. Y. Milman, A. De Vita, and M. J. Gillan, "First-principles study of the atomistic and electronic structure of the niobium- alpha -alumina (0001) interface", Phil. Mag. Lett. **73** [6] 377-83 (1996).

[17] R. Benedek, M. Minkoff, and L. H. Yang, "Adhesive energy and charge transfer for MgO/Cu heterophase interfaces", Phys. Rev. B **54** [11] 7697-700 (1996).

[18] J. P. Perdew and A. Zunger, "Self-interaction correction to density-functional appoloximations for many-electron systems", Phys. Rev. B **23** [10] 5048-79 (1981).

[19] D. M. Bylander, L. Kleinman, and S. Lee, "Self-consistent calculations of the energy bands and bonding properties of $B_{12}C_3$ ", Phys. Rev. B **42** [2] 1394-1403 (1990).

[20] G. P. Kerker, " Efficient iteration scheme for self-consistent pseudopotentianl calculations", Phys. Rev. B **23** [6] 3082-4 (1981).

[21] M. Kohyama, "Ab initio calculations for SiC-Al interfaces: tests of electronic-minimization techniques", Modelling Simul. Mater. Sci. Eng. **4** [4] 397-408 (1996).

[22] N. Troullier and J. L. Martins, "Efficient pseudopotentials for plane-wave calculations", Phys. Rev. B **43** [3] 1993-2006 (1991).

[23] L. Kleinman and D. M. Bylander, "Efficacious form for model pseudopotentials", Phys. Rev. Lett. **48** [20] 1425-8 (1982).

[24] Y. Sugawara, private communication.

[25] C. L. Fu and K. M. Ho, Phys. Rev. B **28**, 5480 (1983); R. J. Needs, R. M. Martin, and O. H. Nielsen, "Total-energy calculations of the structural properties of the group-V element arsenic", Phys. Rev. B**33** [6] 3778-84 (1986).

Grain Boundary Engineering in Ceramics

MOLECULAR DYNAMICS SIMULATION OF FRACTURE TOUGHNESS OF SILICON NITRIDE SINGLE CRYSTAL

Naoto Hirosaki, National Institute for Research in Inorganic Materials, 1-1, Namiki, Tsukuba-shi, Ibaraki, 305-0044, Japan

Shigenobu Ogata and Hiroshi Kitagawa, Department of Adaptive Machine Systems, Osaka University, 2-1, Yamadaoka, Suita-shi, Osaka, 565-0871, Japan

ABSTRACT

The stress distribution near a crack tip and the crack propagation of a β-silicon nitride single crystal is investigated by molecular dynamics (MD) simulations. An atomic model of β-Si_3N_4 crystal with a crack was strained according to a linear elastic calculation, and then relaxed by MD to examine the crack propagation. The crack propagates when K_I is greater than 1.4 MPa$\sqrt{m}$, while it annihilates when K_I is smaller than 1.3 MPa$\sqrt{m}$, indicating that the K_{IC} value is about 1.4 MPa$\sqrt{m}$. The stress distribution for the "critical" crack tip with $K_I = 1.4$ MPa$\sqrt{m}$, is calculated from the MD results, assuming that the stress is the average of atomic stresses. The calculated stress distribution is in good agreement with the linear elastic solution.

INTRODUCTION

Polycrystalline silicon nitride ceramics are composed of β-Si_3N_4 grains and oxynitride grain boundary phases. The mechanical properties of silicon nitride ceramics are governed by the fracture of Si_3N_4 grains and of the grain boundary phases. Many studies have been carried out on fracture in polycrystalline silicon nitride ceramics, in which the fracture of the grain boundary phases is important. Few experimental studies of fracture in monocrystalline Si_3N_4 have been reported because it is difficult to obtain large single crystals. In this study, MD simulations are performed to examine the crack propagation behavior and calculate the fracture toughness. The stress distribution near the crack tip in a β-Si_3N_4 crystal is also examined.

To the extent authorized under the laws of the United States of America, all copyright interests in this publication are the property of The American Ceramic Society. Any duplication, reproduction, or republication of this publication or any part thereof, without the express written consent of The American Ceramic Society or fee paid to the Copyright Clearance Center, is prohibited.

MODEL AND POTENTIALS

The total potential energy of the system (E_T) is a sum of 2-body ($E_{ij}^{(2)}(r_{ij})$) and 3-body ($E_{jik}^{(3)}(r_{ij}, r_{ik}, \theta_{jik})$) interactions as proposed by Vashishta *et al.* [1]:

$$E_T = \sum_{i<j} E_{ij}^{(2)}(r_{ij}) + \sum_{i,j<k} E_{jik}^{(3)}(r_{ij}, r_{ik}, \theta_{jik}) \tag{1}$$

where i, j, and k refer to atoms in the system, r_{ij} is the length of the ij bond, and θ_{jik} is the bond angle between the ij and ik bonds. The 2-body and 3-body functions are

$$E_{ij}^{(2)}(r_{ij}) = A_{ij} \left(\frac{\sigma_i + \sigma_j}{r_{ij}}\right)^{\eta_{ij}} + \frac{Z_i Z_j}{r_{ij}} \exp\left(-\frac{r_{ij}}{r_{s1}}\right) - \frac{\alpha_i Z_j^2 + \alpha_j Z_i^2}{2r_{ij}^4} \exp\left(-\frac{r_{ij}}{r_{s4}}\right) \tag{2}$$

and

$$E_{jik}^{(3)}(r_{ij}, r_{ik}, \theta_{jik}) = B_{jik} \exp\left(\frac{l}{r_{ij} - r_c} + \frac{l}{r_{ij} - r_c}\right) \times (\cos\theta_{jik} - \cos\theta_{jik}^0) \tag{3}$$

where $A, \sigma, \eta, Z, r_{s1}, \alpha, r_{s4}, B, l, r_c, \theta^0$ are parameters. These parameters for Si_3N_4 were determined by Vashishta *et al.* [1].

ELASTIC CONSTANTS

Elastic constants for β-Si_3N_4 are calculated using the interatomic potential of Eq.(1). Tensors of the elastic constants, C_{ij}, are expressed using the total potential energy, E_T, as [2]:

$$C_{ij} = \frac{1}{V} \frac{\partial^2 E_T}{\partial \epsilon_i \partial \epsilon_j} \qquad (i, j = 1, 2, 3, 4, 5, 6 = xx, yy, zz, yz, zx, xy) \tag{4}$$

where ϵ_i and ϵ_j are small strains and V is the volume of the model. This indicates that elastic constants can be calculated by differentiating the interatomic potential with respect to strain.

The MD cell in the β-Si_3N_4 single crystal model to calculate elastic constants involves 224 atoms. The cell dimensions along the x, y, and z axes were 1.52 nm, 1.32 nm, and 1.16 nm, respectively. The cell was deformed numerically by ϵ_i and ϵ_j (about 10^{-3}) and the E_T was calculated by Eq.(1) to find the minimum-energy point. The C_{ij} values were calculated by numerically differentiating E_T in Eq.(1) by ϵ_{ij} and ϵ_{kl} at the minimum-energy point. The compliance constants, S_{ij}, were calculated using the calculated elastic constants. Table 1 shows the calculated elastic and compliance constants for the β-Si_3N_4 single crystal. The MD data calculated by Ching *et al.* [3] using pair potentials obtained by an *ab initio* method and experimentally measured data for single crystals in sintered silicon nitride ceramics by Hay *et al.* [4] are also shown. Our calculated results are in good agreement with the experimentally measured values and Ching's solutions, except for C_{44} and C_{66}, which are

Table I. Elastic constants of β-Si$_3$N$_4$.

	This work	Ching[3]	Hay[4]
C_{11} (GPa)	591	409	343
C_{12} (GPa)	182	271	136
C_{13} (GPa)	162	201	120
C_{33} (GPa)	690	604	600
C_{44} (GPa)	377	108	120
C_{66} (GPa)	205	69	
S_{11} (1/GPa)	0.00194		
S_{12} (1/GPa)	-0.000505		
S_{13} (1/GPa)	-0.000337		
S_{33} (1/GPa)	0.00161		
S_{44} (1/GPa)	0.00265		
S_{66} (1/GPa)	0.00489		

rather higher than the reported values. We used our calculated elastic values to analyze the MD simulations.

The engineering constants of Young's modulus to the x axis (E_x), Poisson's ratio between the x and y axes (ν_{xy}), and the share modulus on the xy plane (G_{xy}) can be related to the compliance constants as follows:

$$E_x = \frac{1}{S_{11}}, \tag{5}$$

$$\nu_{xy} = -S_{12}E_x, \tag{6}$$

$$G_{xy} = \frac{1}{S_{66}}. \tag{7}$$

Using the stiffness constants in Table 1, ν_{xy} and G_{xy} for the β-Si$_3$N$_4$ single crystal were calculated to be 0.26 and 205 GPa, respectively.

CRACK PROPAGATION BEHAVIOR DURING MD RELAXATION

In the MD simulation of crack propagation, we used a cubic cell in which the a and c axes of β-Si$_3$N$_4$ were placed along the x and z directions, respectively, as illustrated in Fig. 1. The basic cell contains 9408 atoms (Si:4032, N:5376). The cell dimensions for the x, y, and z axes were S_x = 9.2 nm, S_y = 9.3 nm, and S_z = 1.2 nm, respectively. Atoms near the cell boundaries, within a width of s/S = 0.06 from the boundaries in the x and y axes, were confined within the cell, whereas those in the z axis were subjected to periodic boundary conditions. A crack of length $S/2$, or half the cell width, was inserted with the crack tip lying along the z axis (*i.e.* x=0, y=0).

Atom displacements in the xy plane caused by introducing the crack were calculated using linear elastic theory based on isotropic materials, because the crystal structure of

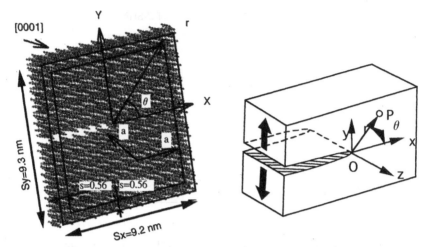

Figure 1. MD model of β-Si3N4   Figure 2. Schematic model of crack tip

β-Si$_3$N$_4$ is orthotropic, *i.e.*, anisotropic in the z axis but isotropic in the xy plane. The displacements (u, v, w) for point P (r, θ, z) $(z = 0)$ for opening mode I shown in Fig. 2 are:

$$u = \frac{K_I}{2G_{xy}}\sqrt{\frac{r}{2\pi}}\cos\frac{\theta}{2}\left(\kappa - 1 + 2\sin^2\frac{\theta}{2}\right)$$

$$v = \frac{K_I}{2G_{xy}}\sqrt{\frac{r}{2\pi}}\sin\frac{\theta}{2}\left(\kappa + 1 - 2\cos^2\frac{\theta}{2}\right) \tag{8}$$

where u, v, and w are the displacements in the x, y, z axes, respectively, and κ is:

$$\kappa = 3 - 4\nu_{xy} \tag{9}$$

under the plain strain condition, and where ν_{xy} is Poisson's ratio. These equations indicate that K_I is related to displacement near the crack tip through macroscopic values, namely, the share modulus and Poisson's ratio.

The crack propagation behavior of the MD model in Fig. 1 was examined. Firstly, the atomic model of a β-Si$_3$N$_4$ crystal with a crack, shown in Fig. 1, was strained according to Eq.(8). All the atoms were then relaxed at a temperature of 100 K except for those within a width of $s/S = 0.06$ from the boundaries, which were held constant to maintain the external stress applied to the MD model. After 1.2 ps, the internal energy became constant, which means that the atoms near the crack tip had relaxed to a stable configuration under the applied external stress. Fig. 3 shows the crack propagation behavior for K_I values from 1.0 to 1.8 MPa $\sqrt{m}$. The crack length decreased in the cases of $K_I = 1.0$ and 1.2 MPa $\sqrt{m}$.

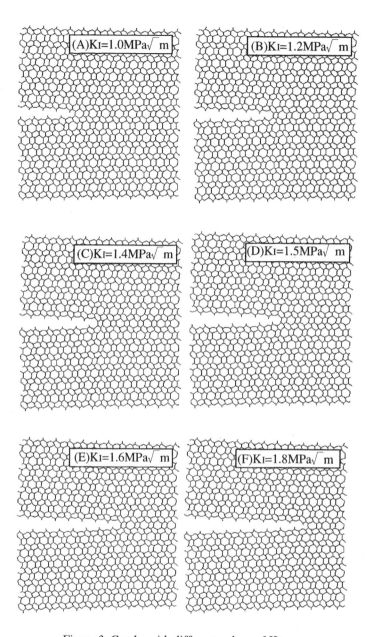

Figure 3. Cracks with different values of K_I.

However, it began to grow for $K_I = 1.4$ MPa $\sqrt{m}$, and increased with further increases in the value of K_I. Assuming that the lowest K_I value at which a crack begins to propagate is the critical stress intensity factor, from the MD simulations, K_{IC}^{sim} is estimated to be about 1.4 MPa$\sqrt{m}$.

STRESS DISTRIBUTION NEAR THE CRACK TIP

Stress distribution in the simulation cell was calculated using the MD results of stress at each atom. For the set of particles i with a weight of m_i at a position x_α^i and x_β^i, and having velocities of v_α^i and v_β^i for the α and β directions, respectively, the stress tensor, $\sigma_{\alpha\beta}$, in a small volume in the MD model is defined on the basis of the virial theorem [5] as:

$$\sigma_{\alpha\beta} = \frac{1}{V_0} \cdot \left(\frac{\partial E_{V_0}}{\partial x_\alpha^i} x_\beta^i + \sum_{i \in V_0} m_i v_\alpha^i v_\beta^i \right) \qquad (\alpha, \beta = 1, 2, 3) \qquad (10)$$

where α, β are the numbers 1,2, or 3 for the x, y, or z axes, V_0 is a small volume in the MD cell, and E_{V_0} is the total potential energy of atoms in the volume V_0, defined as:

$$E_{V_0} = \sum_{i \in V_0, j \neq i} E_{ij}^{(2)}(r_{ij}) + \sum_{i \in V_0, j \neq i, k \neq i, j > k} E_{jik}^{(3)}(r_{ij}, r_{ik}, \theta_{jik}). \qquad (11)$$

An MD caluculation for the tensile test in the y axis of β-Si$_3$N$_4$ without a crack was carried out to investigate the validity of $\sigma_{\alpha\beta}$ defined in Eq.(10). A basic cell contained 9408 atoms, which was the same as that used in the crack propagation simulations, was used for the tensile test. Each atom in this model was loaded directly with a tensile stress of 20 MPa along the y axis and $\sigma_{\alpha\beta}$ was caluculated. The dimensions of V_0 for the x, y, and z axes were $L_x = 0.3$ nm, $L_y = 9.3$ nm, and $L_z = 1.2$ nm, respectively. Fig. 4 shows the σ_{22} (MD result) and the stress distribution function for the linear elastic solution (LES). The σ_{22} values, calculated using atoms within each 0.3 nm step along the x axis, were almost constant and in good agreement with the stress calculated with the LES, suggesting that stress evaluation using the MD simulation is accurate.

Next, the stress distributions for the models containing a crack were calculated. Fig. 5 shows the σ_{11} and σ_{22} values near the crack tip for $K_I = 1.4$ MPa $\sqrt{m}$, where the crack starts to propagate. The LES curves were calculated by:

$$\sigma_{11}(r) = \frac{K_I}{\sqrt{2\pi r}} \cos\frac{\theta}{2} \left(1 - \sin\frac{\theta}{2} \sin\frac{3\theta}{2} \right)$$

$$\sigma_{22}(r) = \frac{K_I}{\sqrt{2\pi r}} \cos\frac{\theta}{2} \left(1 + \sin\frac{\theta}{2} \sin\frac{3\theta}{2} \right). \qquad (12)$$

The volume for the calculation of $\sigma_{\alpha\beta}$ was 0.3 nm by 0.46 nm by 1.2 nm. The MD values are in good agreement with the results of the linear elastic analysis, except for near the crack tip. In contrast, within 0.5 nm from the crack tip, where the LES values increase up to ∞,

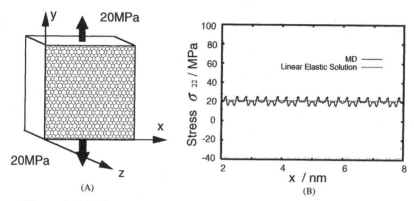

Figure 4. (A)MD model of tensile stress with size of x=9.1 nm, y=9.2 nm, and z=1.2 nm, and (B)stress distribution along x axis according to tensile stress of 20 MPa for y direction.

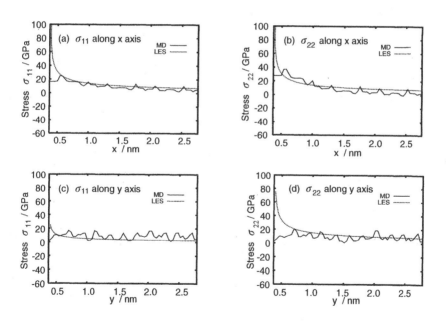

Figure 5. Stress distribution near the crack tip.

the MD values are quite different. In this range, the condition of cracks should be evaluated directly using an atomic MD model because the stress defined according to the continuous model becomes meaningless.

CONCLUSION

Crack propagation behavior in single crystals of β-Si_3N_4 was evaluated by a molecular dynamics simulation using two- and three-body potentials. A crack in the prism plane $(10\bar{1}0)$ propagated when $K_I \geq 1.4$ MPa$\sqrt{m}$. Stress distribution near the crack tip when K_I was 1.4 MPa$\sqrt{m}$, at which level the crack did not grow, was calculated from the MD results based on the virial theorem. The stress distribution 0.5 nm farther from the crack tip was in good agreement with linear elastic solutions. These results indicate that our MD simulation describes the fracture behavior of β-Si_3N_4 remarkably well.

REFERENCES

1. R. Vashishta, R. K. Kalia, A. Nakano, and I. Ebbsjö, "Molecular Dynamics Methods and Large-Scale Simulations of Amorphous Materials," pp. 151-213 in *Amorphous Insulators and Semiconductors*, Ed. by M.F.Thorpe and M.I. Mitkova, Kluwer, Dordrooht, (1996).

2. Y. Shibutani, "Definitions of Elastic Constants," *J. Soc. Mat. Sci., Japan*, **46** [3], 218-227 (1997).

3. W-Y Ching, Y-N Xu, J. D. Gale, and M. Rühle, "*Ab-Initio* Total Energy Calculation of α- and β-Silicon Nitride and the Derivation of Effctive Pair Potentials with Application to Latiice Dynamics," *J. Am. Ceram. Soc.*, **81** [12] 3189-96 (1998).

4. J. C. Hay, E. L. Sun, G. M. Pharr, P. F. Becher, and K. B. Alexander, "Elastic Anisotoropy of β-Silicon Nitride Whiskers," *J. Am. Ceram. Soc.*, **81** [10], 2661-69 (1998).

5. S. Kotake, pp.68-74 in *Bunshi Netsuryuutai (in Japanese)*, Maruzen, Tokyo, Japan, (1990).

VARIABLE-CHARGE MOLECULAR DYNAMICS OF AGGREGATION OF TiO$_2$ NANOCRYSTALS

Shuji Ogata
Department of Applied Sciences, Yamaguchi University, Ube 755-8611, Japan

Hiroshi Iyetomi
Department of Physics, Niigata University, Niigata 950-2181, Japan

Kenji Tsuruta
Department of Electrical and Electronic Engineering, Okayama University
Okayama 700-8530, Japan

Fuyuki Shimojo
Faculty of Integrated Arts and Sciences, Hiroshima University
Higashi-Hiroshima 739-8521, Japan

Aiichiro Nakano, Rajiv K. Kalia, and Priya Vashishta
Department of Physics and Astronomy and Department of Computer Sciences
Louisiana State University, Baton Rouge, Louisiana 70803-4001

ABSTRACT

Molecular dynamics (MD) simulations of both rutile and anatase TiO$_2$-nanoclusters (size 60–80 Å) are performed using an environment-dependent interatomic potential developed recently by Ogata et al. [J. Appl. Phys. **86** [16], 3036-3041 (1999)]. In the potential, atomic charges vary dynamically according to the generalized electronegativity-equalization principle. Formation of double-charge layer and enhanced atomic diffusion are observed in surface regions of the free TiO$_2$-nanospheres by inclusion of the atomic-charge variation. Their effects on aggregation processes of TiO$_2$ nanoclusters are investigated through MD simulations.

INTRODUCTION

Nanophase materials synthesized by consolidating nanometer-sized clusters, draw increasing attention in recent years because of their improved properties such as high strength, high toughness, and high catalytic-reactivity compared to

To the extent authorized under the laws of the United States of America, all copyright interests in this publication are the property of The American Ceramic Society. Any duplication, reproduction, or republication of this publication or any part thereof, without the express written consent of The American Ceramic Society or fee paid to the Copyright Clearance Center, is prohibited.

conventional materials [1]. TiO_2 is one of the fundamental ceramics with a wide variety of applications such as photocatalysts, high dielectrics, and pigments [1,2]. Rutile and anatase are major crystalline phases of TiO_2; the former is the thermodynamically preferred form at all temperatures. Theoretical understanding of physical properties of TiO_2 nanoclusters and nanophase TiO_2 is of scientific interest as well as of technological significance.

Based on the formulation by Streitz and Mintmire [3] we have recently developed a variable-charge interatomic potential for TiO_2 systems [4], in which atomic charges vary dynamically in response to the environmental atoms. The potential reproduces the pressure-dependent static dielectric constants, vibrational density of states, melting temperature, and surface relaxation of the rutile crystal, as well as the cohesive energy and elastic moduli. The potential is applicable to anatase; its experimental cohesive energy, lattice constants, and dielectric constants are reproduced as well. In the first part of the present paper we will briefly describe the variable-charge interatomic potential [4]. Then we will apply the potential to both rutile and anatase nanoclusters with size 60–80 Å to investigate effects of the atomic-charge variation on their aggregation processes.

VARIABLE-CHARGE INTERATOMIC POTENTIAL

In the present variable-charge potential [4], the total-potential energy E_{pot} for a charge-neutral system composed of N_{Ti} titanium and N_O oxygen atoms with charges $\{q_i e\}$ located at $\{\vec{r}_i\}$, is modeled as composed of four terms:

$$E_{pot} = \sum_i E_i^{atom}(q_i) + \sum_{i<j} V_{ij}^{es}(r_{ij}; q_i, q_j)$$
$$+ \sum_{i<j} V_{ij}^C(r_{ij}) + \sum_{i \in \{Ti\},\, j \in \{O\}} \Delta V_{ij}^{TiO}(r_{ij}; q_i, q_j) \qquad (1)$$

The first term in the right-hand side of Eq. (1) represents the atomic energy [3-5]

$$E_i^{atom}(q_i) = E_i^{atom}(0) + q_i \chi_i + \frac{1}{2} J_i q_i^2 \qquad (2)$$

with the electronegativity χ_i and the hardness J_i; they are regarded as free parameters in optimizing the potential formulas. The second term in Eq. (1) is the electrostatic interaction-energy:

$$V_{ij}^{es}(r_{ij}; q_i, q_j) = \int d\vec{r}_1 \int d\vec{r}_2 \frac{\rho_i(\vec{r}_1, q_i)\rho_j(\vec{r}_2, q_j)}{r_{12}} \qquad , \qquad (3)$$

where ρ_i denotes a model charge-density distribution of the Slater form around the i-th atom. The third term in Eq. (1) represents the covalent bonding and the steric repulsion between the atomic cores, which has a modified Rydberg form

$V_{ij}^c(r_{ij}) = -C_{ij}[1 + (r_{ij} - r_{ij}^e)/l_{ij}]\exp[-\alpha_{ij}(r_{ij} - r_{ij}^e)/l_{ij}]$. The last term in Eq. (1), involving only pairs of neighboring Ti and O atoms, is responsible in reproducing relaxation behavior of low-index surfaces (e.g., (110)) and melting temperature of the rutile crystal.

The parameters in the potential were determined through fitting to experimental or first-principles electronic-structure data including the lattice constant, cohesive energy, elastic moduli, static dielectric constants, surface energies of low-index planes [(110) and (100)], melting temperature at ambient pressures, and surface relaxation properties for (110) plane of the rutile [6-13]. Table I compares the calculated values for several physical quantities with the corresponding experimental results.

Table I. Comparison of various physical quantities.

	Present	Experiments
$a(\text{Å})$	4.68	4.59 [8]
c (Å)	2.58	2.60 [8]
Cohesive energy per TiO_2 (eV)	19.8	19.8 [10]
Bulk moludus	228	216 [11]
C_{33} (GPa)	519	484 [11]
C_{44} (GPa)	136	124 [11]
C_{13} (GPa)	123	150 [11]
ε_{xx}	91.8	86 [12]
ε_{zz}	195.9	170 [12]
Melting temperature (°C)	1,700–2,100	~1,800 [13]

We have also calculated the surface energies of (110) and (100) planes of the rutile. We found that the surface energies are 39 meV/Å^2 for (110) and 41 meV/Å^2 for (100), which agree well with the first-principles data [6] 55 meV/Å^2 for (110) and 70 meV/Å^2 for (100). Displacements of atoms from the crystalline positions agree well with the results obtained in the first-principles calculations [6]. Our results for the deviation of the atomic charges from the bulk values show reasonable agreement with those calculated using the tight-binding interatomic potential [7]. In the course of molecular dynamics (MD) simulations with the present potential, the atomic charges, q_i, will be determined dynamically by minimizing E_{pot} with respect to a set of $\{q_i\}$ under the constraint of total-charge conservation.

To test transferability of the present potential, we have investigated the properties of the anatase phase of TiO_2. We have calculated the ground-state energy of anatase to find it larger than that of rutile. The resulting energy difference, 0.09eV per Ti atom, compares favorably the corresponding value (0.068eV per Ti atom) evaluated from the experimantal data on the heat of formation [13]. The lattice constants at zero temperature were obtained to be $a = b = 3.85$ (3.79) Å and $c = 8.78$ (9.51) Å where the numbers in the parentheses are the corresponding experimental values [2]; the density has an error of 4 %. We have also calculated the dielectric constants along the a- and c-axes and the values are $\varepsilon_{xx} = 38.1$ and $\varepsilon_{zz} = 61.6$, respectively. The experimental value [2] of ε for a powder sample of anatase is 48, which is much lower than $\varepsilon = 114$ for rutile. Averaging the present values for anatase over the principal directions yields $\varepsilon = (2 \times 38.1 + 61.6)/3 \approx 46$. This is in excellent agreement with the experimental value.

MOLECULAR DYNAMICS SIMULATIONS

Nanoclusters may be synthesized by several methods [14-16]. Recent development in the sol-gel method [16] has made it possible to synthesize high-purity TiO_2 nanoclusters below 10 nm in size with varying degrees of aggregation by controlling the reaction rates. The anatase nanoclusters obtained in the sol-gel method sintered [17] at relatively low temperatures ~ 600 °C as compared to the case of rutile nanoclusters (~1,000 °C). During the sintering of the anatase nanoclusters, anatase-to-rutile phase transformation has been observed in the vicinity of grain boundaries [17]. Such low sintering-temperatures are desirable for technical applications since grain growth should be significantly suppressed during the sintering. Dispersed nanoclusters are desirable for their use in high-quality nanophase TiO_2. Recently Trentler et al. [18] have obtained loosely aggregated anatase-nanocrystals with diameter 5–9 nm, which exhibited round or irregular shapes. Theoretical investigation for aggregation mechanisms of TiO_2 nanoclusters will offer valuable information for improved processing of the nanoclusters.

Anatase nanoclusters processed in the sol-gel method assumed round shapes with diameter d ~ 70 Å [18]. This observation indicates that the ground state shape of an anatase nanocluster in this range of diameter is spherical (or round) rather than faceted. To theoretically study the morphology of free rutile-nanoclusters in the same size-range we have calculated the potential energies, E_{pot}, of both spherical and faceted nanoclusters. We cut out rutile spheres with $d = 60$ Å and 80 Å from a rutile crystal; the total numbers of atoms are $N = 10,446$ for $d = 60$ Å and $N = 24,870$ for $d = 80$ Å. Then E_{pot} of each nanosphere was obtained through energy minimization by both moving the atoms and changing the atomic charges following the conjugate gradient (CG) method. Faceted rutile-nanoclusters with nearly the same numbers of atoms ($N = 10,164$ and 24,102) were also prepared through the Wulff construction using the surface energies given in Ref. 4; the CG minimization of E_{pot} with respect to the atomic positions and charges were thereon performed. We thus found $E_{pot}/N = -6.609$ eV for the nanosphere with $d = 60$ Å ($N = 10,446$) and $E_{pot}/N = -6.601$ eV for the faceted nanocluster with $N = 10,164$. The $E_{pot}/N = -6.615$ eV for the nanosphere with $d = 80$ Å ($N = 24,870$), $E_{pot}/N = -6.608$ eV for the faceted nanocluster with $N = 24,102$. Comparing those values, we predict that rutile nanoclusters at the ground state with size 60 – 80 Å assume spherical shapes. In the following we thus consider spherical nanoclusters with $d = 60$ Å for both anatase and rutile.

We investigate possible diffusion of surface atoms in anatase and rutile nanospheres by performing MD simulation runs during 18 ps at temperature $T = 1,123$ °C. In the simulations we adopt the fast-multipole method to speed up calculations of the Coulomb interactions. The Nosé-Hoover thermostats-chain is used to control the temperature of the system. The MD time-step is 0.37 fs. By calculating mean-square displacements of atoms $\langle |\Delta \vec{r}|^2 \rangle$, we find that Ti and O atoms at radius $r > 28$ Å are diffusing for both anatase and rutile nanospheres, while the interiors retaining the crystalline structures. The tangential-diffusion constants $D_t = d\langle \Delta r_t^2 \rangle /(4dt)$ in the anatase nanosphere, 7.0×10^{-6} cm^2/s for Ti and 6.0×10^{-6} cm^2/s for O, are 2–3 times larger than the corresponding radial-diffusion constants $D_r = d\langle \Delta r_r^2 \rangle /(2dt)$ as well as D_t and D_r in the rutile nanosphere. Separately we perform MD simulations of anatase nanospheres ($d = 60$ Å) with the atomic charges fixed to their bulk-crystalline values. We thereby find that the variation of atomic charges acts to enhance diffusion of the surface atoms.

Both for anatase and rutile nanospheres, net charges in the surface regions ($r = 28 – 30$ Å) are positive while they are negative in the adjacent inner-regions ($r = 24 – 28$ Å). Such double-charge layer structure are maintained at an elevated temperature $T = 1,123$ °C for both rutile and anatase nanospheres, as demonstrated in Fig. 1 for the rutile nanosphere. We note that similar double-charge layer structure has been found also for low-index surfaces of rutile in the tight-binding calculations by Shelling et al. [7] for slab geometry. Interaction potential between two nanospheres should be affected by the double-charge layer structure in each nanosphere. In fact we have found that the dynamic charge-transfer acts to add repulsive intercluster interaction-potential when their separation distance R between the two centers-of-mass are $63 - 65$ Å for both rutile and anatase nanospheres. Since the surface region ($r > 28$ Å) in each nanosphere has a net positive-charge, Coulomb repulsion between the two nanoclusters results at close proximity.

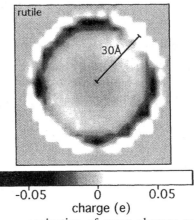

Figure 1: The x-y projection of space charges averaged over 18 ps in a slice $z = [-2\text{Å}, 2\text{Å}]$ of the rutile nanosphere with diameter 60 Å.

Through those analyses we have learned that the effects of atomic-charge variation are to add residual repulsive Coulomb-interaction between the two nanospheres and to enhance diffusion of surface atoms. One may expect that the former effect will hinder aggregation, while the latter will promote it. To understand their combined effects on the aggregation processes, we perform MD simulations for aggregation of two nanospheres ($d = 60$ Å) at $T = 1{,}127$ °C. We put two nanospheres at $R \sim 61$ Å; three combinations of internal-lattice orientations are considered. In the case of anatase nanospheres, they aggregate after ~ 5 ps irrespective of the lattice orientations, which is accompanied by surface atomic-diffusion to form a 'bridge' between the two nanospheres as depicted in Fig. 2 (the left panel). In the case of rutile nanospheres, on the other hand, R increases monotonously until they separate completely as demonstrated in Fig. 2 (the right panel). For reference we separately perform MD simulations of two anatase-nanospheres with the atomic charges fixed to their bulk-crystalline values (i.e., without charge variation). In those rigid-ions runs, the two

Grain Boundary Engineering in Ceramics

nanospheres show no tendency to aggregate during 10 ps and no substantial deformation in the surface regions. Accumulating the results obtained here, we conclude that the degrees of surface atomic-diffusion play a critical role in determining the aggregation processes of TiO_2 nanoclusters.

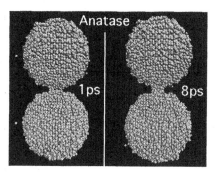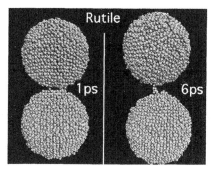

Figure 2: Snapshots of the MD runs for two contacted nanospheres of anatase (left) and those of rutile (right).

CONCLUSION

We have carried out MD simulations of rutile- and anatase-nanoclusters of TiO_2 (size 60–80 Å) using the variable-charge interatomic potential. The atomic-charge variation relatively enhanced diffusion of surface atoms in the single anatase-nanosphere with the double-charge layer structure. The simulation results have clarified difference between the rutile- and anatase-nanoclusters in their aggregation processes, which is in harmony with the experimental observation.

ACKNOWLEGEMENT

This work was performed under the auspices of the US-Japan Research Program (96-03264, MPCR-357) operated by NSF and JSPS, and also supported by Sasaki Environment Tech. Fund.

REFERENCES

[1] *Nanomaterials Synthesis, Properties, and Application*, edited by A.S. Edelstein and R.C. Cammarata (IOP Pub., London, 1996).
[2] *Concise Encyclopedia of Advanced Ceramic Materials*, edited by R.J. Brook (Pergamon, Cambridge, 1991), pp. 486-488.

[3] F.H. Streitz and J.W. Mintmire, "Charge Transfer and Bonding in Metallic Oxides," *J. Adhesion Sci. Technol.* **8** [8], 853-864 (1994).

[4] S. Ogata, H. Iyetomi, K. Tsuruta, F. Shimojo, R.K. Kalia, A. Nakano, and P. Vashishta, "Variable-Charge Interatomic Potentials for Molecular-Dynamcs Simulations of TiO_2," *J. Appl. Phys.* **86** [6], 3036-3041 (1999).

[5] A.K. Rappe and W.A. Goddard, "Charge equilibration for molecular dynamics simulations," *J. Phys. Chem.* **95** [8] 3355-3363 (1991).

[6] M. Ramamoorthy, D. Vanderbilt, and R.D. King-Smith, "First-Principles Calculations of the Energetics of Stoichiometric TiO_2 Surfaces," *Phys. Rev. B* **49** [23], 16721-16727 (1994).

[7] P.K. Schelling, N. Yu, and J.W. Halley, "Self-Consistent Tight-Binding Atomic-Relaxation Model of Titanium Dioxides," *Phys. Rev. B* **58** [3], 1279-1293 (1998).

[8] S.C. Abrahams and J.L. Bernstein, "Rutile: Normal Probability Plot Analysis and Accurate Measurement of Crystal Structure," *J. Chem. Phys.* **55** [7] 3206-3211 (1971).

[9] J.G. Traylor, H.G. Smith, R.M. Nicklow, and M.K. Wilkson, "Lattice Dynamics of Rutile," *Phys. Rev. B* **3** [10], 3457-3472 (1971).

[10] *CRC Handbook of Chemistry and Physics, 79th edn.* (CRC Press, Florida, 1996).

[11] M.H. Manghnani, "Elastic constants of single-crystal rutile under pressures to 7.5 kilobars", *J. Geophys. Res.* **74** [17], 4317-4328 (1969).

[12] C.R.A. Catlow, R. James, W.C. Mackrodt, and R.F. Stewart, "Defect Energetics of α-Al_2O_3 and TiO_2," *Phys. Rev. B* **25** [2], 1006-1026 (1982).

[13] *JANAF Thermodynamic Tables, 3rd edn,* edited by Chase *et al.* (AIP, New York, 1985).

[14] R.W. Siegel, S. Ramasamy, H. Hahn, L. Zongquan, L. Ting, and R. Gronsky, "Synthesis, characterization, and properties of nanophase TiO_2," *J. Mater. Res.* **3** [6] 1367-1372 (1988).

[15] H. Hahn, J. Logas, and R.S. Averback, "Sintering Characteristics of Nanocrystalline TiO2," *J. Mater. Res.* **5** [3] 609-615 (1990).

[16] Q. Xu and M.A. Anderson, "Physical-Chemical Properties of TiO_2 Membranes Controlled by Sol-Gel Processing," *Mat. Res. Soc. Symp. Proc.* **132**, 41-46 (1989).

[17] K.-N.P. Kumar, K. Keizer, A.J. Burggraaf, T. Okubo, H. Nagamoto, and S. Morooka, "Densitification of Nanostructured Titania Assisted by a Phase Transformation," *Nature* **358** [2] 48-51 (1992).

[18] T.J. Trentler, T.E. Denler, J.F. Bertone, A. Agrawal, and V.L. Colvin, "Synthesis of TiO_2 Nanocrystals by Nonhydrolytic Solution-Based Reactions," *J. Am. Chem. Soc.* **121** [7] 1613-1614 (1999).

Diffusion and Transformation

ANISOTROPIC GRAIN BOUNDARY PROPERTIES FOR MODELING GRAIN GROWTH PHENOMENA

David J. Srolovitz[1] and Moneesh Upmanyu[1, 2]

[1]Princeton Materials Institute and Dept. of Mechanical & Aerospace Engineering
Princeton University
Princeton, NJ 08854 USA

[2]Dept. of Material Science & Engineering
University of Michigan
Ann Arbor, MI 48109 USA

ABSTRACT

Anisotropy in grain boundary mobility and energy can be very important in determining microstructural evolution. In this paper, we examine the misorientation and temperature dependence of the boundary mobility and temperature using atomistic simulations. The activation energy and pre-exponential factor describing boundary mobility and the boundary energy vs. misorientation show cusps at misorientations corresponding to low Σ boundaries. The temperature dependence of the mobility is consistent with the compensation effect. The boundary mobility corresponding to low Σ boundaries can correspond to minima or maxima in the actual boundary mobility or can change from one to another as the temperature is changed. The consequences for microstructural evolution are discussed.

INTRODUCTION

Microstructural evolution during all grain growth phenomena occurs by grain boundary migration in response to various driving forces. To describe these phenomena, models must include accurate descriptions of both grain boundary mobility and driving forces. In all cases, the parameters that describe these are anisotropic. These anisotropies arise from both the dependence of the grain boundary structure on the bicrystallography and the crystallographic factors associated with the driving force (e.g., slip within the grains). All models of grain growth phenomena require parameterization of the grain boundary mobility and the grain boundary energy. In the present paper, we review some of our recent results on both the orientation dependence of grain boundary mobility and grain boundary energy.

To the extent authorized under the laws of the United States of America, all copyright interests in this publication are the property of The American Ceramic Society. Any duplication, reproduction, or republication of this publication or any part thereof, without the express written consent of The American Ceramic Society or fee paid to the Copyright Clearance Center, is prohibited.

Many experiments have been performed and theories developed to clarify our understanding of grain boundary mobility in the past few decades. One common approach was to perform grain growth experiments and extract the average grain boundary mobility from the evolution of grain size. However, since this process involved a statistical averaging over grain boundaries, the mobility data extracted from such experiments could not yield information on the misorientation dependence of the grain boundary mobility. Some of the earlier experimental measurements of the mobilities of individual boundaries in metals and ceramics made using curvature driven grain boundary migration can be found in [1-6]. Even in the experiments performed on the highest purity materials available, there is evidence suggesting that the mobility was affected by impurity segregation (see below). Atomistic simulations provide an approach for studying intrinsic boundary mobilities (i.e., without impurities). Several simulations of boundary migration have been performed to date [7-11]. The earlier simulations did not systematically extract grain boundary mobility.

In the present paper, we examine the misorientation dependence of grain boundary mobility and grain boundary energy in a model two-dimensional bicrystal under constant driving force conditions. In the next section, we briefly outline our simulation procedure. Next we present our results on the grain boundary mobility as a function of temperature and misorientation and the boundary energy as a function of misorientation. In the following section, we discuss the consequences on grain growth phenomena.

SIMULATION METHOD

The present simulations were designed to study steady-state, curvature driven grain boundary migration. Shvindlerman and co-workers [6,12] introduced the 2-d U-shaped half-loop bicrystal geometry shown in Fig. 1 in order to study steady-state, curvature-driven boundary migration and it is this geometry that we adopt here. The reduced mobility of the grain boundary M^* can be calculated from the boundary velocity v as $v=MF=M\gamma\kappa=M^*\kappa$, where F is the driving force for boundary migration which reduces to $F=\gamma\kappa$ for curvature driven migration, γ is the boundary energy and κ is the boundary curvature. The rate of change of the area of the half-loop grain is

$$\dot{A} = vw = -(M\gamma)\kappa w = -(M\gamma)(\frac{2}{w})w = -2M\gamma , \qquad (1)$$

w being the half-loop width. The simulations were performed in 2-d using molecular dynamics and simple, empirical (Lennard-Jones) pair potentials (the depth of the potential is ε and the minimum is at $r=r_0$). Three edges of the simulation cell are left free, so as to decrease the stresses produced (and vacancies generated [13]) during grain boundary migration. Three atomic planes are frozen on the fourth edge to maintain the desired grain misorientations. The simulations were all performed at constant temperature. The underlying crystal structure is that of a triangular lattice with a nearest neighbor spacing r_0. For more details, see [11]. Times are reported in units of $\tau=(Mr_0/\varepsilon)^{1/2}$, energies in units of ε, distance in units of r_0, and area in units of the perfect crystal area per atom a_0, where M is the atomic mass. The simulations were performed with bicrystal misorientations corresponding to various general (for e.g. $\Sigma = -30$), singular ($\Sigma = 7 - 38.22$ and $\Sigma = 13 - 32.23$) and vicinal (near $\Sigma = 7$ and $\Sigma = 13$) boundaries. The present study focuses on high angle boundaries. Hence, the misorientations examined here are limited to the range $30° < \theta < 40°$ (the entire range of unique boundary misorientations lies between $30° < \theta < 60°$ or, equivalently, $0° < \theta < 30°$). In order to determine the instantaneous area of the half-loop grain, each atom is assigned to one of the two grains based upon the angles between it and its nearest neighbors relative to a fixed direction in the laboratory frame. The assignment to each grain, is made based on whether the average nearest neighbor angles (modulo $2\pi/6$) are closer to those of the lattices interior to or exterior to the half-loop boundary. Unless otherwise noted, all data is averaged over at least three simulation runs.

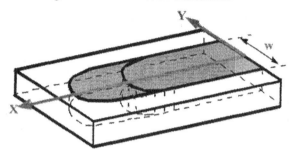

Fig. 1. Schematic illustration of the U-shaped half-loop using in the simulations and experiments [6].

RESULTS AND DISCUSSION

We perform a series of grain boundary migration simulations as a function of half-loop width and temperature. The atomic configuration of the entire system is monitored throughout the grain boundary migration simulations. Fig. 2 shows the temporal evolution of a $w = 21r_0$ bicrystal using a simulation performed at $T = 0.145$ ε/k. Following an initial transient, the half-loop grain settles into a well defined shape that is maintained as the half-loop retracts. During this half-loop retraction, the crystallographic misorientation between the grains is maintained and the sides of the half-loop, far from the highly curved region remain nearly parallel. Careful observations of the half-loop during its retraction show that small fluctuations about what appears to be a steady-state half-loop shape do occur. While the boundary profile seen in Fig. 2 appears to be a steady-state profile, relatively long excursions from this shape are occasionally observed.

The temporal evolution of the size of a half-loop grain is shown in Fig. 3, where we plot grain area A versus time t. The grain area A is approximately equal to the number of atoms in the half-loop grain times the area per atom in the perfect crystal. Following an initial transient, the area of the half-loop grain A decays linearly with time. When the height of the half-loop is a few times that of its width w, the half-loop shape is perturbed by the frozen (3 layers of) atoms at the bottom of the simulation cell; thereby, giving rise to another transient (at late times). Examination of Fig. 3 between the beginning and ending transient regimes shows an overall linear decay of grain size with time with a superimposed high frequency background oscillation. This is consistent with the fast fluctuations in grain shape that occurs during the retraction of the half-loop, as described above. Nonetheless, the rate of change of the half-loop area may easily be extracted from this data.

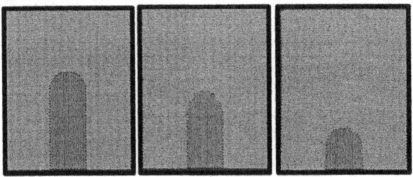

Fig. 2. Temporal evolution of the grain boundary profile at T=0.145 ε/k and a width of w=21 r_o at t=0, 500 and 1500τ.

In order to extract the dependence of the grain boundary migration rate on the grain boundary curvature, we perform a series of simulations at T=0.125 ε/k at different widths and extract values of $\dot{A}$ by performing linear curve fits to the non-transient regions of the $\dot{A}$ versus t curves (e.g., Fig. 3). According to Eq. (1), we should expect $\dot{A}$ to be independent of boundary width if the grain boundary velocity is proportional to curvature. Fig. 4 shows $\dot{A}$ as a function of half-loop width, w. For small half-loop widths, $\dot{A}$ decreases with increasing w and then asymptotes to a constant value $\dot{A}_\infty$ at large widths (see the horizontal line in Fig. 4). Therefore, these simulations show that grain boundary velocity is proportional to grain boundary curvature for sufficiently large grains. We expect that the deviation at small grain size is attributable to elastic interactions between the dislocations that make up the boundary, which should be negligible beyond a few inter-dislocation spacings. In the present T=0.125 ε/k simulations, we find that $\dot{A}_\infty = 0.26$ $(r_o)^2/\tau$, which corresponds to a reduced mobility of $M\gamma=0.13$ $(r_0)^2/\tau$. Deviations from pure curvature driven boundary migration are only significant for half-loop widths below 20 r_0.

The temperature dependence of the reduced grain boundary mobility is determined by performing a series of simulations over a temperature range from T=0.075-0.250 ε/k and with a half-loop width w for which classical curvature driven growth occurs, w=25r_0. These data sets are each well fit by a straight line ($\ln M^* = \ln M_0^* - Q/kT$), the slope of

Grain Boundary Engineering in Ceramics

which yields the activation energy for grain boundary migration. The linearity of the data (on this axes of Fig. 5) demonstrates that the mobility is indeed an Arrhenius function of temperature ($M=M_o e^{-Q/kT}$), consistent with the predictions of absolute reaction rate theory. The activation energies lie between 0.3-0.5 ε, where ε is the bond strength. These activation energies are smaller than those obtained experimentally (normalizing the experimental data by bond strength), likely due to impurity effects in the experimental studies (see [11]). The activation energy and the pre-exponential factor M_o^* vs. θ plots (not shown) exhibit pronounced minima (cusps) at misorientations corresponding to $\Sigma=7$ ($\theta=32.20^\circ$) and $\Sigma=13$ ($\theta=38.23^\circ$). There exists a temperature where all of the Arrhenius curves for misorientations near the same cusp cross (known as the compensation temperature). This is seen in Fig. 5 and in experiments [11]. The reduced mobility vs. misorientation shows a maximum at $\Sigma7$ and a minimum at $\Sigma19$ at all temperatures examined (see Fig. 6). However, the $\Sigma13$ boundary corresponds to a minimum in the reduced mobility at high temperature and a maximum at low temperature. This is consistent with the fact that only the compensation temperature for the $\Sigma13$ boundary falls within the temperature range examined.

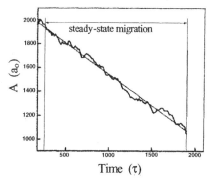

Figure 3. The time dependence of the U-shaped half loop grain at T=0.145 ε/k and a width of w=21 r_o.

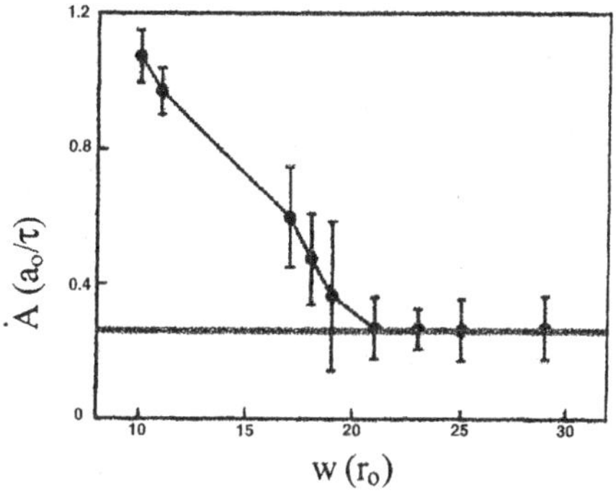

Figure 4. The time rate of change of the area of the U-shaped grain vs. half loop width at T=0.145 ε/k.

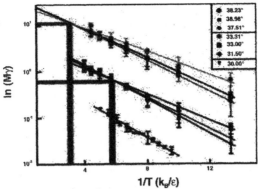

Fig. 5. Arrhenius plot of the reduced mobility for a series of simulations corresponding to simulations performed at or near Σ7 (~38°), Σ13 (~32°) and a 30° boundary.

In curvature driven growth, the driving force is directly proportional to the grain boundary energy. We extracted the grain boundary energy from a plot of the total energy of the system with the shrinking half-loop vs. grain boundary area. Since the width of the half-loop is fixed, the slope of this curve is simply equal to twice the grain boundary energy (actually enthalpy). Figure 7 shows that the boundary energy has similar misorientation dependence as the activation energy for boundary migration, with cusp like minima at the low Σ misorientations. This data can be used to extract the mobility from the reduced mobility ($M=M^*/\gamma$).

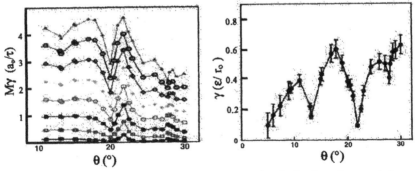

Fig. 6. Reduced mobility vs. boundary misorientation. The curves from bottom to top range from T=0.075 to 0.250 ε/k in units of 0.025 ε/k, respectively.

Fig. 7. Grain boundary enthalpy vs. boundary misorientation at T=0.125 ε/k.

Grain Boundary Engineering in Ceramics

CONSEQUENCES OF ANISOTROPIC BOUNDARY PROPERTIES

Since grain boundary velocity is proportional to the product of the driving force and mobility, anisotropy in either of these will modify microstructural evolution during grain growth. Anisotropy in the misorientation dependence of boundary energy also modifies the evolution by changing the angles at which grain boundaries meet. This, in turn, further modifies the curvature of the boundaries and hence their velocities. Monte Carlo simulation suggests that this is the dominant anisotropy for microstructural evolution during normal grain growth with an initially random texture [14]. In general, anisotropy in the misorientation dependence energy can be expected to lead to texture formation and evolution from one type of texture (e.g., a deformation texture) to another and a change in the relative abundance of different boundary types [14,15]. Anisotropy in mobility alone will not lead to texture development in an initially random polycrystal [14], but is expected to modify texture evolution in an initially textured material. The largest effects are expected for boundaries for which cusps exist in the energy and/or mobility. Since the above data suggests that there are cusps in both the energy and mobility, we expect that the preponderance of the corresponding misorientation will either decrease or increase during the evolution, depending on whether these cusps correspond to maxima or minima. Since raising the temperature can change maxima to minima, the nature of the microstructural evolution may be very different at high and low T (other than the overall rate). In order to make quantitative predictions of microstructural evolution based on these data, it is imperative that approaches be developed to predict the effects of segregation on both boundary mobility and energy. While there has been significant progress in the latter, there has been relatively little progress in the former. Additionally, there is little simulation data on boundary mobility available in 3-d [9].

ACKNOWLEDGEMENT: The authors acknowledge useful discussions with L.S. Shvindlerman and the support of the US Dept. of Energy grant DE-FG02-99ER45797.

REFERENCES

1. K. T. Aust and J. W. Rutter, Misorientation dependence of tilt grain boundaries in ultra-pure Sn and Al. Trans. Am. Inst. Min. Engrs. **215**, 119 (1959).
2. K. T. Aust and J. W. Rutter, Misorientation dependence of tilt grain boundaries in ultra-pure Sn and Al. Trans. Am. Inst. Min. Engrs. **215**, 820 (1959).
3. K. T. Aust and J. W. Rutter, Migration of <100> tilt grain boundaries in high purity lead, Acta metall. **13**, 181 (1965).
4. B. B. Rath and H. Hu, Curvature driven grain boundary migration in zone-refined Al. Trans. Am. Inst. Min. Engrs. **245**, 1577 (1969).
5. R. C. Sun and C. L. Bauer, Measurement of grain boundary mobilities through magnification of capillary forces. Acta metall. **18**, 635-638, (1970).
6. V. Y. Aristov, Y. M. Fridman, and L. S. Shvindlerman, Boundary migration in aluminium bicrystals. Phys. Met. Metall. **35**, 859 (1973).

7. R. J. Jhan and P. D. Bristowe, A molecular dynamics study of grain boundary migration without the participation of secondary grain boundary dislocations. Scripta metall. mater. **24**, 1313 (1990).

8. A. P. Sutton, Atomistic simulation study of grain growth using shrinking cylinders. In: Computer Simulation in Materials Science : Nano/Meso/Macroscopic Space and Time Scales. Eds. H.O. Kirchner, et al. (NATO-ASI series E 308), p. 163 (1996).

9. B. Shönfelder, D. Wolf, S. R. Philpot and M. Furtkamp, Molecular-dynamics method for the simulation of grain-boundary migration. Interface Science **5**, 245 (1997).

10. M. Upmanyu, R. W. Smith, and D. J. Srolovitz, Atomistic simulation of curvature driven grain boundary migration. Interface Science **6**, 41 (1998).

11. M. Upmanyu, D. J. Srolovitz, L. S. Shvindlerman, and G. Gottstein, Misorientation dependence of intrinsic grain boundary mobility: Simulation and experiment. Acta mater. **47**, 3901 (1999).

12. G. Gottstein, D. A. Molodov and L. S. Shvindlerman, Grain Boundary Migration in Metals: Recent Developments, Interface Science **6**, 7 (1998).

13. M. Upmanyu, D. J. Srolovitz, G. Gottstein and L. Shvindlerman, Vacancy Generation During Grain Boundary Migration, Interface Science **6**, 287 (1998).

14. E. A. Holm and G. N. Hassold, unpublished communication.

15. G.S. Grest, D. J. Srolovitz and M.P. Anderson, Computer Simulation of Grain Growth: IV. Anisotropic Grain Boundary Energies, Acta metall. **33**, 509 (1985).

COMPUTATIONAL MODELING OF CERAMIC MICROSTRUCTURE BY MC AND MD METHODS

Hideaki Matsubara, Hiroshi Nomura, Atsushi Honda, Katsuyuki Matsunaga
Japan Fine Ceramics Center
4-6-1 Mutsuno Atsuta-ku, Nagoya, 456-8587, Japan

ABSTRACT

We have successfully developed a computer simulation technique for microstructural design of ceramic materials. The Monte Carlo (MC) and the molecular dynamic (MD) methods have been used for the simulations of the microstructures at the particle size level and atomic level, respectively. The MC simulations were performed for the array of two- or three-dimensional lattices. Plural mechanisms of mass transport were introduced in the MC simulations of sintering and grain growth in ceramic systems, which involve a liquid phase and second solid particles. The MD simulation, which was applied to alumina ceramics, demonstrated the atomic structures and excess energy at grain boundaries and interfaces. The MC and MD simulations for sintering process are useful for microstructural design in ceramic materials.

INTRODUCTION

New techniques in microstructure design are needed to obtain novel functions and efficient fabrication processes in ceramic materials. Although experimental studies are critical in the development of new materials and processes, other methods must be used for complete success. We need to optimize microstructures of ceramic materials so the materials have well interacted and integrated functions for their applications. Computer simulation is a promising technique to design microstructures suitable for required performances in materials. The studies of material design using computer simulation can give us not only a basic understanding about microstructural development, but also new important directions for material technologies.

In this study, the computer simulation techniques by Monte Carlo (MC) and molecular dynamics (MD) methods are studied to design microstructures in ceramics.

To the extent authorized under the laws of the United States of America, all copyright interests in this publication are the property of The American Ceramic Society. Any duplication, reproduction, or republication of this publication or any part thereof, without the express written consent of The American Ceramic Society or fee paid to the Copyright Clearance Center, is prohibited.

The MC simulation is used to design the processes of sintering and grain growth and the MD simulation is investigated to design atomic structures at grain boundaries in ceramics.

SIMULATION METHOD

Fig. 1 shows the construction of our simulation system for microstructure design in ceramic materials. The MC simulation is a probability type method in which sequences proceed in the direction of decreasing total energy of the grain boundary, surface and interface. The MC simulation is applied to some important processes of microstructural development in ceramics, such as grain growth through liquid (Ostwald ripening), sintering of solid state, and sintering under the existence of liquid (liquid phase sintering).

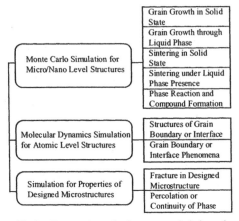

Fig.1　Construction of microstructural design of ceramics by MC and MD simulations.

The array of MC simulation in this study are the two dimensional and triangular lattices, which consist of a matrix solid phase with different crystal orientations, a liquid phase, a second solid phase and a pore (or void) phase.[1-5] There are different energy values depending on the grain boundaries, interfaces and surfaces. When a solid lattice is selected in the array, it tries the mass transfer through grain boundaries and a liquid phase. When a pore lattice is selected, the pore lattice moves due to the four diffusion routes; volume, grain boundary, pore-surface and outside-surface. If the total energy change ≤ 0, this try is in practice.

MD simulation is a well-established technique for examining micro- and nano-scale structures and motion on the atomic level. Using potential energy functions to describe the forces between atoms, we obtain structures, energies and dynamic properties at grain boundaries and interfaces.[6-8] The MD simulation in the three dimensional array under the ensemble of NTP is performed by using the software of MASPHYC (Fujitsu Co.). In this study, we obtain atomic structures and energies of grain boundaries, surfaces and interfaces in three systems of alpha-alumina based ceramics; pure alumina, Mg-, Si- or Ca-doped alumina and an alumina-glass phase system. The two-body potential function in the Al-Si-Ca-Mg-O system was used for the MD simulation.

RESULTS AND DISCUSSION

MC Simulations of Grain Growth and Sintering

Fig. 2 shows the two results of MC simulation of solid state sintering. The case of (a) has the initial configuration of single phase particles and pores (voids). The simulation distinctly shows the behavior of shrinkage accompanied by grain growth. The simulation result at 100 MCS (Monte Carlo steps) is the microstructure including coarsening grains and residual pores. The result at 500 MCS has larger grains and full density. The result of (b) is the simulation of the initial configuration so that second phase particles are distributed in the same initial microstructure of matrix phase particles as in the case of (a). This simulation demonstrates microstructure development under the interaction of three phases of the matrix, second phase particles (dispersoids) and pores. The second phase particles play not only an inhibiting effect on grain growth but also

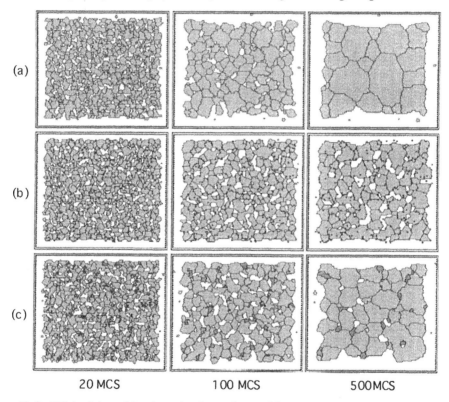

Fig.2   MC simulations of sintering and grain growth at a solid state. (a) solid1, (b) solid 2 + solid 2, (C) solid 1 + solid 2 (growing).

delaying effect on shrinkage. The final simulated structure illustrates a finer grain structure and more residual pores compared to the case of (a). The case of (c) can be treated as a more complex system where the second particles grow together with the matrix phase. It can be seen that the second particles grow through mass transportation in the matrix phase. The inhibiting effect of the second particles on grain growth and sintering in this case is moderate compared to the (b) case.

Fig. 3 shows the quantitative results in the above simulations of sintering and grain growth. Changes in porosity of the three systems are clear as a function of MCS. The grain size vs. MCS plots can indicate change in grain size of the matrix phase in the three cases and the size of the second particles of the (C) case.

Fig. 4 and Fig. 5 show a series of results from the new simulation of liquid phase sintering accompanied by Ostwald ripening of solid particles in the liquid phase. The shrinkage mechanism of this simulation is the combination of liquid wetting on solid surface and rearrangement of solid particles due to capillary force of a liquid. Moreover, this simulation can deal with the Ostwald ripening of solid particles due to a solution and re-precipitation mechanism through a liquid phase. The initial configuration is the mixture of solid particles and the particles that behave as liquid after the simulation starts. The content of liquid phase is varied from 5% to 50% to see the effect of liquid phase content on sintering and grain growth. The microstuctures are shown in Fig. 4 and the changes in porosity and grain size of solid particles are quantitatively indicated in Fig. 5. A low amount of liquid is not enough to obtain a dense material, and the shrinkage is faster with an increase in liquid phase content. Grain growth is enhanced with an increase in liquid phase content, but a large amount of liquid delays Ostwald ripening due to longer diffusion distance. It is noted that Ostwald ripening enhances the rearrangement of solid particles.

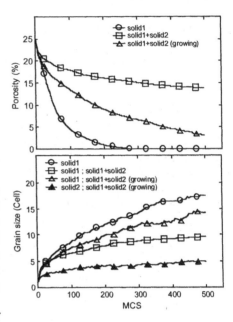

Fig.3 Porosity and grain size changes in the MC simulations of solid state sintering and grain growth.

Grain Boundary Engineering in Ceramics

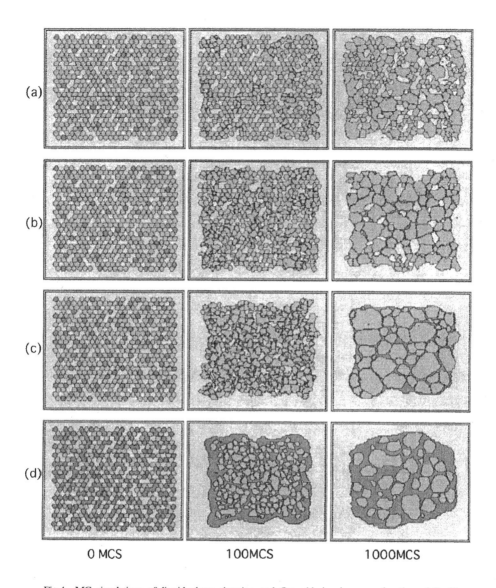

Fig. 6 shows the MC simulations of anisotropic grain growth under the presence of a liquid phase as a function of liquid phase contents. In this simulation, anisotropy in interfacial energy between the solid particles and liquid is introduced to simulate growth behaviors affected by the anisotropy and the liquid phase contents. With a small amount of liquid, grain growth is relatively slow and homogenous. As the liquid phase increases, the grain growth becomes faster. In the limited range of liquid contents from 7.5% to 10%, however, grain growth is not homogenous resulting in heterogeneous microstructures of coarse and fine grains. When the liquid content is above 12%, the microstructure again becomes relatively homogeneous. The compound effect between the anisotropy and the liquid phase contents is a possible mechanism for heterogeneous microstructures or abnormal grain growth in ceramic systems, including systems that contain a liquid phase like silicon nitride.

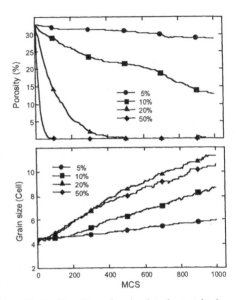

Fig.5 Porosity and grain size changes in the MC simulations of liquid phase sintering and Ostwald ripening as a function of liquid content.

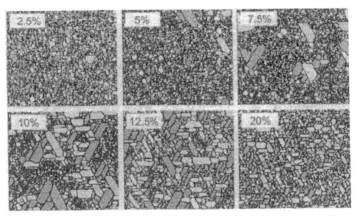

Fig.6 MC simulations of anisotropic grain growth under the presence of a liquid phase as a function of liquid content.

Grain Boundary Engineering in Ceramics

MD Simulations of Grain Boundary and Interface in Alumina System

Fig. 7 shows atomic structures of four kinds of grain boundaries in pure alumina by the MD simulation. It is possible to calculate an excess energy due to the presence of a grain boundary. For the same material, the grain boundary energies depend on the crystal orientation relationship of grain boundaries as shown in Fig. 7. Large differences in grain boundary energies are seen; for example, 2.5 J/m^2 for general boundary and 0.2 J/m^2 for $\Sigma 11$ boundary.

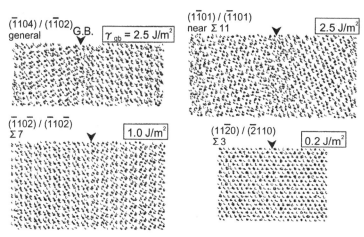

Fig.7 MD simulation of atomic structure and excess energy at grain boundaries of alumina.

The ratio of grain boundary energy (γ_{gb}) vs. surface energy (γ_{sv}) in pure alumina by the MD simulation is listed in Table 1. The ratio of γ_{gb} / γ_{sv} is in reasonable agreement with the experimental results available. The driving force for sintering is the reduction in surface energy when particles come into contact, so the relative surface and grain boundary energies play an important role. These data can also be used as input to MC simulations of alumina, illustrating the complementary nature of the two techniques.

Table 1 Ratio of γ_{gb} / γ_{sv} by MD simulation and experimental study in alumina.

grain boundary / surface	γ_{gb}/γ_{sv}
general / $[(\bar{1}104) + (1\bar{1}02)]/2$	1.1
near$\Sigma 11$ / $(1\bar{1}01)$	1.0
basal twin / (0001)	0.8
$\Sigma 7$ / $(1\bar{1}02)$	0.5
$\Sigma 3$ / $(11\bar{2}0)$	0.1
Experimental	0.7

Such a treatment is available in the systems that include inclusions or sintering additives. Trajectory and γ_{gb} by the MD simulation at grain boundaries of Mg, Ca or Si doped alumina is shown in Fig. 8. The Mg segregated boundary has the lowest energy value among the three boundaries. This means that Mg segregation has a role of stable grain boundary formation in alumina.

Fig. 9 shows the mean square displacement by the MD simulation at grain boundaries of Mg, Ca or Si doped alumina. It is noted that Mg is much faster than Si, Ca and Al at the boundaries. This result indicates that Mg is so easy to move for a uniform distribution at grain boundaries in alumina. The low energy of Mg segregated boundary and the fast mobility of Mg along the grain boundary are possible mechanisms for fine grain structures in the Mg-doped alumina system.

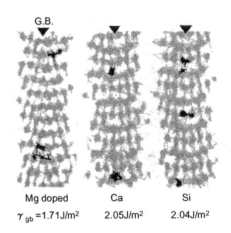

| Mg doped | Ca | Si |
| γ_{gb} =1.71J/m² | 2.05J/m² | 2.04J/m² |

Fig.8   Trajectory and γ_{gb} by MD simulation at grain boundaries of Mg, Ca or Si doped alumina.

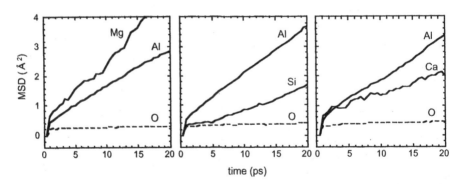

Fig.9   Mean square displacement by MD simulation at grain boundaries of Mg, Ca or Si doped alumina.

Grain Boundary Engineering in Ceramics

The MD simulation can be applied to the system of alumina-glass (or liquid) system. The trajectory at interfaces between alumina and glassy phase with anorthite composition is shown in Fig. 10. The cases of (a) and (b) have thin and thick glassy phases, respectively. The trajectory in the thin film of (a) indicates a lower mobility than that of thick film of (b). The interfacial energy in the thin film is also lower than that of the thick film. These results bring a new understanding of grain boundaries and interfaces in ceramic systems containing crystal solid and glass (liquid) phases. The grain boundaries or interfaces in the system vary the mobility and the energy depending on the thickness of the glassy phase.

In the future, the combination of MC and MD simulations will be a very powerful technique to design ceramic microstructures. Both grain structures and atomic structures obtained by MC and MD are useful and helpful for development of new ceramic materials and systems with novel functions.

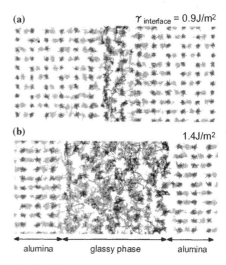

Fig.10   Trajectory and excess energy by MD simulation at interfaces between alumina and glassy phase with anorthite composition.

CONCLUSION

The Monte Carlo (MC) and the molecular dynamic (MD) methods have been used for the simulations of the microstructures at the particle size level and atomic level, respectively.

(1) The MC simulation was applied to the microstructural development of sintering and grain growth at the solid state and under the presence of a liquid phase. The compound effects of the second particle, liquid phase and anisotropy were successfully analyzed by the MC simulations in order to design complex and important microstructures in ceramics.

(2) The MD simulation was applied to three systems of alpha alumina based ceramics; pure alumina, Mg-, Si- or Ca-doped alumina and an alumina-glass phase system. The MD simulations demonstrated the atomic structures, excess energy and mass transportation at the boundaries and interfaces depending on crystal orientation, composition, and film thickness.

REFERENCES

1) H. Matsubara and R. J. Brook, *Amer. Ceram. Soc., Ceramic Transactions,* **71**, 403-418 (1996).
2) H. Matsubara, S. Kitaoka and H. Nomura, *Amer. Ceram. Soc., Ceramic Transactions,* **99**, 51-63 (1998).
3) H. Matsubara, *Key Engineering Materials the Ceramics, Ceram. Soc. Japan,* **166**, 1-8 (1999).
4) M. Tajika, H. Matsubara and W. Rafaniello, *J. Ceram. Soc. Japan,* **105** [11], 928-933 (1997).
5) Y. Okamoto, N. Hirosaki and H. Matsubara, *J. Ceram. Soc. Japan,* **107**, 109-114 (1999).
6) H. Suzuki, H. Matsubara, J. Kishino and T. Kondoh, *J. Ceram. Soc. Japan,* **106** [12], 1215-1221 (1998).
7) H. Suzuki and H. Matsubara, *J. Ceram. Soc. Japan,* **107** [8], 727-732 (1999).
8) A. Honda, K. Matsunaga and H. Matsubara, "Molecular Dynamics Simulation of an Intergranular Glass Phase in Alumina Based Ceramics", *J. Japan Inst. Metals,* to be submitted.

RHEOLOGY OF GRAIN BOUNDARY NETWORK SYSTEMS

Yoshihisa Enomoto and Takashi Mitsuda,
Department of Environmental Technology, Nagoya Institute of Technology,
Gokiso, Nagoya 466-8555, Japan

ABSTRACT

Computer simulations of grain growth kinetics under steady shear strain are performed by means of two-dimensional phase field model. The time evolution of both a single spherical grain and randomly distributed grains undergoing domain coarsening is studied by changing the shear rate. We observe deformation and disappearance of the single grain. For the random grains, some anomalous rheological properties are shown to be due to their topological rearrangement.

INTRODUCTION

Grain growth is a process in which the average grain size of a single-phase polycrystalline material increases as a function of time, driven by the reduction in the total grain boundary energy [1]. Since most advanced materials are polycrystalline, their physical properties, including mechanical and electrical properties, are influenced and very often controlled by the microstructures developed during the grain growth following sintering and/or solidification. Therefore, it is of technological importance to fundamentally understand grain growth kinetics responsible for the temporal microstructural evolution.

Over the last half century, there have been many experimental as well

To the extent authorized under the laws of the United States of America, all copyright interests in this publication are the property of The American Ceramic Society. Any duplication, reproduction, or republication of this publication or any part thereof, without the express written consent of The American Ceramic Society or fee paid to the Copyright Clearance Center, is prohibited.

as theoretical investigations of the grain growth process. Recently, there has been enormous interest in using computer simulations, based on various types of models such as Q-state Potts model [2], vertex model [3], phase field model [4] and their variants. These works have greatly improved our understanding of the grain growth kinetics. However, the mechanical properties, especially of the dynamical or rheological behavior of the grain systems, have been mostly neglected, compared with the grain growth problem. Experimentally, the grain systems have been known to exhibit a complex mechanical response such as viscoelasticity, which differs from that of a simple material, due to their structural complexity [5].

Under these circumstances, we here carry out computer simulations for the rheological behavior of polycrystalline systems under steady uniform shear strains, using the two-dimensional phase field model [4]. Our concern is how the shear strain affects the pattern due to the grain growth kinetics and how this effect appears in the macroscopic properties such as the viscosity.

THE MODEL

In the phase field approach [4], a microstructure is described by a set of spatially continuous and time-dependent field variables (non-conserved order parameter fields); $C_1(\mathbf{r},t),\cdots,C_p(\mathbf{r},t)$ with p being the number of field variables. Each field variable represents grains of a given crystallographic orientation, which is a continuous variable ranging from -1.0 to 1.0. For example, a value of 1.0 (-1.0) for $C_1(\mathbf{r},t)$ means that the material at position $\mathbf{r}$ has the crystallographic orientation labeled as 1 (anti-phase related to orientation 1). At the grain boundary region between orientations 1 and 2, $C_1(\mathbf{r},t)$ and $C_2(\mathbf{r},t)$ have absolute values intermediate between 0.0 and 1.0.

In the diffuse interface theory, the total free energy, F, of an inhomogeneous system containing grain boundaries is given by

$$F = \int d\mathbf{r}\left[f\{C_i(\mathbf{r},t)\}+\frac{k}{2}\sum_{i=1}^{p}(\nabla C_i)^2 \right] \tag{1}$$

Grain Boundary Engineering in Ceramics

with the local free energy density

$$f\{C_i\} = \sum_{i=1}^{p}\left(-\frac{a}{2}C_i^2 + \frac{b}{4}C_i^4\right) + c\sum_{i=1}^{p}\sum_{j>i}^{p}C_i^2 C_j^2 \qquad (2)$$

where k is the gradient energy coefficient, and a, b, and c are phenomenological positive parameters. It may be easily shown that, if $c > b/2$, the above free energy model has $2p$ potential wells (minima) in the p field variable space, which describe the equilibrium free energies of crystalline grains in $2p$ different orientations. It has been known that a finite but large number for p is sufficient for realistically modeling the grain growth kinetics [4].

Microstructural evolution of the polycrystalline system under applied shear deformation is characterized by the extended Ginzburg–Landau–type kinetic equations [4,6]:

$$\frac{\partial C_i}{\partial t} + \mathbf{u}(\mathbf{r},t)\bullet\nabla C_i = -L\frac{\delta F}{\delta C_i} \qquad (3)$$

where L is the kinetic coefficient which describes the grain boundary mobility. The term $\mathbf{u}(\mathbf{r},t)\bullet\nabla C_i$ represents the effect of deformation by an applied strain, using an analogy of the hydrodynamic treatment in binary fluids [6]. Since the elastic interaction is neglected, the deformation field $\mathbf{u}(\mathbf{r},t)$ is equal to the externally imposed strain, which, in the case of shear strain, is given by $\mathbf{u}(\mathbf{r},t) = (Gy,0)$ with the shear rate G. The elastic interaction is, indeed, important in most experimentally accessible situations. Here, however, we deal with eq. (3) as a first step, since inclusion of the elastic interaction is time–consuming.

In our two–dimensional computer simulation, the set of kinetic equations is discretized with respect to space and time, and is solved using the simple Euler technique. The system is a rectilinear form with the size $N \times N$ and with the periodic boundary condition for the two boundaries perpendicular to the strain. For the boundary parallel to the strain we have imposed a sheared periodic boundary condition [6]. All the results reported in this paper are obtained by assuming $a = b = c = 1$, $k = 0.5$, and $L = 1.0$ with lattice spacing 1.0, time interval 0.1, and $N=128$. We also set $p = 4$.

SIMULATION RESULTS

To visualize the microstructures, one may define the following function;

$$\psi(\mathbf{r},t) = \sum_{i=1}^{p} C_i(\mathbf{r},t)^2 .$$ The values of ψ within the grains are close to 1.0,

while those at the boundaries are significantly less. In the following figures, dots are plotted at positions with $\psi(\mathbf{r},t) < 0.7$.

As a first test, we simulate the time evolution of a spherical grain of orientation 1 having radius 32, embedded in a matrix of orientation 2 with small initial randomness. As was expected [4], for the case of $G = 0.0$ (Fig. 1(a)), the spherical grain is shown to shrink and finally disappear. For a small value of G (Fig. 1(b)), the grain is simply elongated to disappear. If we increase the shear rate further, the grain eventually breaks up into many pieces before its disappearance (Fig. 1(c)). In this simulation, disappearance of the grain has been observed for a shear rate larger than 0.05. Such a threshold value of G has been found to be a decreasing function of the initial radius.

Now we study the effects of shear in the grain growth (or coarsening) process. Initially we assign random numbers distributed uniformly between $|C_i(\mathbf{r},t = 0)| < 0.01$ at each lattice point. The time evolution of grains is shown in Fig. 2(a)–(c) for $G = 0$, 0.01 and 0.02, respectively. The microstructures with $G = 0$ (Fig. 2(a)) are remarkably similar to experimentally observed ones of normal grain growth in many single–phase polycrystalline materials [4]. For $G \neq 0$, in the initial regime, the grains are simply elongated by the strain. This elongation, then, causes coalescence and recombination of vertices. In this way, the grain pattern gradually becomes parallel to the strain direction. Another feature of the grain growth is a competition of coarsening and deformation of grains. When the shear rate is small (Fig. 2(b)), the coarsening dominates the deformation. Thus, grains can grow larger compared to those in the strong shear case (Fig. 2(c)) at the same shear strain, $\gamma = Gt$.

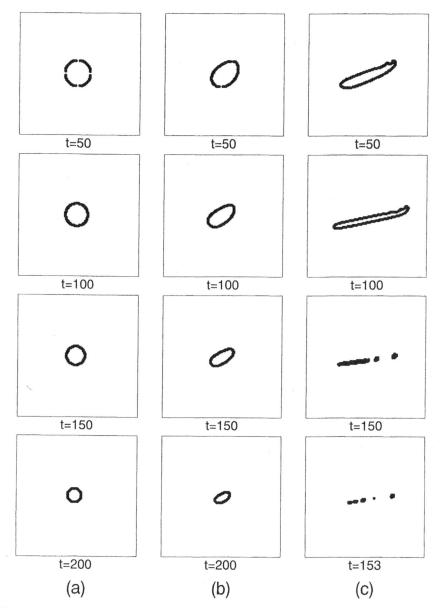

Figure 1   Time evolution of a spherical grain boundary for G =0 (a), 0.01 (b) and 0.05 (c).

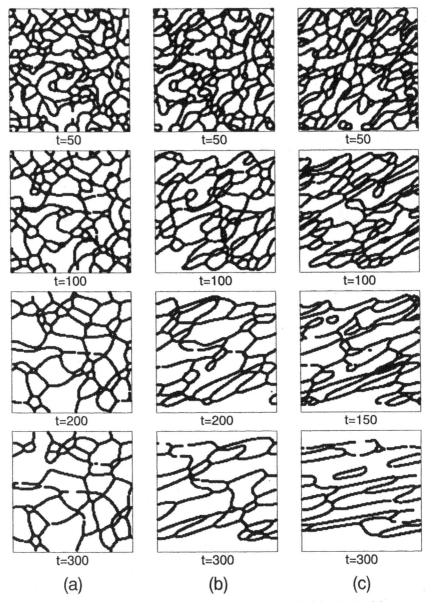

Figure 2   Time evolution of random grains for $G=0$ (a), 0.01 (b) and 0.02 (c).

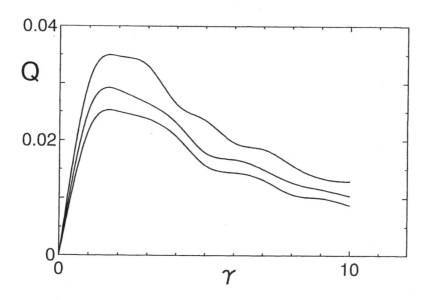

Figure 3   Anisotropy factor Q as a function of the shear strain $\gamma = Gt$ for $G = 0.01$, 0.02 and 0.05 from lower to upper lines.

In studying rheological properties of the system, one of the important quantities is the stress tensor due to grain boundaries. Here, we evaluate the xy component of the macroscopic stress tensor, Q [6],

$$Q\ (t) = \left\langle \frac{\partial \psi}{\partial x} \frac{\partial \psi}{\partial y} \right\rangle \tag{4}$$

where $\langle\ \rangle$ denotes the spatial average. This is a measure of the spatial anisotropy of the grain disturbed by the shear strain. Note that the excess viscosity $\Delta\eta$ is given by $\Delta\eta(t) = Q\ (t)/G$ [6].

In Fig. 3 the values of Q are plotted as a function of the shear strain $\gamma = Gt$ for several values of G. It is found that Q increases rapidly when the strain is small and then starts to decrease slowly at a larger strain. Indeed, Q exhibits a peak at a finite value of the shear strain, i.e., $\gamma \approx 2$, which is insensitive to the shear rate. On the other hand, the peak

height of Q is weakly dependent on the shear rate G as $\max\{Q(t)\} \propto G^{0.2}$.
These values of the peak position and the peak height exponent might depend
on the system size, number of fields, and the shear onset time, whose effects
are now under consideration as well as oscillatory shear effects. It is
interesting to compare these time-dependent behaviors with the grain
boundary structure. The grains are elongated in the time region where
Q increases. Once the grain disappears and the vertex recombination
dominates, Q begins to decrease. This behavior can be explained as follows.
Increase of the grain boundaries due to the elongation of grains results
in a storage of the free energy, and the disappearance of the grains, as
well as the shear enhanced grain disappearance, gives rise to a dissipation
of the energy. These are the causes of viscoelasticity.

CONCLUSION

We have simulated the grain growth kinetics of polycrystalline materials
under steady shear using the two-dimensional phase field model. We have
demonstrated some interesting features of the relationship between the
morphology of grain boundary network systems and their mechanical response.

REFERENCES

[1] D. Weaire and N. Rivier, "Soap, Cell and Statistics", Contemp. Phys.,
     25[1], 55-99, (1984).
[2] J. A. Glazier, M. P. Anderson and G. S. Grest, "Coarsening in the
     Two-dimensional Soap Froth and the Large-Q Potts Model", Philos. Mag. B,
     62[6], 615-645, (1990).
[3] K. Kawasaki, T. Nagai and T. Nakashima, "Vertex Model for
     Two-dimensional Grain Growth", Philos. Mag. B, 60[3], 399-421, (1989).
[4] L. Q. Chen, "A Novel Computer Simulation Technique for Modeling Grain
     Growth", Scripta Metall., 32[1], 115-120, (1995).
[5] T. Okuzono and K. Kawasaki, "Intermittent Flow Behavior of Random Foams",
     Phys. Rev. E, 51[2], 1246-1256, (1995).
[6] T. Ohta, H. Nozaki and M. Doi, "Computer Simulations of Domain Growth
     under Steady Shear Flow", J. Chem. Phys., 93[4], 2664-2675, (1990).

Microstructural Development in α-Al$_2$O$_3$

M. A. Gülgün, R. Voitovych, E. Bischoff, R. M. Cannon[#], and M. Rühle
Max-Planck Institute für Metallforschung, Seestr. 92, D-70174, Stuttgart, Germany
[#]Lawrence Berkeley Laboratory, Berkeley, CA 94720, USA

ABSTRACT

Despite many investigations already performed with respect to sintering, stability and grain growth in α-Al$_2$O$_3$, some fundamental questions are still open. These questions concern mainly the influence of small level of impurities at grain boundaries on the sintering behavior and growth of the grains of the materials, i.e. the microstructure and its stability. Therefore, different α-Al$_2$O$_3$ ceramics were prepared containing a well-defined amount of impurities. The materials were sintered and annealed at high temperatures. Besides conventional microstructural studies, such as grain diameter and grain size distribution, the segregation at different grain boundaries were studied by analytical electron microscopy using a dedicated STEM with high spatial resolution. The results observed so far suggest that abnormal grain growth starts if a certain level of specific impurities is segregated at grain boundaries. Those observations are supported by quantitative studies of the correlation between segregation at grain boundaries, grain size and grain size distribution.

The implications of the different microstructures with respect to microstructural stability are discussed and compared to observations described in the literature.

INTRODUCTION

Impurities (intentional dopants and/or contamination) have dramatic effects on the microstructural evolution, sintering, and properties of polycrystalline α-Al$_2$O$_3$(from hereon called alumina). Coble [1] reported in the early 60's that with small additions of MgO (0.25wt%), it was possible to produce translucent polycrystalline alumina. Over the last 40 years, MgO was shown to inhibit exaggerated grain growth and to help with the densification of the ceramic. A comprehensive review of the role of MgO in alumina was given by Bennison and Harmer [2]. Furthermore, for over more than 50 years, silica additions were used to produce commercial grade sintered alumina ceramics. Over the same period of time, there has been numerous investigations on the effects of impurities, especially Si, Ca,

To the extent authorized under the laws of the United States of America, all copyright interests in this publication are the property of The American Ceramic Society. Any duplication, reproduction, or republication of this publication or any part thereof, without the express written consent of The American Ceramic Society or fee paid to the Copyright Clearance Center, is prohibited.

Ti, Mn, Mg, Zr, and Y on the microstructure [3-6], sinterability [7-8], and mechanical properties of corundum [9-10]. Most of the earlier work has been performed on ceramics produced from commercially available powders. The impurity levels in these powders were well above the tolerable limits for a complete control of the microstructural evolution during sintering as shown elegantly by Baik and coworkers [11-13]. Owing to the co-existence of many impurities at grain boundaries (GBs) in the microstructure there have been a wide spread confusion in the literature over the effects of certain impurities. Lately, purer alumina powders (total cation impurities below 50 ppm) have been available [11]. With these powders it is now possible to isolate the influence of certain impurity ions on the microstructure and properties of the ceramic.

In α-Al_2O_3, in which most impurities (except for Cr and Fe) possess a very low solubility, impurity atoms will strongly segregate to GBs; hence, the impurity effects are mediated through the interfaces in the ceramic. Thus, to understand its microstructure and properties, it is imperative to investigate the structure and the chemistry of the interfaces in polycrystalline alumina, and correlate this to other microstructural parameters. With the advance of the analytical electron microscopy tools, it is now possible to image and analyze the chemical composition at the interfaces on nearly atomic scale. This communication reports on the microstructural and microchemical investigations of alumina with controlled amounts of impurities. Furthermore, the paper relates the interface chemistry to the observed microstructural development and sintering behavior.

EXPERIMENTAL DETAILS

Three different sets of polycrystalline alumina ceramics with varying amounts of impurities were prepared. The first set of samples (GI), were processed through the so-called temperature induced forming method of Bell and Sigmund using AKP53 high purity alumina powders (about 150 ppm Si) [14]. The second and third sets of samples (GII and GIII) were prepared from AKP 3000 high purity powders (with total cation impurity levels < 60 ppm). The powders of GII and GIII were mixed with specified amounts of yttrium in pure isopropanol and ball milled for 1 hour. The GII samples were milled with ZrO_2 ball, and hence they had about 30 ppm additional Zr contamination. The GIII samples were milled with high purity alumina milling balls. These powders did not have any detectable impurity besides the intended dopant. powders were uniaxially compacted into 13 mm diameter discs and the isostatically cold pressed at 800 MPa. All samples were sintered in a bed of their starting powders and closed alumina crucibles at 1550°C for various times in air. Bulk chemical analysis of starting powders and sintered ceramics were done using ICP/OES. Table I shows the as sintered chemical analysis of GI and GII samples. GIII samples had less than 30 wt ppm Si, and <10 wt ppm Zr, and Mg.

For microstructural and microchemical analysis the samples were cut in half perpendicular to the axial directions and analyses were done in these inner surfaces except where indicated otherwise. Polished and thermally etched (1400°C for 1 h) surfaces were observed in optical (OM) and scanning electron microscopes (SEM).

Grain sizes were determined by the linear intercept method at least at two different magnifications [15]. The grain sizes reported here are taken as 3/2 times the mean linear intercept lengths. Conventional (TEM, JEOL 2000FX and Zeiss 912 Omega) and high resolution transmission electron microscopy (HRTEM, JEOL 4000 EX) were used to investigate the structure of the interfaces in the samples. The segregation of impurities to the interfaces were determined using a dedicated scanning transmission electron microscope (STEM, VG STEM HB501 UX). The quantification of the excess impurities at the grain boundary plane was done with the box method described by Ikeda et al. [16]. The yttrium excess values are corrected by using the experimentally determined k-factor. The excess values for other elements were evaluated by the k-factors supplied by the EDS software. In each sample a minimum of 12 grain boundaries with three measurement points on the average were quantified.

Table 1a. Volumetric Concentrations of impurities in GI (ICP/OES).

Sample	Y_2O_3 [wt ppm]	Si [wt ppm]	Ca [wt ppm]	Mg [wt ppm]	Zr[*] [wt ppm]
GI Y0	<3	300 ±50	20 ±10	40 ±10	~50
GI Y150	156 ±6	300 ±50	220 ±20	30 ±10	~50
GI Y300	307 ±12	300 ±50	30 ±10	40 ±10	~50
GI Y500	507 ±20	330 ±50	30 ±10	40 ±10	~50
GI Y1000	1046 ±50	300 ±50	30 ±10	40 ±10	~50
GI Y1500	1409 ±80	320 ±50	30 ±10	30 ±10	~50
GI Y3000	1700 ±90	370 ±50	30 ±10	40 ±10	~50

Table 1b. Volumetric Concentrations of impurities in GII (ICP/OES).

Sample	Y_2O_3 [wt ppm]	Si [wt ppm]	Ca [wt ppm]	Mg [wt ppm]	Zr [wt ppm]
GII Y0	<3	<100	<10	<10	<10
GII Y150	165 ±6	<100	<10	<10	30 ±10
GII Y300	424 ±17	<100	<10	<10	60 ±10
GII Y500	600 ±30	<100	<10	<10	80 ±10
GII Y1000	1270 ±80	<200	<10	<10	80 ±10
GII Y3000	2200 ±190	<200	<10	<10	30 ±10

"safe detection" limit for Zr, Mg and Ca 10 ppm. [*] estimated

RESULTS

Tables Ia and b show the bulk chemical composition of the three sets of samples after sintering. The GI samples were contaminated during processing or firing of the ceramics by Si and Ca. Since they were also milled with zirconia balls for 1 hours we expect them to have also about 50 ppm Zr in their composition. The GII samples contained less Si and Ca, however, also Zr impurities due the same reason. The Si content was less than the limit of detectability of the instrument used at that time. According to the manufacturer the certified silicon contend of the powder batch used in this work is about 6 ppm.

Figures 1 show the microstructure of the GI samples with varying amounts of Y-doping. All samples except undoped alumina were sintered under identical conditions at 1550° C for 2 h. The Y0 (undoped) sample was sintered at

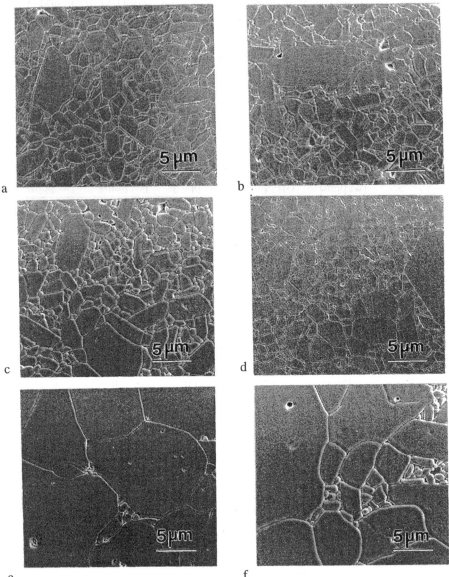

Figure 1. SEM micrographs of GI samples with yttrium contents of a) 0 , b)150, c)300, d) 500, e) 1050, and f) 1410 wt ppm of Y_2O_3.

Grain Boundary Engineering in Ceramics

1500 for 2 h. Up to 500 wt ppm Y_2O_3 doping, the microstructure remained fine grained, although some effects of Si and ca contamination were indicated by the elongated slab-like grain morphologies. For the 1000 wt ppm and higher amounts of Y_2O_3 doped samples the microstructure was unstable and clearly showed a bimodal grain size distribution with exaggeratedly grown grains. The majority of the grains were about 10 to 20 μm. The space between the large grains was filled with fine grains (~1-2 μm). A closer look at these abnormally grown microstructure with TEM revealed that some grain boundaries were wetted with an amorphous film (Figure 2). There were also triple point pockets filled with a glass. Energy filtered images of such boundaries and triple point pockets showed that the amorphous film at the grain boundaries as well as the glass in the pockets contained Si, Ca, Y, and Al [17]. HRTEM images confirmed the existence of approximately 2 nm thick amorphous film at some grain boundaries in these samples with large grains.

Figure 2. TEM dark-field micrograph of GI sample doped with 1500 wt ppm of Y2O3 showing amorphous grain boundary film and triple point pockets.

Figure 3 shows the SEM micrographs from the GII samples with varying Y contents sintered at 1550° C for 2 h. In GII, samples up to 500 wt ppm Y_2O_3 doping levels had a stable microstructure with a uniform distribution of equi-axed 1-2 μm grains. These microstructures did not contain any significant amounts of Si and Ca as confirmed by high spatial resolution EDS analysis. If the Y_2O_3 concentration exceeded 1000 wt ppm the grain size jumped to 4-5 μm. However, the microstructure remained equi-axed and uniform. No abnormal grain growth was observed. SEM analysis of the polished surfaces revealed that with increasing

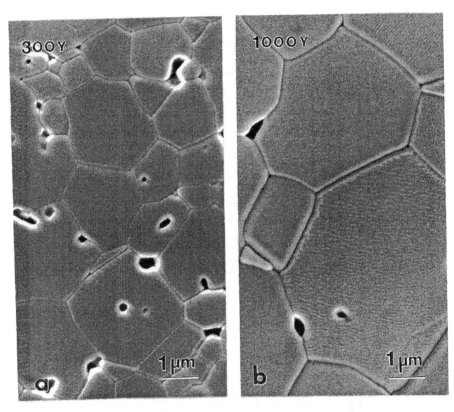

Figure 3. SEM micrographs of GII samples with yttrium contents of a) 300 , b1000, wt ppm of Y_2O_3.

Y content the porosity in the samples decreased. Figure 4 is the plot showing the porosity in the GII samples as a function of yttrium at 1550°C for 2 h. Annealing of GII in a Si rich environment (MoSi$_2$ furnace) resulted in an enormous grain growth. Figure 5 is the SEM picture from this outer region in the sample doped with 500 wt ppm yttrium. The microstructure was still uniform with equi-axed grains. However, the average grain size was about 60 μm. Further AEM analysis of the region indicated that this exaggerated grain growth phenomenon is closely related to the interface chemistry in this outer region.

The microchemistry of the interfaces in all of the samples were analyzed. The results are shown (Figure 6) in a representation suggested by Cannon (the 'Cannon-plot'). Cumulative frequency of analyzed GBs is plotted as a function of excess Y at the interfaces. Open symbols were used for samples where no YAG precipitates [5,18] were observed, and closed symbols for samples that had YAG precipitates. Similar plots were constructed for Si, Ca and Zr excess concentrations. The 0.5

Grain Boundary Engineering in Ceramics

cumulative frequency value for the excess concentration was taken as the average Γ^{excess} at the grain boundaries for each sample.

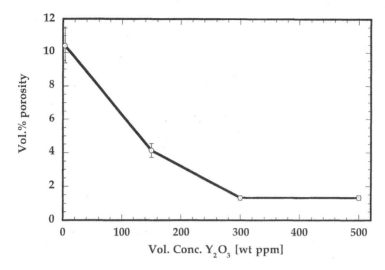

Figure 4. Plot showing volumetric porosity in GII samples with increasing amounts of Y_2O_3 doping.

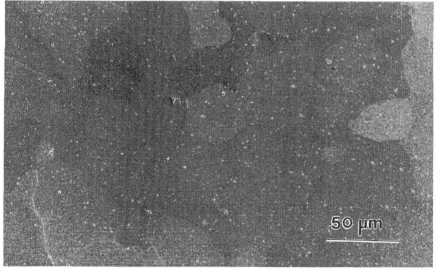

Figure 5. SEM micrograph from the Si contaminated outer region of 500wt ppm Y_2O_3 doped GII sample showing exaggeratedly grown yet equi-axed grains.

Figure 7 is a composite plot of grain size and average excess impurity concentration at the grain boundaries as a function of volumetric yttrium dopant concentration in samples GI. Very similar plots are obtained for GII and GIII samples. In terms of segregation behavior, three distinctly different regimes are delineated on the plot [18]. In regime I, Y ions are existent in the microstructure as segregant at grain boundaries and as solute in the alumina grains. The excess concentration at grain boundaries can be calculated in this regime after McLean-Langmuir from the GB area [19]. This relationship is plotted as hypothetical excess concentration in Figure 7. In regime II the excess concentration of Y reaches its solubility limit, and perhaps even go into supersaturation. As a consequence, precipitation of YAG is expected. In regime III YAG precipitates are observed in the microstructure and the excess Y concentration at the grain boundaries appears to have settled to their saturation value of approximately 6.5 Y atoms/nm^2. A more comprehensive treatment of this saturation precipitation behavior is given in an upcoming publication [19]. The grain growth behavior exhibits a similar delineation on the same plot. In regime I, the mean grain diameter versus volumetric Y concentration is about constant. In regime II, the microstructure becomes unstable. In regime III, a sudden jump in the grain sizes was observed. This jump was more eminent in samples with high Si and Ca contamination (GI) than in the sets of cleaner samples (GII and GIII).

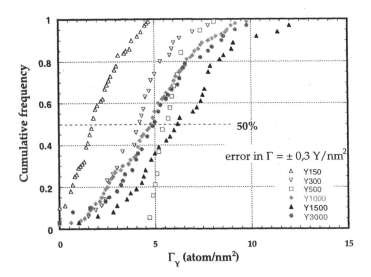

Figure 6. The Cannon plot showing the cumulative distribution of excess yttrium at the grain boundaries of GI samples with varying amounts of volumetric concentration of Y_2O_3.

Grain Boundary Engineering in Ceramics

DISCUSSION

Table I shows that although the GI samples were all contaminated, their contamination amounts in Si, Ca, and Zr are very similar to each other. The only impurity that systematically changed in the composition was Y. Thus, the drastic changes observed in microstructure must be intimately related with the amount of yttrium in the samples. However, the GI samples are not suitable to isolate the effect of Y only. Certainly, other contaminants are contributing to the observed variations in the evolution of the microstructure. GII and GII samples are processed progressively cleaner, and hence will be able to clarify the role of yttrium as well as the synergetic effects of the other impurities in sintering , grain growth and other properties of these doped alumina ceramics [19].

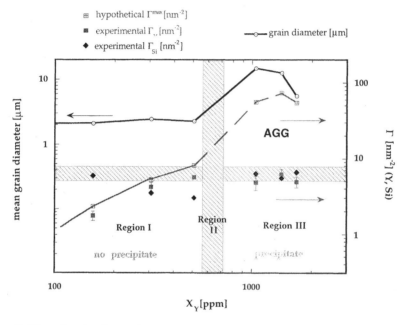

Figure 7. Plot of grain size and average excess impurity concentration at the grain boundaries as a function of volumetric yttrium dopant concentration in samples GI.

From the microstructures shown in Figures 1 and 3, dependence of the porosity on the yttrium content , one can argue that yttrium is enhancing the sintering and densification in alumina. The nominally pure alumina had about 10 % porosity, whereas the porosity levels continuously decreased with increasing yttrium content. At the same time the grain size increased slightly up to the precipitation limit of YAG, and then jumps to higher values after precipitation. This result is in an apparent contradiction to earlier studies about the sintering behavior of yttrium doped alumina [8]. Certainly, there is a need for more in-depth studies on this subject. We address this issue in more detail in the upcoming publication [19]. One important aspect that needs to be clarified is the interplay between the excess yttrium and silicon at the grain boundaries. In the absence of Si impurities, the sudden jump in the grain size in regime III is much milder than the case with silicon presence. Grains after YAG precipitation in sample GI and the outer regions in some of GII samples are about 5 to 30 times larger than the grains right before YAG precipitation. AEM studies showed that the interfaces in the larger grain areas in both sets of samples contained significant amount of silicon (< 6 Si atoms/nm^2) in addition to 5 to 7 yttrium atoms/ nm^2. It is worth noting that in the case GI samples where there was also Ca contamination present, the microstructure show the classic example of abnormally grown alumina with frequent slab-like grain morphology. In GII sample where there was no Ca present, the microstructure was made up of exaggeratedly grown grain with uniform size and equi-axed morphology. A further detailed discussion on the grain growth behavior in these Y doped aluminas is given in [19].

CONCLUSIONS

The segregation behavior of Y at the grain boundaries of an α-Al_2O_3 ceramic have been studied and quantified. At dilute concentrations, the excess yttrium at the grain boundaries can be calculated using a McLean -Langmuir type of relationship. At higher yttrium concentrations the grain boundaries become saturated and YAG particles precipitate. The excess concentration of yttrium after precipitation is about 6.5 yttrium atoms/ nm^2.

The observed microstructural evolution and grain growth behavior were related to the excess concentration of yttrium at the grain boundaries and its precipitation behavior. It appears that yttrium is enhancing sintering an densification in α-Al_2O_3, and simultaneously stimulating grain growth. Grain growth rates are strongly increased when Si is present along with saturation amount of yttrium at the grain boundaries.

References:
1. R. L. Coble, „Sintering of crystalline Solids, II experimental test of diffusion models in porous compacts," J. Appl. Phys., 32 [5], pg. 793-799, 1961, R.L. Coble, „Transparent alumina and method of preparation," U.S. Patent@ 3 026 210 March 1962.
2. S. J. Bennison and M. P. Harmer, „A history of the role of MgO in the sintering of a-Al2O3,"in Ceramic Transactions, Vol. 7, *Sintering of Advanced Ceramics*, ed. by C. A. Handwerker, J. E. Blendel, and W. Q. Kaysser, pg. 13-49, Am. Ceram. Soc., Westerville, OH, 1990.
3. P. A. Morris, „ Impurities in ceramics: Processing and effects on properties," in Ceramic Transactions, Vol. 7, *Sintering of Advanced Ceramics*, ed. by C. A.

Handwerker, J. E. Blendel, and W. Q. Kaysser, pg. 50-85, Am. Ceram. Soc., Westerville, OH, 1990.

4. J. Zhao and M. P. Harmer, „Sintering of high purity alumina doped simultaneously with MgO and FeO," J. Am. Ceram. Soc., 70 [12], pg. 860-866, 1987.

5. M. A. Gulgun, V. Putlayev, and M. Ruhle, „Effects of yttrium doping α-alumina: I, microstructure and microchemistry," J. Am. Ceram. Soc., 82 [7], pg. 1849-1856, 1999.

6. P. Gruffel and C. Carry, „Effect of grain size on yttrium grain boundary segregation in fine grained alumina," J. Europ. Ceram. Soc., 11, pg. 189-199, 1993.

7. R. J. Brook, „Frontiers of sinterability," in Ceramic Transactions, Vol. 7, *Sintering of Advanced Ceramics*, ed. by C. A. Handwerker, J. E. Blendel, and W. Q. Kaysser, pg. 3-12, Am. Ceram. Soc., Westerville, OH, 1990.

8. J. Fang, A. M. Thompson, M. P. Harmer, and H. M. Chan, „Effect of yttrium and lanthanum on the final stage sintering behavior of ultrahigh purity alumina," J. Am. Ceram. Soc., 80 [8], pg. 2005-2012, 1997.

9. S. Lartigue, C. Carry, L. Priester, „Grain boundaries in high temperature deformation of yttria and magnesia co-doped alumina," Colloque de Physique, C1 [1], Tome51, pg. c1-985-c1-990, 1990.

10. H. Yoshida, Y. Ikuhara, and T. Sakuma, High temperature creep resistance in rare earth doped fine graind Al_2O_3," J. Mater. Res., 13 [9], pg. 2597-2601, 1998.

11. J. H. Yoo, J. C. Nam, and S. Baik, „Quantitative evaluation of glass forming impurities in alumina: Equivalent silica concentration (ESC)," J. Am. Ceram. Soc., 82 [8], pg. 2233-2238, 1999.

12. S. I. Bae and S. Baik, „Determination of critical concentration of Silica and calcia for abnormal grain growth in alumina," J. Am. Ceram. Soc., 76 [4], pg. 1065-1067, 1993.

13. I. J. Bae and S. Baik, „Abnormal grain growth in alumina," Materials Sc. Forum, 204-206, pg. 485-490, 1996.

14. N. Bell and W. Sigmund, unpublished work 1998.

15. M. I. Mendelson, „Average grain size in polycrystalline ceramics," J. Am. Ceram. Soc., 52, pg. 443, 1969.

16. J. A. S. Ikeda, Y. M. Chiang, and A. J. Garrath-Reed, and J. B. Vander-Sande, Space charge segregation at grain boundaries in titanium dioxide: II model experiments," J. Am. Ceram. Soc., 76, pg. 2447-2457, 1993.

17. M. A. Gulgun, R. Voitovych, B. Kabius, and M. Ruhle, „Grain boundary structure and chemistry in a polycrystalline α-Al_2O_3: Y, Si, Ca," in preparation, 2000.

18. M. A. Gulgun and M. Ruhle, „Yttrium in polycrystalline α-alumina," Key Engineering Materials, 171-174, *Creep and Fracture of Engineering Materials and Structures*, ed. by T. Sakuma and K. Yagi, pg. 793-800, 2000.

19. M. A. Gulgun, R. M. Cannon, R. Voitovych, and M. Ruhle, „Segregation and grain growth behavior in polycrystalline α-alumina," in preparation, 2000.

GRAIN BOUNDARY FACETING TRANSITION AND ABNORMAL GRAIN GROWTH IN OXIDES

Chan Woo Park and Duk Yong Yoon

Department of Materials Science and Engineering, Korea Advanced Institute of Science and Technology, 373-1, Kusong-dong, Yusong-gu, Taejon 305-701, Korea

ABSTRACT

Grain boundaries in metals and oxides are faceted at low temperatures and become defaceted above certain critical temperatures. The defaceted grain boundaries are expected to have an atomically rough structure and hence nearly isotropic properties. The grain boundary faceting and defaceting can be also induced by additives. When all or some of the grain boundaries are faceted in polycrystals, abnormal grain growth (AGG) occurs, and when all grain boundaries are defaceted, normal grain growth (NGG) occurs. In alumina, for example, all grain boundaries in a specimen prepared from a high purity powder and sintered at either 1620°C or 1900°C are defaceted with smoothly curved shapes, and the grains grow normally. When sintered at 1620°C after adding 100 ppm of SiO_2 and 50 ppm of CaO, some grain boundaries become faceted and AGG occurs. When 600 ppm of MgO is added to this specimen, all grain boundaries become defaceted again and NGG occurs. Similar effects of SiO_2 and MgO addition are observed when sintered at 1900°C. In all of these specimens no liquid phase is found at the triple junctions under TEM. Such a correlation between the grain boundary faceting and AGG is found in a number of metals and some oxides. When the grain boundaries are faceted, some of the facet planes have ordered structures and may migrate by the movement of steps existing on such defects as dislocations or twins, or produced by two dimensional nucleation. The effective mobility of the larger grains will then be larger than that of the smaller grains, causing AGG.

INTRODUCTION

At low temperatures crystals in equilibrium with vapor or liquid have been predicted[1-5] and indeed observed[7-9] to be polyhedral with flat surface planes and sharp edges. The flat low index surface planes are singular with atomically flat or smooth structures corresponding to the cusps in the polar plot of the surface free energy γ against the surface normal. With a temperature increase, the corners and edges of the crystal become atomically rough with macroscopical rounding[5,6,10,11], and the area of the flat surface decreases until it disappears at the roughening temperature. Certain models predict that this roughening transition of a singular surface is an infinite order transition with a discontinuous change of the surface curvature to a finite value[12-14]. The roughening transition can thus be observed from the change of the macroscopic shape or the surface atomic structure using diffraction techniques[15]. The roughening transition corresponds to the blunting of cusps in the γ-plot.

To the extent authorized under the laws of the United States of America, all copyright interests in this publication are the property of The American Ceramic Society. Any duplication, reproduction, or republication of this publication or any part thereof, without the express written consent of The American Ceramic Society or fee paid to the Copyright Clearance Center, is prohibited.

If the surface normal of a crystal with an equilibrium shape changes discontinuously at some edges and corners, there will be missing surface orientations, and if the average orientation of a surface is one of such missing orientations, it will be faceted with a hill-and-valley shape as predicted by Herring[1]. At finite temperatures the facet planes will usually consist of both singular and rough surface segments, and with a temperature increase the total area of the rough surface segments will increase until the entire surface becomes rough. This is the defaceting transition which is a first order type as pointed out by Cahn[16]. Thus a defaceted surface will be atomically rough, as well as a low index surface, above its roughening transition temperature.

It is well known that the growth behavior of a crystal depends on its surface structure. A crystal with a singular surface grows by two dimensional nucleation of surface steps[17-19] or on the steps produced by screw dislocations[17] or on the reentrant edges produced by twins[20]. If, on the other hand, a crystal surface is rough, it grows by a continuous process with the rate determined by atom flux in the surrounding fluid.

In a system of many grains dispersed in a liquid matrix, the grains coarsen to reduce the total interfacial area. If the grains are spherical and hence have a rough surface, they undergo normal Ostwald ripening with the rate determined by diffusion in the liquid matrix[21-25]. If, on the other hand, the grains are polyhedral and hence have singular surfaces, they undergo abnormal growth[26-28]. It has been proposed[29,30] that the abnormal grain growth (AGG) occurred because the grains grow by two dimensional nucleation of surface steps or on the steps produced by screw dislocations. If the liquid volume fraction is small (in the order of a few percent), the grains rarely have the planar surfaces which appear in the equilibrium shape. Then the grain surfaces are faceted[31], and the grain still undergo AGG[32,33].

In single phase polycrystals with the grains separated from each other by grain boundaries, there seems to be an analogous relationship between the grain boundary structure and the grain coarsening behavior. There have been numerous experimental indications that the grain boundaries are singular at low temperatures and become rough at high temperatures[34,35]. Rottman[36] predicted the roughening transition of low angle grain boundaries by making an analogy with stepped surface. Recently, Dahmen and Westmacott [37] observed that some planar grain boundaries of a polyhedral Al grain embedded in another large grain became rounded at a high temperature and planar again at a low temperature. This was the first direct observation of the roughening transition of a grain boundary based on the grain boundary shape change. In polycrystals the grain boundaries rarely have orientations corresponding to the cusps in the polar plot of the grain boundary energy γ_g against the grain boundary normal or the inclination angle. Therefore, they are faceted at low temperatures and their defaceting transitions were observed by Hsieh and Balluffi[38] in Al and Au. The defaceted grain boundaries are atomically rough as pointed out by Cahn[16].

We reported recently that in Ni[39], Ag[40], a Ni-base superalloy[41], alumina[42], and BaTiO$_3$[43,44], AGG occurred when all or some of the grain boundaries were faceted and normal growth occurred when defaceted. In this paper we review the grain boundary faceting in alumina and BaTiO$_3$, the grain boundary structure including the effect of additives, and examine the mechanism for relating the growth behavior to the grain boundary structure. In Ni[39], Ag[40], and a Ni-base superalloy[41], the grain boundary faceting-defaceting and the growth behavior were dependent on the heat-treatment temperature and the atmosphere (the oxygen activity), but in the oxides the additives were found to produce strong effects.

RESULTS AND DISCUSSION

Grain Boundary Engineering in Ceramics

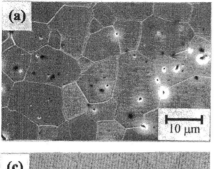
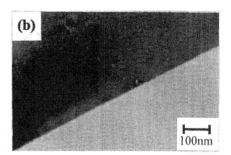

Fig 1. (a) The SEM microstructure of pure Al_2O_3 sintered at 1900°C for 3 h and (b) a grain boundary at a high magnification under TEM and (c) a high resolution image of a grain boundary.

As we reported earlier[42] and observed previously by others[45,46], normal grain growth occurred when pure alumina powder was sintered at either 1620 or 1900°C as displayed in Fig. 1(a). At a relatively low magnification used for Fig. 1(a), the grain boundaries were curved and the transimission electron micrograph in Fig. 1(b) also showed curved grain boundaries without any faceting at a higher magnification. The high resolution image of Fig. 1(c) did not reveal any special boundary structure. The specimen heat-treated at 1620°C also showed similar defaceted grain boundaries[42]. These defaceted grain boundaries must have been rough at the heat-treatment temperatures, and although they could have undergone faceting transitions at low temperatures, the transition kinetics could have been slow enough to freeze some characteristics of the rough structure during the rapid cooling. The normal grain growth with defaceted grain boundaries is consistent with the results observed in Ni[39], Ag[40], and a Ni-base superalloy[41]. The rate of the rough grain boundary movement is expected to be determined by the rate of atomic jump across the boundary and hence increase linearly with the driving force. Then normal growth is expected as shown also by the analysis of Thompson *et al.*[47] and the simulations of Srolovitz *et al.*[48].

When sintered at 1620°C after adding 100 ppm (by mole) of SiO_2 and 50 ppm of CaO and at 1900°C after adding 150, 300, or 500 ppm of SiO_2, AGG occurred with the large grains elongated along their basal planes as shown in Fig. 2(a). The TEM observations of the grain boundaries and their triple junctions at high magnifications did not reveal any liquid phase. The basal planes of the large grains in contact with the small grains were flat at most of the areas as can be seen in Fig. 2(a) and also in Fig. 2(b) at a higher magnification. Such flat grain boundaries have important implications on the structure of the grain boundaries. It appears that the grain boundaries which

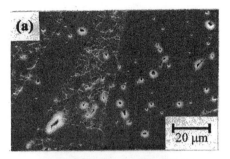
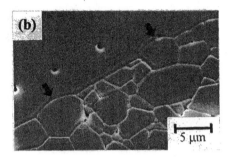
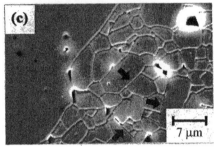

Fig 2. (a) The microstructure of Al_2O_3 doped with 100 ppm SiO_2 and 50 ppm CaO and sintered at 1620°C for 12 h at a low magnification and (b) a flat grain boundary and (c) matrix grains at a high magnification under SEM.

lie on the basal planes of any grain with any orientation of the neighboring grain often have low boundary energies corresponding to cups in the γ_g-plot. The grain boundaries are flat because of an apparently large torque effect as proposed by Herring[49] and it can be shown by using the capillarity vector of Hoffman and Cahn[50] that a large grain that forms low energy grain boundaries on its basal plane with the neighboring grains of any orientation have a flat grain boundary with those neighboring grains as shown in Figs. 2(a) and (b). With some neighboring grains, however, curved grain boundaries were also formed as indicated by arrows in Fig. 2(b), and these grain boundaries were either not singular or singular with relatively shallow cusps in the γ_g-plot. It thus appears that a grain can form singular grain boundaries with its neighbors on its basal plane but the boundary energy can vary with the orientation of each neighboring grain.

The characteristics of these flat grain boundaries is that the flat shape is maintained even at the triple junctions without any indication of bending. Many such flat grain boundaries which run across several neighboring grains were also observed among the fine matrix grains as indicated by arrows for some of them in Fig. 2(c). It thus appears that the formation of singular grain boundaries on the basal planes is quite general.

The addition of Y_2O_3 apparently has the same effect as the addition of SiO_2 and CaO on the grain boundary shape and grain coarsening behavior. Cho et al.[51] and Gülgün et al.[52] observed that an alumina specimen with 1000 ppm (by mole fraction) of Y_2O_3 showed AGG with a bimodal grain size distribution and some long planar grain boundaries lying on $(0001), \{11\bar{2}0\}$, and $\{01\bar{1}2\}$ planes. Some grain boundaries were also faceted along these directions. Bouchet et al.[53] and Fionova et al.[54] also observed that half of the grain boundaries examined in an Y_2O_3-

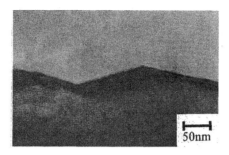

Fig 3. A faceted grain boundary in Al_2O_3 sintered at 1900°C for 3 h doped with 300 ppm SiO_2.

doped alumina lay on (0001) and (01$\bar{1}$2) planes. Lartigue Korinek and Dupau[55] also observed both planar and faceted grain boundaries lying on the basal planes in an alumina doped with 500 ppm (apparently by weight) of Y_2O_3 and 500 ppm of MgO.

The singular grain boundaries lying on the low index lattice planes, which are often contrary to such geometrical models as the coincident site lattice, were also found in a number of metals. Dahmen and Westmacott[37] observed that an Al grain embedded in another large grain with a large misorientation angle had an equilibrium shape with planar tilt grain boundaries, one of which coincided with a (001) plane of one grain and an (110) plane of the other. Merkle and Wolf[56,57] also observed that some faceted tilt boundaries in Au were asymmetric boundaries with low and high index planes apparently corresponding to the cusps in the γ_g-plot. These results were corroborated by their computer simulations of the grain boundary atomic structures[56,57]. While the high resolution observations and the simulations of Merkle and Wolf[56,57] are mainly focused on the pure tilt and twist boundaries, our observations show that in alumina the basal planes can form general low energy boundaries with any misorientation relationship with the neighboring grains. Although the planes of the faceted grain boundaries in polycrystalline Ni[39] and other alloys[41,58] can be the low energy tilt boundaries or rough boundaries, some of them can be low energy general boundaries with low index planes. It thus appears that the low index planes can form general low energy grain boundaries and this possibility needs to be further explored.

In our SiO_2-CaO doped and SiO_2 doped specimens about 10 % of the grain boundaries examined under TEM were faceted as shown in Fig. 3 and the rest of them appeared to be planar at the magnification used. It is possible that many of these planar boundaries are the atomically flat singular boundaries of the basal or other low index planes, but the others can be curved with an atomcially rough structure. Although more extensive high resolution TEM observations are required, it thus appears that these specimens had flat singular grain boundaries lying on the basal and possibly on other low index planes, the faceted grain boundaries with most likely some singular segments, and possibly rough grain boundaries.

The addition of SiO_2 and CaO thus produced singular grain boundaries which had either planar or faceted macroscopic shapes. The occurrence of AGG with singular grain boundaries is similar to those observed earlier in Ni[39], Ag[40], and a Ni-base superalloy[41]. In these metals the singular boundary segments appeared as macroscopically faceted grain boundaries and no flat grain boundaries were found apparently because the cusps in their γ_g-plots were not quite deep. Following Gleiter's suggestion[59,60] we proposed[30,39] that AGG occurred with singular grain boundaries because they moved either by producing or on existing boundary steps.

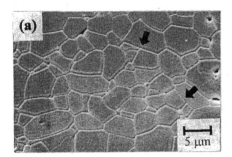 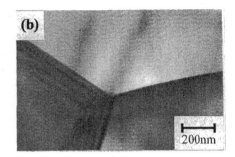

Fig 4. (a) The SEM microstructure of Al_2O_3 doped with 100 ppm SiO_2 + 50 ppm CaO + 600 ppm MgO and sintered at 1620°C for 12 h and (b) grain boundaries at a high magnification under TEM.

As pointed out earlier and shown in Fig. 1, the grain boundaries in the pure alumina specimens were curved and none was found which was flat across a triple junction. The addition of MgO to the specimens doped with SiO_2 and CaO appeared to make the grain boundaries rough again and hence the grain coarsening normal. As shown in Fig. 4(a) the specimen with 100 ppm SiO_2, 50 ppm CaO, and 600 ppm MgO showed normal grain growth when sintered at 1620°C for various periods and the grain boundaries were mostly curved with defaceted shapes as shown in Fig. 4(b). There were a small fraction of grain boundaries which were flat at the triple junctions as indicated by arrows in Fig. 4(a), indicating that the MgO addition might not have made all of the grain boundaries rough. Even if some grain boundaries are singular, the grain coarsening can be almost normal if the free energy of the surface step edges becomes sufficiently small[30]. The roughening effect of the MgO addition can thus be regarded also as that of reducing the step edge energy to zero for most of the grain boundaries and possibly to small finite values for others.

A similar relationship between the grain boundary roughening transition and the grain coarsening behavior was also observed in TiO_2-excess $BaTiO_3$ by Lee *et al.*[43,44]. When $BaTiO_3$ with excess 0.1 mol.% TiO_2 was sintered in air at 1250°C, the grains with (111) double twins grew abnormally and these large grains elongated in the twin directions formed flat grain boundaries on their (111) planes with many neighboring grains. This observation indicates that the (111) planes form general low energy grain boundaries with the neighbors of any orientation. Many grain boundaries between the fine matrix grains were faceted. When the sintering atmosphere was changed to hydrogen, all grain boundaries became curved indicating that they became rough. The flat boundaries with (111) planes thus underwent a roughening transition and the faceted boundaries also became rough by a defaceting transition. Then the grains no longer grew preferentially along the twin directions. Therefore, the twins provide the boundary steps for rapid movement only when they meet the faceted grain boundaries, and when the grain boundaries become rough, all grains grow rapidly without any preference for the twin directions. These results thus provide a definitive evidence that the grains with faceted grain boundaries grow abnormally by a boundary step mechanism.

The possibility of grain growth by either producing or on existing boundary steps was proposed by Gleiter[59,60] based on his observations of the spiral growth of boundary steps[59]. It is likely that the singular boundaries move by the step mechanism and their migration rate is then expected to increase non-linearly with the capillary driving force. Then AGG can occur because the larger grains will grow faster as proposed earlier[30].

CONCLUSIONS

The observations of the grain boundary shapes and atomic structures indicate that general grain boundaries with low index planes can be singular with the minimum boundary energies. Such a possibility needs to be definitively established by further observations and simulation studies. If these grain boundaries move by a step mechanism, AGG can occur. The nature of this step mechanism linking the singular boundary to the AGG behavior is yet to be further clarified. When the grain boundaries become rough by increasing temperature in some metals, by adding MgO to alumina, or by heat-treating $BaTiO_3$ in hydrogen, normal grain growth occurs because the boundary mobility is constant and uniform. The grain boundary roughening transition will obviously have important effects on various boundary properties, including the grain coarsening behavior, as shown in these observations.

REFERENCES

1. C. Herring, "Some Theorems on the Free Energies of Crystal Surfaces," *Phys. Rev.*, **82**, 87-93 (1951).
2. D. S. Fisher and J. D. Weeks, "Shape of Crystals at Low Temperatures: Absence of Quantum Roughening," *Phys. Rev. Lett.*, **50**, 1077-80 (1983).
3. M. Drechsler and J. F. Nicholas, "On the Equilibrium Shape of Cubic Crystals," *J. Phys. Chem. Solids*, **28**, 2609-27 (1967).
4. E. Fradkin, "Roughening Transition in Quantum Interfaces," *Phys. Rev. B*, **B28**, 5338-41 (1983).
5. C. Rottman and M. Wortis, "Statistical Mechanics of Equilibrium Crystal Shapes: Interfacial Phase Diagrams and Phase Transitions," *Phys. Rep.*, **103**, 59-79 (1984).
6. C. Rottman and M. Wortis, "Equilibrium Crystal Shapes for Lattice Models with Nearest- and Next-Nearest-Neighbor Interactions," *Phys. Rev. B*, **B29**, 328-39 (1984).
7. S. Balibar, D. O. Edwards, and C. Laroche, "Surface Tension of Solid ^4He," *Phys. Rev. Lett.*, **42**, 782-84 (1979).
8. J. E. Avron, L. S. Balfour, C. G. Kuper, J. Landau, S. G. Lipson, and L. S. Schulman, "Roughening Transition in the ^4He Solid-Superfluid Interface," *Phys. Rev. Lett.*, **45**, 814-17 (1980).
9. K. O. Keshishev, A. Ya. Parshin, and A. B. Babkin, "Crystallization Waves in He^4," *Sov. Phys. JETP*, **53**, 362-69 (1981).
10. S. Sarian and H. W. Weart, "Factors Affecting the Morphology of an Array of Solid Particles in a Liquid Matrix," *Trans. Metall. Soc. AIME*, **233**, 1990-94 (1965).
11. M. Wortis, "Equilibrium Crystal Shapes and Interfacial Phase Transitions"; pp.367-405 in *Chemistry and Physics of Solid Surfaces VII*. Edited by R. Vanselow and R. F. Howe. Springer-Verlag, Berlin, 1988.
12. J. D. Weeks, "The Roughening Transition"; pp.293-317 in *Ordering in Strongly Fluctuating Condensed Matter Systems*. Edited by T. Riste. Plenum, NY, 1980.
13. J. M. Kosterlitz, "The Critical Properties of the Two-Dimensional xy Model," *J. Phys. C*, **C7**, 1046-60 (1974).
14. J. M. Kosterlitz and D. J. Thouless, "Ordering, Metastability and Phase Transitions in Two-Dimensional Systems," *J. Phys. C*, **C6**, 1181-203 (1973).
15. T. Engel, "Experimental Aspects of Surface Roughening"; pp.407-28 in *Chemistry and Physics of Solid Surfaces VII*. Edited by R. Vanselow and R. F. Howe. Springer-Verlag, Berlin, 1988.

16. J. W. Cahn, "Transitions and Phase Equilibria among Grain Boundary Structures," *J. Phys. Collq.*, **43**, C6-199-C6-213 (1982).
17. W. K. Burton, N. Cabrera, and F. C. Frank, "The Growth of Crystals and the Equilibrium Structure of Their Surfaces," *Phil. Trans. R. Soc., London, Sect. A*, **A243**, 299-358 (1951).
18. S. D. Peteves and R. Abbaschian, "Growth Kinetics of Solid-Liquid Ga Interfaces-Part I Experimental," *Metall. Trans. A*, **22A**, 1259-70 (1991).
19. P. E. Wolf, F. Gallet, S. Balibar, E. Rolley, and P. Nozières, "Crystal Growth and Crystal Curvature near Roughening Transitions in HCP ^{4}He," *J. Physique*, **46**, 1987-2007 (1985).
20. R. C. DeVries, "Observation on Growth of the $BaTiO_3$ Crystals from KF Solutions," *J. Am. Ceram. Soc.*, **42** [11] 547-58 (1983).
21. I. M. Lifshitz and V. V. Slyozov, "The Kinetics of Precipitation from Supersaturated Solid Solutions," *J. Phys. Chem. Solids*, **19**, 35-50 (1961).
22. C. Wagner, "Theory of Precipitate Change by Redissolution," *Z. Elektrochem.*, **65**, 581-91 (1961).
23. A. J. Ardell, " The Effect of Volume Fraction on Particle Coarsening: Theoretical Consideration," *Acta Metall.*, **20**, 61-71 (1972).
24. C. H. Kang and D. N. Yoon, "Coarsening of Cobalt Grains Dispersed in Liquid Copper Matrix," *Metall. Trans. A*, **12A**, 65-69 (1981).
25. S. S. Kang and D. N. Yoon, "Kinetics of Grain Coarsening during Sintering of Co-Cu and Fe-Cu Alloys with Low Liquid Contents," *Metall. Trans. A*, **13A**, 1405-11 (1982).
26. M. Schreiner, Th. Schmitt, E. Lassner, and B. Lux, "On the Origins of Discontinuous Grain Growth during Liquid Phase Sintering of WC-Co Cemented Carbides," *Powder Metall. Inter.*, **16**, 180-83 (1984).
27. J. H. Choi, "Effects of Carbon on the Grain Growth Behavior in TaC-Ni"; M.S. Thesis. Dept. of Materials Science and Engineering , Korea Advanced Institute of Science and Technology, Taejon, Korea, 1997.
28. D. F. K. Hennings, R. Janssen, and P. J. L. Reynen, "Control of Liquid-Phase-Enhanced Discontinuous Grain Growth in Barium Titanate," *J. Am. Ceram. Soc.*, **70** [1] 23-27 (1987).
29. Y. J. Park, N. M. Hwang, and D. Y. Yoon, "Abnormal Growth of Faceted (WC) Grains in a (Co) Liquid Matrix," *Metall. Mater. Trans. A*, **27A**, 2809-19 (1996).
30. D. Y. Yoon, C. W. Park, and J. B. Koo, "The Step Growth Hypothesis for Abnormal Grain Growth"; in *Properties and Applications IV*, Proceedings of the International Workshop on Ceramic Interfaces (Taejon, Korea, September, 1998). Edited by S.-J. L. Kang and H. I. Yoo. The Institute of Materials, London, United Kingdom, 2000, in press.
31. J. E. Blendell, W. C. Carter, and C. A. Handwerker, "Faceting and Wetting Transitions of Anisotropic Interfaces and Grain Boundaries," *J. Am. Ceram. Soc.*, **82** [7], 1889-900 (1999).
32. S. C. Hansen and D. S. Phillips, "Grain-Boundary Microstructures in a Liquid Phase Sintered Alumina (α-Al_2O_3)," *Philos. Mag. A*, **47**, 209-34 (1983).
33. C. A. Powell-Dogan and A. H. Heuer, "Microstructures of 96% Al_2O_3 Ceramics: I, Characterization of the As-Sintered Materials," *J. Am. Ceram. Soc.*, **73** [12] 3670-76 (1990).
34. U. Erb and H. Gleiter, "The Effect of Temperature on the Energy and Structure of Grain Boundaries," *Scripta Metall.*, **13**, 61-64 (1979).
35. L. S. Shvindlerman and B. B. Straumal, "Regions of Existence of Special and Non-Special Grain Boundaries," *Acta Metall.*, **33**, 1735-49 (1985).
36. C. Rottman, "Roughening of Low-Angle Grain Boundaries," *Phys. Rev. Lett.*, **57**, 735-38 (1986).
37. U. Dahmen and K. H. Westmacott, "Studies of Faceting by High Voltage/High Resolution Microscopy"; pp.133-67 in *Interface: Structure and Properties*. Edited by C. S. Pande, B. B. Rath, and D. A. Smith. Trans Tech Publications, Switzerland, 1993.

38. T. E. Hsieh and R. W. Balluffi, "Observations of Roughening /De-faceting Phase Transitions in Grain Boundaries," *Acta Metall.*, **37**, 2133-39 (1989).
39. S. B. Lee, N. M. Hwang, D. Y. Yoon, and M. F. Henry, " Grain Boundary Faceting and Abnormal Grain Growth in Nickel," *Metall. Mater. Trans. A*, in press.
40. J. B. Koo and D. Y. Yoon, "The Dependence of Normal and Abnormal Grain Growth in Ag on Annealing Temperature and Atmosphere," submitted to *Metall. Mater. Trans. A*.
41. S. B. Lee, D. Y. Yoon, and M. F. Henry, "Abnormal Grain Growth and Grain Boundary Faceting in a Model Ni-base Superalloy," submitted to *Acta Mater.*.
42. C. W. Park and D. Y. Yoon, "The Effects of SiO_2, CaO, and MgO Addition on Grain Growth of Alumina," submitted to *J. Am. Ceram. Soc.*.
43. B. K. Lee, "Formation of {111} Twins and Abnormal Grain Growth in $BaTiO_3$"; Ph.D. Thesis. Dept. of Materials Science and Engineering, Korea Advanced Institute of Science and Technology, Taejon, Korea, 2000.
44. B. K. Lee, S.-Y. Chung, and S.-J. L. Kang, "Grain Boundary Faceting and Abnormal Grain Growth in $BaTiO_3$," *Acta Mater.*, in press.
45. S. I. Bae and S. Baik, "Determination of Critical Concentrations of Silica and/or Calcia for Abnormal Grain Growth in Alumina," *J. Am. Ceram. Soc.*, **76** [4] 1065-67 (1993).
46. S. I. Bae and S. Baik, "Sintering and Grain Growth of Ultrapure Alumina," *J. Mater. Sci.*, **28**, 4197-204 (1993).
47. C. V. Thompson, H. J. Frost, and F. Spaepen, "The Relative Rates of Secondary and Normal Grain Growth," *Acta Metall.*, **35**, 887-90 (1987).
48. D. J. Srolovitz, G. S. Grest, and M. P. Anderson, "Computer Simulation of Grain Growth-V. Abnormal Grain Growth," *Acta Metall.*, **33**, 2233-47 (1985).
49. C. Herring, "Surface Tension as a Motivation for Sintering"; pp.143-79 in *The Physics of Powder Metallurgy*. Edited by W. E. Kingston. McGraw-Hill, NY, 1951.
50. D. W. Hoffman and J. W. Cahn, "A Vector Thermodynamics for Anisotropic Surfaces," *Surf. Sci.*, **31**, 368-88 (1972).
51. J. Cho, M. P. Harmer, H. M. Chan, J. M. Rickman, and A. M. Thompson, "Effect of Yttrium and Lanthanum on the Tensile Creep Behavior of Aluminum Oxide," *J. Am. Ceram. Soc.*, **80** [4] 1013-17 (1997).
52. M. A. Gülgün, V. Putlayev, and M. Rühle, "Effects of Yttrium Doping α-Alumina: I, Microstructure and Microchemistry," *J. Am. Ceram. Soc.*, **82** [7] 1849-56 (1999).
53. D. Bouchet, F. Dupau, and S. Lartigue-Korinek, "Structure and Chemistry of Grain Boundaries in Yttria Doped Aluminas," *Microsc. Microanal. Microstruct.*, **4**, 561-73 (1993).
54. L. Fionova, O. Konokenko, V. Matveev, L. Priester, S. Lartigue, and F. Dupau, "Heterogeneities of Grain Boundary Arrangement in Polycrystals," *Interface Sci.*, **1**, 207-11 (1993).
55. S. Lartigue Korinek and F. Dupau, "Grain Boundary Behavior in Superplastic Mg-Doped Alumina with Yttria Codoping," *Acta Metall. Mater.*, **42**, 293-302 (1994).
56. K. L. Merkle and D. Wolf, "Low-Energy Configurations of Symmetric and Asymmetric Tilt Grain Boundaries," *Philos. Mag. A*, **65**, 513-30 (1992).
57. D. Wolf and K. L. Merkle, "Correlation between the Structure and Energy of Grain Boundaries in Metals"; pp.87-150 in *Materials Interfaces*. Edited by D. Wolf and S. Yip. Chapman & Hall, London, United Kingdom, 1992.
58. J. S. Choi, "Effects of Grain Boundary Structure on Grain Growth Behavior in 316L Stainless Steel"; M.S. Thesis. Dept. of Materials Science and Engineering, Korea Advanced Institute of Science and Technology, Taejon, Korea, 1997.
59. H. Gleiter, "The Mechanism of Grain Boundary Migration," *Acta Metall.*, **17**, 565-73 (1969).
60. H. Gleiter, "Theory of Grain Boundary Migration Rate," *Acta Metall.*, **17**, 853-62 (1969).

Grain Boundary Engineering in Ceramics

EFFECTS OF DIFFERENT ADDITIVES ON DENSIFICATION AND α-β PHASE TRANSFORMATION IN Si$_3$N$_4$ CERAMICS

Jian-Feng Yang, Zhen-Yan Deng and Tatsuki Ohji
National Industrial Research Institute of Nagoya,
1-1 Hirate-cho, Kita-ku, Nagoya 462-8510, Japan

Koichi Niihara
Institute of Scientific and Industrial Research, Osaka University,
8-1, Mihogaoka, Ibaraki, Osaka 567-0047, Japan

ABSTRACT

Sintering behavior and microstructure of Si$_3$N$_4$ ceramics with two different additive systems of MgAl$_2$O$_4$-ZrO$_2$ (MA-Z) and Y$_2$O$_3$-Al$_2$O$_3$ (Y-A) were investigated. It was found that, the densification of Si$_3$N$_4$ was enhanced with an increase in additive content. Phase transformation and grain growth were suppressed by increasing the amount of MA-Z, but not affected by Y-A content. This implied that the phase transformation and grain growth were controlled by the diffusion in MA-Z doped Si$_3$N$_4$ and by the interfacial reaction in Y-A doped Si$_3$N$_4$.

INTRODUCTION

Silicon nitride (Si$_3$N$_4$) is difficult to sinter alone, because of its strong covalent bonding between silicon and nitrogen atoms. Usually, sintering additives, such as metal oxides and rare-earth metal oxides, are added to consolidate Si$_3$N$_4$ ceramics by liquid-phase sintering. During the sintering process, the additives react with SiO$_2$ on the surface of the Si$_3$N$_4$ particles and some of the nitrides, and then form an oxynitride liquid. In this way, the sintering behavior of Si$_3$N$_4$, for instance, densification, α-β phase transformation and grain growth, is substantially influenced by the amount and chemistry of the liquid phase. In a way, the onset of densification by particle rearrangement occurs at the temperature of liquid-phase formation, and phase transformation then occurs through a solution–diffusion-reprecipitation process. The reconstructive α-β phase transition seems to provide

To the extent authorized under the laws of the United States of America, all copyright interests in this publication are the property of The American Ceramic Society. Any duplication, reproduction, or republication of this publication or any part thereof, without the express written consent of The American Ceramic Society or fee paid to the Copyright Clearance Center, is prohibited.

the driving force for densification [1], but some studies showed that phase transformation is not related to densification behavior [2, 3].

In this study, the effects of different sintering-additive content on the sintering behavior and resultant microstructure of Si_3N_4 were investigated. Two types of sintering additives were selected: one is to make the consolidation easy, $MgAl_2O_4$-ZrO_2 (MA-Z); the other is to enhance grain growth and is refractory glass phase Y_2O_3-Al_2O_3 (Y-A).

EXPERIMENTAL PROCEDURE

Commercially available Si_3N_4 powder (SN-9S, Denka Co., Ltd., Tokyo, Japan, α phase 92%, average particle size 1.10 μm, specific surface area 7.3 m^2/g, impurities (ppm): Fe 1200, Al 1200, Ca 1200) was used as the starting powder. The total additive content varied from 3.5 to 15 wt%, and the ratios of $MAl_2O_4:ZrO_2$ and $Y_2O_3:Al_2O_3$ were fixed at 5:2 and 1:1, respectively, in order to maintain a constant composition for the grain-boundary glassy phase. Compositions of testing samples are listed in Table 1.

Table 1. Composition of testing materials

Sample	Si_3N_4 (wt%)	$MgAl_2O_4$ (wt%)	ZrO_2 (wt%)	Y_2O_3 (wt%)	Al_2O_3 (wt%)
1.75MA1.75Z	96.5	1.75	1.75		
2.5MA2.5Z	95.0	2.5	2.5		
2.5MA2.5Z	90.0	5.0	5.0		
7.5MA7.5Z	85.0	7.5	7.5		
2.5Y1.0A	96.5			2.5	1.0
5.0Y2.0A	93.0			5.0	2.0
10.0Y4.0A	86.0			10.0	4.0

The procedures to prepare specimens are as follows: the starting mixture powder was ball-milled in ethanol for 24 h, then dried and sieved. The mixture was dry-pressed and CIPed under 200 MPa pressure. The sizes of green samples were 64 mm × 6.5 mm × 5 mm. Sintering was conducted in a graphite resistance furnace (Model No. FVPHP-R-5, Fujidempa Kogyo Co., Ltd., Osaka, Japan), under a nitrogen pressure of 0.6 MPa, at various temperatures from 1300° to 1900°C. The heating rate of 10-20°C/min was used.

The bulk densities of the specimens were measured by the Archimedes method, using distilled water. Theoretical densities were calculated from the starting powder using the rule of mixtures. Relative density was determined from the actual density and the theoretical density. Crystalline phases were identified by X-ray diffraction (XRD; Model No. RU-200B, Rigaku Co., Ltd., Tokyo, Japan), at 50 kV and 150 mA. The weight fraction of β-Si$_3$N$_4$ ($\rho_{\beta/(\alpha+\beta)}$) was calculated from the two highest XRD peaks of the α- and β-Si$_3$N$_4$ phases, following a procedure proposed by Gazzara and Messier [4]. In order to observe the microstructure by scanning electron microscopy (SEM; Model No. S-5000, Hitachi Co., Ltd., Tokyo, Japan), the samples sintered at 1900°C were polished and plasma-etched.

RESULTS AND DISCUSSION

Figure 1 shows the changes in relative density and α-Si$_3$N$_4$ phase content (100% − $\rho_{\beta/(\alpha+\beta)}$) as a function of sintering temperature. It can be seen that densification started at ~1300°C for the MA-Z system and at ~1450°C for the Y-A system, and the phase transformation became active at >1500°C. Densification behavior also is highly affected by additive content, that is, increasing the content enhanced the densification. However, phase-transformation behavior was significantly different for the two systems: it was suppressed by the MA-Z content, but depended very little on the Y-A content. In addition, the density tended to decrease slightly as the temperature increases up to 1650°C, at this temperature point phase transformation was almost complete.

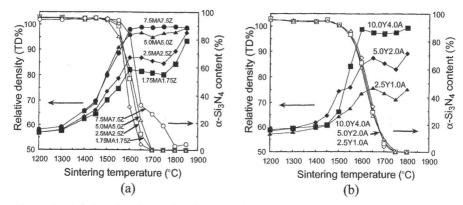

Figure 1. Relative density and weight fraction of α phase as a function of sintering temperature after 2 h sintering. (a) MA-Z; (b) Y-A

The relationship between density and β-Si$_3$N$_4$ content for two types of specimens is shown in Fig. 2. An abrupt density increase with little α–β transformation (<20% for MA-Z system and <40% for Y-A additive system, respectively) could be seen at the beginning of sintering. Further α–β transformation resulted in little densification, and fully dense ceramics could be obtained after the α–β transformation reached 100%.

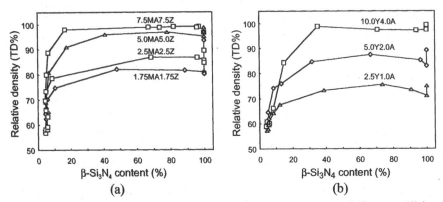

Figure 2. Relative density correlated with β-Si$_3$N$_4$ content at different additive content after 2 h sintering. (a) MA-Z; (b) Y-A

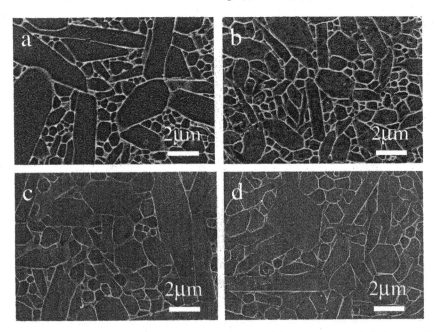

Figure 3. SEM micrographs of Si$_3$N$_4$ ceramics of the samples: (a) 2.5MA2.5Z, (b) 7.5MA7.5Z, (c) 5.0Y2.0A, (d)10.0Y4.0A. Sintering condition: 1900°C/8h

Figure 3 shows the morphology of two types of Si_3N_4 ceramics with the different additives. It is apparent that refined microstructures were obtained by increasing the additive content for the MA-Z system. However, the morphology of Si_3N_4 changes little with the addition of Y-A content. This implies that grain growth of Si_3N_4 is restrained by increasing MA-Z content, but not affected by Y-A content.

Previous studies on the sintering of Si_3N_4 ceramics showed that the densification process was accompanied by the α-β phase transformation and the phase transformation was enhanced considerably by the higher amount of sintering additive [5, 6]. However, in this study, it can be seen that: (1) densification started at relatively lower temperatures than that of phase transformation; (2) phase transformation was active with constant densities; and (3) densification was enhanced by increasing additive content, whereas phase transformation was suppressed or unaffected. These results suggest that the densification and phase transformation have no direct relationship.

The densification of Si_3N_4 is controlled by particle rearrangement and solution–diffusion-reprecipitation, both of which are enhanced by liquid-phase formation at the grain boundary [7]. Particle rearrangement contributes to densification at low temperatures, and solution–diffusion-reprecipitation is dominant at high temperatures. At the same time, phase transformation becomes active at high temperature. In this study, it is believed that the primary densification mechanism is particle rearrangement. Because densification in the present study depended largely on the sintering additive content; the larger amount of boundary liquid phase is more favorable for increasing particle rearrangement. In contrast, if solution–diffusion-reprecipitation is the main mechanism causing densification, the densification would be suppressed or unaffected by the amount of liquid phase, corresponding to the changes in the phase transformation.

An earlier report [8] has shown that grain growth is accelerated by α-β phase transformation, both of which are controlled by solution–diffusion-reprecipitation. Two fundamental mechanisms existed for the mass-transfer: diffusion of Si and N ions through the boundary phase, and reaction (reprecipitation) at the interface between the boundary and the matrix. When solution–diffusion-reprecipitation is slow, the mass-transfer process becomes rate-controlling [9]. In the case of diffusion-controlling, the mass-transfer rate decreases as the boundary liquid film thickness increases if the matrix grain size is the same. However, the mass-transfer rate is independent of such a change in film thickness if the process is controlled by an interfacial reaction [10].

For the present Si_3N_4 ceramics with MA-Z additive, it is suggested that the phase transformation and grain growth are diffusion controlled. The large additive content increases the thickness and viscosity of the liquid phase, due to the constant SiO_2 content on the Si_3N_4 particle surface. This results in the long distance and large difficulty for the diffusion of Si and N along the liquid phase, and therefore the phase transformation and grain growth were retarded.

For the Si_3N_4 ceramics with Y-A additive, solution–diffusion-reprecipitation is controlled by the interfacial reaction, as explained in other works [11]. The occurrence of an interface-controlled mechanism with a high Y_2O_3 content can be explained by the fact that the yttrium ion is primarily a network modifier in a Y-Si-O-N glass [12]. This viewpoint is supported by the results of Walls and Ueki [13], who speculated that the diffusion of nitrogen ions through the liquid phase must be extremely rapid, because they detected negligible nitrogen in the glassy matrix.

In this study, density decreases substantially as the temperature increases after phase transformation is almost complete, particularly in the case of a lower additive content. As we know, the densification of Si_3N_4 during sintering is inhibited by the impingement effect of rodlike β-Si_3N_4 grain growth [14]; so the large decrease in density for a low additive content probably results from the lack of compensation by particle rearrangement. Obviously, this tendency is large for the Y-A system, due to the refractory glass phase and enhanced grain growth.

CONCLUSIONS

Densification was enhanced by an increase in sintering additive, because of the enhanced particle rearrangement. However, phase transformation and grain growth were suppressed by increasing the MA-Z content but not affected by the Y-A content. The phase transformation and grain-growth behavior can be explained by a solution-diffusion-reprecipitation mechanism. In case of the MA-Z system, the above process is controlled by a diffusion mechanism, whereas in the case of the Y-A system, the process is controlled by the interfacial reaction. Moreover, density decreased with an increase in temperature after phase transformation was almost complete, due to the impingement of rodlike β-Si_3N_4 grains.

REFERENCES

1. W. D. Kingery, "Densification during Sintering in the Presence of Liquid Phase. I. Theory," *J. Appl. Phys.*, **30** [3] 301-306 (1959).

2. D. D. Lee, S. J. Kang, G. Petzow, and D. N. Yoon, "Effect of α to β (β') Phase Transformation on the Sintering of Silicon Nitride Ceramics," *J. Am. Ceram. Soc.*, **73** [3] 767-69 (1990).

3. C. Greskovich and S. Prochazka, "Observations on the α–β Si_3N_4 transformation," *J. Am. Ceram. Soc.*, **60** [9-10] 471-72 (1977).

4. C. P. Gazzara and D. R. Messier, "Determination of Content of Si_3N_4 by X-ray Diffraction Analysis," *Am. Ceram. Soc. Bull.* **56** [9] 777-780 (1977).

5. M. Mitomo and K. Mizuno, "Sintering Behavior of Si_3N_4 with Y_2O_3 and Al_2O_3 Addition," *Yogyo-Kyokai-Shi*, **94** [1] 106-111 (1986).

6. A. Kuzjukevics and K. Ishizaki, "Sintering of Silicon Nitride with $YAlO_3$ Additive," *J. Am. Ceram. Soc.*, **76** [9] 2373-75 (1993).

7. G. Wötting, B. Kanka and G. Ziegler, "Microstructural Development, Microstructural Characterization and Relation to Mechanical Properties of Dense Silicon Nitride;" pp. 83-96 in *Non-Oxide Technical and Engineering Ceramics*, Edited by S. Hamshire, Elsevier Appl. Sci. Publishers, Barking, Essex, England, 1986.

8. H. Emoto, H. Hirotsuru and M. Mitomo, "Influence of Phase Transformation on Grain Growth Behavior of Silicon Nitride Ceramics," *Key Engineering Materials*, **159-160**, 215-20 (1999).

9. I. M. Lifzhitz, and V. V. Slyozov, "The Kinetics of Precipitation from Supersaturated Solid Solution," *J. Phys. Chem. Solid*, **19** [1/2] 35-50 (1961).

10. S. M. Han and S. J. Kang, "Phase Transformation and Grain Growth during Liquid Phase Sintering of Si_3N_4 Ceramics," pp. 83-88 in *Proceedings of 1st International Symposium on the Science of Engineering Ceramics* (Koda, Japan, October, 1991). Edited by S. Kimura and K. Niihara, The Ceramic Society of Japan, Tokyo, Japan, 1991.

11. M. Krämer, M. J. Hoffmann, and G. Petzow, "Grain Growth Studies of Silicon Nitride Dispersed in an Oxynitride Glass," *J. Am. Ceram. Soc.*, **76** [11] 2778-84 (1993).

12. M. Kitayama, K. Hirao, M. Toriyama and S. Kanzaki, "Control of β-Si_3N_4 Crystal Morphology and its Mechanism (Part 1)," *J. Ceram. Soc. Japan*, **107** [10] (1999).

13. P. A. Walls, and M. Ueki, "Analysis of the α to β Phase Transformation in Y-Si-Al-O-N Ceramics," *J. Mater. Sci.*, **28**, 2967-74 (1993).

14. O. Abe, "Sintering Process of Y_2O_3 and Al_2O_3-doped Si_3N_4," *J. Mater. Sci.*, **25**, 4018-26(1990).

PRECISE SHAPE FUNCTION FOR INTERPARTICLE NECKS FORMED DURING SOLID-STATE SINTERING

Yukiko Tagami

Institute of Physics, University of Tsukuba
Tsukuba 305-8571, Japan
e-mail address: tagami@cm.ph.tsukuba.ac.jp

ABSTRACT

As a classical problem of inhomogeneous sintering theory, the formation of a neck between two spherical particles by grain boundary and
surface diffusion has been reexamined recently by Svoboda and Riedel.[1]
By introducing similarity analysis they have been able to specify the neck
geometry in closed analytical form without any geometrical assumptions.
Their calculation of the similarity solution, however, contained a drastic
simplification.

The present paper solves directly their nonlinear differential equation
without simplification and gives, for the first time, a precise similarity
solution, along with its equivalent closed form in the form of the Padé
approximant. Several applications of the thus-obtained neck profile function are given and comparison with the corresponding results by Svoboda
and Riedel is made.

INTRODUCTION

It is well-known that formation of a neck between two spherical particles due to
grain boundary and surface diffusion plays a crucial role in the progress of sintering.

To the extent authorized under the laws of the United States of America, all copyright interests in this publication are the property
of The American Ceramic Society. Any duplication, reproduction, or republication of this publication or any part thereof, without
the express written consent of The American Ceramic Society or fee paid to the Copyright Clearance Center, is prohibited.

Recently, Svoboda and Riedel (hereafter we shall call SR)[1] solved by means of similarity analysis a nonlinear differential equation governing the profile function and were able to specify the neck geometry in closed analytical form without any geometrical assumptions.

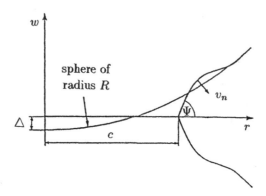

Figure 1: The neck profile $w(r)$.

Figure 1 schematically depicts the geometry of the problem: Two spherical particles with radius R are connected by a circular contact area with radius c. Here $w(r)$ describes the undercutting profile of the neck surface. $\triangle$ is the half a center-to-center approach of the two spheres, Ψ the dihedral angle at the neck tip, and v_n the normal displacement rates of the neck surface. The surface profile $w(r)$ of the neck help evaluate properties characterizing the first stage of sintering. In the similarity analysis of SR, in place of $w(r)$ a dimensionless shape function $W(X)$ of the dimensionless similarity coordinate X is calculated, where

$$w(r) = R(\frac{c}{R})^\alpha W(X) + (\frac{r^2}{2R} - \triangle), \tag{1}$$

and

$$X = \frac{(r-c)/R}{(c/R)^\alpha}. \tag{2}$$

The exponent α is to be evaluated later. The governing equation for $w(r)$, or the nonlinear differential equation satisfied by $W(X)$, is as follows:[1]

$$\frac{d^2}{dS^2}[\frac{W_X''}{(1+W_X'^2)^{3/2}}] - (2\lambda)^3 \frac{W_X'}{(1+W_X'^2)^{1/2}} = 0, \tag{3}$$

Grain Boundary Engineering in Ceramics

with

$$\frac{d}{dS} = (1 + W_X'^2)^{-1/2}\frac{d}{dX}, \tag{4}$$

and λ is a constant. For clarity we hereafter denote by W_X'' the second derivative of W with respect to X. Equation (3) comes from a governing equation derived by Riedel[2] which in turn is based upon the relation[3] between the chemical potential μ of an atom on a curved surface, whose gradient gives the driving force for a diffusive flux of atoms, and the principal curvatures κ_1, κ_2 of the surface

$$\mu = \mu_0 - \gamma_s\Omega(\kappa_1 + \kappa_2). \tag{5}$$

Here μ_0 is the chemical potential of an atom on a flat surface, γ_s the specific surface energy, and Ω atomic volume. The term in the angular brackets in equation (3) is none other than one of the principal curvatures of our neck geometry.

In solving equation (3) with equation (4), SR linearized it by assuming that $W_X'^2 \ll 1$. SR solved the following linear differential equation

$$W_X^{IV} - (2\lambda)^3 W_X' = 0, \tag{6}$$

and obtained a closed form solution

$$W(X) = \frac{\Psi}{2\lambda\sin(2\pi/3 - \phi)}e^{-\lambda X}\sin(\sqrt{3}\lambda X - \phi), \tag{7}$$

where Ψ should properly be replaced by tan Ψ. On the other hand we directly solve equation (3) with equation (4) without assuming that $W_X'^2 \ll 1$, as shown in the present paper.

SIMILARITY SOLUTIONS

Calculating equation (3) with equation (4) we obtain an equation for $W = W(X)$

$$\frac{W_X^{IV}}{(1 + W_X'^2)^2} - 10\frac{W_X'W_X''W_X'''}{(1 + W_X'^2)^3} - 3\frac{(W_X'')^3}{(1 + W_X'^2)^3}$$
$$+ 18\frac{(W_X')^2(W_X'')^3}{(1 + W_X'^2)^4} - (2\lambda)^3 W_X' = 0. \tag{8}$$

Let us introduce a new function $f(X)$ by

$$f(X) = \frac{W_X'}{(1 + W_X'^2)^{1/2}}. \tag{9}$$

Equation (8) then becomes

$$(1 - f^2)f_X''' - ff_X'f_X'' - (2\lambda)^3 f = 0. \tag{10}$$

For convenience we introduce a parameter

$$z = 2\lambda X. \tag{11}$$

Equation (10) now becomes an equation for $f = f(z)$:

$$(1 - f^2)f_z''' - ff_z'f_z'' - f = 0. \tag{12}$$

Since the geometry in Fig. 1 indicates $W_X'(0) = \tan \Psi$, the definition of f in equation (9) leads to its initial value

$$f(0) = \sin \Psi. \tag{13}$$

Altogether we have a nonlinear differential equation, equation (12), for $f = f(z)$ with an initial value, equation (13), to be solved. We further expect

$$f(\infty) = f_z'(\infty) = f_z''(\infty) = \cdots = 0, \tag{14}$$

as a physical requirement.

First we expand f in power series of z:

$$f(z) = \sin \Psi \sum_{n=0}^{\infty} a_n z^n, \tag{15}$$

with $a_0 = 1$. Then we obtain, for arbitrary a_1 and a_2, writing $\sigma \equiv \tan^2 \Psi$,

$$a_3 = \frac{1}{6}[2\sigma a_1 a_2 + (1 + \sigma)], \tag{16}$$

$$a_4 = \frac{1}{12}[3(3\sigma + 1)a_1 a_3 + 2\sigma a_2^2], \tag{17}$$

$$a_5 = \frac{1}{60}\{2\sigma a_1[2\sigma a_1^2 a_2 + (3 + 5\sigma)a_2^2 + 30 a_4] \\ + (1 + \sigma)[(1 + 5\sigma)a_2 + 2\sigma a_1^2]\}, \tag{18}$$

$$a_6 = \frac{1}{30}\sigma(20 a_2 a_4 + 35 a_1 a_5 + a_2^3 + 9 a_1^2 a_4) + \frac{1}{180}\sigma^2(19 + 5\sigma)a_1^2 a_2^2 \\ + \frac{1}{36}\sigma(1 + \sigma)(2 + \sigma)a_1 a_2 + \frac{1}{720}(1 + \sigma)^2(1 + 5\sigma). \tag{19}$$

Likewise we have another power series in z (cf. equation (9)):

$$W_X' = \frac{f}{\sqrt{1 - f^2}} = \tan \Psi \sum_{n=0}^{\infty} b_n z^n, \tag{20}$$

with $b_0 = 1$ and

$$b_1 = (1 + \sigma)a_1, \tag{21}$$

$$b_2 = (1+\sigma)(a_2 + \frac{3}{2}\sigma a_1^2), \qquad (22)$$

$$b_3 = (1+\sigma)[\frac{10}{3}\sigma a_1 a_2 + \frac{1}{2}\sigma(1+5\sigma)a_1^3 + \frac{1}{6}(1+\sigma)], \qquad (23)$$

$$b_4 = (1+\sigma)[\frac{1}{24}(1+\sigma)(1+15\sigma)a_1 + \frac{1}{12}\sigma(19+105\sigma)a_1^2 a_2$$
$$+\frac{5}{3}\sigma a_2^2 + \frac{5}{8}\sigma^2(3+7\sigma)a_1^4], \qquad (24)$$

and similarly for $b_5, b_6, \cdots$. Since $w(c) = 0$ we have from equation (1)

$$W(0) = -\frac{1}{R}(\frac{R}{c})^\alpha(\frac{c^2}{2R} - \triangle). \qquad (25)$$

As the center-to-center approach of the two spheres $2\triangle$ in Fig. 1 has been assumed to be small, we note that $W(0)$ given in equation (25) is likely to be negative.

Now, juxtaposition of $\{a_n\}$'s in equations (16) to (19) with $\{b_n\}$'s in equations (21) to (24) reveals that if we put

$$a_n = \frac{k_n}{n!}a_1^n, \qquad (26)$$

and assume that k_n's do not depend on σ as long as n remains small, we arrive at the following remarkable sets of relations that simultaneously satisfy equations (16) to (19) as well as equations (21) to (24):

$$k_1 = 1, \qquad k_2 = 3, \qquad k_3 = k_4 = -3,$$

$$k_5 = -9(1+6\sigma), \qquad k_6 = 9(1 - 24\sigma - 42\sigma^2), \cdots$$

in conjunction with

$$a_1 = -3^{-1/3}, \qquad a_2 = \frac{1}{2}3^{1/3}, \qquad a_3 = \frac{1}{6}, \qquad a_4 = -\frac{1}{24}3^{-1/3},$$

$$a_5 = \frac{1}{120}3^{1/3}(1+6\sigma), \qquad a_6 = \frac{1}{720}(1 - 24\sigma - 42\sigma^2), \cdots .$$

Hence

$$b_1 = -3^{-1/3}(1+\sigma), \qquad (27)$$

$$b_2 = \frac{3}{2}b_1^2, \qquad (28)$$

$$b_3 = (\frac{5}{2}b_1^2 - 3^{1/3})b_1, \qquad (29)$$

$$b_4 = (\frac{35}{8}b_1^3 - 3^{4/3}b_1 - \frac{3}{2})b_1, \qquad (30)$$

$$b_5 = \left(\frac{63}{8}b_1^4 - \frac{15}{2}3^{1/3}b_1^2 - \frac{87}{20}b_1 + \frac{1}{6}3^{2/3}\right)b_1, \tag{31}$$

$$b_6 = \left(\frac{231}{16}b_1^5 - \frac{35}{2}3^{1/3}b_1^3 - \frac{85}{8}b_1^2 + \frac{67}{30}3^{2/3}b_1 + \frac{139}{90}3^{1/3}\right)b_1. \tag{32}$$

With the help of equation (20) together with equations (27) to (32) we now have the ascending series for the similarity solution $W(X)$:

$$\begin{aligned}
W(X) - W(0) &= \tan\Psi \cdot X \sum_{n=0}^{\infty} \frac{b_n}{n+1}(2\lambda X)^n \\
&= \tan\Psi \cdot X[1 - A + 2A^2 - (5 - 6\sec^{-4}\Psi)A^3 \\
&\quad + 14(1 - \frac{72}{35}\sec^{-4}\Psi + \frac{36}{35}\sec^{-6}\Psi)A^4 \\
&\quad - 42(1 - \frac{20}{7}\sec^{-4}\Psi + \frac{58}{35}\sec^{-6}\Psi + \frac{4}{21}\sec^{-8}\Psi)A^5 \\
&\quad + 132(1 - \frac{40}{11}\sec^{-4}\Psi + \frac{170}{77}\sec^{-6}\Psi + \frac{536}{385}\sec^{-8}\Psi \\
&\quad - \frac{1112}{1155}\sec^{-10}\Psi)A^6 + \cdots],
\end{aligned} \tag{33}$$

where $A = 3^{-1/3}\lambda \sec^2\Psi \cdot X$ and $W(0)$ was given by equation (25).

CLOSED FORM SOLUTIONS

In view of the fact that the solution (33) constitutes the first several terms in the ascending series for $W(X)$, we now construct the simplest of the Padé approximants corresponding to equation (33) that should satisfy the physical requirements that $W(X)$ does not diverge for any $X > 0$ and that $W(X)$ together with its derivatives with respect to X approach zero for $X \to \infty$ (see also equation (14)). We obtain

$$\begin{aligned}
W(X) - W(0) &= 3\tan\Psi \cdot X[3 + 3^{2/3}\sec^2\Psi \cdot \lambda X \\
&\quad - 3^{1/3}\sec^4\Psi \cdot \lambda^2 X^2 + 2\sec^2\Psi(\sec^4\Psi - 3)\lambda^3 X^3]^{-1}.
\end{aligned} \tag{34}$$

The requirement of non-divergence of $W(X) - W(0)$ for $X > 0$ limits the applicable range of σ (and consequently of Ψ) for equation (34) as

$$\sigma = \tan^2\Psi > \frac{9\sqrt{2}}{7} - 1 = 0.81827\cdots, \tag{35}$$

or

$$\Psi > 42.132°. \tag{36}$$

That equation (34) is a Padé approximant resulted in the fact that $W(X)$ given by equation (34) does not apply for smaller values of Ψ all the way down to $0°$.

Grain Boundary Engineering in Ceramics

Hereafter condition (35), or condition (36), will be assumed. Equation (34) can be integrated with respect to X. After a lengthy calculation we arrive at

$$\int_0^\infty [W(X) - W(0)]\, dX = \frac{3^{2/3}}{\lambda^2} \frac{\sin\Psi\,\cos^{1/3}\Psi}{[u^2 + v^2 - 2\sec^{4/3}\Psi(5\sec^4\Psi - 18)]}$$

$$\times \{(u - \sec^{8/3}\Psi)\log\frac{2(u - \sec^{8/3}\Psi)^2}{[u^2 + v^2 + 2u\sec^{8/3}\Psi + 12\sec^{4/3}\Psi(\sec^4\Psi - 3)]}$$

$$+ \frac{\sqrt{3}(u^2 + v^2 + 2u\sec^{8/3}\Psi)}{v}[\frac{\pi}{2} + \tan^{-1}(\frac{u + 2\sec^{8/3}\Psi}{\sqrt{3}v})]\}, \tag{37}$$

where

$$u = (\sqcap + \sqcup)^{1/3} + (\sqcap - \sqcup)^{1/3}, \tag{38}$$

$$v = \left|(\sqcap + \sqcup)^{1/3} - (\sqcap - \sqcup)^{1/3}\right|, \tag{39}$$

and

$$\sqcap = 62\sec^8\Psi - 351\sec^4\Psi + 486, \tag{40}$$

$$\sqcup = 9(\sec^4\Psi - 3)[(\sec^4\Psi - 2)(49\sec^4\Psi - 162)]^{1/2}. \tag{41}$$

APPLICATION OF THE SOLUTIONS

The above-obtained similarity solutions $W(X)$ provide us with aspects of interest on quantities of materials during sintering, of which we state only a few here.

(i) The case of very slow grain boundary diffusion
In this case we put $\triangle = 0$ and $r \approx c$. Assuming $r/R \ll 1$ we have $W(0) \sim 0$. The expression for $W(X)$ for $X \sim 0$ is given by equation (34) as

$$W(X) \sim 3\tan\Psi \cdot X(3 + 3^{2/3}\sec^2\Psi \cdot \lambda X)^{-1}$$

$$\sim \tan\Psi \cdot X(1 - 3^{-1/3}\sec^2\Psi \cdot \lambda X), \tag{42}$$

which gives

$$\int_0^\infty W(X)\, dX \sim \int_0^{3^{1/3}\cos^2\Psi\cdot\lambda^{-1}} W(X)\, dX \qquad \text{as } X \text{ is small}$$

$$= \frac{3^{2/3}\sin\Psi\,\cos^3\Psi}{6\lambda^2}. \tag{43}$$

The equivalence of the volume added to the neck between $r = 0$ and $r = c$ and the volume removed from the surface at $r > c$ becomes

$$\frac{\pi c^4}{4R} = 2\pi c R^2 (\frac{c}{R})^{2\alpha} \int_0^\infty W\, dX, \qquad \alpha = \frac{3}{2}$$

$$\sim \frac{2\pi c^4}{R} \frac{3^{2/3}\sin\Psi\,\cos^3\Psi}{6\lambda^2}.$$

Hence

$$\lambda = 2 \cdot 3^{-1/6} \sin^{1/2} \Psi \cos^{3/2} \Psi. \tag{44}$$

As in SR we now have

$$w(r) = R(\frac{c}{R})^{3/2} W(X), \tag{45}$$

with $W(X)$ given by equation (42). Use of equation (44) in Equations (18) and (20) of Ref. 1 gives

$$(\frac{c}{R})^{11/2} = 352 \cdot 3^{-1/2} \sin^{3/2} \Psi \cos^{9/2} \Psi \cdot \tau, \tag{46}$$

where τ is the non-dimensionalized time defined by Equation (18) of Ref. 1, *i.e.* $\tau \propto t$. Thus we have the same power law as in SR:

$$c \propto t^{2/11}. \tag{47}$$

(ii) The case of slow surface diffusion

In this case we have $\triangle = c^2/(4R) \ll 1$. As in SR we have $\alpha = 2$, so that equation (25) yields $W(0) = -1/4$, in coincidence with the result by SR. Our $W(X) - W(0)$ given in equation (33) or equation (34), however, will not provide a relationship between λ and Ψ. Nevertheless we have from Equation (20) of Ref. 1

$$(\frac{c}{R})^7 = 7(2\lambda)^3 \tau \propto t,$$

therefore the power law $c \propto t^{1/7}$ holds in this case as in SR.

Since the present paper focuses primarily on the derivation of precise similarity solution $W(X)$ (and thus $w(r)$), more detailed application of our results will be presented elsewhere.

CONCLUSIONS

A solution to the nonlinear differential equation, equation (3), posed by Svoboda and Riedel, for the profile function of a neck formed between two spherical particles during the early stage of sintering has been obtained for the first time. Results are given by equation (33) in the form of an ascending series, and by equation (34) in closed-form. As an example of appllications of our results, the time dependence of the neck radius in the two limiting ranges of fast surface diffusion (or very slow grain boundary diffusion) and of slow surface diffusion (or fast grain boundary diffusion) was examined, and we reproduced the power laws $\frac{2}{11}$ and $\frac{1}{7}$, respectively, for the neck radius calculated by Svoboda and Riedel using their closed-form similarity solution, equation (7), which was derived using a linearized form for equation (3), viz. equation (6).

It is practical to assume the dihedral angle Ψ at the neck to be as large as 60°. While the use of linearization by Svoboda and Riedel will limit the validity of their neck profile function to very small values of Ψ, our similarity solutions which have been derived directly from their nonlinear differential equation without simplification, will remain valid through larger values of Ψ. In fact, our use of the Padé approximant in equation (34) requires $\Psi > 42.132°$. Altogether our similarity solutions are to describe properly the neck growth when the neck radius remains small compared to the neighboring necks during the early stage of sintering.

References

[1] Svoboda, J. and Riedel, H.," New Solutions Describing the Formation of Interparticle Necks in Solid-State Sintering," *Acta metall.mater.* **43** [1] 1-10 (1995).

[2] Riedel, H. *Fracture at High Temperatures* (Springer-Verlag, Berlin, 1987).

[3] Herring, C. in *The Physics of Powder Metallurgy*, ed. W. E. Kingston (McGraw-Hill, New York, 1951) p.143.

DIFFUSIONAL RELAXATION AROUND 9R Cu MARTENSITE PARTICLES IN AN Fe MATRIX

R. Monzen
Department of Mechanical Systems Engineering, Kanazawa University
2-40-20 Kodatsuno, Kanazawa 920-8667, Japan

ABSTRACT

High-resolution electron microscopy and transmission electron microscopy experiments have been carried out to investigate the relaxation process of elastic fields created by the b.c.c.$\rightarrow$9R martensitic transformation of Cu particles and the lattice relaxation processes of twinned 9R martensite particles towards final stable structures, in an Fe–Cu alloy aged at 823 K. The relaxation process of the elastic fields was examined by observing the rotation of particle $(009)_{9R}$ planes due to annealing. From the particle size and annealing time dependence of the rotations, it has been concluded that diffusion on the Cu particle–Fe matrix interface causes the relaxation. The activation energy of the interfacial diffusion was evaluated as 1.6×10^5 J/mol. Following the $(009)_{9R}$ plane rotation during aging, the migration of twin boundaries initially occurred in 9R particles greater than about 13 nm. At the same time, 9R segments underwent a partial transformation to a 3R structure. Particles of size 26–40 nm consisted of two variants with stable f.c.c. and f.c.t. structures. The f.c.t. structure had larger lattice constants $a=b=0.369$ nm and $c=0.366$ nm than that for bulk f.c.c. Cu ($a=0.361$ nm). Larger particles were f.c.c. or f.c.t. single crystals which exhibit the Kurdjumov–Sachs orientation relationship to the Fe matrix.

INTRODUCTION

It has long been accepted that small Cu-rich precipitate particles in thermally-aged Fe–Cu model alloys are initially coherent with the b.c.c. Fe matrix, while large overaged particles are f.c.c. single crystals with a Kurdjumov–Sachs orientation relationship to the Fe matrix [1, 2]. Recently, high-resolution electron microscopy

To the extent authorized under the laws of the United States of America, all copyright interests in this publication are the property of The American Ceramic Society. Any duplication, reproduction, or republication of this publication or any part thereof, without the express written consent of The American Ceramic Society or fee paid to the Copyright Clearance Center, is prohibited.

(HREM) of Cu particles in Fe−Cu and Fe−Cu−Ni model alloys aged at 823 K has revealed that the coherent Cu particles in these alloys do not undergo a direct b.c.c.→f.c.c. transformation, but first transform martensitically into a 9R structure [3]. The smallest 9R particles observed had a size of about 4 nm and consisted of only two twin-related variants. Larger particles, with sizes up to about 13 nm, consisted of several twinned segments. It was considered that at a size above about 18 nm a second transformation to a more stable 3R structure occurs. 3R is a non-cubic distorted f.c.c. structure obtained when the regular stacking faults on the $(009)_{9R}$ basal planes are removed, but where the resulting $(003)_{3R}$ basal planes retain a spacing characteristic of 9R basal planes. However, a detailed examination of the 9R→3R→f.c.c. transformation processes has not been carried out.

We have recently shown that the close-packed $(009)_{9R}$ planes in each twin segment rotate to align more closely with matrix $\{110\}_{Fe}$ planes during aging at 823 K or by annealing, and that the particle size at which the b.c.c.→9R martensitic transformation occurs depends on the ambient temperature and is about 12 nm at 823 K and about 4 nm at 213 K [4]. It has also been found that the plane rotations are associated with diffusional relaxation of elastic strains produced by the martensitic transformation. In this study, the mechanism and kinetics of the diffusional relaxation are examined. Moreover, the lattice relaxation processes in 9R particles towards an f.c.c. or f.c.t. structure during aging are revealed in detail.

EXPERIMENTAL

Discs of 3-mm diameter of an Fe−1.5 wt% Cu alloy were punched from 1-mm thick strips and ground to a thickness of about 120 μm. The discs were then solution-treated under vacuum at 1103 K for 10.8 ks, quenched into water and aged at 823 K for 14.4 ks−1.26 Ms. Then part of the specimens were annealed at 373−548 K for 3.6 or 18 ks, or at 673 K for 3.6 ks. The discs were electropolished in a Struers twin-jet electropolisher, using a solution of 5% perchloric acid in methanol cooled to 213 K, and with an applied potential of 20−25 V. Microscopy was carried out on an Akashi 002B operating at 200 kV. All HREM and TEM images were obtained with the beam parallel to a $<111>_{Fe}$ direction in the Fe matrix.

RESULTS AND DISCUSSION

Relaxation Process of Elastic Strains in and around 9R Particles

Figure 1(a) shows a high-resolution electron micrograph of a 9R martensite particle (of diameter 6.6 nm), which consists of five twin-related bands, in a specimen aged at 823 K for 28.8 ks and cooled to 213 K during electropolishing. Figure 1(b)

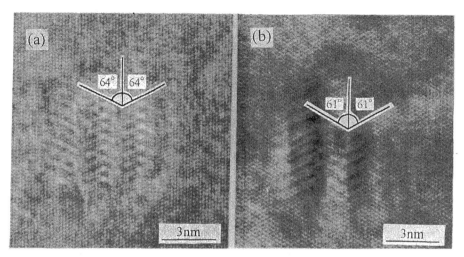

Figure 1. HREM images of 9R particles in Fe–Cu specimens aged at 823 K and cooled to 213 K: (a) unannealed and (b) annealed at 673 K for 3.6 ks.

shows a 9R particle (of diameter 5.5 nm) in a specimen received the same initial heat treatment, and subsequently annealed at 673 K for 3.6 ks. In (a) and (b), the structure of these particles is easily recognized from the characteristic herring-bone fringe pattern [3]. The fringes have spacings of about three times the expected $(009)_{9R}$ close-packed plane spacing (i.e. about 0.6 nm). The boundary between adjacent twin bands lies parallel to a set of matrix $\{110\}_{Fe}$ fringes. This boundary plane is the $(\bar{1}\bar{1}4)_{9R}$ plane of mirror symmetry. The orientation relationship between the Fe matrix and the 9R particles is thus $(011)_{Fe} // (\bar{1}\bar{1}4)_{9R}$ and $[1\bar{1}1]_{Fe} // [\bar{1}10]_{9R}$ [3]. In (a), the angle α between the $(009)_{9R}$ basal and $(\bar{1}\bar{1}4)_{9R}$ twin planes is 64°. This was the case for 9R particles of size 4–12 nm in specimens cooled to 213 K. On the other hand, $\alpha = 61°$ in (b). This was again typical of most particles in this size range. The values of α for particles unannealed and annealed at 673 K for 3.6 ks are in good agreement with those previously reported [4]. That is, the $(009)_{9R}$ planes in each twin band rotated to align more closely with $\{110\}_{Fe}$ planes.

In Fig. 2(a), the average $\Delta\alpha$ of the change in α in 3.6 and 18 ks isochronal annealings is plotted against temperature for 9R martensite particles of average size 8.0 nm, which consist of six twin-related bands, for specimens cooled to 213 K before annealing. Figure 2(b) shows 3.6 ks isochronal annealing curves against temperature for 3.8 nm (two twin bands) and 8.0 nm particles, again for specimens cooled to 213 K before annealing. About 10 particles were examined for each experimental point. The scatter of the data was about ±0.5°. As annealing temperature increases, $\Delta\alpha$ increases and saturates to the same value of about 3.3° regardless of the

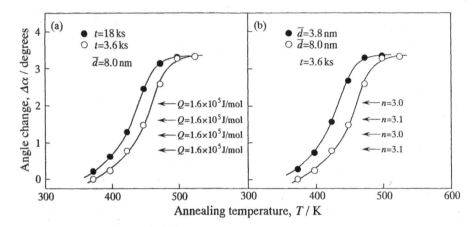

Figure 2. Temperature dependence of the average angle change $\Delta\alpha$ induced by annealing: (a) 8.0 nm particles annealed for 3.6 ks and 18 ks; (b) 3.8 nm and 8.0 nm particles annealed for 3.6 ks.

annealing time and the particle size. At a given temperature, $\Delta\alpha$ is larger for the longer annealing time and the smaller particle size.

We have shown that the $(009)_{9R}$ plane rotations are connected with diffusional relaxation of elastic fields, created upon the b.c.c.$\rightarrow$9R martensitic transformation [4]. Theoretically [5] and experimentally [6], the diffusional relaxation of an elastic energy due to a uniformly stressed particle has been shown to occur with first order kinetics. Therefore, the kinetics in the present work is probably expressed as

$$\Delta\alpha=\Delta\bar{\alpha}\{1-\exp(-t/\tau)\} \tag{1}$$

where t is the annealing time and $\Delta\bar{\alpha}$ is the saturation value (3.3°) of $\Delta\alpha$. τ is the relaxation time and is written as [5]

$$\tau\propto d^3T/D \tag{2}$$

when the pass of diffusion is the particle–matrix interface. Here d is the particle diameter, D is the interfacial diffusivity and T is the temperature. Equations (1) and (2) indicate that the activation energy Q for interfacial diffusion can be obtained by the cross-cut method, using the results of two isochronal annealings for the same size particles. That is, Q can be determined using

$$t_1T_1^{-1}\exp(-Q/RT_1)=t_2T_2^{-1}\exp(-Q/RT_2) \tag{3}$$

where t_1 and t_2 are the times to achieve the same rotation at the annealing temperatures T_1 and T_2, respectively. As shown in Fig. 2(a), Q is 1.6×10^5 J/mol.

Grain Boundary Engineering in Ceramics

This value essentially agrees with that previously determined by the internal friction method, for diffusion along an α-Fe particle–Cu matrix interface [6].

Equations (1) and (2) also indicate that if the same change $\Delta\alpha$ is achieved for the identical annealing time, the following relation holds:

$$T_1^{-1}d_1^{-n}\exp(-Q/RT_1)=T_2^{-1}d_2^{-n}\exp(-Q/RT_2) \tag{4}$$

where $n=3$. T_1 is the annealing temperature for the particle with $d=d_1$ and T_2 is that for the particle with $d=d_2$. The value of n was determined using Fig. 2(b) and $Q=1.6\times10^5$ J/mol, and was found to be nearly 3. Therefore, we conclude that the elastic strains relax by atomic diffusion along the interfaces between the Fe matrix and Cu particles.

The possible existence of cusps on the grain-boundary energy γ against misorientation angle θ curves for f.c.c. metals has been examined using the crystallite rotation method [7]. This method has utilized specimens in which a great number of small-size particles (crystallites) are sintered to flat single-crystal substrates of the same material at random misorientations. A grain boundary exists in the neck region of each sintered crystallite and, when γ of the boundary changes with θ, the crystallite tends to rotate during annealing in a direction which reduces the energy of the boundary. From this, one might expect that the $(009)_{9R}$ plane rotation is attributed to a decrease of the particle–matrix interfacial energy, which would become minimum when $\alpha=120°$. In this case, the rotation requires diffusion of atoms along the twin boundaries in 9R particles. However, a simple analysis shows that the kinetics in this case should not be first order. Therefore this mechanism is ruled out. The observed plane rotations are thought to be caused directly by the relaxation of elastic fields created upon the b.c.c.-to-9R martensitic transformation.

Lattice Relaxation in 9R Particles towards a Final Stable Structure

We have previously revealed that when coherent b.c.c. particles reach the critical diameter of about 12 nm during aging at 823 K, the transformation to 9R occurs and immediately the $(009)_{9R}$ basal planes in each twin segment rotate to align more closely with $\{110\}_{Fe}$ planes [4].

Particles of size above about 13 nm usually showed complicated aspects. The motion of twin boundaries was ordinary, in agreement with the previous observation by Othen et al. [3]. This was sometimes accompanied by a partial transformation to a 3R structure. An example is given in Fig. 3. The structure seen in the left region of the particle is 9R. The herring-bone fringes due to the 9R stacking sequence are visible but the motion of the twin boundary is apparent. The spacing

$d_{(009)_{9R}}$ and angle α of the lattice fringes are right for $(009)_{9R}$; $d_{(009)_{9R}} = 0.200 \pm 0.002$ nm and $\alpha = 61°$. In the right segment of the particle, on the other hand, the herringbone fringes are missing. This segment has the 3R structure. The angle α between the $(003)_{3R}$ and $(\bar{1}\bar{1}4)_{9R}$ $(//(011)_{Fe})$ planes is $61°$ and the $(003)_{3R}$ fringe spacing is 0.200 ± 0.002 nm, the same values as for 9R. Figure 3 provides direct evidence for the 3R structure for the first time. Other particles had still more complex structures intermediate between 3R and f.c.c. or, as will be seen in the following, f.c.t..

Cu particles in a size range from about 26 nm to 40 nm were composed of two alternating variants with f.c.c. and f.c.t. structures. The transformation from 9R to f.c.c. or f.c.t. via an intermediate 3R structure occurred without completely detwinning of the 9R particles. Figure 4 shows a particle (of size about 27 nm) which consists of two variants labeled A and B. The segments A and B show W-shaped diffuse fringes. TEM micrographs taken under some two-beam diffraction conditions revealed that the W-shaped fringes on segments A and B arise from two types of moiré fringe formed by interferences between the 110_{Fe} and $11\bar{1}_{Cu}$ reflections and the $10\bar{1}_{Fe}$ and $\bar{1}\bar{1}1_{Cu}$ reflections, respectively. The boundary between two abutting variants lies parallel to both the matrix $(011)_{Fe}$ and particle $(111)_{Cu}$ fringes. Thus, both the variants exhibit the Kurdjumov–Sachs orientation relationship to the Fe matrix: $(011)_{Fe} // (111)_{Cu}$ and $[1\bar{1}1]_{Fe} // [\bar{1}10]_{Cu}$. It is also clear from Figs. 3 and 4 that the $(111)_{Cu}$ plane describing the Kurdjumov–Sachs relationship is inherited from the $(\bar{1}\bar{1}4)_{9R}$ twin plane.

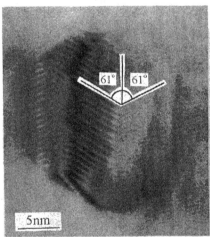 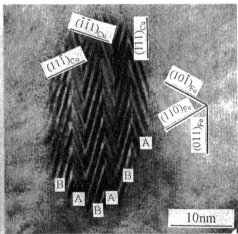

Figure 3. HREM image of a particle which has partially transformed to a 3R structure.

Figure 4. HREM image of a particle which consists of two variants with f.c.c. and f.c.t. structures.

The $(\bar{1}\bar{1}1)_{Cu}$ plane spacing $d_{(\bar{1}\bar{1}1)_{Cu}}$ can be calculated as [8]

$$d_{(\bar{1}\bar{1}1)_{Cu}}=d_{(10\bar{1})_{Fe}}D\left(d^2_{(10\bar{1})_{Fe}}+D^2-2d_{(10\bar{1})_{Fe}}D\cos\phi\right)^{-1/2} \qquad (5)$$

where $d_{(10\bar{1})_{Fe}}$ ($=0.2027\,\mathrm{nm}$) is the spacing of $(10\bar{1})_{Fe}$, D is the observed moiré fringe spacing and ϕ is the angle between $(10\bar{1})_{Fe}$ and the moiré fringes. The angle ψ between $(10\bar{1})_{Fe}$ and $(\bar{1}\bar{1}1)_{Cu}$ can then be calculated as [8]

$$\cos\psi=(d_{(\bar{1}\bar{1}1)_{Cu}}/d_{(10\bar{1})_{Fe}})\sin^2\phi+\{1-(d_{(\bar{1}\bar{1}1)_{Cu}}/d_{(10\bar{1})_{Fe}})^2\sin^2\phi\}^{1/2}\cos\phi. \qquad (6)$$

From (5) and (6), the values of $d_{(\bar{1}\bar{1}1)_{Cu}}$ and the angle α between the $(111)_{Cu}$ (// $(011)_{Fe}$) and $(\bar{1}\bar{1}1)_{Cu}$ planes were estimated. In addition, the angle α and the angle β between the $(111)_{Cu}$ and $(002)_{Cu}$ planes were obtained from measurements on $[1\bar{1}1]_{Fe}$ and $[1\bar{1}0]_{Cu}$ diffraction patterns. The spacing of $(111)_{Cu}$ fringes was also measured directly. Similar analyses and measurements were made for about 30 particles of analogous sizes, which consisted of the same two variants.

Table I summarizes these experimental results for the variant B. The spacings of $(11\bar{1})_{Cu}$ and $(111)_{Cu}$ planes and the values of the angles α and β for the variant A were in good agreement with the expected values for bulk f.c.c. Cu, shown in parentheses. The planar spacings and angles of the variant B are best fitted by an f.c.t. structure with lattice parameter values of $a=b=0.369$ nm and $c=0.366$ nm. Although the degree of tetragonality is small, the magnitudes of the lattice parameters are significantly larger than for f.c.c. Cu. Comparison with the result of Fig. 3 shows that the lattice plane rotations and plane spacing changes in 3R variants have taken place towards the f.c.c. and f.c.t. structures. These changes in angles and plane spacings require diffusion of Cu atoms, presumably along the intervariant boundaries.

Table I. Measured Spacings of $(11\bar{1})_{Cu}$ and $(111)_{Cu}$, Measured Angles α and β, and Calculated Lattice Constants for the Variant B. These Values for Bulk f.c.c. Cu are Indicated in Parentheses.

Plane Spacing (nm)		α (°)	β (°)	Lattice Constant (nm)
$(11\bar{1})_{Cu}$	$(111)_{Cu}$			
0.215 ± 0.002	0.215 ± 0.002	71.5 ± 0.5	53.5 ± 1.0	$a=b=0.369\pm0.002$
(0.209)	(0.209)	(70.5)	(54.7)	$c=0.366\pm0.002$
				(0.361)

In larger particles, alternating variant bands further widened at the expense of their neighbors, and eventually the particles became f.c.c. or f.c.t. single crystals with the lattice parameter values given in Table I when they reached a size of about 40 nm. The single-crystal particles, which were needle-like along $<110>_{Cu}$, existed in approximately equal numbers.

ACKNOWLEDGEMENT

Thanks are due to Professor S. Ikeno and Dr. K. Matsuda of Toyama University whose transmission electron microscope was used for the microscopic observation.

REFERENCES

[1] E. Hornbogen and R. C. Glenn, "A Metallographic Study of Precipitation of Copper from Alpaha Iron," Trans. Metall. Soc. AIME, **218**, [11], 1064-1070, (1960).

[2] S. R. Goodman, S. S. Brenner and J. R. Low, "An FIM-Atom Probe Study of the Precipitation of Copper from Iron–1.4 At. Pct Copper," Metall. Trans., **4**, [10], 2363-2369, (1973).

[3] P. J. Othen, M. L. Jenkins and G. D. W. Smith, "High-Resolution Electron Microscopy Studies of the Structure of Cu Precipitates in α-Fe," Phil. Mag. A, **70**, [1], 1-24, (1994).

[4] R. Monzen, M. L. Jenkins and A. P. Sutton, "The bcc-to-9R Martensitic Transformation of Cu Precipitates and the Relaxation Process of Elastic Strains in an Fe–Cu Alloy," Phil. Mag. A, **80**, (2000), in press.

[5] T. Mori, M. Okabe and T. Mura, "Diffusional Relaxation around a Second Phase Particle," Acta Metall., **28**, [3], 319-325, (1980).

[6] R. Monzen, K. Suzuki and T. Mori, "Internal Friction Caused by Diffusion around a Second Phase Particle --- Cu–Fe Alloy," Acta Metall., **31**, [4], 519-524, (1983).

[7] G. Herrmann, H. Gleiter and G. Bäro, "Investigation of Low Energy Grain Boundaries in Metals by a Sintering Technique," Acta Metall., **24**, [3], 353-359, (1976).

[8] P. B. Hirsh, A. Howie, R. B. Nicholson, D. W. Pashley and M. J. Whelan, "Diffraction and Contrast Effects from Two-Phase Materials," p. 343, Electron Microscopy of Thin Crystals, 2nd ed., Krieger Publishing, Florida, (1977).

ISOTHERMAL T-TO-M TRANSFORMATION NUCLEATED AT GRAIN BOUNDARIES IN ZIRCONIA-YTTRIA CERAMICS

Harushige TSUBAKINO, Natsuki MATSUURA* and Yoshihide KURODA*
Department of Materials Science and Engineering, Faculty of Engineering,
Himeji Institute of Technology, Himeji 671-2201, Japan
*Graduate Student, Himeji Institute of Technology

ABSTRACT

The tetragonal-to-monoclinic (T-M) phase transformation in zirconia containing 0.5-4 mol% yttria during isothermal aging at various temperatures between 75 and 300℃ was studied by means of X-ray diffraction, SEM observations and atomic force microscopy (AFM). The amount of transformation increases sigmoidally with an increase in aging time in all specimens, which indicates that the transformation proceeds with a typical nucleation and growth mode. The rate of sigmoidal curves depends on the grain size and environments, i.e., a shift to shorter aging time occurs with a larger grain size and also after aging in a water environment. The kinetics analysis indicates that these factors have an effect on the nucleation stage of transformation but not on the growth stage. AFM observations suggest that the transformation products nucleate at grain boundaries and advance into grain interiors.

INTRODUCTION

Zirconia ceramics containing a small amount of yttria have excellent mechanical properties, such as high bending strength and high fracture toughness. These excellent mechanical properties are related to the stress-induced martensitic transformation from metastable tetragonal to stable monoclinic phases (T-M transformation) [1]. It is well known that the transformation is complicated because it is affected by many factors such as grain size, water environment, yttria content, cyclic annealing, annealing at low temperature and so on [2]. Furthermore, it is not clear whether this transformation proceeds by athermal mode or isothermal mode in nature [3,4].

In this study, the T-M transformation during isothermal aging in zirconia ceramics

To the extent authorized under the laws of the United States of America, all copyright interests in this publication are the property of The American Ceramic Society. Any duplication, reproduction, or republication of this publication or any part thereof, without the express written consent of The American Ceramic Society or fee paid to the Copyright Clearance Center, is prohibited.

containing various yttria contents was studied by X-ray diffraction and atomic force microscopy (AFM).

EXPERIMENTAL PROCEDURE

High purity zirconia-yttria powders produced by a co-precipitation method were used. The chemical compositions of powders are shown in Table I. These powders were pressed unidirectionally at a pressure of about 100 MPa and sintered in an air at two conditions, at 1650 °C for 54 ks and at 1350 °C for 18 ks. The sintered body was aged isothermally at temperatures between 75 and 530 °C in three environments, open air (40-60% humidity), 20% humidity and vacuum (about 7 mPa). The amount of M-phase was determined using relative XRD intensities of the sum of two monoclinic peaks $((111)_M + (11\bar{1})_M)$ and the sum of the tetragonal, cubic and two monoclinic peaks $((111)_T + (111)_C + (111)_M + (11\bar{1})_M)$. Thermal expansion was measured at a given heating and cooling rate at temperatures ranging from room temperature to 800 °C.

The surface of the specimens aged isothermally was observed using AFM. Before the aging treatment, the specimen surface was polished mechanically using emery paper and diamond paste and then etched thermally in an air at a temperature of 1500 °C for 10 min to form thermal grooves of grain boundaries.

Table I   Chemical compositions of original powders (mass%).

Sample	Y_2O_3	Al_2O_3	SiO_2	Fe_2O_3	Na_2O	ZrO_2
0.5Y	0.93	0.005	0.005	<0.002	0.020	bal.
1Y	2.23	0.006	0.007	<0.002	0.023	bal.
1.5Y	2.73	0.008	0.005	<0.002	0.023	bal.
2Y	3.66	<0.005	0.007	<0.002	0.018	bal.
3Y	5.18	<0.005	0.006	0.004	0.019	bal.
4Y	7.03	<0.005	<0.002	<0.002	0.030	bal.

RESULTS AND DISCUSSION

Dilatation Curves
Typical dilatation curves of 1Y, 2Y and 3Y are shown in Fig. 1. The curves during the heating stage show the expansion at around 330 °C and the contraction at 650 °C and, during the cooling stage, clear expansion at 330-530 °C (Ms). These results are similar to those in 3Y ceramics containing much impurities [5]. These expansions during both stages are attributed to the T→M transformation and the contradiction is due to the reverse transformation, i.e. M→T transformation. The total expansion during both stages, even in the specimens sintered at the same temperature, becomes greater as the

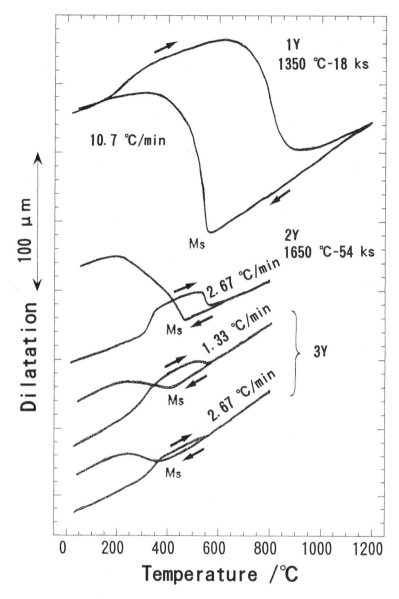

Figure 1. Typical dilatation curves in 1Y, 2Y and 3Y at various
heating-cooling rates.

heating rate decreases, as shown in Fig. 1. These results indicate that the expansion phenomena at about 330 ℃ during the heating stage are related to the isothermal transformation mode.

Furthermore, Fig. 1 indicates that the transformation rate will increase as the yttria content decreases, because the amount of expansion increases as the yttria content decreases. This result is due to the larger grain size in the smaller yttria content specimen, as shown in Fig. 2, i.e., mean grain size decreases from about 3.5 μ m in 0.5Y to 1.5 μ m in 4Y sintered at 1650 ℃ for 54 ks, and from 0.7 μ m in 0.5Y to about 0.3 μ m in 4Y sintered at 1400 ℃ for 18 ks. All specimens can be sintered to high density, but cracks are only observed in 0.5Y even though the sintering temperature is low (1400 ℃).

Isothermal Transformation Curves

Typical transformation curves during isothermal aging at various temperatures are shown in Fig. 3. In all specimens containing various contents of yttria, the transformation increases sigmoidally as aging time increases. These results indicate clearly again that the T → M transformation takes place by an isothermal mode. 1Y has transformed even in the as-sintered specimen up to about 83%. But in that specimen, the transformation increases isothermally

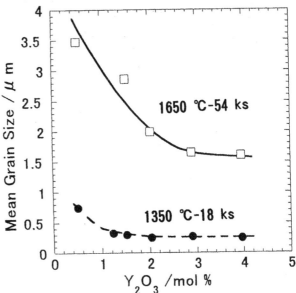

Figure 2.   Mean grain size in various zirconia-yttria ceramics.

over 83 %. It has been reported that the isothermal transformation occurs in specimens with higher yttria content beyond about 2 mol% [2,6,7]. However, this study shows that the T→M transformation takes place isothermally even in the specimens containing much lower yttria content. The isothermal transformation curves shift to a shorter aging time as the yttria content decreases. This means the isothermal transformation becomes faster as the yttria content decreases Furthermore, the saturation values of the transformed monoclinic phase after fully aging decrease as the yttria content increases,

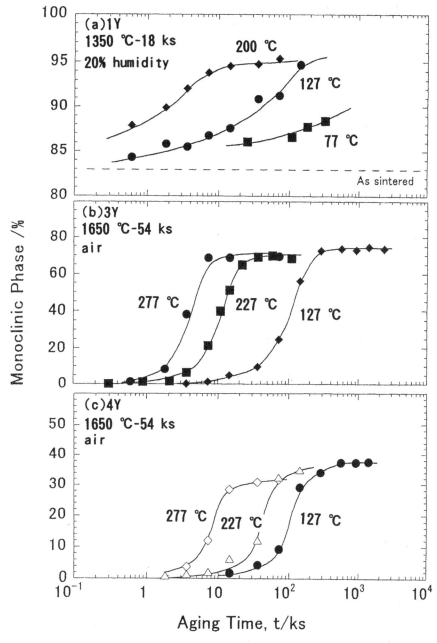

Figure 3. Relationship between monoclinic phase (%) and
aging time at various temperatures.

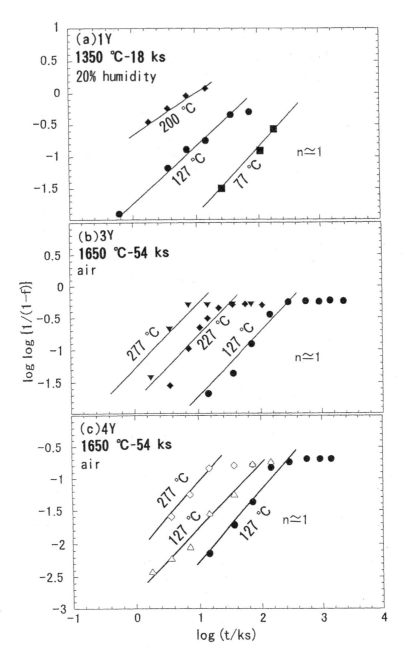

Figure 4. Relationship between loglog[1/(1-f)] and logt.

Grain Boundary Engineering in Ceramics

which is related to the smaller grain size in the specimen with less yttria content.

Many time-transformation isotherms in alloy systems have been described by the Johnson-Mehl equation [8]:

$$f = 1 - exp(-bt^n) \tag{1}$$

$$loglog[(1/(1-f)) = nlogt + log(b/2.3) \tag{2}$$

where f is the fraction transformed, t is the aging time, and b and n are constants. The plots of $loglog[1/(1-f)]$ against $log\ t$ are shown in Fig. 4. In this study, f was taken as the fraction in the saturated amount of monoclinic phase minus the amount of monoclinic phase in the as-sintered specimen. An approximately linear relationship between them was found. This fact suggests that a certain diffusional reaction has taken place in this T→M transformation. From the slope of the straight lines, about 1 is obtained as the n value, irrespective of aging temperature, humidity environment and yttria content. From these results, it can be concluded that the transformation mechanism will be the same under different humidities, i.e., the growth mechanism will be the same.

The value of n in the kinetics law of eq.(1) varies with the shape of transformation products and nucleation sites. $n = 1$ means the grain boundary surface nucleation in a heterogeneous reaction or the thickening of cylinders or needles in a homogenous reaction [8]. As the factors such as aging temperature and yttria content decrease, the transformation kinetics decrease, as shown in Fig. 4. This result indicates that the b-value in eq.(1) decreases as the values of such factors decrease. Therefore, the nucleation rate of the transformation product would decrease with the decrease in such factors.

Figure 5 shows the relationship between the amount of transformation in the as-sintered (dotted line) and fully aged (solid line) specimens and yttria content. This figure shows that the monoclinic phase increases the values of dotted line to those of solid lines with the isothermal aging treatment. The amount of monoclinic phase in the as-sintered specimens increases with a decrease in yttria content, and increases steeply at around 1.5 mol% yttria. This is due to the faster isothermal transformation rate in the specimens containing yttria less than 1.5 mol%, as shown in Fig. 3. So far, it has been believed that the T→M transformation in these zirconia ceramics containing less than 1.5mol% yttria proceeds athermally [4], i.e., diffusionless transformation mode. This will be due to the much faster rate of isothermal transformation. Furthermore, an interesting result obtained in this study was that the monoclinic phase in the 0.5Y specimen sintered at much lower temperature (1200 ℃) increases from 96% (as-sintered specimen) to 100% when the specimen was aged at 250 ℃, as shown in ◇ mark of Fig. 5.

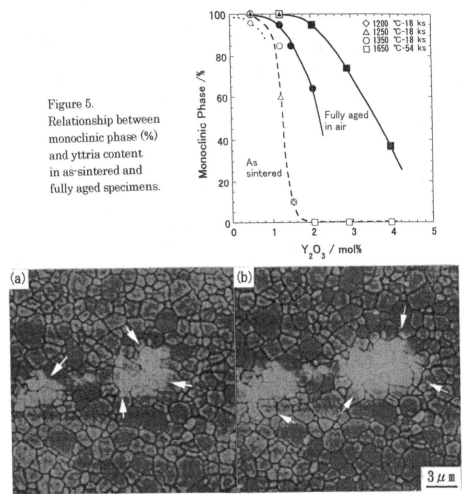

Figure 5.
Relationship between monoclinic phase (%) and yttria content in as-sintered and fully aged specimens.

Figure 6.   AFM micrographs of the surface of 3Y aged at 200°C for (a)11 ks and (b) 14 ks.   (Aged environment is open air)

AFM Observations

Typical AFM micrographs of 3Y aged in an air for various aging times are shown in Fig. 6. Thermally etched grain boundary grooves can be clearly visible. Figs. 6(a) and (b) show the same area of the specimen surface. With an increase in aging time, the bright area of the roughly round shape, which is the surface raised due to the isothermal T→M transformation, spreads out to the adjacent surrounding grains. Detailed observations of each bright area indicate that the bright area start from grain boundaries to grain interiors and then spread to adjacent grains, as shown by the arrows in Figs. 6(a) and (b).

CONCLUSIONS

(1) The T→M transformation proceeds isothermally in all zirconia-yttria systems containing from 0.5 to 4 mol% yttria.

(2) The transformation kinetics are faster as the yttria content decreases.

(3) AFM observations suggest that with isothermal aging, the transformation advances from grain boundaries to grain interiors and then propagates to adjacent grains.

ACKNOWLEGEMENTS

The authors wish to thank Prof. M. Niibe (LASTI, Himeji Institute of Technology) for his guidance in taking the AFM micrographs. This study was partly supported by a Grant-in-Aid for Science Research on Priority Research (11123234), the Ministry of Education, Science, Sports and Culture in Japan.

REFERENCES

[1] D. J. Green, R. H.. J. Hannnink and M. V. Swain, "Mechanics and Mechanisms of Toughening," 57-95, Transformation Toughening of Ceramics, CRC Press, Florida, (1989).

[2] H. Tsubakino, M. Hamamoto and R. Nozato, "Tetragonal-to-Monoclinic Phase Transformation during Thermal Cycling and Isothermal Aging in Yttria -Partially Stabilized Zirconia," J. Mater. Sci., 26 [20] 5521-5526 (1993).

[3] H. Tsubakino, Y. Kuroda and M. Niibe, "Surface Relief Associated with Isothermal Martensite in Zirconia-3 mol% Yttria Ceramics Observed by Atomic Force Microscopy," J. Am. Ceram. Soc., 82[10] 2921-23 (1999).

[4] M. Hayakawa, K. Nishio, J. Hamakita and T. Onda, "Isothermal and Athermal Martensitic Transformations in a Zirconia-Yttria Alloy," Mater. Sci. Eng., A273-275 213-217 (1999).

[5] H. Tsubakino, R. Nozato and M. Hamamoto, "Effect of Alumina Addition on the Tetragonal-to-Monoclinic Phase Transformation in Zirconia-3mol% Yttria," J. Am. Ceram. Soc., 74[2] 440-443 (1991).

[6] T. Sato and M. Shimada, "Crystalline Phase Change in Yttria-Partially Stabilized Zirconia by Low-Temperature Annealing," J. Am. Ceram. Soc., 67[10] C212-213 (1984).

[7] H. Tsubakino, T. Fujiwara, A. Yamamoto, K. Satani and S. Ioku, "Isothermal Transformation Behavior of Martensite in Zirconia-3mol% Yttria Containing Alumina," Solid→Solid Phase Transformations, TMS, 749-754 (1994).

[8] J. W. Christian, "Formal Theory of Transformation Kinetics," 471-495, The Theory of Transformation in Metals and Alloys, Pergamon Press, Oxford, (1965).

Interface Characterization of α-β Phase Transformation in Si$_3$N$_4$ by Transmission Electron Microscopy

Tomohiro Saito and Yuji Iwamoto
Japan Fine Ceramics Center, 2-4-1, Mutsuno, Atsuta-ku, Nagoya 456-8587, Japan

Yoshio Ukyo
Toyota Central R&D Labs.,Inc, Nagakute, Aichi, 480-1192, Japan

Yuichi Ikuhara
Engineerring Research Institute, The University of Tokyo,
2-11-16, Yayoi, Bunkyo-ku, Tokyo 113-8656, Japan

ABSTRACT

High-purity α-Si$_3$N$_4$ powders were heated without sintering additives at 1900°C in a nitrogen gas atmosphere for 10 h to control the α/β ratio to ~50%. The α-β interfaces in a grain were observed mainly by high-resolution transmission electron microscopy (HREM). It was found that the interface of two phases is directly connected without any amorphous layer and has the epitaxial orientation relationship of [0001]α//[0001]β, (10$\bar{1}$0)α//(10$\bar{1}$0)β. The α-β transformation mechanism was discussed in detail on the basis of the experimentally observed results.

To the extent authorized under the laws of the United States of America, all copyright interests in this publication are the property of The American Ceramic Society. Any duplication, reproduction, or republication of this publication or any part thereof, without the express written consent of The American Ceramic Society or fee paid to the Copyright Clearance Center, is prohibited.

INTRODUCTION

It is well known that Si_3N_4 ceramics show α-β phase transformation at high-temperatures during sintering. The mechanism of the phase transformation has been considered to be due to the solution precipitation mechanism.[1] According to the mechanism, α-grains initially dissolve into the grain boundary liquid phase which is formed from sintering additives, and subsequently nucleate precipitates on the β–grains. So far, many reports have been proposed to support the solution precipitation mechanism for the α-β transformation in Si_3N_4, and thus this mechanism is generally accepted. However, Šajgalík and Shimada et al. pointed out that the α-β transformation occurred even without sintering additives.[2,3] This indicates that a solid-solid transformation is possible to occur for the α-β transformation in Si_3N_4.

In this sturdy, high-purity α-Si_3N_4 powders were heated without sintering additives at 1900°C in nitrogen gas atmosphere for 10 hours so that the α/β ratio is controlled to be ~50%. Then the α-β interfaces in a grain were observed mainly by high-resolution transmission electron microscopy (HREM) to reveal the nature of the transformation. It was found that the interface of two phases is directly connected without any amorphous layer and has the epitaxial orientation relationship of $[0001]\alpha//[0001]\beta$, $(10\bar{1}0)\alpha//(10\bar{1}0)\beta$. The α-β transformation mechanism will be discussed in detail on the basis of the experimentally observed results.

EXPERIMENTAL PROCEDURE

High-purity Si_3N_4 powders (UBE E-10) with an average grain size of about 0.2 μm were used as the starting materials. The powders were heated at 1900°C in a nitrogen gas pressure of 1 MPa for 10 h, and then measured by XRD to confirm that the α/β ratio is ~50%.

Cross-sectional HREM specimens were prepared by the following procedure. First the powders were mixed by epoxy resin to obtain a epoxy block in which the powders were dispersed. The block was polished to a thickness of about 100 μm, and dimpled to a thickness of ~20 μm, and finally thinned by ion beam sputtering at a voltage of 5 kV,

Cross-sectional HREM observations were performed using high-resolution electron microscope (Topcon EM-002B) operating at 200 kV.

RESULTS AND DISCUSSION

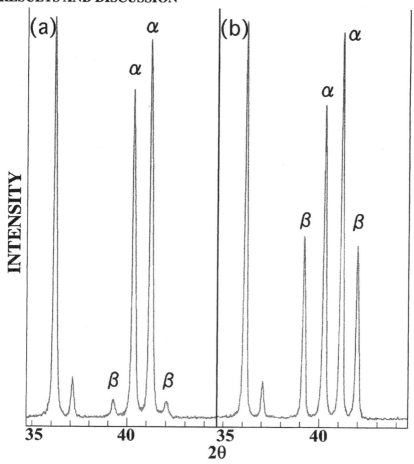

Figure 1. XRD spectrum obtained from Si_3N_4 powders heated at (a) 1750°C for 50 h and (b) 1900°C for 10 h. It is noted that the content of β-phase increases with increasing temperature.

Figure 1 shows the XRD spectrum obtained from the powders heated at (a) 1750°C for 50 h and (b) 1900°C for 10 h. It is clear that the content of β–phase increases with raising temperature, and about half of the powders were transformed to be β–phase by annealing at 1900°C for 10 h. Figure 2(a) shows a bright-field image of a grain with the size about 0.7 μm. In the grain an area with the different contrast can be seen at the edge of the grain as indicated by (A). The micro-beam diffraction patterns taken from region A and region

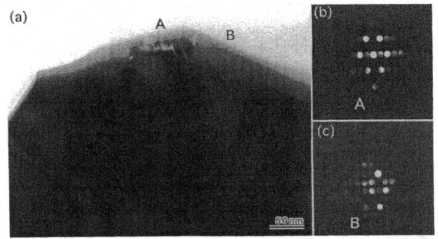

Figure 2. (a) Bright field image of a grain, in which two different contrast are observed as indicated by A and B. Micro-beam diffraction pattern from (b) area A and (c) from area B.

B are inset in (b) and (c), respectively.

According to the diffraction patterns, the area A and area B are confirmed to be α-phase and β-phase, respectively. Figure 3 is high-resolution electron micrograph of the α/β interface with the incident beam parallel to $[11\bar{2}0]\alpha//[11\bar{2}0]\beta$. It is found that the α/β interface is very clear and formed without any amorphous layer although there is some steps along the interface. The interface has the epitaxial orientation relationship of $[0001]\alpha//[0001]\beta$, $(10\bar{1}0)\alpha//(10\bar{1}0)\beta$. The interface is considered to be formed during the α-β transformation. This indicates that solid-solid transformation may occur for the α-β transformation in Si_3N_4. Figure 4 shows atomic structure of α and β phase projected from the $[11\bar{2}0]\alpha//[11\bar{2}0]\beta$ direction. Here, the lattice parameters of the α-phase are $a = 0.7813$nm and $c = 0.5591$nm, and those of the β-phase are $a = 0.7595$nm and $c = 0.2902$nm. That is, the length of c-axis in α-phase is almost twice as that in β-phase. Stacking sequence of the α-phase on the c plane are constructed of - - ABCDABCD - - and the β-phase is - - ABABAB - -. It is noted that the position of silicon atoms is almost the same between $(AB)_\alpha$ and $(AB)_\beta$ planes, whereas $(CD)_\alpha$ planes have different structures from those in $(AB)_\beta$. Thus the α-β phase transformation is considered to occur by rebuilding from $(CD)_\alpha$ to $(AB)_\beta$ structure, that is, β-phase can be obtained by rotating $(CD)_\alpha$ planes by 180° in the α planes. Figure 5 shows the detailed structure in α/β interface, in which the irregular contrasts are observed at very narrow region as indicated

Grain Boundary Engineering in Ceramics

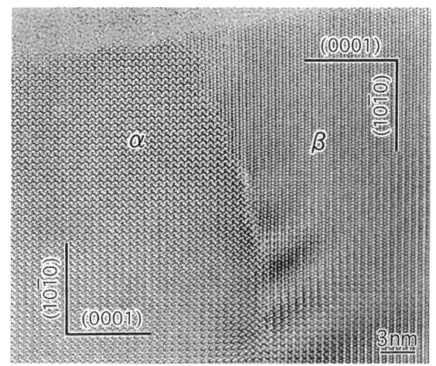

Figure 3. High-resolution electron micrograph of the α/β interface, indicating that the two phases are directly joined without liquid phase, and there is a epitaxial orientation relationship between α and β phase.

by arrows. If the $(CD)_\alpha$ planes is rebuilt to $(AB)_\beta$ planes in the local area, α-phase is transformed to β-phase in the narrow region at the α/β interface. It is also considered that the energy barrier is not so high to change the atomic configuration at the narrow region. The α/β transformation is, thus, likely to occur by the continuous reaction mentioned above at the interface.

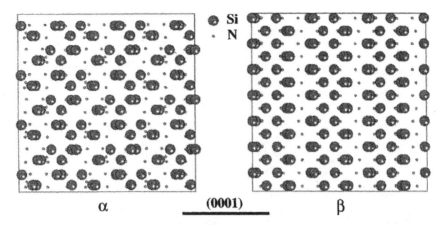

Figure 4. Atomic structure of α and β phase projected from the [11$\overline{2}$0] direction.

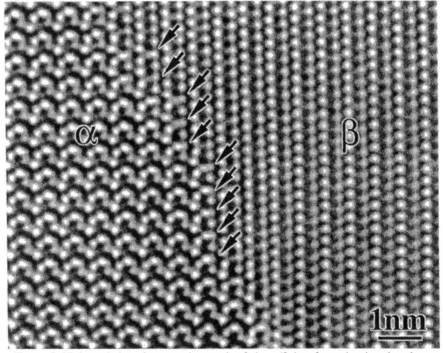

Figure 5. High-resolution electron micrograph of the α/β interface, showing that there are periodic distinct contrast as indicated by the arrows.

CONCLUSIONS

The α/β interface structures in Si_3N_4 grains were characterized by mainly HREM, and the following results were obtained.

(1) Half of high-purity α-Si_3N_4 powders were transformed to β phase at 1900°C for 10h in 1MPa N2 gas atmosphere to form α/β interface in a Si_3N_4 grain.

(2) The interface between α and β phase was directly connected without any liquid phase and has the epitaxial orientation relationship of $[0001]\alpha//[0001]\beta$, $(10\bar{1}0)\alpha//(10\bar{1}0)\beta$.

(3) These results suggest that the solid-solid transformation would occur for the mechanism of α-β transformation in Si_3N_4.

REFERENCES

[1] D. R. Messier, F. L. Riley and R. J. Brook, "The α/β silicon nitride phase transformation," *J. Mater. Sci.,* **13**, 1199 (1978).

[2] P. Šajgalík, "α/β-Phase Transformation of Si_3N_4 without Sintering Additives," *J. Eur. Ceram. Soc.,* **8**, 21-27 (1991).

[3] T. Yamada, M. Shimada and M. Koizumi, "Densification of Si_3N_4 by High Pressure Hot-Pressing," *Am. Ceram. Soc. Bull.,* **60** [12] 1281-1288 (1981).

[4] P.-O. Olsson, "Crystal defects and coherent intergrowth of α- and β -crystals in Y-Ce doped sialon materials," *J. Mater. Sci.,* **24**, 3878-3887 (1989).

[5] S. L. Hwang and I-W. Chen, "Nucleation and Growth of β'–SIALON," *J. Am. Ceram. Soc.,* **77** [7] 1719-28 (1994).

[6] Y. Goto, G. Thomas, "Phase transformation and microstructural chages of Si_3N_4 during sintering," *J. Mater. Sci.,* **30**, 2194-2200 (1995).

AEM STUDY OF INTERFACE STRUCTURES RELATED TO CUBIC-TO-TETRAGONAL PHASE TRANSITION IN ZIRCONIA CERAMICS

N.Shibata, J.Katamura, Y.Ikuhara and T.Sakuma*
Department of Materials Science, School of Engineering, The University of Tokyo,
7-3-1 Hongo, Bunkyo-ku, Tokyo, 113-8656, Japan

*Graduate school of Frontier Sciences, University of Tokyo,
7-3-1 Hongo, Bunkyo-ku, Tokyo, 113-8656, Japan

ABSTRACT

The interface structures formed by cubic-to-tetragonal phase transition were analyzed using a nano-probe analytical transmission electron microscope. In ZrO_2-Y_2O_3 system, TEM-EDS analysis revealed that the periodical lamellar structure, which was formed during heat treatments in the cubic/tetragonal two-phase field, accompanied compositional fluctuation of yttrium ions. This result supports the idea that the modulated structure is formed by the spinodal decomposition. In addition, antiphase domain boundary(APB), which was formed by diffusionless cubic-to-tetragonal phase transition, was examined by TEM-EDS and TEM-EELS analyses. It was found that APBs have a tendency to be segregated by yttrium ions during annealing in a single t-phase field in ZrO_2-Y_2O_3-TiO_2 system. The segregation of yttrium ions can be explained in terms of stabilization of APB structure, and the introduction of oxygen vacancies seems to be very effective for the stabilization.

INTRODUCTION

Mechanical properties of zirconia ceramics are closely related to their grain interior interface structures[1]. In most cases, these interfaces are the products of several solid-solid phase transformations inherent in zirconia ceramics[2,3]. In order to improve the properties in ZrO_2, it is important to understand the nature and the origin of these interface structures including phase transformation mechanisms. In this study, interface structures formed by cubic-to-tetragonal(c-t) phase transition in zirconia ceramics are examined by nano-probe analytical transmission electron microscopy (AEM), and these results are compared and discussed with theoretical approach using computer simulation based on time-dependent Ginzburg-Landau(TDGL) kinetic model.

EXPERIMENTAL PROCEDURE

High-purity ZrO_2-6mol%Y_2O_3 powders (99.9%; Tosoh Co. Ltd., Japan), ZrO_2-3mol%Y_2O_3 powders

To the extent authorized under the laws of the United States of America, all copyright interests in this publication are the property of The American Ceramic Society. Any duplication, reproduction, or republication of this publication or any part thereof, without the express written consent of The American Ceramic Society or fee paid to the Copyright Clearance Center, is prohibited.

(99.9%; Tosoh Co. Ltd., Japan) and TiO_2 powders (99.9%; Fuji Titan Industry Co. Ltd., Japan) were used for starting materials. These powders were mixed in a ball mill for 24h and pressed into green compacts with a diameter of about 10mm. They were arc-melted in an Ar atmosphere and then kept on a water-cooled copper hearth after melting. The arc-melted samples were annealed in air for several conditions. X-ray diffraction (XRD) analyses were carried out on crushed powders with a Rigaku RINT2500 operated at 40kV and 200mA with CuKα radiation. Thin foils for transmission electron microscopy (TEM) were prepared by standard procedure using ion thinning method. Conventional TEM(C-TEM) observations were performed with a HITACHI H-800, operated at 200kV. High-resolution electron microscopy (HREM) observation and microanalyses were carried out using a TOPCON EM-002B, and a TOPCON EM-002BF (field emission type) equipped with Noran Voyager energy-dispersive X-ray spectroscopy (EDS) system with a probe size of less than 1nm, respectively. Both of them were operated at 200kV. TEM-electron energy-loss spectroscopy (EELS) analyses were also carried out with a TOPCON EM-002BF with a parallel EELS detector (Gatan model 666).

RESULTS AND DISCUSSION

Modulated structure in ZrO_2-Y_2O_3 system

It has been reported that a periodical lamellae structure, originally called the modulated structure, is formed during heat treatments in the c-ZrO_2/t-ZrO_2 two-phase field in ZrO_2-Y_2O_3 system[4-7]. This microstructure generates in early stages of isothermal aging or continuous cooling from c-ZrO_2 field, and consists of bright and dark lamellae having a habit plane close to {111}[4,5,7], which is normal to the elastically soft direction of c-ZrO_2[8]. The present author's group proposed that this microstructure is formed by spinodal decomposition between t- and c-ZrO_2 on the assumption that the diffusionless c-t phase transition is a second-order type, although the initial compositional fluctuation in a very early stage occurs in t-ZrO_2[9,10]. In this study, chemical composition of the modulated structure in ZrO_2-6mol%Y_2O_3 was analyzed by TEM-EDS. The samples were annealed in air at 1200°C for 1∼ 126h to observe time dependence of the microstructure evolution.

Figure 1(a)(b) shows C-TEM and HREM images of the modulated structure in ZrO_2-6mol%Y_2O_3

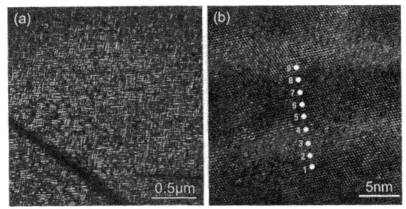

Figure 1: (a) Dark field image of ZrO_2-6mol%Y_2O_3 annealed at 1200°C for 126h. (b) HREM image of the modulated structure in (a).

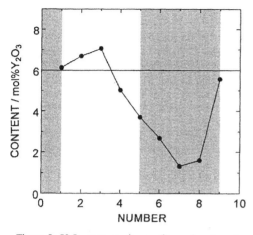

Figure 2: Y$_2$O$_3$ concentration profile analyzed on the spots indicated by white dots in Fig.1(b).

annealed at 1200°C for 126h. Fig.1(a) is a dark field image using a 112 reflection, which is characteristic reflection of t-ZrO$_2$. Lamellae structure with bright and dark contrast can clearly be seen. Fig.1(b) shows HREM image of single bright and dark lamellae. A faint bright and dark contrast seen in the figure corresponds to the single lamellae structure. The wavelength of the lamellae is estimated to be about 15nm[11]. It is found that lattice image is continuous between the bright and dark lamellae. There are no clear boundaries between the lamellae, and the interfaces look like diffuse. Figure2 is the result of TEM-EDS analyses on the spots indicated by the dots in Fig.1(b). The result shows that the content of yttrium ions gradually changes between the bright and dark lamellae. This result indicates that the origin of this lamellae structure is the compositional fluctuation of yttrium ions. Figure3 shows the comparison of concentration distribution at the center of several bright and dark lamellae with different annealing time, 1h and 126h. The horizontal axis describes Y$_2$O$_3$ content and the vertical axis describes frequency. In the sample annealed for 1h, Y$_2$O$_3$ content of the center of each lamella localize around 6mol%Y$_2$O$_3$, which is initial composition of this sample. On the other hand, in the sample annealed for 126h, Y$_2$O$_3$ content of the center of each lamella separate into two regions; higher Y$_2$O$_3$ region with about 8mol%Y$_2$O$_3$ and lower Y$_2$O$_3$ region with 2~4mol%Y$_2$O$_3$. These results clearly indicate that this type of phase separation is different from nucleation-and-growth type whose composition profile will have rectangular feature[12]. The present experimental results; diffuse interface, continuous compositional fluctuation and time dependence of compositional fluctuation, support the idea that the modulated structure is formed by spinodal decomposition.

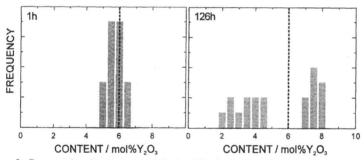

Figure 3: Concentration distribution of ZrO$_2$-6mol%Y$_2$O$_3$ annealed at 1200°C for 1h and 126h, respectivly.

It is possible to simulate the microstructure evolution in ZrO$_2$-Y$_2$O$_3$ system using a thermodynamic model based on the assumption that the diffusionless c-t phase transition is a second-order type[13-15].

We adopted TDGL kinetic model based on the diffuse interface theory for the simulation[16]. The details for the calculation have been reported previously[13-15]. Figure 4(a)(b) shows the result of the simulation on the condition that ZrO_2-6mol%Y_2O_3 is isothermally annealed at 1200°C. Fig.4(a) shows the calculated composition field against τ, which means arbitral time in this simulation. The lamellae structure due to compositional fluctuation is well simulated and the habit planes of the lamellae are likely to be <11>, which is elastically soft direction of this system in two-dimensional field. Fig.4(b) shows composition profiles of Y^{3+} ions along the diagonal line in Fig.4(a). In the profiles, the content of Y^{3+} ion continuously changes between the lamellae, and the existence of diffuse interfaces are predicted. It is also noted that the amplitude of compositional fluctuation becomes greater as annealing proceeds, although the wavelength of fluctuation does not change in this stage. These results are well consistent with the present experimental results. It is thus supported both experimentally and theoretically that the modulated structure must be formed by spinodal decomposition.

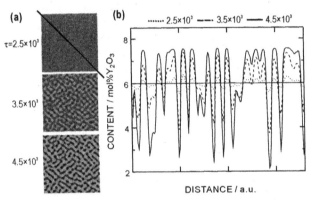

Figure 4: Results of computer simulation for ZrO_2-6mol%Y_2O_3 annealed at 1200°C. (a) shows composition field image and (b) shows composition profiles along the diagonal line in (a).

Domain structure in tetragonal zirconia

In t-ZrO_2 transformed diffusionlessly from c-ZrO_2, the domain structure, which consists of microdomains with a size of about 20-200nm separated by boundaries with curvilinear features, is observed in a dark field image taken with a reflection of *odd odd even* type, which is characteristic reflection of t-ZrO_2[17,18]. The domain boundaries are now generally accepted to be antiphase domain boundaries (APBs) formed by the reversal of oxygen ion displacement along [001] axis of t-ZrO_2. APBs in t-ZrO_2 are thus derived from the phase change in anion sublattice.

Recently, our group has predicted that, in ZrO_2-Y_2O_3 system, APBs in t-ZrO_2 have a tendency to be wetted by yttrium ions from the computer simulation of the microstructure evolution[13]. This simulation predicts that the displacement of oxygen ions becomes smaller near the APB and is zero at the boundary, i.e. APBs have c-ZrO_2 like structure. The segregation of yttrium ions is reasonably explained in terms of stabilization of c-ZrO_2 like APB structure because yttrium ion is a typical cubic-stabilizing oxide.

Grain Boundary Engineering in Ceramics

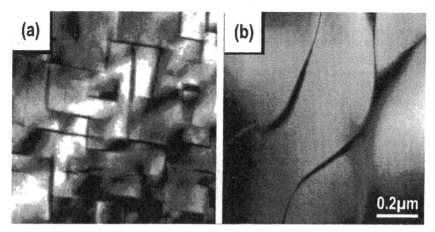

Figure 5: TEM dark field images of (a) ZrO₂-3mol%Y₂O₃-12mol%TiO₂ and (b) ZrO₂-20mol%CeO₂ annealed at 1500°C for 4h.

To examine microstructure evolution in a single t-phase field experimentally, ZrO₂-3mol% Y₂O₃-12mol%TiO₂ (3Y-12T) and ZrO₂-20mol%CeO₂ (20Ce) samples were selected. XRD analyses revealed that 3Y-12T and 20Ce samples were fully tetragonal after annealing at 1500°C. These results are consistent with the current phase diagrams[19,20]. XRD analyses ensure the microstructure evolution in a single-t-phase field in these two systems.

Figure 5(a)(b) shows dark field images of (a) 3Y-12T and (b) 20Ce annealed at 1500°C for 4h. In both cases, the incident beam directions are close to [110] and the micrographs were taken using a [112] reflection. Well facetted APBs can be seen in 3Y-12T, while rather isotropic APBs are seen in 20Ce. The habit planes of the facetted APBs in 3Y-12T are determined to be close to {111} planes by conventional trace analysis, which are normal to elastically soft direction of c-ZrO₂[21]. Figure 6 shows concentration profiles across the respective APB in (a) 3Y-12T and (b)20Ce obtained from TEM-EDS analyses. Yttrium ions segregate in the APB, while titanium ion's content is almost unchanged across the APB in 3Y-12T. This result is consistent with the idea derived from computer simulation that segregation occurs to stabilize c-ZrO₂ like APB structure, because Y₂O₃ has cubic-stabilizing effect while TiO₂ has no such effect[22].

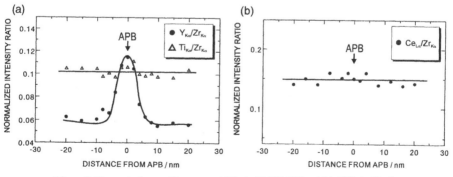

Figure 6: Concentration profiles across APBs in (a) 3Y-12T and (b) 20Ce in Fig.5.

It is reported that CeO_2 has cubic-stabilizing effect although its effect is weaker than that of trivalent oxides such as Y_2O_3[23]. However, as shown in Fig. 6(b), cerium ion's content seems to be constant across the APB and noticeable segregation of cerium ions could not be observed. This result may suggest that we need to clarify the atomic or electronic structure at the interface to understand the nature of stabilizing mechanism of APB structure in t-ZrO_2. TEM-EELS analysis was, then, carried out on the APB in 20Ce. Figure 7(a) is one of the results obtained from TEM-EELS analysis exactly at the APB and domain interior 10nm off from it. The solid line shows the spectrum from APB and dotted one from domain interior, respectively. These spectra are the energy-loss near-edge structure (ELNES) of Ce M_4, $_5$-edges which consist of two strong peaks, so-called white lines. These white lines are dominated by 3d →4f excitonic transitions, and energy difference between the M_4 and M_5 transitions is a result of spin-orbit coupling in the d-shell[24]. The white line ratio, M_5/M_4, is much influenced by 4f occupancy, and therefore it is thought to be possible to distinguish valence state of cerium ions by comparing white line ratios[25,26]. Figure 7(b) shows comparison of white line ratio between on and off APBs. The white line ratios were estimated from second-derivative of the spectra, which is suitable for estimation of white line ratio[27]. It is noted that there is a tendency that white line ratio at APB is higher than that of domain interior. It is reported that the higher the white line ratio is, the lower the valency of cerium ion will be[28]. The present result indicates that valence state of cerium ions may be lower at APBs than domain interior. It is generally accepted that the addition of lower valent rare-earth oxides than Zr^{4+} is effective for stabilizing high temperature cubic phase. In addition, Ce^{3+} must have much larger cubic-stabilizing effect than Ce^{4+} inferring from current phase diagram of ZrO_2-$CeO_{1.5}$ system[29]. APBs in ZrO_2-CeO_2 system may be stabilized by lower valent cerium ions.

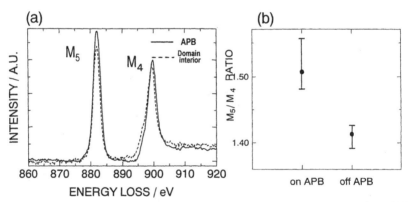

Figure 7: (a) An example of Ce M_{45}-edge ELNES obtained from APB and domain interior in 20Ce. (b) Comparison of white line ratio between APB and domain interior.

Oxygen vacancies seem to be concentrated around APBs to compensate local charge neutrality in the both cases of ZrO_2-3mol%Y_2O_3- 12mol%TiO_2 and ZrO_2-20mol%CeO_2, because lower valent cations are segregated around them. APBs are the boundary where oxygen ions are forced to become closer each other than domain interior. This closeness will cause electric repulsive force between oxygen ions, which may destabilize APB structure. Introduction of oxygen vacancies seems very effective to reduce repulsive force between closer oxygen ions. Oxygen vacancies may play a critical role to maintain APB structure in tetragonal zirconia as well as cubic-stabilizing cations.

Grain Boundary Engineering in Ceramics

CONCLUSIONS

The modulated structure and the domain structure in zirconia ceramics are examined using an analytical electron microscope with a probe size less than 1nm. It was revealed that the modulated structure consists of diffuse interfaces and is derived from compositional fluctuation of Y_2O_3. The modulated structure must be formed by spinodal decomposition. It was also revealed that APBs in t-ZrO_2 have a tendency to be wetted by yttrium ions in ZrO_2-Y_2O_3-TiO_2 system. APBs are expected to have cubic-like structure by computer simulation, and to be stabilized by the segregation of cubic -stabilizing dopants. On the other hand, in ZrO_2-CeO_2 system, APBs seem to be stabilized by change in valency of cerium ions. These results may suggest that introduction of oxygen vacancies should be very important to maintain APB structure in tetragonal zirconia.

ACKNOLEDGEMENT

The authors wish to thank Mr. Takashi Yanaka (TOPCON Co. Ltd.) and Dr. Takahisa Yamamoto (The University of Tokyo) for their experimental assistance. This work was financially supported by the Grand-in-Aid for Scientific Research on the Priority Area (A) 09242206, 10136206, 11123205, from the Ministry of Education, Science and Culture, Japan.

REFERENCES

[1] A.H.Heuer, R.Chaim and V.Lanteri,"Reviw: Phase transformation and microstructure and microstructural characterization of alloys in the system in the system Y_2O_3-ZrO_2." In S.Somiya, N.Yamamoto and H.Yanagida, editors, Advance in Ceramics, Vol.24A, *Science and Technology of Zirconia III*, American Ceramic Society, OH,(1986).

[2]T.Sakuma and H.Hata, "Diffusionless cubic-to-tetragonal transformation and microstructures in ZrO_2-Y_2O_3,"*J. Jpn. Inst. Metals*,**53**[9](1989)972-979.

[3]T.Sakuma,"Microstructural aspects on the cubic-tetragonal transformation in zirconia,"*Key Eng. Mater.*,Vols.**153-154**(1998)75-96.

[4]T. Sakuma, Y. Yoshizawa and H. Suto, "The Modulated Structure Formed by Isothermal Aging in ZrO_2-5.2mol%Y_2O_3 Alloy," *J. Mater. Sci.*, **20**[3], (1985), 1085-92..

[5]T. Sakuma, Y. Yoshizawa and H. Suto, "The Metastable Two-Phase Region in the Zirconia-Rich Part of the ZrO_2-Y_2O_3 System," *J. Mater. Sci.*, **21**[4], (1986), 1436-40.

[6]M. Hayakawa, K. Adachi and M. Oka, "Tweed Contrast with (223) Habit in Arc-melted Zirconia-Yttria Alloys," *Acta metall. mater.*, 38[9], (1990), 1761-7.

[7]M. Doi and T. Miyazaki, "On the Spinodal Decomposition in Zirconia-Yttria (ZrO_2-Y_2O_3) Alloys," *Phil. Mag. B*, **68**[3], (1993), 305-15.

[8] H.M. Kandil, J.D. Greiner and J.F. Smith, "Single-crystal elastic constants of yttria-stabilized zirconia in the range $20°$ to $700°C$," *J. Am. Ceram. Soc.*, **67**[5], (1984), 341-6.

[9] M. Hillert and T. Sakuma, "Thermodynamic modeling of the c$\rightarrow$ t transformation in ZrO_2 alloys," *Acta metall mater.*, 39, (1991)111-115.

[10] J. Katamura and T. Sakuma, "Thermodynamic Analysis of the Cubic-Tetragonal Phase Equilibria in the System ZrO_2-$YO_{1.5}$," *J. Am. Ceram. Soc.*, **80**[10], (1997), 2685-88.

[11] J.Katamura, N.Shibata, Y.Ikuhara and T.Sakuma, "Transmission electron microscopy-energy-dispersive X-ray spectroscopy analysis of the modulated structure in ZrO_2-6mol%Y_2O_3 alloy,"*Philos. Mag. Lett.*,**78**[1](1998)45-49.

[12] J.W.Cahn,"Spinodal decomposition, " *Trans.Metal.Soc.AIME*,**242**[2] (1968) 166- 180.

[13] J.Katamura and T. Sakuma, "Computer simulation of the microstructure evolution during the diffusionless cubic-tetragonal transition in the system ZrO_2-Y_2O_3," *Acta Mater.*, **46** (1998) 1569-1575.

[14] Junji Katamura, Yuichi Ikuhara and Taketo Sakuma, "Theoretical investigation of the cubic-to-tetragonal transition in ZrO_2-based alloys," Proceedings of Third Pacific Rim International Conference on Advanced Materials and Processing, The Minerals, Metals and Materials Society, Warrendale, Pa.(1998),1399-1404.

[15] N.Shibata, J.Katamura, Y.Ikuhara and T.Sakuma, "Microanalysis of modulated structure in zirconia ceramics", The third pacific international conference on advanced materials and processing,The Minerals, Metals and Materials Society, Warrendale, Pa.(1998),1405-1410.

[16] J.W. Cahn and J.E. Hilliard, "Free energy of a nonuniform system. I . Interfacial free energy," *J. Chem. Phys.*, **28**, (1958), 258-67.

[17]T.Sakuma, "Development of domain strucuture associated with the diffusionless cubic-to-tetragonal transition in ZrO_2-Y_2O_3 alloys," *J.Mater.Sci.*,**22**(1987)4470-4475.

[18]A.H.Heuer, R.Chaim and V.Lanteri, "The displacive cubic $\rightarrow$ tetragonal transformation in ZrO_2 alloys,"*Acta Metall.*,**35**[3](1987)661-666.

[19]E.Zshech, P.N.Kountouros, G.Petzow, P.Behrens, A.Lessmann and R.Frahm, "Synchrotron radiation Ti-*K* XANES study of TiO_2-Y_2O_3-stabilized tetragonal zirconia polycrystals,"*J.Am.Ceram.Soc.*, **76**[1](1983)197-201.

[20] E.Tani, M.Yoshimura and S.Somiya,"Revised phase diagram of the system ZrO_2-CeO_2 below 1400ºC,"*J.Am.Ceram.Soc.*,**66**[7](1983)506-510.

[21] H.Ogawa, A.Yasuda, N.Shibata, Y.Ikuhara and T.Sakuma, "Segregation of Yttrium Ions to Domain Boundaries of Tetragonal Zirconia,"*Phil. Mag. Lett.*, **77**[4](1998) 199-203.

[22] V.C.Pandofelli, J.A.Rodriguew and R.Stevens,"Effect of TiO_2 addition on the sintering of ZrO_2·TiO_2 composites and on the retention of the tetragonal phase of zirconia room temperture," *J.Mater.Sci.*.**26**[19] (1991)5327-5334.

[23]S.Trong, K.Miyazawa and T.Sakuma, "The diffusionless cubic-to-tetragonal phase transition in near-stoichiometric ZrO_2-CeO_2,"*Cramic International*,**22**(1996)309-315.

[24] G.Kaindl, G.Kalkowski, W.D.Brewer, B.Perscheid and F.Holtzberg, "*M*-edge x-ray absorption spectroscopy of 4*f* instabilities in rare-earth systems (invited)," *J.Appl.Phys.*,**55**[6](1984)1910-1915.

[25] B.T.Thole and G.van der Laan,"Branching ratio in x-ray absorption spectroscopy," *Physical Review B*, **38**[5](1988)3158-3171.

[26] R.F.Egerton. Electron energy-loss spectroscopy in the electron microscope. Plenum press, New York and London, second edition(1996).

[27] C.C.Appel, G.A.Botton, A.Horsewell and W.M.Stobbs,"Chemical and structural changes in manganese-doped yttria-stabilized zirconia studied by electron energy loss spectroscopy combined with electron diffraction," *J.Am.Ceram.Soc.*,**82**[2](1999)429-435.

[28] J.A.Fortner, E.C.Buck, A.J.G.Ellison and J.K.Bates,"EELS analysis of redox in glasses for plutonium immobilization,"*Ultramicroscopy*,**67** (1997)77-81.

[29] H.Y.Zhu,"$CeO_{1.5}$-stabilized tetragonal ZrO_2," *J.Mater.Sci.*, **29**(1994) 4351-4356.

Electronic Ceramics

POSSIBLE CENTER FOR POLAR CLUSTER IN LEAD MAGNESIUM NIOBATE Pb(Mg$_{1/3}$Nb$_{2/3}$)O$_3$

H. Z. Jin[1,2], Jing Zhu[1,2], Shu Miao[1,2], X. W. Zhang[2] and Z. Y. Cheng[1]

1. Electron Microscopy Laboratory, School of Materials Science and Engineering, Tsinghua University, Beijing 100084, P. R. China
2. Department of Materials Science and Engineering, Tsinghua University, Beijing 100084, P. R. China

ABSTRACT

Analytical electron microscopy and image processing techniques have been used for studying the compositional distribution and the correlation between the ordered domain and the polar cluster in lead magnesium niobate Pb(Mg$_{1/3}$Nb$_{2/3}$)O$_3$(PMN). The results show that there is a strong composition fluctuation of Mg/Nb ratio in either the ordered or disordered regions, and the dispersive of fluctuation distribution of Mg/Nb is stronger in the boundary area than in the ordered and the disordered regions. Antiphase domain boundaries and dislocations are found. The fringe image of Nb and (Nb+Mg) {111} planes are distorted randomly, which may be caused by the composition fluctuation of Mg/Nb ratio and local atomic layer displacement. The boundary and dislocation area, with strong composition fluctuation and lattice strain, may play an important role in forming a center of polar cluster.

INTRODUCTION

Lead magnesium niobate (PMN, Pb(Mg$_{1/3}$Nb$_{2/3}$)O$_3$), is a relaxor type ferroelectric with diffuse phase transition (DPT) and a strong frequency dispersion behavior. At room temperature, PMN has a cubic perovskite structure with space group Pm3m and a$_0$=0.404 nm.[1] In the Curie region the PMN is treated as "super-paraelectrics" consisting of a mass of polar clusters.[2,3,4] Meanwhile a short range ordering was identified[5] and characterized by the presence of *111/2* superlattice

To the extent authorized under the laws of the United States of America, all copyright interests in this publication are the property of The American Ceramic Society. Any duplication, reproduction, or republication of this publication or any part thereof, without the express written consent of The American Ceramic Society or fee paid to the Copyright Clearance Center, is prohibited.

reflections.[6] The ordered regions are about 2-5 nm in size.[6] A space charge model was suggested by Chen et al[7] that the ordered regions in PMN had a local Mg/Nb of 1:1, deviated from the average ratio of 1:2, and the doubling of the unit cell is due to a non-stoichiometric 1:1 ordering of Mg and Nb atoms between two B-site cation sublattice. By using high-resolution electron microscopy (HREM) and image simulation and Madelung electro-static energy calculations, a set of structural models had been suggested by Bursill et al[8] for possible ordered and disordered distribution of Nb and Mg over the B sites of the PMN, including a $6a_0 \times 6a_0 \times 6a_0$ cubic unit cell with a Mg/Nb=1:1 ordered inner region and a Nb-rich disordered outer shell. They modeled the chemical domain textures using the next-nearest-neighbor Ising model. The atomic-resolution Z-contrast imaging technique[9] was used to determine the two different cation sites occupancies (B' and B") of Nb and Mg. The results showed that the ordered domain structure in $Pb(Mg_{1/3}Nb_{2/3})O_3$ is in agreement with the charge-balanced random-layer model.[10] By means of microdiffraction and microanalysis methods with 1-nm probe size, Miao et al found that the composition inhomogeneity is severe in both the ordered and disordered regions in PMN.[11] It is reported by Viehland et al[12] that the strong fluctuations in the B-site cation ratio between ordered and disordered regions didn't exist; the possibility of weak fluctuations is consistent with their observed lattice images in the $(Pb_{0.4}Ba_{0.6})(Mg_{1/3}Nb_{2/3})O_3$.

By observation of the superlattice diffraction peaks in x-ray spectrum from PMN single crystal Zhang et al suggested that the doubling of the unit cell in the superlattice regions involved significant ionic displacements in addition to a chemical ordering.[13] N. de Mathan et al's study on PMN at 5K by x-ray and neutron diffraction supposed a model that the local symmetry is rhombohedral with atomic shifts along the <111> direction of the cubic, i.e. the rhombohedral axis.[14] Recently, the effect of the substitution rate on the distribution of the physical sizes of heterogeneity, thus on the relaxor properties, in PMN ceramics had been studied by Bidault et al.[15] Note that the following two topics are not well understood up to now.

(1) What type of chemical ordering exists in the ordered domain of PMN?

(2) What is the correlation between the ordered regions and the polar clusters?

In present work the composition fluctuation and lattice strain have been studied at the nanometer scale in ordered, disordered domain and interdomain areas between ordered and disordered regions of PMN.[16] The chemical ordering in the ordered domain and the correlation between the ordered domains and polar clusters are discussed.

EXPERIMENTAL PROCEDURE

The samples of PMN used in this study were achieved by first synthesizing $MgNb_2O_6$ using a columbite precursor as described by Swartz et al.[17] Then high-purity oxide powders Pb_3O_4 were properly weighed and mixed with $MgNb_2O_6$ by ball milling. After drying and sieving, the powder was calcined in an alumina crucible at 800°C for 2 h. After remilling and drying, the powders were then cold-pressed into disks of 1.2 mm thickness and 10 mm in diameter. The disks were sintered at 1280°C in a closed alumina crucible and $PbZrO_3$ pellets were used to prevent the excessive loss of PbO. Then the surfaces were removed to expose the interior section for further experiments.

The phase content of the samples were examined by X-ray diffractometry using Cu Kα radiation generated at 40 kV and 100 mA (Rigaku, Dmax-RB). Samples for transmission electron microscopy study were first mechanically polished to obtain a thickness of ~30 μm and then be placed on a copper support. The pellets were then thinned by ion milling (Gatan 600) till perforation in the center, using argon ions at an acceleration voltage of 4 kV and an incident angle of 12°. Suitable specimens for TEM examination were finally obtained via ion milling at small angles of 10° for about 30 minutes. Microscopic examination was conducted using a TEM with a field emission gun operated at 200 kV (JEM-2010F, JEOL). Compositional analysis in the nanoscale region was performed using X-ray energy dispersive spectroscopy (EDS) (Link ISIS 300) attached to the TEM. Taking the advantages of the Gatan Image Filter system (GIF200) attached to the TEM, digital images were acquired directly into the computer through a slow scan CCD (charge-coupled device) camera, and image filtering and processing were conducted in this study using the software of DigitalMicrograph (Gatan).

RESULTS

Compositional analysis

Computer simulation results indicated that the images observed in HREM were strongly affected by experimental conditions, such as the thickness of samples and objective lens defocus value.[18,19] Furthermore, the overlapping between matrix and ordered domains along the electron beam direction will also affect the results of compositional analysis. Because of the size of the ordered domains of PMN is only 2~5 nm, the border area of the sample which is thin enough was used to collect EDS data. In order to identify the structure of the local position being investigated, the nano-beam diffraction (NBD) technique was employed in this study.

First a thin section of the sample is selected for analysis. Then the sample is tilted until the electron beam is incident along the [110] axis. After that the TEM is switched from the normal mode to the NBD mode with a beam spot size of 1.6 nm. Electron diffraction was then carried out and a NBD pattern corresponding to the selected area was obtained. Due to the small beam spot compared with the size of the ordered domains, it is possible that only ordered domains or disordered matrix would be selected exclusively. If the superlattice diffraction spots appear in the NBD pattern and there is no obvious intensity difference between all of the diffraction spots, the selected area is considered to be an ordered domain. On the contrary, the selected area is considered as the matrix when no superlattice diffraction spots exist. Figure.1(a) and (b) are NBD patterns from ordered domains and disordered matrix, respectively.

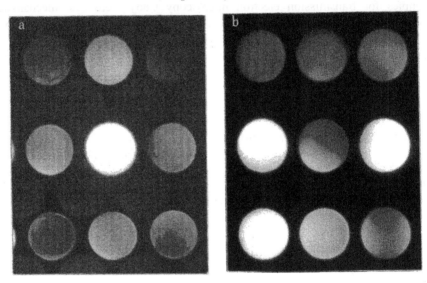

Figure.1 Nano-beam diffraction patterns from ordered domain (a) and disordered matrix (b).

Once the structure of the selected area was identified by means mentioned above, the corresponding EDS spectrum was collected. More than 20 spectra were carefully acquired from both the ordered domains and the disordered matrix, respectively. The acquisition time for each spectrum was about 100 s in order to collect enough counts under the low illumination since the TEM is operated in the NBD mode where a very small condenser aperture was used.

Grain Boundary Engineering in Ceramics

Quantitative data was obtained and the compositional ratio of Mg/Nb was calculated for each spectrum. Finally, a histogram of the ratio distribution was produced, followed by a normal curve fit. The normal fit curve is characterized by two parameters, i.e. the mean value and the standard deviation. They are 0.55 and 0.11 for the ordered and 0.55 and 0.10 for the disordered regions, respectively. No distinct differences can be revealed between these two cases, indicating that no Mg:Nb=1:1 ordering exists, which is inconsistent with the suggestions of the space charge model.[7]

It should be noted that there are no sharp interfaces between the ordered and disordered regions. Transition regions exist between them. It is difficult to determine the composition at the boundary between the ordered and disordered area. Though the function of line-scan is supplied by the EDS, it can not fulfill the needs in this case because the region is too small. The composition information in this region was acquired by recording a series of EDS spectra while moving the beam spot step by step manually from the ordered region to the matrix. The specimen was kept static after the suitable region had been selected, and the electron beam was moved instead to limit the specimen drift. The suitable area was decided by checking the NBD patterns so that the beam spot was located in the ordered domain for the first point and moving towards and into the disordered matrix at the sixth point. Six EDS spectra were acquired as a series for each of thus area. The spot size is 1.6 nm and the step size is 0.7 nm.

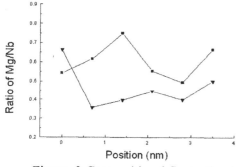

Figure. 2 Compositional fluctuation in regions from the ordered domain towards the matrix. Two lines are given as examples.

Quantitative analysis was performed on each spectrum and the variation of the Mg/Nb ratio versus the position was calculated. Altogether ten groups of spectra were carefully acquired and two of them are shown in Figure.2. Fluctuations of the ratio in such a nanoscale region can obviously be revealed, indicating that nanoscale Mg or Nb rich areas may exist.

There are altogether sixty spectra contained in the ten series. A histogram of the Mg/Nb ratio was calculated based on these 60 data followed by a normal curve fit. The mean and standard deviation for this curve is 0.57 and 0.13, respectively. This third curve is similar to the other two that have been discussed above and no major differences can be found between them. This third curve, conformable with the other two, is also considered as proof to the validity of the data. The standard deviation value of this third line is a bit larger than the other two, indicating that the distribution of the Mg/Nb ratio is more dispersive in the interdomain or boundary area than in the ordered domain and the matrix region.

Lattice strain and displacement

Figure.3a is a Fourier-filtered HREM image of PMN, with the electron beam incident along the [110] axis, recorded through the slow scan CCD camera. The fast Fourier transform (FFT) of (a) is shown in the inset. As we know, the Fourier transform yields a diffraction pattern similar to the electron diffraction pattern so similar indices can be assigned to the spots of the FFT pattern (for a brief reference of the image processing techniques used in this paper, see [20]). In this case the FFT image is the diffraction pattern of [110] zone axis. The *111/2* diffraction spots can be revealed, which resulted from the nanoscale ordered domain where B-site ordering exists.

Figure.3b is also a filtered image where only superlattice *111/2* reflections were included in the Fourier space as shown in the inset. The bright areas of this image refer to the ordered domain and the dark ones refer to the disordered matrix, since the *111/2* reflections were caused by the B-site ordering. This method had been successfully utilized to distinguish the ordered domain from the disordered matrix and the mean size of the ordered domain was acquired.[18] From Figure.3b severe distortions and strains are found.

Figure.3c is a one-dimensional image achieved by including only one pair of the *111/2* reflections in the Fourier space. The intensity of the *111/2* reflection originated solely from the scattering of the ordered Nb and Mg atoms; thus, the fringe contrast in Figure.3c should represent the Nb and (Nb+Mg) {111} atom planes. The curvature of the fringes in Figure.3c characterizes the displacement of local Nb and Mg atoms layer. Figure.3d is given for a comparison where the *111* matrix reflections are selected instead of the corresponding superlattice reflections. Antiphase domain boundaries and dislocations can be found in Figure.3c, while in Figure.3d the fringes are much more straight and countinuous even when the fringes pass through the boundary area.

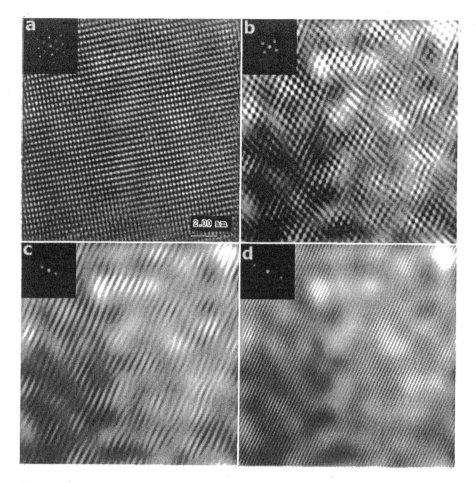

Figure.3 (a) A Fourier-filtered HREM image of PMN with the electron beam incident along the [110] direction; (b) A Fourier-filtered image where only superlattice *111/2* reflections are included; (c) A one-dimensional image, where one pair of superlattice *111/2* spots were included, shows a severe displacement of Nb and Mg atom layers; (d) A one-dimensional image where one pair of *111* spots was included;

DISCUSSION AND CONCLUSION

1. The results of compositional analysis show a strong composition fluctuation in both the ordered and disordered regions of PMN. The mean value of the

Mg/Nb ratio and the standard deviation are 0.55 and 0.11 for the ordered region, 0.55 and 0.10 for the disordered region, and 0.57 and 0.13 for the whole region. The large fluctuation of the Mg/Nb ratio that is often observed between two neighboring local area does not support the Chen et al's space charge model,[7] which requires the 1:1 (Mg:Nb) stoichiometric ordered domain with a Nb-rich outer shell.[8]

2. The fringes of the {111} planes obtained by only including the matrix reflections are basically continuous. However, the lattice image of Nb and (Nb+Mg) obtained by superlattice *111/2* reflections are curved and distorted. This implies that there is a significant local atom layer displacement and lattice strain of the Nb or (Nb+Mg) on the {111} atomic plane. In addition, antiphase domain boundaries and local dislocations exist. All of these boundaries are the areas with strong composition fluctuation and severe strain.

3. In an area with strong composition fluctuation and lattice strain, the local symmetry inevitably breaks down, then the polar cluster has to be formed. The random interdomain – boundary area may play an important role in forming a center of polar cluster.

ACKNOWLEDGEMENTS

This work was supported by the National Natural Science Foundation of China and made use of the resources of Electron Microscopy Lab of Tsinghua University and the State Key Lab of New Ceramics and Fine Processing.

REFERENCES

[1] V.A. Pokov and I.E. Mglinkova, "Electrical and Optical Properties of Single Crystals of Ferroelectrics with Diffused Phase Transition," *Sov. Phys. –Solid State (Engl. Transl.)* **3** [3] 613-623 (1961).

[2] G. A. Smolenskii, "Physical Phenomena in Ferroelectrics with Diffused Phase Transition (Review)," *J. Phys. Soc. Jpn.*, **28**, 26-37 (1970).

[3] L.E. Cross, "Relaxor Ferroelectrics: An Overview," *Ferroelectrics,* **151**,305-320 (1994).

[4] S.P. Li, J.A.Eastman, R.E. Newnham and L.E. Cross, "Diffuse Phase Transition in Ferroelectrics with Mesoscopic Heterogeneity: Mean- Field Theory," *Physical Rev. B,* **55** [18] 12067-12078 (1997-II).

[5] H. B. Krause, J. M. Cowley, and J. Wheatley, "Short-Range Ordering in $Pb(Mg_{1/3}Nb_{2/3})O_3$," *Acta Cryst.*, **A 35** 1015-17 (1979).

[6] C. Randall and A. Bhalla, "Nanostructure-Property Relations in Complex Lead Provskites," *Jpn. J. Appl. Phys.*, **29** 327 (1990).

[7] J. Chen, H. M. Chan, and M. P. Harmer, "Ordering Structure and Dielectric Properties of Undoped and La/Na-Doped $Pb(Mg_{1/3}Nb_{2/3})O_3$," *J. Am. Ceram. Soc.*, **72** [4] 593-98 (1989).

[8] L.A. Bursill, Hua Qian, Julin Peng, X.D. Fan, "Observation and Analysis of Nanodomain

Textures in the Dielectric Relaxor Lead Magnesium Niobate," *Physica B* **216** 1-23 (1995).

[9] Y. Yan, S. J. Pennycook, Z. Xu, and D. Viehland, "Determination of the Ordered Structure of $Pb(Mg_{1/3}Nb_{2/3})O_3$ and $Ba(Mg_{1/3}Nb_{2/3})O_3$ by Atomic-Resolution Z-Contrast Imaging," *Appl. Phys. Lett.*, **72** [24] 3145-47 (1998).

[10] M. A. Akbas and P. Davies, "Domain Growth in $Pb(Mg_{1/3}Ta_{2/3})O_3$ Perovskite Reslaxor Ferroelectric Oxides," *J. Am. Ceram. Soc.*, **80** [11] 2933-36 (1997).

[11] Shu Miao, X.W. Zhang and Jing Zhu, "Structure and Compostion Determination of Ordered Domain in $Pb(Mg_{1/3}Nb_{2/3})O_3$ and $Pb_{1-x}La_x(Mg_{1/3}Nb_{2/3})O_3$ by Analytical Electron Microscopy Techniques," *J. Am. Ceram. Soc.*, accepted.

[12] D. Viehland, N. Kim, Z. Xu, and D. A. Payne, "Structural Studies of Ordering in the $(Pb_{1-x}Ba_x)(Mg_{1/3}Nb_{2/3})O_3$ Crystalline Solution Series," *J. Am. Ceram. Soc.*, **78** [9] 2481-89 (1995).

[13] Q.M. Zhang, H. You, M. L. Mulvihill and S.J. Jang, "An X-ray Diffration Study of Superlattice Ordering in Lead Magnesium Niobate," *Solid State Communications,* **97** [8] 693-698 (1996).

[14] N de Mathan, E. Husson, G. Calvarin, J.R. Gavarri, A.W. Hewat and A. Morell, "A Structural Model for the Relaxor $Pb(Mg_{1/3}Nb_{2/3})O_3$ at 5 K," *J. Phys.: Condens. Matter,* **3** 8159-71 (1991).

[15] O.Bidault, E. Husson and P. Gaucher, "Substitution Effects on the Relaxor Properties of Lead Magnesium Niobate Ceramics," *Phil. Mag. B,* **79** [3] 435-48 (1999).

[16] H.Z. Jin, Jing Zhu, Shu Miao, X.W. Zhang, Z.Y. Cheng, "Ordered Domain and Polar Cluster in Lead Magnesium Niobate $Pb(Mg_{1/3}Nb_{2/3})O_3$," *J. Appl. Phys.* Accepted.

[17] S. L. Swartz and T. R. Shrout, "Fabrication of Perovskite Lead Magnesium Niobate," *Mater. Res. Bull.,* **17** 1245-50 (1982).

[18] X. Q. Pan, W. D. Kaplan, M. Rühle, R. E. Newnham, "Quantitative Comparison of Transmission Electron Microscopy Techniques for the Study of Localized Ordering on a Nanoscale," *J. Am. Ceram. Soc.*, **81** [3] 597-605 (1998).

[19] K. Park, L. Salamanca-Riba, M. Wutting, and D. Viehland, "Ordering in Lead Magnesium Niobate Solid Solutions," *J. Mater. Sci.,* **29** 1284-89 (1994).

[20] John C. Russ, *"The Image Processing Handbook",* Second Edition, CRC Press, Charper 5, 283-346 (1995).

SEGREGATION OF BaZrO$_3$ IN MELT TEXTURED YBa$_2$Cu$_3$O$_{7-x}$

Fatih Dogan and Jeffrey D. Reding
University of Washington, Department of Materials Science and Engineering,
Roberts Hall, Box 352120
Seattle, WA 98195, U.S.A.

Masanobu Awano
National Industrial Research Institute of Nagoya
Shidami Human Science Park, 2268-1, Simo-Shidami, Moriyama-Ku
Nagoya 463-8687, JAPAN

ABSTRACT

Microstructural development of melt textured YBa$_2$Cu$_3$O$_{7-x}$ and BaZrO$_3$ composites was investigated. Melt texturing by continuous undercooling resulted in macrosegragation of nanosized BaZrO$_3$ particles ($\cong$ 70 nm) at the solid-liquid interface during solidification. Processing at higher solididification rates through large undercooling of a semisolid sample was proposed to trap nanosized particles within the textured YBa$_2$Cu$_3$O$_{7-x}$ phase.

INTRODUCTION

Achieving a high critical current, J$_c$, in high temperature superconducting materials requires precise control of the microstructural during material processing. It is necessary to eliminate any grain boundaries that act as weak links for the circulating superconducting current and to introduce effective flux pinning sites dispersed uniformly throughout the matrix. Melt textured YBa$_2$Cu$_3$O$_{7-x}$ (Y123) contains many defects such as non-superconducting particles, twin boundaries, stacking faults, cracks, pores and dislocations which may act as pinning centers. Y$_2$BaCuO$_5$ (Y211) are typical non-superconducting inclusions that are trapped in melt textured Y123. A uniform distribution of entrapped Y211 particles appears to be the case in most Y123 bulk materials that are doped with platinum to prevent the particles from coarsening in the melt.

To the extent authorized under the laws of the United States of America, all copyright interests in this publication are the property of The American Ceramic Society. Any duplication, reproduction, or republication of this publication or any part thereof, without the express written consent of The American Ceramic Society or fee paid to the Copyright Clearance Center, is prohibited.

However, segregation of these inclusions has also been observed as distinguishable patterns in seeded crystals [1-3]. Other additives such as CeO_2, SnO_2 and ZrO_2 in Y-Ba-Cu-O system lead to inhomogeneous distribution of fine particles pushed at the solidification front during the melt growth process. Endo et al. [4] investigated the segregation of Y211 particles dependent on the undercooling (ΔT), growth rate (R) and growth direction. Takao et al. [5] synthesized Y123-$BaZrO_3$ composite powders by spray drying and obtained very fine inclusions of $BaZrO_3$ in sintered materials. Higher critical current, J_c, values were measured in composite superconductors due to the effective flux pinning by $BaZrO_3$ inclusions.

The objective of this work is to investigate segregation of $BaZrO_3$ in platinum doped and melt textured Y123. Pushing and entrapment of particles at the solid/liquid interface are discussed dependent on the size of non-superconducting inclusions.

EXPERIMENTAL PROCEDURE

Two $YBa_2Cu_3O_{7-x}$ powder compositions containing, 10 mol% $BaZrO_3$ and 0.2 wt.% Pt, with and without addition of 10 mol% Y_2BaCuO_5 phase, were prepared by a coprecipitation method. Pt was added to reduce the size of Y211 particles during melt texturing process. Stoichiometric amounts of Y_2O_3, $BaCO_3$, CuO and Pt powders were first dissolved in nitric acid to prepare a stock solution. The precipitation reaction was carried out by adding stock solution into the oxalic acid $((COOOH)_2\ 2H_2O)$ solution. Both solutions were mixed with 50 vol% ethanol prior to the reaction in order to achieve a well dispersed precipitate. The slurry was then spray dried to obtain a fluffy precursor powder, which was calcined at 870°C for 12 hours. The calcined powder was compacted first by uniaxial pressing followed by cold isostatic pressing at 200 MPa. The compacts were heated on MgO single crystals above the melting temperature of Y123 to 1050°C for 2 hours and cooled to 1010°C at a heating/cooling rate of 2°C/min. Melt texturing was achieved by slow cooling of the samples from 1010 to 980°C at a rate of 1°C/hr.

The phase purity of the powders was analyzed by X-ray diffraction (XRD). The microstructure of the powders and melt-textured samples was characterized by scanning electron microscopy (SEM) and transmission electron microscopy (TEM).

RESULTS AND DISCUSSION

The XRD pattern of calcined powders, shown in Fig. 1, reveals the presence of both $YBa_2Cu_3O_{7-x}$ and $BaZrO_3$ phases. The TEM image in Fig. 2

shows BaZrO$_3$ particles less than 100 nm in size segregated at the surface of Y123 particles, whose size is typically on the order of several micrometers.

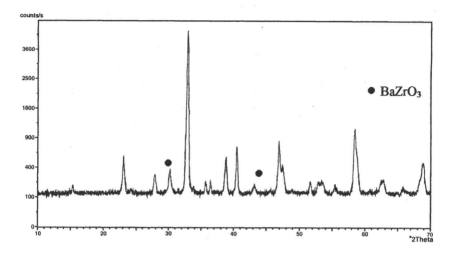

Figure 1. XRD pattern of YBa$_2$Cu$_3$O$_{7-x}$ - BaZrO$_3$ composite powders.

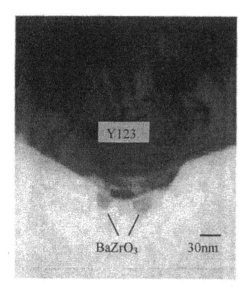

Figure 2. TEM image of the powder showing YBa$_2$Cu$_3$O$_{7-x}$ and BaZrO$_3$.

In sintered polycrystalline Y123 samples, fine $BaZrO_3$ particles are uniformly distributed throughout the matrix resulting in a nanostructural composite with enhanced flux pinning properties [5]. It is known that the grain boundaries in polycrystalline Y123 samples act as weak links for supercurrents. Formation of high angle grain boundaries can be greatly reduced by melt texturing techniques to further improve the superconducting properties of Y123.

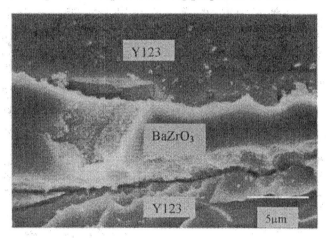

Figure 3. SEM micrograph of melt textured sample showing segregation of $BaZrO_3$ particles along the solidification front of $YBa_2Cu_3O_{7-x}$ crystals.

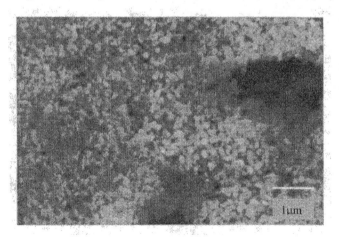

Figure 4. SEM micrograph of segregated nanosized $BaZrO_3$ particles in melt textured $YBa_2Cu_3O_{7-x}$.

Figure 3. shows the SEM image of a selected area from the microstructure of melt textured Y123-BaZrO$_3$. The fracture surface of the sample reveals a typical crystal growth pattern of Y123 from the melt. Note that BaZrO$_3$ particles are almost completely separated from the Y123 matrix. This indicates that nanosized particles are not trapped within the Y123 phase but rather pushed and segregated during the melt texturing process. Segregation of BaZrO$_3$ particles leads to their local accumulation at the solidification front of Y123 crystals as shown in Fig. 4. Particle segregation prevents continuous growth of the superconducting phase. The relationship between the critical size of a particle and the critical growth rate of the crystals has been discussed in the Y123-Y211 system based on the pushing/trapping theory [6]. In order to be able to trap nanosized particles, the growth rate should be increased. This can be achieved at a large undercooling (ΔT). However, the extent of ΔT is limited due the spontaneous nucleation of Y123 phase in the melt [7].

The relation ship between the critical growth rate (R), critical size of a particle (r) and interfacial energy ($\Delta\sigma_0$) can be approximated by:

$$R \propto \Delta\sigma_0 / \eta r$$

where η is the viscosity of the melt and

$$\Delta\sigma_0 = \sigma_{SP} - \sigma_{LP} - \sigma_{SL}$$

where σ_{SP}, σ_{LP}, and σ_{SL} are the solid particle, liquid-particle and solid-liquid interfacial energies, respectively [6].

We have observed particle pushing by continual cooling of the samples at a rate of 1°C/min. Further experiments will verify whether entrapment of nanosized BaZrO$_3$ particles can be achieved at constant growth of highly undercooled samples where ΔT approaches to a maximum prior to the spontaneous nucleation of Y123. Recent studies by Awano et al. show that elimination of grain boundaries and uniform distribution of fine pinning centers, such as BaZrO$_3$ inclusions, in superconducting Y123 matrix can be achieved by solid-state diffusion at elevated temperatures under applied pressure [8].

CONCLUSION

Composite powders of Y123 and BaZrO$_3$ were synthesized by the oxalate coprecipitation technique. Powder compacts were melt textured by continuous cooling of the semisolid melt. It was found that nanosized BaZrO$_3$ inclusions ($\cong$70 nm) were pushed at the solidification front of Y123 and segregated as locally concentrated particles. Melt processing at higher solidification rates and larger undercooling were proposed to trap nanosized $BaZrO_3$ inclusions within the superconducting Y123 matrix.

ACKNOWLEDGMENT

One of the authors (FD) would like to acknowledge the generous support from AIST-MITI Japan during his visit at NIRIN.

REFERENCES

[1] A. Endo, H. S. Chauhan, T. Egi, and Y. Shiohara, "Macrosegregation of Y_2BaCuO_5 Particles in $YBa_2Cu_3O_{7-x}$ Crystals Grown by an Undercooling Method," *J. Mater. Sci.*, **11** [4] 795-803 (1996).

[2] S. Honjo, M. J. Cima, M. C. Flemings, T. Ohkuma, H. Shen, K. Rigby, and T. H. Sung, "Seeded Crystal Growth of $YBa_2Cu_3O_{6.5}$ in Semisolid Melts," *J. Mater. Res.*, **12** [4] 880-890 (1997).

[3] C. Varanasi, M. A. Black, and P. J. McGinn, " Demonstration of Y_2BaCuO_5 Particle Segragation in Melt Processed $YBa_2Cu_3O_{7-x}$ through a Computer Visualization Method," *J. Mater. Res.*, **11** [3] 565-571 (1996).

[4] A. Endo, H. S. Chauhan, Y. Nakamura, and Y. Shiohara, "Relationship between Growth Rate and Undercooling in Pt-added $YBa_2Cu_3O_{7-x}$," *J. Mater. Sci.*, **11** [5] 1114-1119 (1996).

[5] Y. Takao, M. Awano, and H. Takagi, "Preparation and Properties of $Ba_2YCu_3O_{7-x}$-$BaZrO_3$ Superconductive Composite by Spray Drying Method," *J. Cer. Soc. Jap.*, **102** [3] 237-240 (1994).

[6] Y. Shiohara and A. Endo, "Crystal Growth of Bulk High-T_c Superconducting Oxide Materials," *Mater. Sci. & Eng.*, **R19** [1-2] 1-86 (1997).

[7] F. Dogan and J. D. Reding, "Effect of Y_2BaCuO_5 Particle Size on Undercooling and Solidification Rate of $YBa_2Cu_3O_{7-x}$," in *Impact of Recent Advances in Processing of Ceramic Superconductors, Ceramics Transactions*, **84**, Eds. U. Balachandran, W. Wong-Ng and A. Bhalla, The American Ceramic Society, Westerville, OH, 23-30, (1998).

[8] M. Awano, Y. Fujishiro, J. Moon, H. Takagi, S. Rybchenko, and S. Bredikhin, "Microstructure Control of an Oxide Superconductor on Interaction of Pinning Centers and Growing Crystal Surface," to be published in *Physica C.*

MICROSTRUCTURE CONTROL IN BaTiO$_3$ SINTERS BY A SMALL AMOUNT OF DOPANTS

T. Yamamoto[*], Y. Ikuhara[*], K. Hayashi[+] and T. Sakuma[‡]
[*]Engineering Research Institute, School of Engineering, The University of Tokyo
2-11-16 Yayoi Bunkyo-ku Tokyo, 113-8656
[‡]Department of Advanced Materials Science, Graduate School of Frontier Science,
University of Tokyo, 7-3-6 Hongo Bunkyo-ku Tokyo, 113-8656
[+]Graduate student

ABSTRACT

Grain growth characteristics in BaTiO$_3$ polycrystals were examined with a special interest in grain boundary structures. The grain growth behavior in BaTiO$_3$ was very sensitive to dopant addition in a level of 0.1mol%. The compounds doped with B-site type dopant (B-excess compounds) exhibit abnormal grain growth while normal grain growth takes place in the compounds doped with A-site type dopant (A-excess compounds). Precise investigation of grain growth behavior revealed that the abnormal grain growth observed in B-excess compound is due to the suppression of the normal grain growth. The inhibition of the normal grain growth was found to be closely related to the formation of extra bonding at grain boundaries in B-excess compound. High-resolution transmission electron microscopy (HRTEM) study revealed that grain boundaries in B-excess compounds tend to facet with {210} type and periodical structure generates along the facetted grain boundaries. Oxygen K-edge ELNES (electron energy-loss near edge structure) clearly showed the difference in the chemical bonding state between the grain interior and the facetted grain boundary. Comparing with molecular orbital calculation, TiO$_6$ octahedral linkage changed from corner-sharing at grain interior to edge-sharing at facetted grain boundaries. Consequently, the extra Ti-O bonding is formed along the boundaries. It could be concluded that the suppression of the grain growth observed in B-excess compounds is due to the formation of extra Ti-O bonding.

INTRODUCTION

BaTiO$_3$-based materials are often used for practical electroceramics. In order to obtain good electroceramic components, it is necessary to control the grain size of sinters uniformly so that the abnormal grain growth during sintering must be suppressed from the viewpoint of practical use. It is known that the grain growth behavior in BaTiO$_3$ is

To the extent authorized under the laws of the United States of America, all copyright interests in this publication are the property of The American Ceramic Society. Any duplication, reproduction, or republication of this publication or any part thereof, without the express written consent of The American Ceramic Society or fee paid to the Copyright Clearance Center, is prohibited.

affected with type and amount of various dopants[1-2]. In addition, it was reported that the grain growth behavior was very sensitive to a small deviation from cation stoichiometric composition. Yamamoto et al. showed that the compound with 0.03mol% excess Ba exhibits normal grain growth while the abnormal grain growth takes place in the compound doped with 0.03mol%Ti[3]. He insisted that the abnormal grain growth observed in such compounds is due to the strong suppression of the normal grain growth. However, it is not cleared yet what stops the normal grain growth. In this study, the grain boundary structure was examined with a special interest in the chemical bonding state by HRTEM and EELS analysis.

MICROSTRUCTURE EVOLUTION IN BaTiO₃ WITH A SMALL ADDITION OF DOPANTS

The grain growth behavior in $BaTiO_3$ is very sensitive to a small change of Ba/Ti ratio. A typical example is shown in Figure 1. Fig. 1 shows optical micrographs in the compounds whose Ba/Ti ratio is varied from 0.9987 to 1.0010 in as-annealed state at 1300°C after sintering. Ba-excess compound has uniform-grained structure of a grain

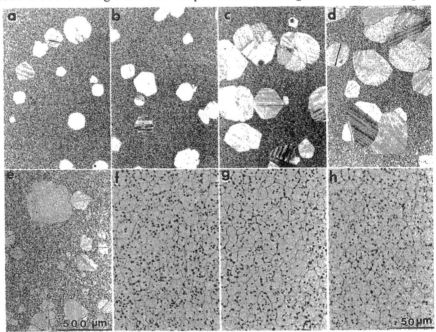

Figure 1   Optical micrographs in $BaTiO_3$ compounds ; (a) Ba/Ti=0.9987, (b) 0.9990, (c) 0.9993, (d) 0.9997, (e) 1.0000, (f) 1.0003, (g) 1.0007, and (h) 1.0010. In the figure, the magnification of (a)~(e) and (f)~(h) are indicated in (e) and (h), respectively.

size of around 10μm. On the other hand, abnormally coarse grains of more than 100μm are formed in fine-grained matrix of about 2μm in Ti-excess compound. The similar microstructure change caused by the variation of Ba/Ti ratio is also observed in co-doped compounds. For example, abnormal grain growth takes place in 0.1mol%Y-0.2mol%Ti compound (A/B ratio < 1.000) while 0.2mol%Ba-0.1mol%Nb compound (A/B ratio > 1.000) exhibits normal grain growth[4], where A and B mean cation sites in a perovskite structure of ABO_3, and Y and Nb ions are A-site and B-site type dopants, respectively. These facts clearly indicate the sensitivity of the microstructure to a small change in A/B ratio.

HIGH RESOLUTION TRANSMISSION ELECTRON MICROSCOPY STUDY

Figure 2 shows bright field images in Ba-excess and Ti-excess compounds in as-sintered state. Comparing the grain boundaries in the two compounds, the grain boundaries are facetted in Ti-excess compound as in (a) while Ba-excess compound exhibits round and smooth grain boundaries as in (b). The facet boundary is a distinct feature in the compound whose A/B ratio is under 1.000, i.e., the grain growth behavior is a type of abnormal one.

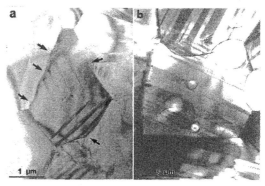

Figure 2   Bright field images in (a) Ti-excess and (b) Ba-excess compounds.

Figure 3 shows a facetted grain boundary of Ti-excess compound in as-sintered state. In the figure, the result obtained from TEM-trace analysis is also shown. As seen in the figure, the habit planes are indexed as $(\bar{2}10)$ and $(\bar{1}20)$ from the grain B, and $(\bar{1}3\ 42)$ and $(\overline{12}13\bar{1}0)$ from the grain A, respectively. Although the other habit planes such as (310) or (410) type are sometimes observed, however, most of habits examined in this study

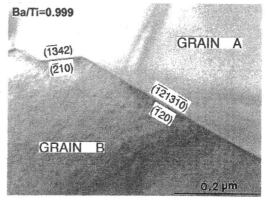

Figure 3 A facetted grain boundary in Ti-excess compound.

could be estimated to be {210} type.

Figure 4 is a HRTEM image in the vicinity of the faceted grain boundary. In the image, the (210) habit is set at the edge-on condition and [001] direction of the grain B with a (210) facet is parallel to the electron beam direction. The adjacent grain A is slightly inclined from the [110] direction. It is noted that the array of white dots appears periodically along the facetted grain boundary as indicated by the arrows. This fact suggests the existence of an extra ordered structure along the facetted grain boundary.

EELS ANALYSIS

Figure 5 shows oxygen K-edge ELNES obtained from grain boundaries in A-excess compounds and in B-excess compounds, respectively. It is noted that the shape of the spectra is clearly different among the two types of compounds. In the case of ELNES of the A-excess compounds, which exhibit normal grain growth, the spectra are clearly split while the spectra in the B-excess compounds, which exhibit abnormal grain growth, show plateau-like features as indicated by the arrows. The difference in the ELNES must be caused by the change in the chemical bonding state, i.e., the atomic structure of grain boundaries in the B-excess compounds may be different from that in the A-excess ones.

It has been proposed that a first principle molecular orbital calculation using DV-Xα cluster method is useful for interpretation of experimental ELNES[5]. In this study, we

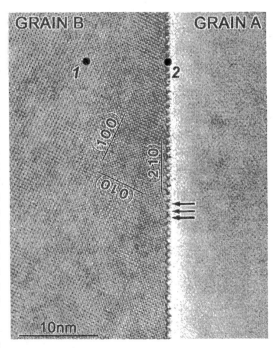

Figure 4 HRTEM image of a facetted grain boundary in Ti-excess compound. EELS analysis is carried out at the area as shown with the dots. The arrows show periodical structures along the facetted grain boundary.

adopted this method to understand complicated ELNES as shown in Fig. 5. The clusters of $(Ba_8TiO_6)^{8+}$ for $BaTiO_3$ and $(TiO_6)^{18-}$ for rutile type TiO_2 are used for DV-Xα calculation. Both structures have TiO_6 octahedrons in the structure but their linkage manner is different between the two structures. In a perovskite structure of $BaTiO_3$, TiO_6 octahedrons are linked at corners of the octahedrons while partially at edges in rutile structure of TiO_2. Figure 6 shows calculated PDOS (partial density of states) of O-2p for $BaTiO_3$ in (b) and TiO_2 in (c) respectively. In the figure, experimental ELNES obtained in Ti-excess compound are also shown for comparison. The oxygen K-edge ELNES corresponds to PDOS of O-2p. The main peaks in the PDOS are alphabetically indexed in the figure. There are three peaks of A, B, and C in $BaTiO_3$, and two peaks of D and E in TiO_2 in the energy-loss range from 520eV to 550eV. According to the analysis by Tanaka et al., the peaks of D and E are identified as t_{2g} and e_g bands in the rutile structure which results from O-2p orbital antibonding with Ti-3d orbital[5]. In addition, the peaks D and E are explained to be closely related to the linkage manner of TiO_6 octahedrons. Tanaka et al. insisted that the height ratio of peaks A and B or D and E strongly depend on the bonding angle of Ti-O-Ti in each structure. The bonding angle is different between a perovskite and a rutile structures because the perovskite structure has corner-shared TiO_6 octahedrons while rutile has partially edge-shared TiO_6 ones. As a result, a shoulder emerges at the peak B in $BaTiO_3$ so that the intensity of the peak B is reduced comparing with that of the peak A in the perovskite structure (Fig. 6(b)) while the peak D is lower than the peak E in the rutile structure (Fig. 6(c)). On the other hand, the experimental ELNES from the facetted grain boundary shows the plateau-like feature. This fact suggests that the

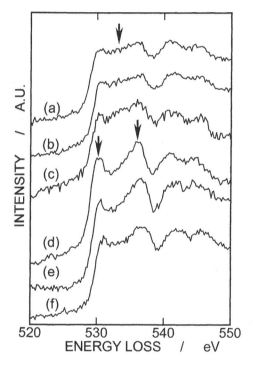

Figure 5 Oxygen K-edge ELNES taken from grain boundaries in (a) Ti, (b) Zr, (c) Nb, (d) Ba, (e) Y, and (f) Ca-excess compounds.

intensity of the spectra as indicated by the arrow in Fig. 6(a) increases. This can be interpreted in terms of the overlapping of the two PDOS of BaTiO$_3$ (Fig. 6(b)) and TiO$_2$ (Fig. 6(c)). Namely, the TiO$_6$ octahedral linkage changes from corner-sharing in the grain interior to edge-sharing at the grain boundary in the B-excess compounds.

Figure 7(a) is a model of the facetted boundary. The extra Ti-O$_2$ (210) layers are set on the Ti-O$_2$ (210) plane of BaTiO$_3$ structure as an ordered structure. In the case of this model, TiO$_6$ octahedral linkage changed from corner-sharing to edge-sharing at the ordered structure. Fig. 7(b) shows a simulated image obtained from the atomic model shown in Fig. 7(a). It is noted that the particular contrast for the ordered structure appears in the simulated image as indicated by the arrows in the figure.

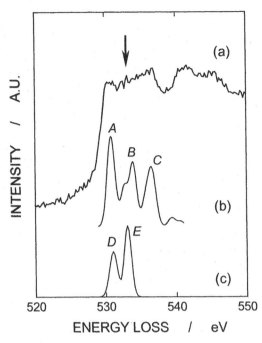

Figure 6 Theoretical ELNES calculated with DV-Xα method; (a) experimental ELNES as shown in Fig. 5(a), (b) theoretical ELNES of BaTiO$_3$, and (c) TiO$_2$.

GRAIN GROWTH IN BaTiO$_3$

The abnormal grain growth observed in Ti-excess BaTiO$_3$ is often discussed in terms of the eutectic liquid formed between BaTiO$_3$ and Ba$_6$Ti$_{17}$O$_{40}$ at 1332°C [6]. In contrast, there are several reports that abnormal grain growth takes place even in solid state [7-8]. One of the present authors showed that the initial grain growth is strongly suppressed in Ti-excess compounds below the eutectic temperature and the suppression result in abnormal grain growth[1]. As mentioned above, the distinct feature in the abnormal grain growth in the B-excess compounds is the strong suppression of normal grain growth. The suppression mechanism can be concluded to be the formation of extra Ti-O bonding at the grain boundaries in the B-excess compounds.

Meanwhile, the abnormal grain growth behavior in B-excess compounds is found to be a

Grain Boundary Engineering in Ceramics

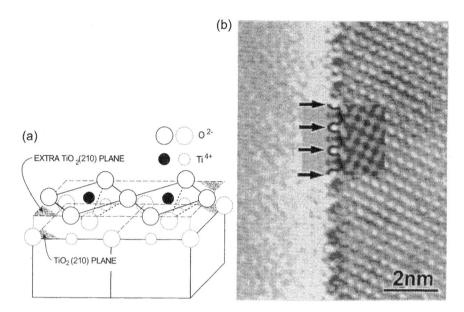

Figure 7　A model for a facetted grain boundary in (a) and a simulated image calculated from the model in (b). Note that the periodical contrast appears in the simulated image as indicated by the arrows.

type of site-saturation from a kinetical point of view[9]. The number of nucleation sites of abnormal grains is depending on not annealing temperature but the content of dopants. In addition, the migration rate of abnormal grains increases with an increase in A/B ratio in the region of A/B<1.000. Therefore it is possible to control the size of abnormal grains by changing the A/B ratio. We have succeeded in fabrication of coarse-grained polycrystals with a grain size of ～1mm by conventional sintering method using the compound whose A/B ratio is 0.9995 [10].

CONCLUSIONS

The grain boundary structure of $BaTiO_3$ doped with a small amount of dopants was examined by high-resolution transmission electron microscopy and electron energy-loss spectroscopy. The following conclusion could be obtained.
1. Grain boundaries in the B-excess compounds are facetted with {210} habits. An ordered structure exists along the habits from HRTEM observation.
2. EELS analysis clearly showed the difference in the chemical bonding state between grain interiors and facetted grain boundaries in the B-excess compounds.
3. Comparing with the theoretical ELNES obtained from molecular orbital method, an ordered structure along the facetted grain boundaries results from the change in the

TiO$_6$ octahedral linkage, i.e., from corner-sharing in grain interiors to edge-sharing in grain boundaries.

4. The suppression of the normal grain growth observed in the B-excess compounds is due to the formation of the extra ordered structure along the grain boundaries.

REFERENCES

[1] D. Hennings, "Control of Liquid-Phase-Enhanced Discontinuous Grain Growth in Barium Titanate" *J. Am. Ceram. Soc.*, **70** [1] 23-27 (1987).

[2] M. Drofenik, "Oxygen Partial Pressure and Grain Growth in Donor-Doped BaTiO$_3$," *J.Am.Ceram.Soc.*, **70**[5] 311-14 (1987).

[3] T. Yamamoto, "Influence of Small Ba/Ti non-stoichiometry on Grain Growth behavior in Barium Titanate," *Br. Ceram. Trans.*, **94** [5] 196-200 (1995).

[4] T. Yamamoto and T. Sakuma, "Influence of Small Cation Nonstoichiometry on the Grain Growth of BaTiO$_3$", *Euro. J. Solid State Inorg. Chem.*, **32**[7-8] 731-40 (1995).

[5] I. Tanaka, I. Nakajima, J. Kawai, H. Adachi, H. Gu, and M. Ruhle, "Dopant-Modified Local Chemical Bonding at A Grain Boundary in SrTiO$_3$," *Phil. Mag. Lett.*, **75**[1] 21-27 (1997).

[6] K. W. Kirby and B. A. Wechsler, "Phase Relations in the Barium Titanate-Titanium Oxide System," *J. Am. Ceram. Soc.*, **74** [8] 1841-47 (1991).

[7] G. Kastner, R. Wagner and V. Hilarius, "Nucleation of twins by grain coalescence during the sintering of BaTiO$_3$ ceramics," *Phil. Mag. A*, **69** [6] 1051-71 (1994).

[8] H. Schmeltz and A. Mayer, "The Evidence of Anomalous Grain Growth below the Eutectic Temperature in BaTiO$_3$ Ceramics," *Ber. Dtsch. Keram. Ges.*, **59** [8/9] 436-40 (1982).

[9] P. R. Rios, T. Yamamoto, T. Kondo and T. Sakuma, "Abnormal Grain Growth Kinetics of BaTiO$_3$ with an Excess TiO$_2$", *Acta Mater* **46**[5] 1617-1623 (1998).

[10] K. Hayashi, T. Yamamoto and T. Sakuma, "Grain Orientation Dependence of PTCR Effect in Niobium-Doped Barium Titanate", *J. Am. Ceram. Soc.*, **79** [6] 1669-1672 (1996).

ACKNOLEDGEMENT

A part of this study was financially supported by Proposal-Based R&D Program of New Energy and Industrial Technology Development Organization (NEDO) Project No. 98Y28-019.

GRAIN BOUNDARIES IN STRONTIUM TITANATE

O. Kienzle, S. Hutt, F. Ernst, and M. Rühle
Max-Planck-Institut für Metallforschung, Seestraße 92, 70174 Stuttgart, Germany

ABSTRACT

This article summarizes experimental studies on the atomistic structure and the composition of *grain boundaries* in $SrTiO_3$ by advanced techniques of transmission electron microscopy (TEM). Furthermore, we report on first experiments of fabricating $SrTiO_3$ bicrystals for fundamental studies by diffusion bonding of $SrTiO_3$ single crystals in high vacuum.

INTRODUCTION

Polycrystalline strontium titanate ($SrTiO_3$) serves as electroceramics in important technical applications, for example in capacitors and in varistors. The unique and technically useful electrical properties of the material originate from *segregation* of charged point defects to the inherent *grain boundaries*. In general, excess segregation of either positively or negatively charged point defects leads to a net charge on the boundary plane. This charge attracts further point defects of opposite charge from the adjacent bulk material, and these build up *space charge* layers on both sides of the boundary. Since a charged grain boundary with a space charge layer corresponds to a barrier in the electrostatic potential, segregation of charged point defects to grain boundaries affects the charge transport properties of $SrTiO_3$ ceramics and thus controls the technically relevant, macroscopic properties. On the other hand, the extent to which point defects segregate to a grain boundary depends on the arrangement of the atoms (ions) at this particular boundary [1, 2]. This implies that the macroscopic electric properties of $SrTiO_3$ ceramics depend on the *atomistic structure* of the inherent grain boundaries.

In order to understand the interrelation between grain boundary structure and segregation of point defects (dopants) we have studied grain boundaries in Fe-doped $SrTiO_3$ ceramics and Fe-doped $SrTiO_3$ bicrystals by both, high-resolution TEM (HRTEM) and analytical TEM (ATEM) [3]. Under the conditions we consider, Fe doping introduces Fe^{3+} ions on Ti sites, Fe'_{Ti}, and in order to preserve

To the extent authorized under the laws of the United States of America, all copyright interests in this publication are the property of The American Ceramic Society. Any duplication, reproduction, or republication of this publication or any part thereof, without the express written consent of The American Ceramic Society or fee paid to the Copyright Clearance Center, is prohibited.

electroneutrality, every two dopant ions cause the formation of an oxygen vacancy $V_O^{\bullet\bullet}$ [3]. The following sections summarize the results of these studies, obtained by advanced methods of TEM. Furthermore, we report on first experiments of fabricating SrTiO$_3$ bicrystals of well-defined crystallography by diffusion bonding of two SrTiO$_3$ single crystals.

GRAIN BOUNDARY SEGREGATION OF DOPANTS

Experimental Method

To image the atomistic structure of grain boundaries in SrTiO$_3$ ceramics and in SrTiO$_3$ bicrystals we employed a JEM 4000 EX (JEOL) high-resolution electron microscope with a point resolution of 0.175 nm. For ATEM we have employed a dedicated scanning transmission electron microscope (HB 501, Vacuum Generators), which can produce an electron probe with a diameter as small as 0.22 nm (full width at half maximum) and thus allows us to carry out x-ray energy-dispersive spectroscopy (XEDS) and electron energy-loss spectroscopy (EELS) with very high lateral resolution. With this instrument, we have

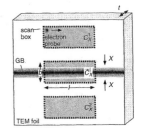

Fig. 1. The 'box method' of analytical TEM.

determined the composition of various SrTiO$_3$ grain boundaries by applying the 'box technique' [2]. This technique works by scanning the electron beam over three equi-sized rectangles, one centered on the grain boundary and one in each adjacent crystal (Fig. 1). While continuously scanning the electron beam over each rectangle, we integrate the XEDS or EELS signal to obtain the average composition of the underlying 'box' of material. Over the 'spot technique', where the electron beam is at rest, the box technique has the major advantage that one obtains an image of the specimen *during* the analysis, which allows the microscope operator to compensate for specimen drift. Comparing the spectra of the interface box (index 0) with the spectra of box 1 and box 2 in the bulk we obtain information about the excess Γ_X of a particular atomic species X at the grain boundary in the following way. In each box i ($i = 0, 1, 2$), the absolute number N_X of X atoms amounts to

$$N_X = C_X^i \cdot b \cdot l \cdot t,$$

where C_X^i denotes the (average) volume concentration of element X in box i, and b, l, and t represent the width, length, and depth of the box. The box depth, t, equals the thickness of the TEM foil, and we assume that t does not vary over the entire region of interest. Thus, all three boxes have the same dimensions. If the grain boundary were decorated with an excess of Γ_X^{gb} X atoms per unit area, the analysis would indicate an excess concentration

$$C_X^0 = C_X^0 - \frac{1}{2}(C_X^1 + C_X^2) = \frac{\Gamma_X^{\text{gb}}}{b}$$

of X atoms in box 0 over the average concentration of X in box 1 and box 2. In reality, of course, segregants will not exclusively accumulate in the boundary plane but also in regions adjacent to the grain boundary. Therefore, the quantity $\Gamma_X = C_X^0 \cdot b$, which we can determine by measuring C_X^0, corresponds to the excess of X atoms in the grain boundary *after projecting their positions onto the boundary plane.*

Grain Boundaries in SrTiO₃ Ceramics

In the following, we describe the results we have obtained concerning the correlation between grain boundary structure and grain boundary composition for six different SrTiO₃ grain boundaries. Table I summarizes the data we have obtained for the composition of these boundaries. The column labeled N indicates how many measurements at different locations along the boundary have been carried out in each case. The columns labeled Γ_{Fe} and Γ_{Ti} show the average excess of Fe and Ti, respectively, in box 0 when projecting onto the boundary plane. These data were obtained by XEDS with a box width b of typically ≈ 5 nm. The columns labeled 'D_{Fe}' and 'D_{Ti}' indicate the average minimum detectable concentration Γ_0 of the respective species under the experimental conditions. The minimum detectable concentrations are determined by the requirement that excess counts from an element at the grain boundary can be considered as significant only if they exceed three times the standard deviation of the counts recorded in the bulk. When calculating the average concentrations Γ_{Fe} and Γ_{Ti} we assumed the values of D_{Fe} and D_{Ti} for those measurements where the Fe concentration or the deviation of the Ti concentration from stoichiometry were below the respective minimum detectable amount. Therefore, some of the values for Γ_{Fe} and Γ_{Ti} in Table I actually constitute upper limits for the true values and, consequently, carry the prefix '$<$'. The column labeled Q, finally, indicates how the ratio of the Ti and Sr cation concentrations change at the grain boundary with respect to the average of their value in the two bulk crystals

$$Q := \frac{C_{\text{Ti}}^0/C_{\text{Sr}}^0}{\frac{1}{2}\left(C_{\text{Ti}}^1/C_{\text{Sr}}^1 + C_{\text{Ti}}^2/C_{\text{Sr}}^2\right)}$$

With this definition, $Q = 1$ indicates no difference between the cation ratio at the grain boundary and the cation ratio in the bulk.

Table I. *Composition of Grain Boundaries in SrTiO₃ as Determined by Analytical TEM.*

Σ	Fig.	N	Γ_{Fe}	D_{Fe}	Γ_{Ti}	D_{Ti}	Q
3A	2a	4	$< 0.30 \pm 0.00$	0.28	4.38 ± 0.65	2.13	1.08 ± 0.01
3B	2a	4	0.55 ± 0.12	0.20	2.68 ± 1.02	2.30	1.06 ± 0.02
?	2b	4	4.03 ± 0.24	0.20	7.83 ± 0.52	2.60	1.11 ± 0.03
13	3a	7	0.28 ± 0.04	0.14	$< 2.03 \pm 0.52$	1.57	1.04 ± 0.02
5	3b	6	$< 0.15 \pm 0.02$	0.15	$< 2.18 \pm 0.62$	1.51	1.06 ± 0.03
3	3c	6	$< 0.18 \pm 0.02$	0.18	$< 3.47 \pm 0.56$	3.47	1.00 ± 0.02

Figure 2 presents two HRTEM images of grain boundaries in SrTiO₃ ceramics. In Fig. 2a the orientation relationship of the two grains approximately corresponds to $\Sigma=3$. The grain boundary in this image consists of two different types of facets, labeled 'A' and 'B'. The A facets lie parallel to {111} planes in both grains. The regular HRTEM image pattern with a short period indicates that these facets correspond to 'special', low-energy grain boundaries (note that in SrTiO₃ a $\Sigma=3$, (111) boundary does not correspond to a 'coherent twin boundary' because the perovskite structure of SrTiO₃ is *primitive* cubic). The B facets, in contrast, follow a different plane and appear with a diffuse contrast. This indicates that the disturbance of the perovskite structure is more severe than in the case of the A facets.

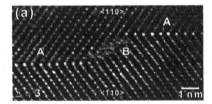 

Fig. 2. *HRTEM images of grain boundaries in SrTiO₃ ceramics. (a) Near $\Sigma=3$ boundary. (b) 'Random' orientation relationship.*

According to the data of Table I, the different grain boundary structures A and B of Fig. 2a lead to significant differences in Fe and Ti segregation: The Fe segregation caused by the disturbed atom configuration of the B facets amounts to nearly twice the upper limit of the Fe segregation caused by the $\Sigma=3$, (111) boundary of the A facets. On the other hand, the A facets exhibit a stronger segregation of Ti. For both types of facets, we observe that the Ti/Sr ratio is slightly higher than in the bulk material.

Fig. 2b shows a HRTEM image of a 'non-special' grain boundary—a boundary with a 'random' orientation relationship between the two grains. For this reason,

Grain Boundary Engineering in Ceramics

this boundary is not well-suited for HRTEM imaging and the image exhibits only one set of lattice fringes in each grain. Along the boundary plane, the image exhibits a speckle contrast, which indicates the presence of an amorphous film between the two grains. The data obtained by ATEM supports this hypothesis: Table I indicates substantial segregation of both, Fe and Ti to this boundary. The Fe segregation corresponds to about eight times the amount observed for the B facets of the $\Sigma=3$ boundary, and the Ti segregation amounts to about three times the value of the B facets.

Grain Boundaries in SrTiO$_3$ Bicrystals

In the SrTiO$_3$ ceramics considered so far, it is a matter of chance to locate grain boundaries suitable for solving their atomistic structure by HRTEM imaging. For this reason, we have studied not only SrTiO$_3$ ceramics but also grain boundaries in SrTiO$_3$ *bicrystals*. The latter have a well-defined crystallography, suited to solve the atomistic structure by HRTEM, and a macroscopic extension, which allows us to prepare TEM specimens with the boundary in regions sufficiently thin for HRTEM.

Figure 3a shows a HRTEM image of a near-$\Sigma=13$, (510) grain boundary in SrTiO$_3$. This boundary constitutes a $\langle 100 \rangle$ tilt boundary with a tilt angle of $(23.6 \pm 0.1)°$ between the two crystal lattices (an exact $\Sigma=13$ orientation relationship would require a tilt angle of 22.62°). Owing to the deviation from the exact $\Sigma=13$ crystallography, the boundary exhibits facets parallel to (100) planes in one grain and parallel to (310) planes in the other. These facets occur about every 90 nm along the boundary. Between the facets, as shown in Fig. 3a, the boundary plane is not atomically flat but exhibits some roughness; the boundary 'core' extends over about 0.6..0.9 nm normal to the boundary plane. Moreover, the core does *not* consist of a periodic atom configuration. Moreover, the HRTEM image reveals small sections with a speckle contrast similar to the contrast of the amorphous interlayer in Fig. 2b, indicating a heavy disturbance of the perovskite structure.

The ATEM data in Table I indicates substantial segregation of Fe at this grain boundary. Only in one location out of seven we have analyzed along this boundary the segregation of Fe was below D_{Fe}. The Ti signal remains below the detection limit of Ti segregation, however the slightly increased parameter of the cation stoichiometry, $Q = 1.04$, does indicate an accumulation of Ti at this grain boundary.

Figure 3b presents a HRTEM image of the $\Sigma=5$, (310) boundary. By evaluating HRTEM images we find that the misorientation of the two grains in this bicrystal corresponds to a $(36.8 \pm 0.1)°$ rotation around a common $\langle 100 \rangle$ axis (the viewing direction of Fig. 3b). Within the limits of error, the tilt angle equals the value of 36.87° required for an exact $\Sigma=5$ orientation relationship. Compared to the

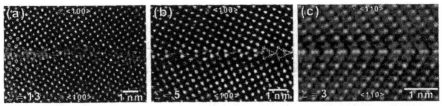

Fig. 3. HRTEM images of grain boundaries in SrTiO₃ bicrystals. (a) Σ=13 boundary. (b) Σ=5, (310) boundary. (c) Σ=3, (111) boundary.

near-Σ=13, (510) boundary of Fig. 3a, the Σ=5, (310) boundary has a regular pattern with pronounced intensity maxima. This image indicates an ordered atom arrangement with a unique, small repeat unit, in agreement with the results of Browning *et al.* [4]. The interface itself is atomically flat and precisely follows a common (310) plane of the two grains. As indicated in Table I, the segregation of Fe to this boundary remains below the detectability limit. While three out of six measurements indicated an enrichment of Ti, these measurements also indicated the presence of impurities: Ca and Zn. Therefore, we assume that the Ti-segregation in the respective cases constitutes a case of co-segregation with impurities and not an intrinsic feature of the Σ=5, (310) boundary. Accordingly, the average Ti/Sr cation ratio was found to exceed that of the bulk material, however only by a small amount.

Figure 3c, finally, presents a HRTEM image of the Σ=3, (111) in SrTiO₃. Experimentally, we do not detect any deviation from the ideal crystallography. The boundary plane is sharp and atomically flat over extended distances. Among all grain boundaries studied in this work, this one has the shortest repeat unit and constitutes the weakest disturbance of the perovskite structure. This becomes apparent not only from the HRTEM image but also from the data we obtained by ATEM. As indicated in Table I, the Σ=3, (111) boundary does not cause any detectable segregation of Fe and Ti, and the Ti/Sr cation ratio at the boundary precisely equals the value we find in the bulk: $Q = 1.00$.

In summary, we obtained the following results: 'Random' grain boundaries in Fe-doped SrTiO₃ tend to develop an amorphous thin film, and such a film has a large capacity for segregants. Among the above examples, the grain boundary of Fig. 2b exhibits the most intense Fe and Ti segregation. The grain boundaries without an amorphous film feature substantially less segregation. The extent of Fe and Ti segregation systematically decreases with increasing structural perfection of the boundary. The near Σ=13, (510) boundary with its irregular atom configuration exhibits the largest amount of segregants, while for the Σ=3, (111) boundary and for the Σ=5, (310) boundaries with their high structural perfection we cannot detect any segregation at all. Accordingly, the near Σ=13, (510) boundary con-

Grain Boundary Engineering in Ceramics

stitutes an intermediate case between structurally perfect boundaries ($\Sigma=3$, (111) and $\Sigma=5$, (310)) and 'random' boundaries like the one in Fig. 2b.

The tendency of grain boundaries with 'random' orientation relationships to develop an amorphous film can be rationalized by the consideration of Clarke [5], Fig. 4: Suppose that the interface between an amorphous film and one of the crystal grains has the energy γ_{sf}—irrespective of the orientation of the crystal. Then a grain boundary with an amorphous film possesses an interface energy of $\gamma = 2\gamma_{sf}$, which corresponds to the dashed horizontal line in Fig. 4. The solid graph schematically describes the interface energy for a boundary without an amorphous film. In this case the energy typically features deep cusps for 'special' orientation relationships (special tilt angles θ around a particular common axis). For those tilt angles θ for which the energy of a film-free boundary (solid line) exceeds the energy $2\gamma_{cf}$ of

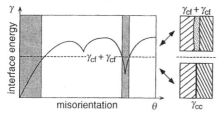

Fig. 4. Dependence of the grain boundary structure on the interface energies γ_{cf}, according to Clarke [5].

the dashed line, the stable $2\gamma_{cf}$, a boundary with an amorphous film constitutes the stable configuration. Vice versa, for those orientation relationships where the energy without amorphous film lies below $2\gamma_{cf}$—either for small angle boundaries or for the energy cusps of 'special' grain boundaries—the stable configuration corresponds to a periodic repetition of a unique building block of ions.

FABRICATION OF SrTiO₃ GRAIN BOUNDARIES BY DIFFUSION BONDING

While SrTiO$_3$ bicrystals are much more suitable than SrTiO$_3$ ceramics for fundamental studies of the correlation between grain boundary structure and grain boundary composition, commercial bicrystals are available with only a few different orientation relationships and few inclinations of the boundary plane. For this reason, we have recently begun to fabricate our own SrTiO$_3$ bicrystals by diffusion bonding of (undoped) SrTiO$_3$ single crystals. After cutting and polishing the surface of two single-crystals according to the desired bicrystallography we align the appropriate in-plane crystallographic directions by means of a specially designed sample holder and bond the two single crystals under high vacuum (10^{-6} mbar), applying a temperature between 1350 and 1500 °C and a pressure of 1 MPa. After diffusion bonding we anneal the bicrystal in a re-oxidizing atmosphere (24 h at 1100 °C in O$_2$ or 60 h at 1480 °C in air).

As a first result, Fig. 5 presents a HRTEM image of the $\Sigma=5$, (310) grain boundary in a diffusion-bonded bicrystal. Owing to the high quality of the surface polishing prior to bonding and to the special design of the specimen holder by which we align the in-plane directions the misorientation versus the desired Σ=5 orientation relationship is particularly small: $(0.24 \pm 0.09)°$ in tilt and $(0.16 \pm$ 0.055$)°$ in twist. The HRTEM image reveals a structurally intact, atomically flat grain boundary with no amorphous film. As to the composition, however, XEDS analysis of this boundary has revealed a significant increase of the Ti/Sr cation ratio with respect to the bulk: $Q = 1.12$ (similar to the 'random' grain boundary of Fig. 2b). Therefore, gaining better control of the grain boundary composition will be an important subject of further experimental work on diffusion-bonded $SrTiO_3$ bicrystals.

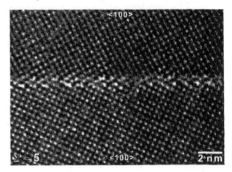

Fig. 5. HRTEM micrograph of a Σ=5, (310) boundary in a diffusion-bonded bicrystal.

CONCLUSION

For grain boundaries in $SrTiO_3$ ceramics and in $SrTiO_3$ bicrystals we observe a pronounced correlation between (i) the extent to which the grain boundary disturbs the perovskite structure of $SrTiO_3$ and (ii) the intensity of Fe and Ti segregation to the boundary. $SrTiO_3$ bicrystals for further fundamental studies of this correlation can be fabricated with high accuracy and with any desired crystallography by diffusion bonding.

REFERENCES

1. J. S. Ikeda and Y. M. Chiang: *Space charge segregation at grain boundaries in titanium dioxide: I, Relationship between lattice defect chemistry and space charge potential*, J. Am. Ceram. Soc. **76** (1993) 2437.

2. J. S. Ikeda and Y. M. Chiang: *Space charge segregation at grain boundaries in titanium dioxide: II, Model experiments*, J. Am. Ceram. Soc. **76** (1993) 2447.

3. O. Kienzle: *Atomistische Struktur und chemische Zusammensetzung innerer Grenzflächen von Strontiumtitanat (Atomistic Structure and Composition of Grain Boundaries in Strontium Titanate)*, doctoral thesis, Universität Stuttgart (1999).

4. N. D. Browning, S. J. Pennycook, M. F. Chisholm, M. M. McGibbon and A. J. McGibbon: *Observation of structural units at symmetric (001) tilt boundaries in SrTiO₃*, Interf. Sci. **2** (1995) 397-423.

5. D. R. Clarke: *On the equilibrium thickness of intergranular glass phases in ceramic materials*, J. Am. Ceram. Soc. **70** (1987) 15-22.

ELECTRONIC STRUCTURE CALCULATION OF SYMMETRIC TILT BOUNDARIES IN ZnO

Fumiyasu Oba, Shigeto R. Nishitani and Hirohiko Adachi
Department of Materials Science and Engineering, Kyoto University,
Sakyo, Kyoto 606-8501, Japan

Isao Tanaka
Department of Energy Science and Technology, Kyoto University,
Sakyo, Kyoto 606-8501, Japan

ABSTRACT

Electronic structure of the [0001] /($\bar{1}\bar{2}30$) $\Sigma 7$ symmetric tilt boundary in ZnO has been investigated by a first-principles molecular orbital method. Atomic arrangements at the grain boundary are modeled by the static lattice calculations. The electronic structure of the derived geometries are discussed comparatively with those of the bulk and the $(10\bar{1}0)$ surface of ZnO. In the configuration with the lowest grain boundary energy, there are small open channels at the boundary core but all of ions have preserved the four-fold coordination. The electronic structure resembles that of the bulk regardless of the presence of the small open channels and no remarkable interfacial electronic states form. The configuration with the second lowest energy has relatively large open channels and some of the ions adjacent to the channels have dangling-bonds. Local electronic states at the ions are similar to the surface states in the valence and the conduction bands. However, interfacial states are not found within the band gap even for the configuration with dangling-bonds.

INTRODUCTION

Non-linear current-voltage characteristics exhibited in polycrystalline ZnO are one of the most interesting phenomena originating from grain boundaries[1-3]. These characteristics have been attributed to the formation of the double Schottky barrier at the grain boundaries[4,5]. Although the generation of the barrier is known to be associated with the presence of some impurities and excessive oxygen[6-8], it should be a significant step to understand the electronic structures of grain boundaries in undoped ZnO. In the present study, we investigate the electronic structures of the [0001] /($\bar{1}\bar{2}30$) $\Sigma 7$ symmetric tilt boundaries in ZnO. Atomic arrangements at the grain boundary are modeled by the static lattice calculations and the electronic structures of the derived geometries are calculated by a first-principles molecular orbital method. Electronic

To the extent authorized under the laws of the United States of America, all copyright interests in this publication are the property of The American Ceramic Society. Any duplication, reproduction, or republication of this publication or any part thereof, without the express written consent of The American Ceramic Society or fee paid to the Copyright Clearance Center, is prohibited.

states specific to the boundary geometries are discussed in comparison with those of the bulk and the $(10\bar{1}0)$ surface of ZnO.

ATOMIC STRUCTURE

Static lattice calculations for the grain boundary have been performed employing the computer code MARVIN[9]. The total internal energy of a simulation cell is iteratively minimized using the Quasi-Newton method under two-dimensional periodic boundary conditions parallel to an interface. The lattice energy is decomposed into a long-range electrostatic energy and a short-range repulsive energy. Two-body potentials of the Buckingham form were used with the parameters reported by Lewis *et al.*[10] in order to model the short-range repulsive interactions. The potentials were confirmed to reproduce the bulk properties of ZnO satisfactorily as shown elsewhere[11]. According to the CSL theory, the $[0001]/(\bar{1}\bar{2}30)$, $38.21°$ $\Sigma 7$ symmetric tilt boundary in ZnO was modeled assuming the stoichiometry. The other $\Sigma 7$ symmetric tilt boundary with the orientation of $[0001]/(\bar{1}\bar{4}50)$, $21.79°$ exists owing to its definition, which is not reported here. Calculations were performed from a number of initial configurations in which the one half-crystal cell was translated relative to the other three-dimensionally. The boundary energies were calculated as the energies relative to the bulk lattice. Although a number of stable configurations were obtained, the geometries with the lowest and the second lowest boundary energies were selected in the present study.

The calculated atomic structures are depicted in Fig. 1. The translations in the [0001] direction are the same as half the length of the lattice constant c for both of the geometries: the basal planes of the two half-crystals are common at the boundary regions. In this direction, all of the atomic displacements from the initial positions are within 0.1Å. Some types of core structures are simultaneously observed for the boundaries with the same misorientation in metal oxides such as NiO[12] and ZnO[13], and that may result from the presence of some structures with similar boundary energies. The present two boundaries may be examples of this and hence we deal with them equally regardless of the energy difference. Boundary A has the lowest energy of 1.37 J/m^2. This should be explained by the fact that all the ions have preserved the four-fold coordination only in this boundary. In boundary B, large open channels along the [0001] direction can be seen at the boundary core. The upper and lower ions of the channels, numbered 1 in Fig. 1(b), have lost one of the first nearest neighbors. However, the boundary energy of 1.42 J/m^2 is only slightly higher than that of the boundary A. More detailed discussion on the atomic structure has been done elsewhere[11].

ELECTRONIC STRUCTURE

Electronic structure calculations have been carried out by the discrete variational (DV)-Xα method employing the computer code SCAT[14]. The electronic structures of model clusters are self-consistently obtained by solving one-electron Schrödinger equations within the local density approximation. Molecular orbitals (MOs) are represented by the linear combination of atomic orbitals. The atomic orbitals are numerically obtained by solving a radial part of a Schrödinger equation for each atom. In the present study, 1s-2p for O and 1s-5sp for Zn were used as the basis sets. In order to discuss electronic structure, we use partial density of states (PDOS) for each atom. The PDOSs were obtained by replacing the atomic orbital populations[15] at discrete MOs with Gaussian function of 0.7 eV FWHM. The analysis in an unit of the atomic orbital population enables detailed investigation of electronic structure.

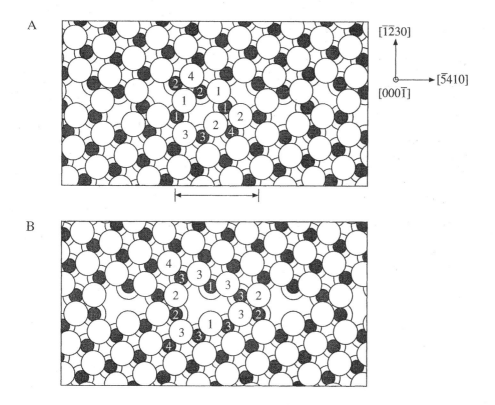

Figure 1. View of the calculated geometries from the $[000\bar{1}]$ direction. The scale indicates a periodicity unit. Filled and open circles denote Zn and O ions, respectively. Three of the four basal planes in the ZnO unit cell can be seen in the order of O-Zn-O.

For grain boundary calculations, we employed model clusters containing the numbered ions and their first and second nearest neighbors in Fig. 1. The electronic structures of the numbered ions were investigated discriminately because they have different surroundings. The ions numbered 1 are the most characteristic to the boundaries. For the boundary A, they are adjacent to the two open channels, whereas they face to the channel and one of the first nearest neighbors is absent in the boundary B. The numbered 2 and 3 have the complete four nearest neighbors but the numbered 2 are more open to the channels owing to the distortion of the bond angle. The numbered 4 are further from the channels. In order to take electrostatic potential into account and minimize the termination effects of the finite-sized clusters, the model clusters were embedded in about 10,000 point charges with the charge of ±2 located at the external atomic sites. Calculations for the bulk and the (10$\bar{1}$0) surface of ZnO were also performed for comparative discussion of the electronic structure. In this case, the model clusters constructed according to relaxed geometries by the static lattice calculations for the bulk and the surface were employed.

The PDOSs for the bulk and the (10$\bar{1}$0) surface model clusters are shown in Fig. 2. For the bulk model in Fig. 2(a), the valence band is composed mainly of Zn-3d and O-2p orbitals, and they are considerably admixed here. Over the band gap around 1.0 eV, the conduction band is constructed mainly by Zn-4sp and -5sp orbitals. The gap between the highest occupied molecular orbital (HOMO) and the lowest unoccupied molecular orbital (LUMO) is 1.8 eV although this is hardly recognized in the figure owing to the Gaussian broadening. This value is smaller than an experimental band gap of 3.3 eV[16] due to the local density approximation. For the surface model cluster in Fig. 2(b), electronic states specific to the surface can be recognized near the conduction band edge: 1.9-4.8 eV above the HOMO. The states are constructed mainly of Zn-4sp and -5sp orbitals on the surface layer and therefore they are often referred to as surface states. The origin of the states is attributed to both the presence of dangling-bonds and the change of electrostatic potential at the surface. The surface effects are also recognized in the valence band. Zn-3d and O-2p components separate from each other relatively to the bulk. Similar features to the surface are expected to be recognized at the grain boundaries modeled in the present study since they have structures with open channels.

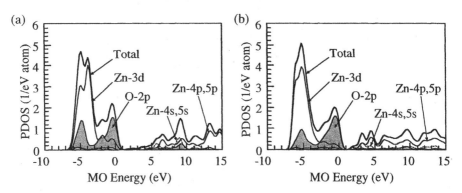

Figure 2. Partial density of states (PDOS) for (a) the bulk and (b) the (10$\bar{1}$0) surface model clusters.

Figure 3 shows PDOSs for the numbered ions in the boundary model clusters (See Fig. 1). For the boundary A in Fig. 3(a), the PDOSs resemble that of the bulk model at all the numbered sites. A significant mixture of Zn-3d and O-2p components is recognized in the valence bands. Moreover, localized states are not clearly present near the conduction band edges. In this boundary, all ions have the four-fold coordination, which may be the

Grain Boundary Engineering in Ceramics

reason for the similarity in electronic structure to the bulk. The presence of the small open channels does not change the electronic structure remarkably.

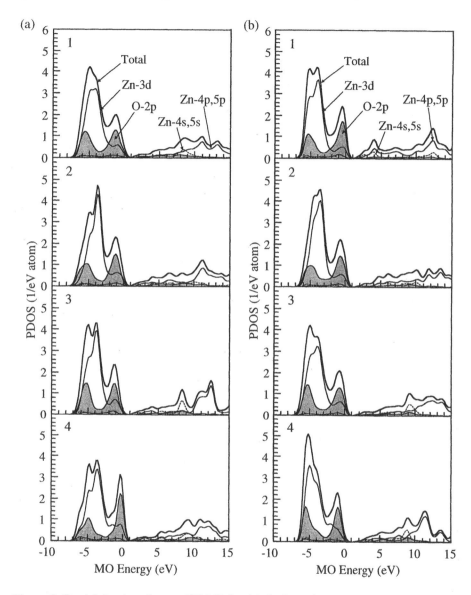

Figure 3. Partial density of states (PDOS) for (a) the boundary A and (b) the boundary B model clusters.

On the other hand, the model cluster for the boundary B in Fig. 3(b) shows particular electronic structure for the ions numbered 1. There are states similar to the surface states near the conduction band edge: 1.8-4.5 eV above the HOMO. They are considered to be states in the conduction band according to the band gap of 1.8 eV for the bulk model. The energy positions are expected to be influenced very slightly by the increase of a cluster size because of the spatially localization on Zn ions numbered 1: the states are not recognized clearly in the PDOSs for ions numbered 2, 3 and 4. The valence structure for the numbered 1 also resembles that of the $(10\bar{1}0)$ surface rather than the bulk. This may be easily understood from the similar environment to the surface. The PDOSs for the other ions resemble that of the bulk rather than the surface owing to the complete four-fold coordination. Deep interfacial electronic states have not been recognized for the grain boundaries of Al_2O_3[17] and TiO_2[18] either.

In ZnO varistors, the non-linear current-voltage characteristics are considered to be generated by localized interface states deep from the conduction band edge (>0.6 eV)[6,7,19-21] Such deep states are not recognized in the present stoichiometric boundaries. Although they are just two examples of the boundaries in ZnO, local atomic relaxation at least preserving three of the four first nearest neighbors is expected to occur at other boundaries. This may also result in the electronic structure without the deep interface states. The nature of the varistor property is not likely to be explained by atomic arrangements at stoichiometric boundaries but by the effects of intrinsic point defects and/or impurities .

CONCLUSION

The electronic structure of the [0001] / $(\bar{1}\bar{2}30)$ $\Sigma7$ symmetric tilt boundary in ZnO has been investigated by a first-principles molecular orbital method. In the boundary A with the lowest boundary energy, all ions are four-fold coordinated as in the case of the bulk ZnO. The electronic structure resembles that of the bulk regardless of the presence of the small open channels. In the boundary B with the second lowest boundary energy, there are large open channels at the boundary cores. Ions adjacent to the channels have dangling-bonds and the local electronic states are similar to the surface. However, deep interface states are not observed either. Local atomic relaxation preserving at least three of the four-fold coordination is expected to be similar at other boundaries. The nature of the non-linear properties of polycrystalline ZnO may therefore be attributed not to the atomic arrangements at stoichiometric boundaries but to the segregation of intrinsic point defects and/or impurities at boundary regions.

ACKNOWLEDGMENTS

We are grateful to Ben Slater and David H. Gay for their support and helpful discussions. This study was performed through Special Coordination Funds of the Science and Technology Agency of the Japanese Government.

Grain Boundary Engineering in Ceramics

REFERENCES

[1] M. Matsuoka, "Nonohmic Properties of Zinc Oxide Ceramics," *Jpn. J. Appl. Phys.*, **10** [6] 736-46 (1971).

[2] K. Mukae, K. Tsuda and I. Nagasawa, "Non-ohmic Properties of ZnO-Rare Earth Metal Oxide-Co$_3$O$_4$ Ceramics," *Jpn. J. Appl. Phys.*, **16** [8] 1361-68 (1977).

[3] D. R. Clarke, "Varistor Ceramics," *J. Am. Ceram. Soc.*, **82** [3] 485-501 (1999).

[4] G. E. Pike and C. H. Seager, "The dc Voltage Dependence of Semiconductor Grain-Boundary Resistance," *J. Appl. Phys.*, **50** [5] 3414-22 (1979).

[5] G. E. Pike, S. R. Kurtz, P. L. Gourley, H. R. Philipp and L. M. Levinson, "Electroluminescence in ZnO Varistors: Evidence for Hole Contributions to the Breakdown Mechanism," *J. Appl. Phys.*, **57** [12] 5512-18 (1985).

[6] K. Tsuda and K. Mukae, "Characterization of the Interface States in ZnO Varistors by DLTS Method (in Japanese)," *J. Ceram. Soc. Jpn.*, **97** [10] 1211-18 (1989).

[7] F. Greuter, G. Blatter, M. Rossinelli and F. Stucki, "Conduction Mechanism in ZnO-Varistors: An Overview"; pp. 31-53 in *Advances in Varistor Technology*. Edited by L. M. Levinson. American Ceramic Society, Westerville, 1989.

[8] F. Stucki and F. Greuter, "Key Role of Oxygen at Zinc Oxide Varistor Grain Boundaries," *Appl. Phys. Lett.*, **57** [5] 446-8 (1990).

[9] D. H. Gay and A. L. Rohl, " MARVIN: A New Computer Code for studying Surfaces and Interfaces and its Application to Calculating the Crystal Morphologies of Corundum and Zircon," *J. Chem. Soc. Faraday Trans.*, **91** [5] 925-36 (1995).

[10] G. V. Lewis and C. R. A. Catlow, "Potential Models for Ionic Oxides," *J. Phys. C: Solid State Phys.*, **18**, 1149-61 (1985).

[11] F. Oba, I. Tanaka, S. R. Nishitani, H. Adachi, B. Slater and D. H. Gay, "Geometry and Electronic Structure of [0001] / ($\bar{1}\bar{2}30$) Σ=7 Symmetric Tilt Boundary in ZnO," *Phil. Mag. A*, in press.

[12] K. L. Merkle and D. J. Smith , "Atomic Structure of Symmetric Tilt Grain Boundaries in NiO," *Phys. Rev. Lett.*, **59** [25] 2887-90 (1987).

[13] A. N. Kiselev, F. Sarrazit, E. A. Stepantsov, E. Olsson, T. Claeson, V. I. Bondarenko, R. C. Pond and N. A. Kiselev, " High Resolution Electron Microscopy of ZnO Grain Boundaries in Bicrystals Obtained by the Solid-Phase Intergrowth Process," *Phil. Mag. A*, **76** [3] 633-56 (1997).

[14] H. Adachi, M. Tsukada and C. Satoko, "Discrete Variational Xα Cluster Calculations. I. Application to Metal Clusters," *J. Phys. Soc. Jpn.*, **45** [3] 875-83 (1978).

[15] R. S. Mulliken, "Electronic Population Analysis on LCAO-MO Molecular Wave Function. I, *J. Chem. Phys.*, **23** [10] 1833-40 (1955).

[16] V. Srikant and D. R. Clarke, "On the Optical Band Gap of Zinc Oxide," *J.Appl. Phys.*, **83** [10] 5447-51 (1998).

[17] S. -D. Mo, W. Y. Ching and R. H. French, "Electronic Structure of a Near Σ 11 *a*-axis Tilt Grain Boundary in α-Al$_2$O$_3$," *J. Am. Ceram. Soc.*, **79** [3] 627-33 (1996).

[18] I. Dawson, P. D. Bristowe, M. -H. Lee, M. C. Payne, M. D. Segall and J. A. White, "First-Principles Study of a Tilt Grain Boundary in Rutile," Phys. Rev. B, **54** [19] 13727-33 (1996).

[19] J. P. Gambino, W. D. Kingery, G. E. Pike, H. R. Philipp and L. M. Levinson, "Grain Boundary Electronic States in Some Simple ZnO Varistors," *J. Appl. Phys.*, **61** [7] 2571-4 (1987).

[20] R. A. Winston and J. F. Cordaro, "Grain-Boundary Interface Electron Traps in Commercial Zinc Oxide Varistors," *J.Appl. Phys.*, **68** [12] 6495-500 (1990).

[21] K. Tsuda and K. Mukae, "Characterization of Interface States in ZnO Varistors using Isothermal Capacitance Transient Spectroscopy (in Japanese)," *J. Ceram. Soc. Jpn.*, **100** [10] 1239-44 (1992).

ATOMIC AND ELECTRONIC STRUCTURE ANALYSIS OF COINCIDENCE BOUNDARIES IN β-SiC

Koji TANAKA and Manori KOHYAMA

Department of Material Physics, Osaka National Research Institute, AIST, 1-8-31 Midorigaoka, Ikeda, Osaka, 563-8577

ABSTRACT

The atomic structures of $\Sigma=3$, 9 and 27 coincidence boundaries, and multiple junctions in β-SiC were studied by high-resolution electron microscopy (HREM). The existence of the variety of structures of $\Sigma=3$ incoherent twin boundaries and $\Sigma=27$ boundary were observed. The structures of $\Sigma=3$, 9 and 27 boundary were explained by structural unit models. Electron energy-loss spectroscopy (EELS) was used to investigate the electronic structure of grain boundaries. The spectra recorded from the grain, the $\{111\}\Sigma=3$ coherent twin boundary (CTB) and the $\{211\}\Sigma=3$ incoherent twin boundary (ITB) did not show significant differences. The energy-loss corresponding to carbon 1s-to π^* transition was not found. This indicates that C atoms exist at grain boundaries on the similar condition of C atoms in grains.

INTRODUCTION

There has been increasing interest in SiC as high-temperature devices or high–performance components. It is very important to investigate the structures and properties of grain boundaries (GB) of SiC, because grain boundaries in sintered ceramics or polycrystalline films have significant effects on the bulk properties. HREM of GB is now one of the main tools for understanding materials. A rigid body translation of one grain relative to the other is an important part of the atomic relaxation of a grain boundary. The presence of a rigid body translation along the common <111> direction at $\{211\}\Sigma=3$ ITB was found in Si [1] and Ge [2], whereas no rigid body translation was found under different conditions in Si [1] or in diamond [3]. It was also pointed out that the translation state was sensitive to the environment of the boundary [4].

To the extent authorized under the laws of the United States of America, all copyright interests in this publication are the property of The American Ceramic Society. Any duplication, reproduction, or republication of this publication or any part thereof, without the express written consent of The American Ceramic Society or fee paid to the Copyright Clearance Center, is prohibited.

In the present study, {211}Σ=3 ITB and its junction with {111}Σ=3 CTBs, {221}Σ=9 boundary and its triple junction with {111}Σ=3 CTBs, and {552}Σ=27 boundary and its quadruple junction with {111}Σ=3 CTBs have been investigated by HREM and compared with a theoretical study.

EXPERIMENT

The transmission electron microscope (TEM) specimen of chemical vapor deposition (CVD) β-SiC was prepared by mechanical thinning and ion milling. CVD techniques can easily provide dense materials of high purity and interfaces that are well-defined even in covalent materials. Moreover, CVD specimens often exhibit preferred orientation during growth and this enhances the probability of coincident site lattice (CSL) grain boundary formation. HREM and EELS analysis was performed on CVD β-SiC with a preferred orientation of {220} and grain size of approximately 10 μm using the field emission (FE) TEM (JEOL, JEM-3000F) equipped with a Gatan Imaging Filter. Atomic models were derived based on the structural unit model.

RESULTS

In the case of Σ=3
Figures 1 (a) ~ (d) show the atomic models of the non-polar interfaces of {211}Σ=3 ITB in β-SiC. They include suitable reconstructions, in which arrows indicate the <011> bonds. These are symmetric and asymmetric models, respectively. Type A and B are constructed for the respective models. Si atoms are reconstructed in Type A, and C atoms in Type B. The relaxed atomic positions in Fig. 1 are calculated by the self-consistent tight-binding (SCTB) method.

The wrong bonds exist at the two types of positions. These are the intergranular bonds and the <011> bonds. The intergranular bonds exist on the {022} plane and cross the interface. The <011> bonds that are reconstructed connect the two atoms on the neighboring {022} planes, and double the periodicity along the <011> direction. The kinds and positions of the wrong bonds in Type A are inverted in Type B.

The rigid-body translations and grain boundary energies of Fig. 1 (a) ~ (d) are shown in Table 1. The symmetric Type A model is the most stable and the symmetric Type B model has the largest energy. The energy difference between the symmetric Type A model and the asymmetric Type A model is small indicating both structures can occur.

Figures 2 (a) and (c) are HREM images of the Σ=3 ITB in β-SiC. The length of incoherent twin is 6 {111} layers in Fig. 2 (a), 18 {111} layers in (c). [The lengths are indicated by the numbers in Fig. 2 (a) ~ (d).] As can be easily recognized, the {111} planes on the different sides of ITB reveal no shift in Fig. 2 (a), whereas there is no shift near the junctions with {111}Σ=3 CTB and is a shift of about one fifth of the {111} plane distance near the center of {211}Σ=3 ITB in (c). This indicates that the

Grain Boundary Engineering in Ceramics

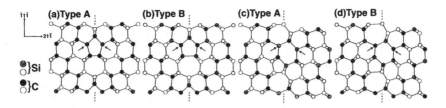

Figure 1 Atomic models of {211}Σ=3 incoherent twin boundary in β-SiC. (a) and (b) are symmetric. (c) and (d) are asymmetric. Arrows indicate the <011> reconstruction bonds. Si atoms are reconstructed in Type A, and C atoms in Type B.

Table I. Rigid Body Translations and Grain Boundary Energies

Structure	Translation Along [21$\bar{1}$] (nm)	Translation Along [$\bar{1}$1$\bar{1}$] (nm)	Grain Boundary Energy (J/m²)
Fig. 1 (a)	0.001	0	1.71
Fig. 1 (b)	0.024	0	2.37
Fig. 1 (c)	0.016	0.072	1.39
Fig. 1 (d)	0.032	0.068	2.42

rigid body translation along the <111> is zero in Fig. 2 (a), and is zero near the junctions and about 0.05 nm near the center in (b).

Atomic models for Σ=3 ITB in Fig. 2 (a) and (c) were given on the basis of the results of the theoretical calculations. Figures 2 (b) and (d) are structural unit models superimposed on HREM images corresponding to Fig. 2 (a) and (c), respectively. White circles represent the reconstruction along the <011> direction. The structural unit consists of symmetric 5-7-6 membered rings aligning along the boundary in Fig. 2 (b). The symmetric structural units are filled in near the junction between CTB and ITB, and asymmetric 5-7-6 membered rings are filled in at the middle of the ITB in Fig. 2(d). All structural unit models fit with the HREM images very well. It should be noted that the positions of all white circles correspond to the Si site and it was shown by the theoretical calculation that the grain boundary energy was lower when Si atom was reconstructed than when C atom was.

The obtained <111> translation was approximately 0.05 nm in experimental result and 0.072 nm in the asymmetric Type A model in theoretical calculation, respectively. It should be noted that the structure of {111}Σ=3 CTB acts like an anchor because it is very stable and rigid, and a {211}Σ=3 ITB boundary always stay with a {111}Σ=3 CTB in real materials. This situation is essentially different from that of ideal bicrystal in the simulation. At the short boundary, it is difficult to translate along the common <111> direction because the top and bottom of the {211}Σ=3ITB are anchored by the {111}Σ=3 CTB. However, the <111> translation can occur at the relatively long boundary because the restriction from the coherent boundary is not so severe near the center of the ITB. The reason why

Figure 2 (a) and (b) : HREM images of the $\Sigma=3$ incoherent twin boundary in β-SiC. (c) and (d) : Structural unit models superimposed on HREM images corresponding to (a) and (c), respectively. White circles represent the reconstruction along <011> direction.

the experimental value is smaller than the theoretical calculation one might be because the atoms can not be fully relaxed due to the anchor effect from $\{111\}\Sigma=3$ CTBs.

Figure 3 shows the C-K edge of EELS spectra from grain, $\Sigma=3$ CTB and ITB in β-SiC. The π^* peak was found at $\Sigma=3$ ITB in diamond [3], but not found in β-SiC as seen in Fig. 3. It indicates that C atoms exist at grain boundaries on the similar condition of C atoms in grains.

As discussed above, the structures of $\Sigma=3$ ITB in β-SiC were explained very well by the prediction of SCTB calculation.

In the case of $\Sigma=9$ and triple junction

Figure 4 (a) shows the atomic model of the ideal triple junction (TJ) between one $\{221\}\Sigma=9$ and two $\{111\}\Sigma=3$ boundaries in diamond structure along [011] direction. It is seen from Fig. 4 (a), atoms

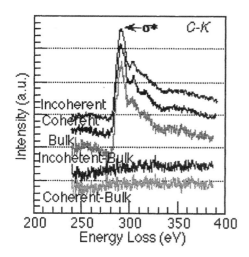

Figure 3 EELS spectra from bulk, Σ=3 coherent and incoherent twin boundary in β-SiC.

can not have proper bonds at {221}Σ=9 because of the height differences in {022} stacking. By introducing 1/4[011] translation, symmetric 5-6-7-6 membered rings structure can be achieved at Σ=9 boundary. And by introducing 1/9[41$\bar{1}$] translation, zigzag 5-7-5-7 membered rings structure can be achieved. It was shown that the zigzag 5-7-5-7 structure has lower energy than symmetric 5-6-7-6 structure in Si [5]. However, introducing 1/4[011] translation at GB result in introducing screw dislocation at TJ and introducing 1/9[41$\bar{1}$] translation result in introducing edge dislocation. Therefore, both cases cause increasing total GB energy around TJ. On the other hand, zigzag 5-7-5-7 structure can be achieved by shifting GB plane by one atomic column as seen in Fig. 4 (b). In this case, no extra dislocation is introduced, therefore total GB energy around TJ seems lower than the case of introducing translation.

Figure 5 (a) is a HREM image of TJ and (b) is a structural unit model superimposed on HREM images corresponding to Fig. 5 (a). It is clear that Σ=9 boundary start from one atomic column away form the ideal TJ and stable zigzag 5-7-5-7 structure is achieved. It should be noted that choosing GB position has the same effect as a translation, and a rigid body translation at Σ=9 boundary is difficult to occur because adjacent Σ=3 CTBs play like an anchor. It may be the reason why the changing of GB position is prefeable to the rigid body translation.

In the case of Σ=27 and quadruple junction

Figure 6 (a) is a HREM image of the quadruple junction (QJ) between one {552}Σ=27 and three {111}Σ=3 boundaries in β-SiC and (b) is a structural unit model superimposed on HREM images corresponding to Fig. 6 (a). It is seen that Σ=27 does not start from the ideal QJ and symmetric 5-7-6-5-6-7 membered rings structure is achieved near QJ but zigzag 5-7-5-7-6 membered rings structure is achieved far from QJ. The GB energy of the zigzag model is smaller than the symmetric model according to the SCTB calculation for β-SiC [6].

Figures 6 (c) and (d) are the atomic models of the {552}Σ=27 boundary. Figure 6 (c) is a symmetric model and (d) is a co-existence model of symmetric and zigzag structure. As indicated in Fig. 6 (c), the zigzag 5-7-5-7-6 structure like Fig. 6 (d) is achieved if the GB position shifts one atomic column from the symmetric structure position. To achieve the zigzag structure, atomic relaxation is necessary.

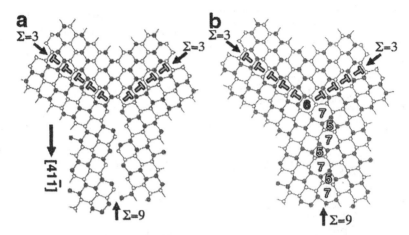

Figure 4 Atomic model of the triple junction between one {221}Σ=9 and two {111}Σ=3 boundaries in diamond structure along [011] direction. Open and closed circles indicate the atoms with different height. (a) : Ideal position. (b) : One atomic column off.

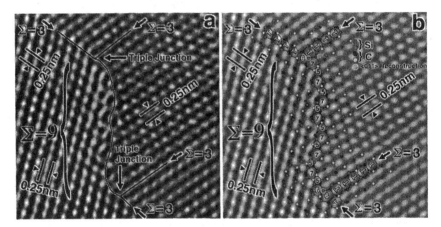

Figure 5 HREM image of triple junction (a) and structural unit model superimposed on it (b). White circles represent the reconstruction along <011> direction.

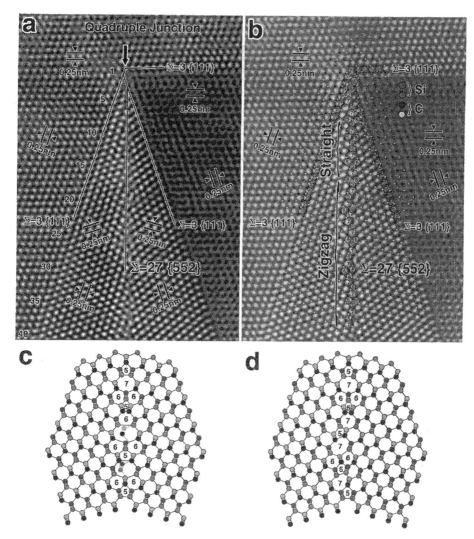

Figure 6 (a) : HREM image of the quadruple between one {552}Σ=27 and three {111}Σ=3 boundaries in β-SiC. (b) : Structural unit model superimposed on (a). (c) : Symmetric model. (d) : Co-existence model of symmetric and zigzag structure.

Also, the restriction from the {111}Σ=3 boundary is so strong near QJ that there is no space for atomic relaxation. This might be the reason why the symmetric structure is achieved near QJ and the zigzag structure is achieved far from QJ.

CONCLUSION

HREM and theoretical calculations showed that the atomic structures of Σ=3, 9 and 27 boundaries, and the multiple junction in β-SiC, were well explained by structure unit models. The translation state was sensitive to the environment of the boundary. A shift of GB position has the same effect as a rigid body translation. C atoms exist at grain boundaries on the similar condition of C atoms in grains.

REFERENCES

1. H. Ichinose, Y. Tajima, and Y. Ishida, "Structures Analysis of Σ=3 CSL Boundary in Polysilicon by HREM Lattice Imaging", *Grain Boundary Structure and Related Phenomena,* Supple. to Trans. of the Japan Institute of Metals, (Sendai, Japan: The Japan Institute of Metals), **27**, 253-260, 1986.
2. A. Bourret, and J. J. Bacmann, "Defect Structure in the Σ=27, 11 and (21$\bar{1}$) Σ=3 Symmetrical Grain Boundaries in Germanium", *Grain Boundary Structure and Related Phenomena,* Supple. to Trans. of the Japan Institute of Metals, (Sendai, Japan: The Japan Institute of Metals), **27**, 125-134, 1986 .
3. Y. Zhang, H. Ichinose, Y. Ishida, K. Ito, and M. Nakanose, "Atomic and Electronic Structures of Grain Boundary in Chemical Vapor Deposited Diamond Thin Film", *Diamond for Electronic Application,* Mater. Res. Soc. Proc., edited by D. L. Dreifus, A. Collins, T. Humphreys, K. Das, and P. E. Pehrsson,(Pittsburgh, Pennsylvania: Material Research Society), **416**, 355-360, 1996.
4. K. Tanaka, M. Kohyama, and M. Iwasa, "The Variety of Structures of the Σ=3 Incoherent Twin Bounday in β-SiC", Matl. Sci. Forum, **294-296**, 187-190, 1999.
5. M. Kohyama, "Theoretical Study of Grain Boundaries in Silicon: Features of Atomic and Electronic Structures", Matl. Sci. Forum, **207-209**, 265-268, 1996.
6. M. Kohyama and K. Tanaka, to be published.

EVALUATION OF OPTICALLY ACTIVE Cr ION IN Cr:Al₂O₃ CERAMICS BY SNOM

Hiroshi Murotani[1], Moriaki Wakaki[1], Yoshihito Narita[2] and Susumu Teruyama[2]

[1] Department of Electro Photo Optics, Tokai University, 1117 Kitakaname, Hiratsuka-Shi, Kanagawa, 259-1292 Japan

[2] JASCO Corp.,2967-5 Ishikawa-Cho, Hachioji, Tokyo, 192-8537 Japan

ABSTRACT

A single crystal is typically used as a laser material for solid state lasers. However, large single crystals are difficult to fabricate and doping large concentration of optically active elements is also difficult to achieve. To avoid these limitations, it is proposed to use polycrystalline ceramics based in Cr:Al₂O₃ for the laser. In this study, we evaluated the spatial distribution of Cr ions, which were doped as optically active elements, by measuring the emission spectrum of the element. These measurements were performed with the SNOM (Scanning Near-field Optical Microscope) installed with a spectrum photometer. The fluorescence spectra of Cr ions were measured using this equipment with the spatial resolution on the order of hundreds of nm. The variation of the spatial distribution of optically active Cr ions was observed for different Cr concentrations. It was found that the spatial distribution of the optically active Cr ion depended on the doped Cr concentration according to the difference in the segregation behavior of Cr.

INTRODUCTION

Solid state lasers are replacing gas lasers since they are smaller and require less maintenance. Applications for the solid state laser are also expanding due to its high output power and good stability. Therefore, the research and development of solid state laser materials has been focused on achieving higher output power, higher efficiency, larger wavelength tunability and wider wavelength coverage. [1-5] Currently, single crystals are

To the extent authorized under the laws of the United States of America, all copyright interests in this publication are the property of The American Ceramic Society. Any duplication, reproduction, or republication of this publication or any part thereof, without the express written consent of The American Ceramic Society or fee paid to the Copyright Clearance Center, is prohibited.

mainly used as laser materials for solid state lasers. However, large single crystals are difficult to fabricate and doping large concentration of optically active elements is also difficult to achieve. To avoid these limitative factors, it is proposed to use polycrystalline ceramics for the laser.[6-9] Therefore, we are studying the feasibility of sintered materials (polycrystalline materials) for the application to the solid state laser medium.

In this study, we evaluated the spatial distribution and the emission spectra of a Cr ion which was doped as an optically active element.[10] The spectral and spatial measurements were performed by using the SNOM (Scanning Near-field Optical Microscope) equipped with a spectrum photometer.[11,12] The fluorescence spectra of Cr ions were measured using this equipment with a spatial resolution on the order of hundreds of nm. The spatial distribution of the optically active Cr ions at different doping concentrations of Cr was also examined.

EXPERIMENTAL

Sample preparation
An Al_2O_3 powder with a purity of 99.99% was used as a starting material for ceramics fabrication. A Cr_2O_3 powder was used as a material for doping optically active element Cr.

After mixing these raw powder materials, a green body was made using fabricated cold isostatic press (CIP). During CIP, a glycerin was used as a hydraulic fluid and a pressure of about 5500 kg/cm^2 was applied for a minute, followed by a pressure of about 5000 kg/cm^2 for 5 minutes. The green body was sintered under a vacuum of 10^{-1} Pa at a temperature of 1800 ℃ to obtain a specimen. The sample was polished to an optical grade for the observation of surface topography. The Cr concentrations of the samples were estimated using an atomic absorption method as 0.66, 0.76, 3.09 and 5.35 wt.%. The crystal structure of the sample was characterized using powder X-ray diffraction (MAC Science Co. Ltd.: MXP18VA-SRD).

Topography and Photoluminescence Measurements by SNOM
The distribution of the optically active Cr ion was observed by using the SNOM. The typical probe used for the SNOM has an optical fiber with a sharpened end. The metallic film is coated around the fiber except for the small aperture ($\sim$ 100 nm) at the sharpened end. The evanescent light is guided into the probe with this structure. The scattering light occurs from the interaction between the evanescent field and the probe by inserting the probe into the field. The distribution of the evanescent field (the profile of the surface structure in a smaller scale than the wavelength used for the SNOM) is obtained by

detecting the scattering light and scanning the probe over the sample surface. The resolution of the system is restricted by the probe size and the separation between the probe and the sample surface. In this study, an illumination-collection mode (i-c mode) was used for the optical arrangement of the SNOM. In the arrangement, the evanescent light is illuminated through the aperture of the probe end and the scattered light from the sample is collected by using the same probe as shown in Fig.1. [11,12]

The microscopic distribution of the PL intensity from a Cr^{3+}: Al_2O_3 (sintered ruby) with different Cr concentrations was observed with the same spatial resolution as the aperture of the probe. The measurements were carried out using the system for more than five areas selected arbitrarily. A He-Ne laser (632.8 nm) was used as an excitation light source for the PL. [13-15] The topographical image of the sample surface was also obtained using the change of an atomic force acting to the probe. The distribution of the Cr atom and the surface morphology of the sample were observed with an EPMA and a SEM (JEOL Ltd.: JXA-840), respectively.

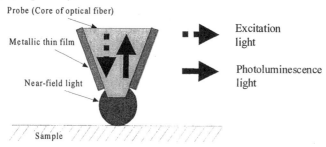

Fig.1 Principle of SNOM(Scanning Near-field Optical Microscope) for photoluminescence measurement.

RESULTS AND DISCUSSION

A corundum structure was obtained for the samples as a result of the X-ray measurement and no other structural phases were found.

The relation between the microscopic distribution of the Cr ion, which was doped as an optically active ion, and the doped Cr concentration was discussed in this study. The inter-relation between the distribution of the PL intensity of the R line and the surface topographical image was considered. An almost homogeneous distribution of the optically active Cr ion was observed for the sample with the low concentration of Cr (0.76 wt.%) as shown in Fig.2. However, a small fluctuation of the PL intensity was observed which might come from the surface profile like a pore and the scanning of the probe.[16-19] It was found that the Cr in the sample with the Cr concentration of 0.66 wt.% showed almost the same behavior.

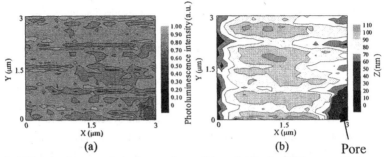

(a)   (b)   Pore

Fig.2 (a)Photoluminescence intensity mapping at 694 nm (R1 line) and (b)topographical
mapping of sintered Al_2O_3 doped with 0.76 wt% Cr observed with SNOM.

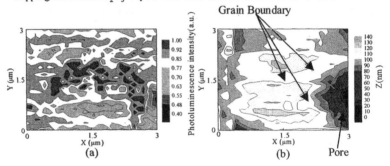

Grain Boundary

(a)   (b)   Pore

Fig.3 (a)Photoluminescence intensity mapping at 694 nm (R1 line) and (b)topographical
mapping of sintered Al_2O_3 doped with 3.09 wt% Cr observed with SNOM.

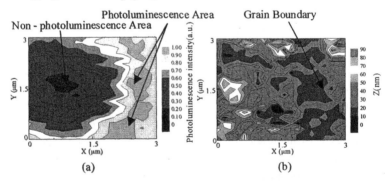

Non - photoluminescence Area   Photoluminescence Area   Grain Boundary

(a)   (b)

Fig.4 (a)Photoluminescence intensity mapping at 694 nm (R1 line) and (b) topographical
mapping of sintered Al_2O_3 doped with 5.35 wt% Cr observed with SNOM.

The topographical image and the spatial distribution of the PL intensity for the sample with
the Cr concentration of 3.09 wt.% are shown in Fig.3. [16-19]   The decrease of the PL

intensity was observed due to the surface profile like a pore. The enhancement of the intensity was observed around grain boundaries and pores, which suggested that optically active Cr ions existed near the grain boundaries. The topographical image and the distribution of the PL intensity for the sample with a higher Cr concentration (5.35 wt.%) is shown in Fig.4. Optically non-active regions which didn't show a PL emission were observed. [16-19] Furthermore, the segregation of the optically active Cr ions was observed around the grain boundaries. It was found from the topographical image that there were nearly smooth surface areas smaller than several 10 nm. This implies that the decrease of the PL intensity does not come from the surface profile but from the existence of an optically non-active material phase. These results are obvious from the comparison of the PL spectra at the optically active region with that at the non-active region as shown in Fig.5. Furthermore, the spectral profile of the sample is very different from that of the sample with smaller Cr concentration as shown in Fig.6. It is considered to be the effect of concentration quenching caused by the decrease of the separation among adjacent Cr ions as the Cr concentration increases.

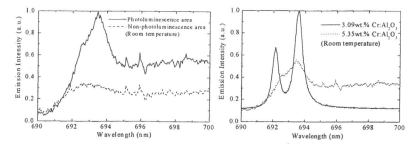

Fig.5 Photoluminescence spectra of sintered Al_2O_3 doped with 5.35 wt% Cr observed with SNOM for photoluminescence area and non-photoluminescence area.

Fig.6 Photoluminescence spectra of sintered Al_2O_3 doped with 3.09 and 5.35 wt% Cr observed with SNOM.

The displacement of an Al ion with a Cr ion causes strain in the lattice, because the ion radius of Cr is larger than that of Al. The maximum amount of Cr displacement within the Al_2O_3 lattice is limited to a certain level. But the strain may be relaxed by the grain boundaries, which allows the extra displacement of Cr ions. From these reasons, it is thought that the segregation of the optically active Cr ions was observed around the grain boundaries for the samples with the Cr concentrations of 3.08 and 5.35 wt.%.

The areas around the pores and the grain boundaries were observed by EPMA and SEM for the sample with the Cr concentration of 5.35 wt.%, where the distribution of optically active Cr ions was observed as shown in Fig.7. A slight difference in the Cr concentration was observed, but it was within the measurement error. EPMA and SEM observations were performed for the segregated material phase region, where almost the same morphology as the optically non-active region detected by the SNOM was recognized as shown in Fig.8. As a result, a larger amount of Cr was detected and Al was not detected in the segregated region. These segregated materials are assumed to be an oxide or a metallic phase of Cr.

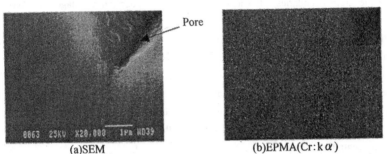

Pore

(a)SEM   (b)EPMA(Cr:kα)

Fig.7 Distribution of Cr in sintered Al$_2$O$_3$ doped with 5.35 wt.% Cr observed with a SEM(a) and an EPMA(b) for Cr:kα in the same field.

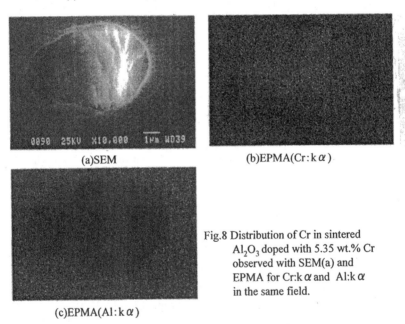

(a)SEM   (b)EPMA(Cr:kα)

Fig.8 Distribution of Cr in sintered Al$_2$O$_3$ doped with 5.35 wt.% Cr observed with SEM(a) and EPMA for Cr:kα and Al:kα in the same field.

(c)EPMA(Al:kα)

CONCLUSION

The behavior of the optically active Cr ions was observed in the microscopic area by using a new method, SNOM. These results are summarized as follows.

(1) In the case of the low doping concentration of Cr (0.66 and 0.76 wt.%), an almost homogeneous distribution of the PL intensity was observed, which indicated the homogeneous displacement of the doped Cr with the Al in Al_2O_3 lattice.

(2) The segregation of the optically active Cr ions was observed around grain boundaries for the Cr doped concentration of 3.09 and 5.35 wt.%. It is thought that the segregation of the optically active Cr ions is observed around the grain boundaries, because the strain may be relaxed by the grain boundaries, which allows the extra displacement of Cr ions.

(3) An optically non-active segregated phase was observed for the sample with the Cr concentration of 5.35 wt.%. These segregated materials are assumed to be an oxide or a metallic phase of Cr.

The spatial distribution of optically active element in the polycrystalline medium was observed with the high resolution by using SNOM. As a result, a basic knowledge was obtained for heavy doping of optically active elements and achieve high efficiency operation of solid state laser medium.

Acknowledgments

We thank Y. Kondo and A. Fukuda of PILOT Precision CO. LTD. for his cooperation in sample preparation.

References

1) J. Harrison, A. Finch, D. M. Rines, G. A. Rines and P. F. Moulton, "Low-threshold, CW, All-solid-state Ti : Al_2O_3 Laser," *Opt. Lett.*, **16** [-8] 581-583 (1991).

2) J. C. Walling, O. G. Peterson, H. P .Jenssen, R. C. Morris and E. W. Odwell, "Tunable Alexandrite Lasers," *IEEE J. Quant. Electron.*, **16** [12] 1302-1315 (1980) .

3) T. Y. Fan and R. L. Beyer, "Diode Laser-pumped Solid-state Lasers," *IEEE J. Quant. Electron.*, **24** [6] 895-912 (1988).

4) R. Scheps, B. M. Gately, J. F. Myers, J. S. Krasinski and D. F .Maller, "Alexandrite Laser Pumped by Semiconductor Lasers," *Appl. Phys. Lett.*, **56** [23] 2288-2290 (1990).

5) R. Scheps, J. F. Myers, H. B. Serreza, A. Rosenberg, R. C. Morris and M. Long,

"Diode-pumped Cr : LiCaAlF₆ Laser," *Opt. Lett.*, **16** [11] 820-822 (1991).

6) A. Ikesue, I. Frusato and K. Kamata, "Fabrication of Polycrystalline, Transparent YAG Ceramics by a Solid State Reaction Method," *J. Am. Ceram. Soc.*, **78** [1] 225-228 (1995).

7) A. Ikesue, T. Kinosita, "Fabrication and Optical Properties of High Performance Polycrystalline Nd : YAG Ceramics for Solid-state Lasers," *J. Am. Ceram. Soc.*, **78**[4] 1033-1040 (1995).

8) A. Ikesue, K. Kamata and K. Yosida, "Effects of Neodymium Concentration on Optical Characteristics of Polycrystalline Nd : YAG laser materials," *J. Am. Ceram. Soc.*, **79** [7] 192-1926 (1996).

9) H. Murotani, T. Mituda, M. Wakaki and Y. Kondou, "Optical Characteristics of Al₂O₃ Ceramics With Highly Doped Cr," *Proc. Sch. Eng. Tokai Univ.*, **39** [1] 95-99 (1999) [in Japanese]

10) T. P. Jones, R. L. Coble and C. J. Mogab, "Defect Diffusion in Single Crystal Aluminum Oxide," *J. Am. Ceram. Soc.*, **52** [6] 331-334 (1969).

11) Y. Narita, T. Tadokoro, T. Ikeda, T. Saiki, S. Mononobu and M. Ohtsu, "Near-field Raman Spectral Mesurement of Polydiacetylene," *Appl. Spectroscopy*, **52**[9]1141-1144 (1998)

12) S. Kawata, "Sensing Technology for Near-field Optical Imaging," *Med. Img. Technolgy*, **14**[1]3-8(1996) [in Japanese]

13) Donald E. McCarthy, "Transmittance of Optical Materials from 0.17 μ to 3.0 μ, " *Applied Optics*, **6**[11] 1896-1898 (1967).

14) S. Sugano, Y. Tanabe, "Absorption Spectra Cr³⁺ in Al₂O₃ Part A," *J. Phys. Soc. Jpn.*, **13**[8] 880-898 (1958).

15) S. Sugano, I. Tsujikawa, "Absorption Spectra Cr³⁺ in Al₂O₃ Part B," *J. Phys. Soc. Jpn.*, **13**[8] 899-910 (1958).

16) Y. Aoki, N. T. My, H. Takesita, S. Yamamoto, P. G. Langer and H. Naramoto, "Luminescent Properties of Cr Doped Al₂O₃ under Ion Bombardment," *JAERI-Review*, **94**[5] 113-115 (1994).

17) I. Wieder and L. R. Sarles, "Stimulated Optical Emission from Exchange-coupled Ions of Cr³⁺ in Al₂O₃."*Phys. Rev. Lett.*, **6**[3] 95-96 (1961).

18) A. L. Schawlow and G. E. Devilin "Simultaneous Optical Maser Action in Two Ruby Satellite Lines" *Phys. Rev. Lett.*, **6**[3] 96-98 (1961).

19) D. F. Nelson and M. D. Sturge. "Relation between Absorption and Emission in The Region of R Lines of Ruby," *Phys. Rev. A*, **137**[4A]1117-1130(1965)

ELECTRONIC CONDUCTION THROUGH A GRAIN BOUNDARY IN BaTiO₃ POSITIVE TEMPERATURE COEFFICIENT THERMISTORS

K. Hayashi, T. Yamamoto, Y. Ikuhara and T. Sakuma,
Department of Materials Science, The University of Tokyo, 7-3-1 Hongo, Bunkyo-ku, Tokyo 113-8656, Japan

ABSTRACT

The conduction mechanism through a single grain boundary in positive temperature coefficient thermistor was investigated by a combination of direct measurements using bicrystals of niobium-doped barium titanate and a device simulation analysis. The potential barrier model with a *n-i-n* structure can consistently interpret the observed electric properties rather than the double Schottky barrier model. The absence of a sub-ohmic behavior in the current-voltage characteristics and excess low-frequency capacitance at zero-bias are caused by the spatially distributed traps at the insulating layer.

INTRODUCTION

The electronic conduction through a grain boundary in positive temperature coefficient (PTC) thermistor has generally been understood from Heywang's theory [1] on the basis of a double Schottky barrier (DSB) model. Since some discrepancies between experimental observations and theoretical predictions have been found, several alternative models have been extended to explain the discrepancies by introducing an insulating layer formed by an intergranular second phase in grain boundaries [2, 3] or by enrichment of Ba-vacancies near the boundaries [4].

Mader *et al.* showed that the current-voltage (*J-V*) characteristics in the PTC range of a polycrystalline sample can be calculated by assuming a spatial extension of acceptor states [5]. We found a temperature-dependent *J-V* characteristic in a single grain boundary in 0.1mol% Nb-doped BaTiO₃, which cannot be explained from the DSB model [6]. The previous attempts to account for the nonlinear *J-V* characteristics in PTC thermistors have not been successful to explain the overall conduction properties. In this study, we tried to make a systematic analysis of the potential barrier by combining direct measurements at single grain boundaries and a semiconductor device simulation. This approach provides a far suitable explanation for the complex response of the potential barrier to changes in temperature and applied voltage.

To the extent authorized under the laws of the United States of America, all copyright interests in this publication are the property of The American Ceramic Society. Any duplication, reproduction, or republication of this publication or any part thereof, without the express written consent of The American Ceramic Society or fee paid to the Copyright Clearance Center, is prohibited.

METHOD

Direct Measurement of a Single Grain Boundary

Polycrystalline 0.1 mol% Nb-doped $BaTiO_3$ with a nominal [Ba]/[Ti, Nb] ratio of 0.9995 was fabricated by a conventional sintering technique. The materials were annealed at 1350°C for 6 h to obtain coarse grains by abnormal grain growth and then cooled down to 900°C with a cooling rate of 50°C/h followed by furnace-cooling to room temperature in air. Bicrystals or single crystals with $0.7 \times 0.25 \times 0.2$ mm in size were machined from the wafers. Electrodes were attached to both sides of the specimens using an ohmic In/Ga-Ag paste and Pt wires as shown in Fig 1. The resistance-temperature (R-T) and J-V characteristics were measured using a current source and a digital multimeter. The applied voltage was controlled to be 0.02 V in the R-T measurement to avoid nonlinear characteristics. The capacitance-voltage (C-V) characteristics were measured at 1.0 and 31.6kHz with a LCR meter. The AC modulation level was 0.02 V rms.

Modeling of Potential Barrier

The general semiconductor device simulation technique [7] for a one-dimensional problem was employed to analyze the measured electric properties. We examined two specific potential barrier models at a grain boundary; one is the double Schottky barrier (DSB) model and the other is the semiconductor-insulator-semiconductor (n-i-n) model, where an acceptor-rich insulating layer is introduced between the depletion layers of n-type semiconductors. The schematic band diagrams for the two models are depicted in Table I. The parameters for the potential barrier models were obtained by fitting the calculated R-T characteristics to the experimental one shown in the next section.

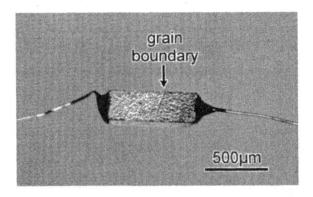

Figure 1   The optical micrograph of a bicrystal sample connected with electrodes.

Table I. Parameters Used for the Device Simulation

Potential barrier model	Energy of acceptor states, $E_C - E_A$ (eV)	Width of insulating acceptor region (nm)	Total effective density of acceptor states (cm^{-2})	Effective donor density (cm^{-3})
DSB MODEL 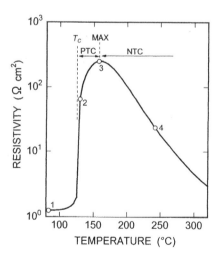	1.01	0	1.7×10^{14}	2.0×10^{18}
n-i-n MODEL	0.90	300	1.4×10^{14}	$2.0 \times 10^{18†}$

† The value was set to be zero at the acceptor rich insulating region.

RESULTS

Figure 2 shows the resistance-temperature (*R-T*) characteristic in a bicrystal of 0.1 mol% Nb-doped BaTiO$_3$ with a large PTC effect. The open circles indicate the temperature points at which *J-V* characteristics were measured. Each point is selected from the specific temperature ranges: (1) below Curie temperature T_C, (2) at the positive temperature coefficient (PTC) range, (3) at the maximum resistivity temperature (MAX) and (4) from the negative temperature coefficient (NTC) range.

Figure 2   Experimental *R-T* characteristics for a 0.1% Nb-doped BaTiO$_3$ bicrystal.
Open circles show the points where *J-V* characteristics are estimated.

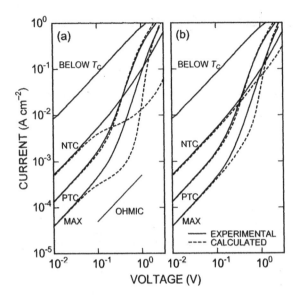

Figure 3 Solid lines show experimental J-V characteristics for the same specimen in Fig. 2. Dashed lines show J-V characteristics calculated with the DSB (a) and n-i-n (b) models

The effective donor density N_D was set to be $2.0 \times 10^{18}\,\mathrm{cm}^{-3}$ for the device simulation from the conductivity measurements of a single crystal examined in this study. The Curie-Weiss law has been adopted to describe the temperature dependence of the permittivity in the calculation. The Curie-Weiss constant C_0 was estimated to be 1.6×10^5 from the capacitance measurement of insulating BaTiO$_3$ ceramics.

The experimental J-V curves at the four temperature ranges are plotted by the solid lines in Fig. 3. The experimental J-V characteristics below T_C show ohmic behavior with $\alpha = 1$, where the nonlinearity coefficient α is defined as $d\log J/d\log V$. The nonlinear behavior is observed above T_C. The maximum α value of about 3 is obtained at 128°C. While we examined more than 20 specimens, no sub-ohmic behavior with $\alpha < 1$ was found above the Curie temperature.

The calculated J-V characteristics for DSB and n-i-n models are also plotted in Figs. 3(a) and (b), respectively. They are calculated from the device simulation technique using the parameters shown in Table I. A calculated curve of the DSB model in PTC range shows a fairly good agreement with the experimental observations, but not at the maximum resistivity temperature and NTC range due to the appearance of sub-ohmic behavior in the calculated J-V curves. Contrary to the DSB model, the J-V curves calculated from the n-i-n model show a good agreement with experimental curves for the whole temperature

Grain Boundary Engineering in Ceramics

ranges. The *n-i-n* model is preferable to describe the observed *J-V* characteristics rather than the DSB model.

Figure 4(a) shows the *C-V* characteristics measured at the same grain boundary in Figs. 2 and 3. At each temperature, the value of *C* is higher at 1 kHz than at 31.6 kHz especially for the DC bias of less than 1V. The capacitance difference between two frequencies can be regarded as excess low-frequency capacitance as observed in a Si grain boundary [8]. However, there is a clear difference in the *C-V* characteristic between Si and BaTiO$_3$ grain boundaries; the value of *C* at zero bias changes with frequency in BaTiO$_3$ in contrast to Si [8].

Figures 4(b) and (c) show the *C-V* characteristics calculated from DSB and *n-i-n* models. Direct current capacitance C_{dc} was calculated on the assumption that the occupation in the acceptor states follows completely the AC modulation. The charge trapping effect at the acceptor states is, therefore, involved in this calculation. The change in the occupation of acceptor states is not accounted for the calculation of high-frequency capacitance C_{hf}, which corresponds to geometric capacitance [8] contributed only from electron depletion layers in the potential barrier. A capacitance at an infinite frequency is considered to lie within the two limiting values of C_{dc} and C_{hf}.

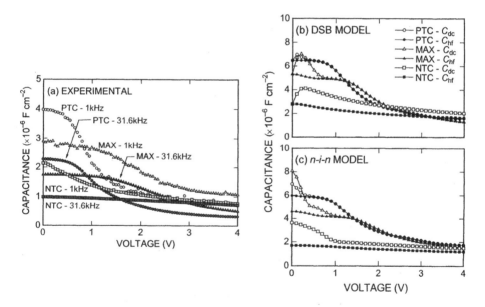

Figure 4   Measured *C-V* characteristics for the same specimen in Figs. 2 and 3 (a), and corresponding calculated characteristics in the DSB (b) and *n-i-n* (c) model.

In Figs. 4(b) and (c), the excess low-frequency capacitance appears at a DC-bias of less than about 1.5 V. Hence, the measured excess capacitance in Fig. 4(a) can be explained in terms of an interaction between electrons and acceptor states. It should be noted that the *n-i-n* model in Fig. 4(c) exhibits different values of C_{dc} and C_{hf} at zero-bias as found in the experimental *C-V* characteristics This is not expected from the DSB model as shown in Fig. 4(b).

DISCUSSION

The potential barrier of the *n-i-n* structure can reasonably describe the observed *J-V* and *C-V* characteristics in comparison with that of DSB. The difference in *J-V* characteristics between the two models are characterized by the appearance of sub-ohmic behavior with $\alpha < 1$ [6, 9].

Temperature and applied voltage dependence of the voltage drop at an insulating layer and depletion layers in the *n-i-n* model were analyzed to clarify the conduction mechanism through the potential barrier, and are given in Fig. 5. At lower external voltage, the voltage drop for the depletion layers is dominant in the PTC range, and the voltage drop for insulating layer rises as the temperature increases. At higher external voltage, the voltage drop at the depletion layers become dominant again. It was clarified from the analysis that the transitions are triggered by the nearly complete occupation of the acceptor states at the insulating layer. If the voltage drop at the depletion layers is dominant, the *n-i-n* potential barrier behaves as a DSB because the DSB consists of a double depletion layer. This can be confirmed by the fact that both DSB and *n-i-n* models show almost the same *J-V* characteristics for a PTC range as shown in Fig. 3. As the influence of acceptor-rich insulating layer becomes dominant, the conduction mechanism through the *n-i-n* potential barrier will show the characteristics of an insulator.

Nonlinear *J-V* characteristics in an insulator have often been discussed in terms of space charge limited current (SCLC). Nemoto and Oda measured *J-V* characteristics at single grain boundaries in PTC thermistor, and insisted that SCLC through an intermediate second phase is responsible for nonlinear current-voltage characteristics [3]. According to their interpretation, the observed *J-V* relationships with Child's law ($\alpha = 2$) [10] is the evidence for the SCLC. However, Child's law does not always appear in *J-V* characteristics as demonstrated in the present results.

It should be pointed out that the Child's law is not the evidence for SCLC through a grain boundary, because the assumptions of the original theory [10] are not necessarily satisfied in the potential barrier at a grain boundary. Although the Child's law is not observed in the present results, we believe that a similar effect to SCLC must be found in the current conduction through the *n-i-n* structure. The absence of sub-ohmic behavior in the *n-i-n* structure will be understood from the analogy with SCLC, where *J-V* curves never lie below the ohmic line [9].

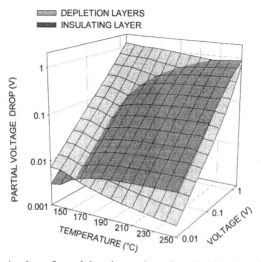

Figure 5   A plot of partial voltage dorp for double depletion layers and acceptor-rich insulating layer as functions of temperature and applied voltage.

As demonstrated in the previous section, the difference in *C-V* characteristics between two potential barrier models can be characterized by the frequency dependence of the capacitance at zero bias. The main contribution to the excess capacitance in Si has been explained from the interaction between conduction electrons and the grain boundary trap states [8]. However, this explanation is insufficient to understand the disagreement between C_{dc} and C_{hf} at zero-bias because no zero-bias excess capacitance was observed in a Si grain boundary.

The disagreement between C_{dc} and C_{hf} in the *n-i-n* model will be interpreted by the spatially distributed acceptor states at the insulating layer [9]. Since the occupation probability of the acceptor states whose energy level is close to the Fermi level can be drastically changed even by the small fluctuation of external voltage, spatial charge distribution at the insulating layer will be easily altered by the application of the small AC signal. The whole acceptor states at the insulating layer, therefore, act as a large dipole. The difference in observed capacitance with frequency must be a direct evidence of the spatial distribution of acceptor states.

According to Gerthsen and Hoffmann, a grain boundary has an intergranular second phase with a thickness of 100nm to 1μm and a dielectric constant of 40 to 400, because the depletion layers in DSB lead to much higher values of capacitance in comparison with the measured value [2]. However, the calculated capacitance in Fig. 4(c) agrees with the experimental values in Fig. 4(a) within a factor of about 2. Hence, it is not necessary to introduce such intergranular layer with relatively low dielectric constant to explain the experimental C-V characteristics. It seems reasonable that the insulating layer is attributed from BaTiO₃ bulk enriched with acceptor point defects [4], as has already been proposed.

We examined various widths of acceptor-rich insulating layers in the present calculation. The agreement between experimental and calculated J-V curves is essentially the same for the width from 200nm to 600nm. The width of the insulating layer must be on the order of 100nm. This value coincides with the prediction of thermodynamic and kinetic estimation [4].

CONCLUSION

The potential barrier with an acceptor-rich insulating layer on the order of 100nm can reasonably explain both J-V and C-V characteristics measured at a single grain boundary in 0.1 mol% Nb-doped BaTiO₃. The temperature dependent transition from conduction through DSB to SCLC in J-V characteristics can be interpreted from the n-i-n model. The low-frequency excess capacitance observed even at zero-bias is also attributed to the spatial distribution of the acceptor states in the insulating layer.

REFERENCES

[1] W. Heywang, "Resistivity Anomaly in Doped Barium Titanate," *J. Am. Ceram. Soc.*, **47** [10] 489-90 (1964).
[2] P. Gerthsen and B. Hoffmann, "Current-Voltage Characteristics and Capacitance of Single Grain Boundaries in Semiconducting BaTiO₃ Ceramics," *Solid State Electron.*, **16** [6] 617-22 (1973).
[3] H. Nemoto and I. Oda, "Direct Examinations of PTC Action of Single Grain Boundaries in Semiconducting BaTiO₃," *J. Am. Ceram. Soc.*, **63** [7-8] 398-401 (1980).
[4] J. Daniels and R. Wermicke, "New Aspects of an Improved PTC Model," *Philips Res. Rep.*, **31** 554-9 (1976).
[5] G. Mader, H. Meixner and P. Kleinschmidt, "Mechanism of Electrical Conductivity in Semiconducting Barium Titanate Ceramics," *Siemens Forsch.- u. Entwickl. -Ber.*, **16** [2] 76-82 (1987); **16** [4] 131-5 (1987).
[6] K. Hayashi, T. Yamamoto and T. Sakuma, "Grain Boundary Potential Barrier in Barium Titanate PTC Ceramics," *Key. Eng. Mater.*, **157-158** 199-206 (1999).
[7] S. Selberher, *Analysis and Simulation of Semiconductor Devices*, Springer-Verlag, Wien, 1984.
[8] C. H. Seager and G. E. Pike, "Anomalous Low-Frequency Grain-Boundary Capacitance in Silicon," *Appl. Phys. Lett.*, **37** [8] 747-9 (1980).
[9] K. Hayashi, T. Yamamoto and T. Sakuma, "Grain Boundary Electrical Barriers in Positive Temperature Coefficient Thermistors," *J. Appl. Phys.*, **86** [5], 2909-13 (1999).
[10] M. A. Lampert and P. Mark, *Current Injection in Solids*, Academic Press, New York, 1970.

INTERFERENCE OF ELECTRON WAVES AND ITS APPLICATION TO VISUALIZE ELECTROMAGNETIC MICROFIELDS

Tsukasa Hirayama, Kimiya Miyashita and Tomohiro Saito
Japan Fine Ceramics Center
2-4-1 Mutsuno, Atsuta-ku, Nagoya 456-8587, Japan

ABSTRACT

Electromagnetic microfields, related to electromagnetic functions of materials and devices, cannot be observed by conventional transmission electron microscopy. However, electron-wave interferometry, such as electron holography or three-electron-wave interference, enables us to visualize electromagnetic fields in microscopic regions. This paper presents their principles and some experimental results.

INTRODUCTION

Modern quantum mechanics requires the concept of duality for the existence of elementary particles such as electrons and photons. Although it is impossible to predict the behavior of one electron exactly, statistical behavior of large numbers of electrons can be considered as a complex wave function having a phase term and an amplitude term. When an electron wave passes through an electromagnetic microfield, only the phase term of the wave function is changed. This phase change, unfortunately, cannot be observed by conventional transmission electron microscopy, because the image is the square of the absolute value of the wave function i.e., the information of the phase change is lost. However, the phase change, having information of the electromagnetic fields, can be observed by interference of electron waves. Since electromagnetic functions are related to electromagnetic fields in materials and devices, this type of observation is important not only for pure physics but also for industrial applications.

To the extent authorized under the laws of the United States of America, all copyright interests in this publication are the property of The American Ceramic Society. Any duplication, reproduction, or republication of this publication or any part thereof, without the express written consent of The American Ceramic Society or fee paid to the Copyright Clearance Center, is prohibited.

Electron holography [1] is a useful technique for detecting the phase change of the electron wave and to visualize electromagnetic microfields. We have been using electron holography to study magnetic domain states of particles and thin films [2-4]. Furthermore, we recently developed a new method, three-electron-wave interference, which enables us to directly visualize electromagnetic microfields in real time [5,6]. In this paper we would like to describe these methods and present some of our experimental results.

ELECTRON HOLOGRAPHY

Brief history of electron holography

Electron holography was invented by Dennis Gabor [7] in 1948 to correct spherical aberration of electron lenses and to improve the resolution of electron micrographs. An application of electron holography to the observation of electromagnetic fields was proposed by Cohen [8] in 1967. Experimental studies were started in the early 1970's [9,10] and various magnetic substances were observed successfully in the 1980's [11-13]. New methods of electron holography were developed in 1990's and now electron holography is being used for some practical applications to study many types of materials or devices [14-16].

Experimental procedure of electron holography

Electron holography is a two-step imaging method. The first step is hologram formation and the second step is image reconstruction. Figure 1 shows the hologram formation in the first step realized by using a transmission electron microscope equipped with a field emission electron gun and an electron biprism [1]. An electron beam illuminates the specimen and the image is formed, through magnifying lenses, in the image plane. A reference beam passing though the vacuum beside the specimen also reaches the image plane. These two beams are superposed by an electron biprism to form an interference pattern. The interference pattern is recorded on film as a hologram.

Figure 2 shows an optical system, Mach-Zehnder interferometer, for the image reconstruction in the second step [1]. In this system, a hologram is illuminated with a laser beam and an original wave function beneath the specimen is reconstructed in the observation plane. At the same time, this laser beam is split by a half mirror and sent to the observation plane. These two beams interfere at the observation plane and form an interference micrograph, where the electric field is displayed as equipotential lines and the magnetic field is visualized as magnetic flux lines. Recently, the Fourier-transform method, digital processing by a computer equivalent to the optical reconstruction system, is also available.

Studies on magnetic domain states of barium ferrite particles

To observe magnetic materials, the magnetic field of the objective lens should be controlled to prevent specimens from changing their magnetic states.

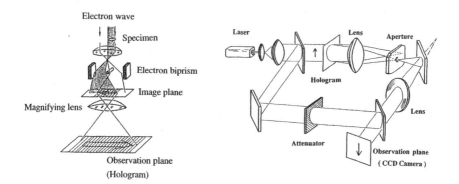

Figure 1. Formation of hologram   Figure 2. Optical reconstruction system

In the present experiment, for observing particles about 1 μm in size, the objective lens was turned off before specimens were installed into the electron microscope, and the image was focused using the first intermediate lens. Under this condition, the maximum direct-magnification of the electron microscope is about 2,000×, which is large enough to observe the interference area of more than 3 μm required to make holograms of such particles. To observe smaller particles, we employed a specially designed low magnetic-field pole piece [2,17] as shown in Fig. 3. The specimen is set above the lens gap and this objective lens is used with an electric current of about 1 A. Under this condition, the magnetic field at the specimen position is 0.55 mT and the focal length is 8.6 mm. Using this objective lens, the maximum magnification of the electron microscope is about 500,000×.

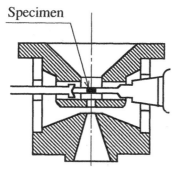

Figure 3. Cross sectional view of the low magnetic-field lens. The specimen is set above the lens gap. The focal length is 8.6 mm.

In general, magnetic substances have magnetic-domain structures to minimize their magnetic energy, which is the total of the magnetic domain-wall energy and magnetostatic energy. The magnetic domain-wall energy is proportional to the area (the square of diameter) of the domain wall. On the other hand, the magnetostatic energy is proportional to the volume (the cube of diameter) of the particle. Therefore, as the particle size decreases, the relative contribution of the domain-wall energy to the total magnetic energy of the particle increases. Finally, the single-domain state becomes the most stable below a certain critical size. The idea of the single-domain state in a small particle was first proposed by Frenkel and Dorfman [18] in 1930 and the critical size of the single-domain particles was, then, theoretically studied by Kittel [19] in 1946, Néel [20] in 1947 and Brown [21] in 1957. Proving the existence of single domain particles is very important for industry as well as scientific research of magnetism, because the maximum coercive forces are expected theoretically for all permanent magnets if these magnets are composed only of the single-domain particles of the materials. Although those theoretical studies were done many years ago, the existence of single-domain particles was not confirmed experimentally until Goto [22-24] et al. successfully observed domain structures of barium ferrite particles in 1980 by the application of the colloid-SEM method. In the present work, we first studied magnetic domain states of barium ferrite particles about 1 μm in size.

Barium ferrite (BaO · 6Fe2O3) is known to be a magnetoplumbite-type oxide having a hexagonal structure (a_0=0.5876 nm, c_0=2.317 nm) and it is widely used for permanent magnets. The present sample was prepared by Goto et al.[22-24]. The sample was confirmed by X-ray powder diffraction to be (BaO · 6Fe2O3) having the hexagonal structure. These barium ferrite particles were dispersed onto a thin carbon film and carbon was then deposited on the particles by vacuum evaporation in order to prevent charging up. Although most particles were sticking to each other to form large clusters of a few tens of micrometers because of their magnetic force, isolated particles were occasionally found. In the present experiment, some completely isolated particles were selected to study the domain structure of individual particles.

Figures 4(a), 4(b) and 4(c) show an ordinary electron micrograph, an hologram and an interference micrograph of a barium ferrite particle, respectively [2-4]. This particle is about 1 μm in size and has a well-defined crystal habit. The particle is so thick that its image contrast is entirely black. In Figure 4(b) information of the magnetic field around the particle is recorded as fringe-bending. Figure 4(c) shows that magnetic flux lines emerging from an N-pole and converging into an S-pole are very clearly seen. The shape of the magnetic flux lines indicates that this particle is in the single-domain state.

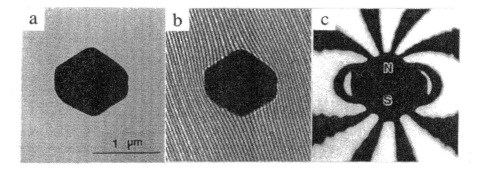

Figure 4. Barium ferrite particle (a) ordinary electron micrograph, (b) hologram, (c) reconstructed interference micrograph

THREE-WAVE INTERFERENCE

Experimental set up for three-wave interference

Electron holography is a useful method for observing electromagnetic microfields as described above. However, since holography is a two step imaging method, it was previously impossible to observe electromagnetic fields in real time. To overcome this problem, real-time observation systems were developed. In particular, an on-line real-time reconstruction system realized observation of magnetic phenomena dynamically by employing a liquid crystal panel to combine an electron holographic microscope and an optical reconstruction system [25,26]. Nevertheless, the dual operation of the electron holographic microscope and the reconstruction system was not an easy task.

Recently, we have developed a new method, realized by interference of three electron waves, for directly visualizing electromagnetic microfields. With this method, those fields are visualized only by an electron holographic microscope. In the present experiment, we used an electron holographic microscope, Hitachi HF-2000, equipped with a cold-type field-emission electron gun and two electron biprisms. Figure 5 shows an experimental set-up for the three-electron-wave interference. The two biprisms should be set in the same horizontal plane as illustrated in Fig. 5, but such an experimental set up is technically difficult. Therefore, one biprism (the upper biprism) was set between the objective lens and the first intermediate lens. The other (the lower biprism) was set between the first and second intermediate lenses. The objective lens was turned off before the experiment. The second intermediate lens and the two projector lenses were fully excited. The focus was adjusted by controlling the electric current of the first intermediate lens.

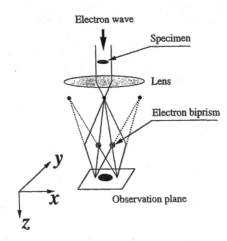

Figure 5. Schematic drawing of three-electron-wave interference using two electron biprisms. The z axis is the direction of electron-wave propagation. The image of the sample is formed in the $x - y$ plane. The two biprisms are set above the $x - y$ plane along the y axis. Thus the object wave is not deflected by the biprisms. Reference waves, however, are deflected in the $+x$ or $-x$ direction. The object wave and the two reference waves are superposed to form a three-wave interference pattern.

Visualization of electric field by three-wave interference

Latex is a type of electrically insulating organic material, and its particles, with uniform diameter, are often used for calibrating the magnification of electron microscopes. Figures 6(a) and 6(b) show an electron micrograph and a three-wave interference pattern of a latex particle on a thin carbon film, respectively [5,6]. The particle is a sphere of about $0.5 \mu m$ diameter, as seen in Fig. 6(a). In Fig. 6(b), the interference fringes are observed. In addition to the fringes, almost concentric equipotential lines of the electric field induced by charging up of the latex particle are observed as the intensity modulation of the fringes.

Observation of magnetic fields by three-wave interference

Figures 7 (a) and 7(b) show the three-wave interference pattern of a barium ferrite particle and its illustration, respectively [5,6]. In Fig. 7(a), magnetic flux lines emerging from the N-pole of single magnetic-domain particle and converging into the S-pole can be observed, as illustrated in Fig. 7(b).

Grain Boundary Engineering in Ceramics

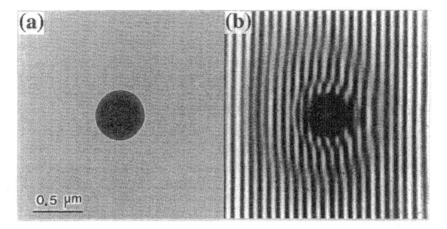

Figure 6. Latex particle on a thin carbon film. The particle is charged up by electron-beam irradiation and an electric field is produced around the particle: (a) electron micrograph, (b) interference pattern of three electron waves. Note that the concentric equipotential lines of the electric field are visualized as the intensity modulation of interference fringes.

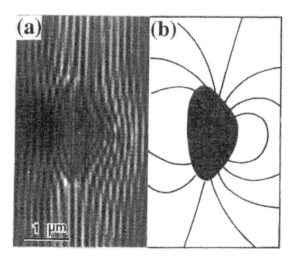

Figure 7. Single magnetic-domain particle of barium ferrite: (a) interference pattern of three electron waves, (b) illustration of the particle and magnetic flux lines.

CONCLUSIONS

We have presented two types of electron interferometry to visualize electromagnetic microfields; electron holography and three-electron-wave interference. Electromagnetic fields in microscopic regions are visualized by these methods. We believe that these techniques are useful for studying materials and /or devices having electromagnetic functions.

ACKNOWLEDGMENTS

We would like to thank Dr. Akira Tonomura of Hitachi Advanced Research Laboratory, Prof. Takayoshi Tanji of Nagoya University and Prof. Yuichi Ikuhara of Tokyo University for their encouragement and valuable suggestions throughout this work. We also thank Dr. Qingxin Ru for image reconstruction of electron holography. We are also grateful to Prof. Kimiyoshi Goto and Mr. Tomoaki Sakurai of Tohoku University for providing barium ferrite samples and for informative discussion. Prof. Guanming Lai of Shizuoka University and Prof. Nobuo Tanaka of Nagoya University are greatly acknowledged for discussion about spatial coherency of electron waves. This work was partially supported by New Energy and Industrial Technology Development Organization (NEDO).

REFERENCES

[1]   A. Tonomura, *Electron holography*, (Springer, Berlin, 1993).

[2]   T. Hirayama, Q. Ru, T. Tanji, and A. Tonomura, "Observation of magnetic domain states of barium ferrite particles by electron holography," *Appl. Phys. Lett.*, **63**,[3], 418-420 (1993).

[3]   T. Hirayama, J. Chen, Q. Ru, K. Ishizuka, T. Tanji, and A. Tonomura, "Observation of single magnetic-domain particles by electron holographic microscopy," *J. Electron Microsc.*, **43**[4], 190-197 (1994).

[4]   T. Hirayama, N. Osakabe, Q. Ru, T. Tanji and A. Tonomura, "Electron holographic interference micrograph of a single magnetic-domain particle," *Jpn. J. Appl. Phys.* **34**[6A], 3294-3297 (1995).

[5]   T. Hirayama, T. Tanji, and A. Tonomura, "Direct visualization of electromagnetic microfields by interference of three electron waves," *Appl. Phys. Lett.*, **67**[9], 1185-1187 (1995).

[6]   T. Hirayama, G. Lai, T. Tanji, N. Tanaka and A. Tonomura, "Interference of three electron waves by two biprisms and its application to direct visualization of electromagnetic fields in small regions," *J. Appl. Phys.*, **82**[2], 522-527 (1997).

[7]   D. Gabor, "A new microscopic principle," *Nature*, 161, 777-778 (1948).

[8]   M. Cohen, "Wave-optical aspects of lorentz microscopy," *J. Appl. Phys.* **38**[13], 4966-4976 (1967).

[9]   A. Tonomura, "The electron interference method for magnetization measurement of thin films," *Jpn. J. Appl. Phys.* **11**[4], 493-502 (1972).

[10] G. Pozzi and G. F. Missiroli, Interference electron microscopy of magnetic domains," *Journal de Microscopie*, **18**, 103-108 (1973).

[11] A. Tonomura, T. Matsuda, J. Endo, T. Arii and K. Mihama, "Direct observation of fine structure of magnetic domain walls by electron holography," *Phys. Rev. Lett.* **44**[21], 1430-1433 (1980).

[12] N. Osakabe, K. Yoshida, Y. Horiuchi, T. Matsuda, H. Tanabe, T. Okuwaki, J. Endo, H. Fujiwara and A. Tonomura, "Observation of recorded magnetization pattern by electron holography," *Appl. Phys. Lett.* **42**[8], 746-748 (1983).

[13] T. Matsuda, A. Tonomura, R. Suzuki, J. Endo, N. Osakabe, H. Umezaki, H. Tanabe, Y. Sugita and H. Fujiwara, "Observation of microscopic distribution of magnetic fields by electron holography," *J. Appl. Phys.* **53**[8], 5444-5446 (1982).

[14] M. R. McCartney, D. J. Smith, R. Hull, J. C. Bean, E. Völkl, and B. Frost, "Direct observation of potential distribution across Si/Si p-n junctions using off-axis electron holography," *Appl. Phys. Lett.* **65**[20], 2603-2605 (1994).

[15] M. Mankos, M. R. Scheinfein, and J. M. Cowley, "Absolute magnetometry at nanometer transverse spatial resolution: Holography of thin cobalt films in a scanning transmission electron microscope," J. Appl. Phys. **75**[11], 7418-7424 (1994).

[16] T. Tanji, Q.Ru, and A. Tonomura, "Differential microscopy by conventional electron off-axis holography," *Appl. Phys. Lett.*, **69**[18], 2623-2625 (1996).

[17] K. Shirota, A. Yonezawa, K. Shibatomi and T. Yanaka, "Ferro-magnetic material observation lens system for CTEM with a eucentric goniometer," *J. Electron Microsc.* **25**[4], 303-304 (1976).

[18] J. Frenkel and J. Dorfman, "Spontaneous and induced magnetisation in ferromagnetic bodies," *Nature* , **126**[3173], 274-275 (1930).

[19] C. Kittel, "Theory of the Structure of Ferromagnetic Domains in Films and Small Particles," *Phys. Rev.* **70**[11], 965-971 (1946).

[20] L. Néel, "MAGNÉTISME-Propriétés d'un ferromagnétique cubique en grains fins," *CR Acad. Sci.* **224**, 1488-1490 (1947).

[21] W. Brown, "Criterion for uniform micromagnetization," *Phys. Rev.* **105**[5], 1479-1482 (1957).

[22] K. Goto and T. Sakurai, "A colloid-SEM method for the study of fine magnetic domain structures," *Appl. Phys. Lett.* **30**[7], 355-356 (1977).

[23] K. Goto, M. Ito and T. Sakurai, "Studies on magnetic domains of small particles of barium ferrite by Colloid-SEM method," *Jpn. J. Appl. Phys.* **19**[7], 1339-1346 (1980).

[24]   K. Goto, "Experimental studies on verification of single-domain particles of ferromagnetic materials and on their critical sizes," *J. Jpn. Soc. Powd. Powd. Metall.* **36**[6], 761-769 (1989).

[25]   J. Chen, T. Hirayama, G. Lai, T. Tanji, K. Ishizuka, and A. Tonomura, "Real-time electron-holographic interference microscopy with a liquid-crystal spatial light modulator," *Opt. Lett.* **18**[22], 1887-1889 (1993).

[26] T. Hirayama, J. Chen, T. Tanji, and A. Tonomura, ""Dynamic observation of magnetic domains by on-line real-time electron holography," Ultramicroscopy **54**, 9-14 (1994).

ELECTRON TRANSPORT ACROSS BOUNDARIES IN Nb-DOPED SrTiO$_3$ BICRYSTALS

T. Yamamoto[*], Y. Ikuhara[*], K. Hayashi[+] and T. Sakuma[‡]
[*]Engineering Research Institute, School of Engineering, The University of Tokyo
2-11-16 Yayoi Bunkyo-ku Tokyo, 113-8656
[‡]Department of Advanced Materials Science, Graduate School of Frontier Science,
The University of Tokyo, 7-3-1 Hongo Bunkyo-ku Tokyo, 113-8656
[+]Graduate student

ABSTRACT

Current-voltage (I-V) characteristics in Nb-doped SrTiO$_3$ bicrystals were investigated. Bicrystals having two types of $\Sigma 1$ boundaries were fabricated by hot-joining technique. The boundary planes and the misfit angles of the bicrystals are (001) and $\approx 0.47°$, and near (012) and $\approx 1.7°$, respectively. High-resolution transmission electron microscopy (HREM) study associated with X-ray energy dispersive spectroscopy (EDS) revealed that the two single crystals contacted perfectly without any amorphous films in all of bicrystals. $\Sigma 1$ boundary of (001) type exhibited linear I-V relation which was similar to that of a single crystal. But non-linearity in I-V behavior clearly appeared even in the same boundary when cobalt ions concentrated in the vicinity of the boundary. On the other hand, the other type boundary exhibited a small non-linearity. In addition, the non-linearity became remarkable by a reduction of cooling rate after joining process. It could be concluded that the electrical transport across the grain boundaries in Nb-doped SrTiO$_3$ bicrystals depends on the distribution behavior of impurities and charged point defects in the vicinity of the grain boundaries.

INTRODUCTION

Electrical properties of polycrystalline electroceramic components such as boundary layer capacitors, thermistors, varistors are closely related to that of potential barriers at grain boundaries[1-3]. Therefore it is important to investigate the electrical transport behavior directly across a single grain boundary for understanding the electrical properties of a polycrystalline material. There are some reports about direct examinations at single grain boundaries and it has been revealed that the characteristics of the potential barriers are sensitive to the grain orientation relationship between adjacent grains[4-6]. Hayashi et al.

To the extent authorized under the laws of the United States of America, all copyright interests in this publication are the property of The American Ceramic Society. Any duplication, reproduction, or republication of this publication or any part thereof, without the express written consent of The American Ceramic Society or fee paid to the Copyright Clearance Center, is prohibited.

reported that resistivity jump in PTCR (positive temperature coefficient of resistivity) effect decreased with an increase in coherency at boundaries. They showed that no PTCR effect could be obtained in highly coherent $\Sigma 3$ boundary[4]. However, it is not cleared yet why electron transport changes by a variation of grain orientation relationship. This is because that the change in grain orientation relationship affects various grain boundary characteristics such as segregation of impurities, lattice mismatch, defect chemistry and so on, which have an influence on the electrical properties across grain boundaries. Therefore, it must be necessary to separate and to reveal each effect from the experimental point of view. For this purpose, it is very useful to use a model boundary such as a bicrystal.

In this study, we investigated the current-voltage (*I-V*) characteristics in the nearly perfect $\Sigma 1$ boundaries and small angle boundaries of semiconducting SrTiO$_3$ bicrystals.

EXPERIMENTAL PROCEDURE

Commercial 0.2at% Nb-doped SrTiO$_3$ single crystals with a size of 10x10x3mm^3 (Earth Chemical Co. Ltd.) were used for preparation of bicrystals. Four types of grain boundaries were prepared in this study as shown in Fig. 1 and Table I, respectively. The contact planes were ground and polished to a mirror-state finally with 0.25μm diamond slurry. As for the bicrystals having boundaries with cobalt ions, metallic cobalt was evaporated on the polished surface of a single crystal before joining. Then the two single crystals were joined by hot-pressing method at 1400°Cx10h under a pressure of 0.4MPa in air after adjusting the edges of both single crystals so as to have the aimed orientation relationship as shown in Fig. 1. As for the bicrystals with small angle boundaries, two kinds of cooling rates after contacting were used, i.e., 200°C/h and 50°C/h, respectively. After joining, plates were machined from the bicrystals parallel to (001) with a

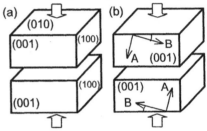

Figure 1 Schematics of SrTiO$_3$ bicrystals prepared in this study, (a) $\Sigma 1$ boundary and (b) a small angle boundary. The direction A is [010] and B [100] in (b). B direction is inclined at an angle of 22.5° from (010) and 67.5° from (100).

Table I   Characteristics in bicrystals prepared in this study

Boundary name	$\Sigma 1$ boundary as in (a) of Fig. 1		Small angle boundary as in (b) of Fig. 1	
Misfit angle*	0.47°		1.7°	
Characteristics	Cobalt evaporation		Cooling rate	
	Not evaporated	Evaporated	200°C/h	50°C/h

*measured values (see text)

thickness of 1mm. Current-voltage (*I-V*) characteristics were measured with the plates by a computerized system consisting of a current source (Keithley, model 220) and a voltage meter (Keithley, model 2010) at room temperature. After then, the boundaries were examined by a high-resolution transmission electron microscope (HRTEM) with a field emission type gun (TOPCON, EM-002BF) operated at 200kV. Specimen for HRTEM observation were prepared by a conventional ion-milling method with foils mechanically thinned to about 50μm thickness. Chemical analysis was carried out with a probe size of 1nm by X-ray energy dispersive spectroscope (Noran, Voyager System) attached to the microscope.

RESULTS AND DISCUSSION

Σ1 Boundary

Figure 2 shows a HRTEM image taken in the vicinity of a boundary in a cobalt-evaporated bicrystal. The observation direction is parallel to [001] direction of the plate and the boundary is set at the edge-on condition. A diffraction pattern inset in the figure is a selected area diffraction pattern including both crystals. It is noted that the boundary is free from secondary phases such as amorphous films and the two single crystals contact each other perfectly in an atomic scale. The coherency of the boundary is very high as indicated in the diffraction pattern. But one crystal is slightly inclined with respect to the other crystal so that grain boundary dislocations are introduced periodically along the boundary as indicated by the arrows in the figure. The interval of grain boundary

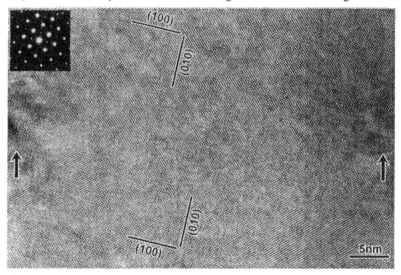

Figure 2   HRTEM image in Co-evaporated Σ1 boundary. In the figure the arrows show grain boundary dislocations.

dislocations is estimated to be about 48nm. As the coherency of the boundary is very high, the location of the boundary is determined with an array of such dislocations. In the case of the high coherent boundary such as the boundary as seen in the figure, the misfit angle between the two grains can be calculated from the interval of grain boundary dislocations. According to Frank's equation, a misorientation angle of a symmetric tilt boundary can be written as

$$d = |b| / 2\sin(\Delta\theta / 2),$$

where d is a spacing between adjacent grain boundary dislocations, $|b|$ magnitude of a Burgers vector of a grain boundary dislocation, and $\Delta\theta$ a misorientation angle, respectively[7]. Assuming that a Burgers vector is [001] type which is the largest one in $SrTiO_3$ lattice, the misorientation angle of the boundary is calculated to be about 0.47° from the equation. Thus the boundary is extremely near a

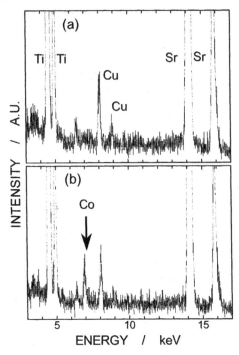

Figure 3　EDS profiles taken from (a) 100nm apart from a boundary and (b) a boundary in Co-evaporated Σ1 bicrystal.

perfect Σ1 boundary. Meanwhile, a similar orientation relationship could be observed in the Σ1 boundary without evaporation of cobalt.

Figure 3 shows EDS spectra taken from (a) the grain interior of 100nm apart from the boundary and (b) the boundary in the cobalt-evaporated bicrystal. The clear peak of Co Kα line clearly appears at the boundary as in (a) while it disappears at the grain interior as shown in (b). This result indicates that the evaporated cobalt distributes in the vicinity of the boundary after annealing for contacting. Since no secondary phases are observed at the boundary as seen in Fig. 2, cobalt ions can be considered to solve and to form a "chemical boundary" in the cobalt-evaporated bicrystal.

Figure 4 shows I-V relations across the boundaries in Σ1 and the cobalt-evaporated bicrystals, respectively. Σ1 boundary exhibits Ohmic relationship of α=1 as in the line A, where α is defined as a non-linearity coefficient of $(\partial \log J / \partial \log V)$. On the other hand, the I-V behavior clearly deviates from linear relationship to non-linear behavior after Ohmic relation in the cobalt-evaporated boundary as in the curve B. The result of Fig. 4 clearly

Grain Boundary Engineering in Ceramics

shows that the *I-V*
characteristic strongly depends
on the impurity segregation
irrespective of coherency at the
boundary.

Small angle boundary

Figure 5 shows a HRTEM
image of a small angle
boundary in as-joint state. In
the figure, observation
direction is nearly parallel to
[001] direction. This boundary
also contacts perfectly each
other and is free from
secondary phases. As seen in
the inset diffraction pattern, the
two crystals have a larger
misorientation angle
comparing with Σ1 boundary
as shown in Fig. 2. But the
angle is so small that the grain
boundary structure is
constructed with grain

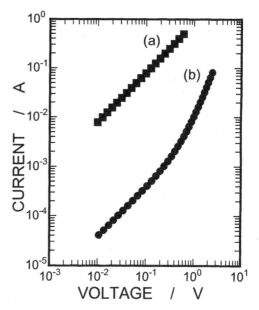

Figure 4 *I-V* relationship in (a)Σ1 boundary
and (b) Co-evaporated Σ1 boundary. Note that
clear non-linearity appears in Co-evaporated
Σ1 boundary.

boundary dislocations as indicated by the arrows in the figure. In the case of this
boundary, the boundary has a twist component in addition with a tilt one. The
misorientation angle of the boundary is estimated to be about 1.7°.

Figure 6 is a plot of the voltage dependence of the current in small angle boundaries
with two kinds of cooling rates. For comparison, the result obtained from Σ1 boundary
is also shown in the figure. *I-V* relationship is not largely different from Ohmic relation
in the boundary with a cooling rate of 200°C/h. But the slowly cooled boundary exhibits
clear non-linearity above the applied voltage of about 0.2V. As shown in Fig. 5, this
boundary is a type of a small angle one. Therefore, the boundary structure is constructed
with a dislocation network, whose structure is determined by a misfit angle between the
two crystals. Namely, its grain boundary structure should not change by a reduction of a
cooling rate because such process does not vary the misfit angle. Therefore, the
transition from near Ohmic to clear non-linear relation in *I-V* behavior as in Fig. 6 is not
caused by the change in the grain boundary structure, i.e., a dislocation structure.

The appearance of non-linearity in *I-V* behavior is closely related to the formation of
potential barriers, which generate when electrons in a conduction band are trapped at an
effective acceptor state in a band gap[8]. Two mechanisms can be considered for the

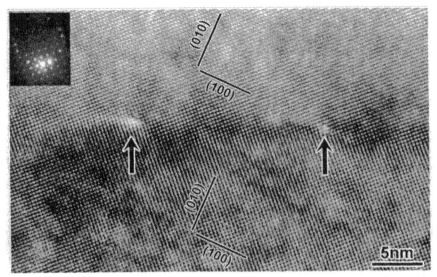

Figure 5   HRTEM image in small grain boundary. In the figure the arrows show grain boundary dislocations.

formation of the acceptor state, i.e., a lattice mismatch[9] and charging up of acceptor type point defects[10]. Since lattice mismatch takes place at grain boundaries because of a discontinuity of lattice arrangement, irregular bondings such as dangling bonds generate for the accommodation of the mismatch. Some of the irregular bondings often serve acceptor states. On the other hand, potential barriers are also formed when acceptor type point defects negatively charge up. This charging up takes place at the occurrence of the unbalance between acceptor and donor type defects. This unbalanced state is closely related to the annealing process. In general, point defects generate mainly at grain boundaries and they diffuse into grain interiors during heating up and annealing. On the contrary, they diffuse to grain boundaries and sink there during cooling. At this time, the defects whose diffusivity is higher will recover preferentially[11]. In the case of $SrTiO_3$, the diffusivity of an anion, i.e., an oxygen ion, can be considered to be higher than that of cations, i.e., strontium and titanium ions. Thus the layer enriched mainly with cation type defects will be formed in the vicinity of grain boundaries. If these defects charge up, electrons are trapped and potential barriers generate. In this study, as seen in Fig. 6, the non-linearity of I-V behavior in a small angle boundary became remarkable only by a reduction of a cooling rate. Thus the change in I-V behavior observed in a small angel boundary is not due to the change in the lattice mismatch but the distribution of defects in the vicinity of grain boundaries.

It has been revealed that the electrical properties at grain boundaries have dependency on the grain orientation relationship. Considering with the fact as mentioned above, it may be reasonable that such dependency results from grain orientation dependence of

Grain Boundary Engineering in Ceramics

the distribution of point defects. In other words, the generation/recover process of point defects in itself has the dependency of the grain orientation relationship.

CONCLUSIONS

The electron transport across single grain boundary was examined for $\Sigma1$ boundary with and without cobalt ions and small angle boundaries in 0.2at%Nb-doped $SrTiO_3$ bicrystals. The following conclusions could be obtained.

1. Two single crystals contact perfectly by hot-pressing method at 1400°C for 10h under a pressure of 0.4MPa in air. The grain boundaries were free from secondary phases such as amorphous films or participate.
2. The $\Sigma1$ boundary with cobalt ions exhibits clear non-liner I-V behavior while Ohmic relation can be obtained in $\Sigma1$ boundary without cobalt ions. Potential barriers are formed only by concentration of cobalt ions, which serve effective acceptor states in a band gap, even if the boundary has extremely high coherency.
3. Non-linearity in I-V relations of a small angle boundary became remarkable only by a reduction of cooling rate. It can be considered that the change in I-V relations caused by the difference in the cooling rate is related to the change in the generation/recover process of point defects in the vicinity of the boundary.

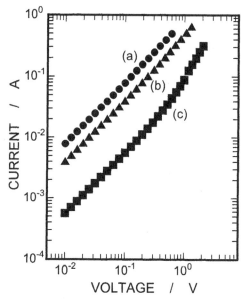

Figure 6 A plot of I-V behavior in small grain boundaries with a cooling rate of (b) 200°C/h and (c) 50°C/h. In the figure, I-V relation obtained in $\Sigma1$ boundary is also shown in (a). Note that the non-linearity becomes remarkable with a decrease in cooling rate.

REFERENCES

[1]G. Goodman, "Grain Boundary Phenomena in Electronic Ceramics," Advances in Ceramics, vol. 1, ed. by L. M. Levinson and D. C. Hill (1981).

[2]B. Huybrechts, K. Ishizaki and M. Tanaka, "The Positive Temperature Coefficient of Resistivity in Barium Titanate," *J. Mat. Sci.*, **30** 2463-74 (1995).

[3]T. K. Gupta, "Application of Zinc Oxide Vristors," *J. Am. Ceram. Soc.*, **73** [7] 1817-40 (1990).

[4]K. Hayashi, T. Yamamoto and T. Sakuma, "Grain Orientation Dependence of PTCR Effect in Niobium-Doped Barium Titanate", *J. Am. Ceram. Soc.*, **79** [6] 1669-1672 (1996).

[5]M. Kuwabara, K. Morimo, T. Matsunaga, "Single Grain Boundaries in PTC Resistors," *J. Am. Ceram. Soc.*, **79** [4] 997-1001 (1996).

[6]M. Nomura, N. Ichinose, H. Haneda and J. Tanaka, "Joining of Strontium Titanate Single Crystals and Their Electrical Characteristics," *Key Engineering Materials* **157-158**, 207-212 (1999).

[7]"Interfaces in Crystalline Materials," ed. by A. P. Sutton and R. W. Balluffi, Clarendon press, Oxford (1995)

[8]C. H. Seager and G. E. Pike, "Grain Boundary States and Varistor Behavior in Silicon Bicrystals," *J. Appl. Phys. Lett.*, **35** [9] 709-711 (1979).

[9]G. E. Pike and C. H. Seager , "The dc Voltage Dependence of Semiconductor Grain-Boundary Resistance," *J. Appl. Phys.*, **50**[5] 3414-3422 (1979).

[10]F. Oba, I. Tanaka, Shigeto R., H. Adachi, B. Slater, and David H. Gay, "Geometry and Electronic   Structure of [0001]/(1230) Σ=7 Symmetric Tilt Boundaries in ZnO", *Phil. Mag.* A **80** 1567-80 (2000).

[11]J. Daniel and R. Wermicke, "New Aspects of an Improved PTC Model," *Philips Res. Rep.*, **31** 554-59 (1976).

[12]Y. M. Chiang and T. Takagi, "Grain-Boundary Chemistry of Barium Titanate and Strontium Titanate: I, High-Temperature Equilibrium Space Charge", *J. Am. Ceram. Soc.*, **73** [11] 3278-85 (1990).

ACKNOWLEDMENT

A part of this study was financially supported by Nippon Sheet Glass Foundation for Materials Science and Engineering.

Structural Ceramics

HIGH TEMPERATURE PROPERTIES AND GRAIN BOUNDARY STRUCTURE IN SILICON NITRIDE BASED CERAMICS

M. Mitomo, T. Nishimura and Y. Kitami
National Institute for Research in Inorganic Materials
1-1, Namiki, Tsukuba-shi, Ibaraki, 305-0044, Japan.

ABSTRACT

The high temperature strength and creep resistance of an α-sialon ceramic was investigated before and after crystallizing the YAG($Y_3Al_5O_{12}$) phase. The crystallization has been carried out at lower temperatures than the sintering temperature. Dewetting of some α/α and α/YAG boundaries during crystallization resulted in improved mechanical properties. The high temperature properties of the silicon nitride ceramics have been also increased by crystallizing the $Yb_4Si_2O_7N_2$(J) phase at grain boundaries by sintering with Yb_2O_3 and SiO_2 additives. Although the simultaneous addition of alumina accelerated crystallization, creep resistance was impaired by the thickening grain boundary films and possibly by lowering viscosity. Results indicate that both grain boundary crystallization and the control of nanostructures at grain boundaries are important for optimizing high temperature properties of the ceramics.

INTRODUCTION

The sintering of silicon nitride ceramics has been investigated using oxide additives which form liquid phase at grain boundaries. The viscosity of the solidified glassy phase decreased at high temperature, which results in poor high-temperature mechanical properties. A number of attempts have been made to improve the high temperature properties of the ceramics, including the use of refractory additives (1,2) and crystallization of a grain boundary glassy phase by annealing (3,4). Although the partial crystallization generally confirmed by TEM investigations (4-6) increases the high temperature mechanical properties up to 1200 °C, these properties at higher temperature depend largely on the amount and the composition of residual glassy

To the extent authorized under the laws of the United States of America, all copyright interests in this publication are the property of The American Ceramic Society. Any duplication, reproduction, or republication of this publication or any part thereof, without the express written consent of The American Ceramic Society or fee paid to the Copyright Clearance Center, is prohibited.

phase. High temperature strength and creep resistance were quite recently improved by crystallization of refractory oxides, i.e. lanthanide disilicates, in silicon nitride ceramics (7,8). However, the presence of a thick glassy phase was still observed at the phase boundaries between silicon nitride and lanthanide disilicates. The complete crystallization of multi-grain junctions has not been successful as the chemical compositions are heterogeneously distributed in the glassy phases (9).

By controlling the chemical composition at grain boundaries in the α/β-sialon ceramics, the crystallization of YAG phase by low temperature annealing changed the grain boundary morphology and resulted in better creep resistance (10). The reaction of residual glassy phase with an α-sialon to form a β-sialon would be possible in such a phase composition, because the chemical composition of α-sialon is more nitrogen rich than β-sialon. Although a few investigations have tried to characterize nanostructures at grain and phase boundaries, the present work was conducted to investigate the relation between the high temperature properties and the nanostructure at the grain boundaries in α-sialon and silicon nitride ceramics. Special attention was paid to controlling the chemical composition at the grain boundaries to minimize the amount of residual glassy phase.

EXPERIMENTAL

The general formula for the Y-α-sialon is $Y_{m/3}(Si_{12-(m+n)}Al_{m+n})(O_nN_{16-n})$. The S-N bonds in the α-Si_3N_4 are replaced by m(Al-N) bonds and n(Al-O) bonds together with the interstitial dissolution of Y. The starting mixture in Si_3N_4-AlN-Al_2O_3-Y_2O_3 system for the Y-α-sialon with m=1.35 and n=0.675 and 10 wt% YAG phase was hot-pressed at 1750 °C for 1 h under 20 MPa. The materials were also subsequently annealed at 1450 °C for 72 h to crystallize the grain boundary glassy phase (11,12). Silicon nitride ceramics were hot-pressed with 4.7 mol% Yb_2O_3 and 1.1 mol% SiO_2 additives at 1750 °C for 4 h, and crystallized the $Yb_4Si_2O_7N_2$ (J phase) during sintering (13,14), which is isostructural to $Y_4Al_2O_9$ (YAM). In order to examine the effect of small difference in the grain boundary chemistry on the crystallization and high temperature properties, 0.5 wt % Al_2O_3 was also added simultaneously (15).

The crystalline composition and microstructural characterization were examined by X-ray diffraction (XRD) and scanning electron microscopy (SEM), respectively. The microstructures were also investigated by transmission electron microscopy (TEM). Bright-field (BF) as well as dark-field (DF) images were compared to know crystallization behavior. DF images will be shown in the present work. The grain boundary structures were observed by high resolution transmission electron microscopy (HREM). Some of the chemical compositions were investigated by energy dispersive X-ray

Grain Boundary Engineering in Ceramics

spectroscopy (EDX) and electron energy loss spectroscopy (EELS). Bending strength and creep resistance were measured by four-point bending tests with 10/20 mm spans and compressive tests, respectively.

RESULTS AND DISCUSSION

α-SIALON CERAMICS

The XRD results indicated that the as-sintered α-sialon ceramics were composed of an α-phase only, implying that the grain boundary phase is glassy. After low temperature annealing, the YAG phase crystallized together with a trace amount of β-sialon (12). The changes in the strength of α-sialon ceramics with temperature are shown in Figure 1(a). The strength was nearly constant up to 1400 °C in the annealed sialon whereas it decreases shapely in the as-sintered material. The better creep resistance for the annealed materials is shown in Figure 1(b). Figure 2 presents the general micro-structure of the α-sialon before and after annealing. The glassy phase in the as-sintered material was uniformly distributed at the grain boundaries and penetrated completely between the two α-grains. The crystalline YAG phase is more heterogeneously distributed as in Figure 2(b). Clearly, the dihedral angle between the α-grains is higher than for the as-sintered material. The thick intergranular glassy

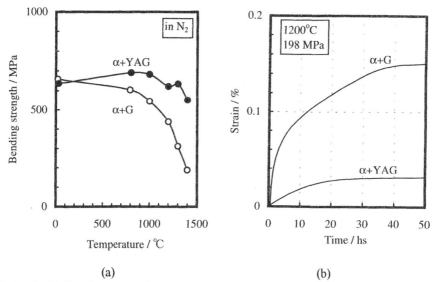

(a)   (b)

Figure 1. (a) Bending strength, and (b) compressive creep rate at 1200 °C under 198 MPa, of as-sintered and annealed α-sialon ceramics.

phase of about 2 nm is visible in the HREM in Figure 3(a) in between the α-grains of as-sintered material. However, complete elimination of the glassy phase from grain boundaries is clearly seen in Figure 3(b). The EELS analysis also indicated the elimination of Y from the boundaries. The coalescence and spherical morphology of

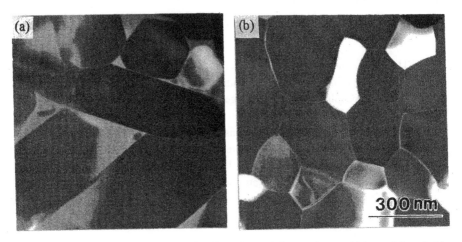

Figure 2. Dark-field TEM micrograph of as-sintered (a), and annealed α-sialon (white grains are YAG) (b).

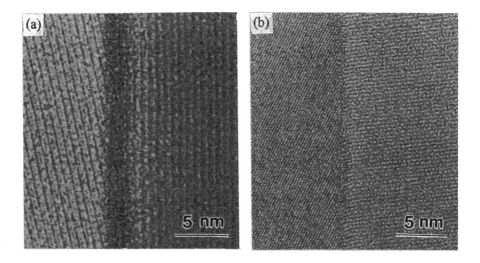

Figure 3. HREM micrograph of α/α boundary in as-sintered (a), and annealed α-sialon ceramics (b).

Grain Boundary Engineering in Ceramics

YAG might be related to higher interface energy in α/YAG boundaries. In α/β composite sialons (10), the reaction

$$\alpha\text{-sialon} + \text{glass} = \beta\text{-sialon} + \text{YAG} \tag{1}$$

might occur during low temperature annealing. Lower thermodynamic stability of α-sialon in presence of oxide glass might be the reason for the absence of intergranular films at grain boundaries. In the present materials, however, appreciable chemical reaction to form β-sialon is not detected. It suggests that the chemical composition at grain boundaries remains nearly the same during low temperature annealing. It is attributed to the control of grain boundary chemistry into (α+YAG) composition (11). It is also shown that large YAG grains are covered by thick glassy phase of about 10 nm as in Figure 4(a), whereas there is no intergranular phase in between α and small YAG grains (Figure 4(b)). These results indicates that annealing the α-sialon ceramics changed the grain boundary morphology from a continuous glass to isolated YAG. The dewetting of α/α and α/small YAG boundaries resulted in improved high temperature strength and creep resistance.

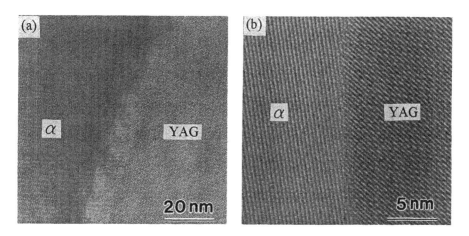

Figure 4. HREM micrograph of α/YAG boundary around large YAG (a), and small spherical YAG grain (b).

SILICON NITRIDE CERAMICS

The influence of the small change in the grain boundary chemistry was examined for the silicon nitride ceramics with $Yb_4Si_2O_7N_2$ (J) at the grain boundaries. This grain boundary phase is not compatible with silicon nitride in most lanthanide oxide systems.

The J phase only crystallizes during sintering in the Si_3N_4-Yb_2O_3-SiO_2 system (13). The results of the XRD reveal that the addition of a small amount of Al_2O_3 does not influence the crystallization behavior of the J phase. It is expected that Al_2O_3 dissolves into J phase to form the solid solution, J'. High temperature strength was similar for both types of materials, as shown in Figure 5(a). The slow crack growth rate was not influenced by the small difference in grain boundary chemistry. However, the creep rate at 1370 °C was significantly different, as shown in Figure 5(b). The accelerated creep rate due to adding alumina (at 1400 °C) has also been reported by Rendtel et al (16). They sintered silicon nitride with additions of 10 wt% Yb_2O_3 and 0.5 wt % Al_2O_3. Although crystallization of the J' phase was detected in their materials, the creep rate was about two order of magnitude higher than that in the present work. This suggests differences in the grain boundary structure due to different chemical compositions have a great effect on grain boundary diffusion.

The TEM micrographs of the materials with J phase show that most of the grain boundary phase in very thin sections was amorphous, although it was crystalline in thick sections. This suggests that the J phase easily vitrifies during polishing or ion etching. The vitrification has even been observed under electron beam irradiation. Crystalline J is shown in Figure 6(a). A higher amount of crystalline J' phase is visible in Figure 6(b), which indicates that the addition of a small amount of alumina

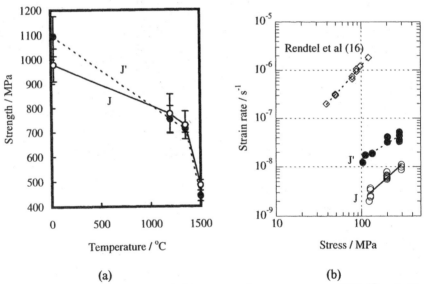

(a) (b)

Figure 5. (a) Bending strength, and (b) compressive creep rate at 1370 °C, of silicon nitride ceramics with J and J' phase at grain boundaries (data(16) were at 1400 °C).

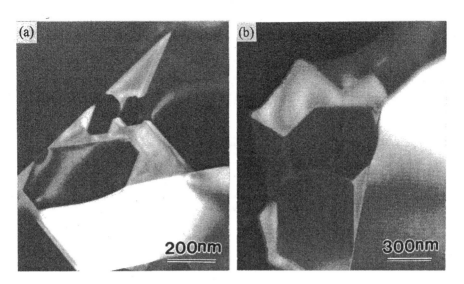

Figure 6. TEM micrograph of silicon nitride ceramic with J phase (a), and J' phase (b).

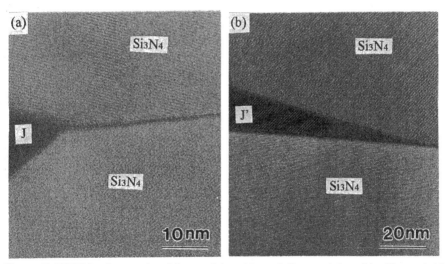

Figure 7. HREM micrograph of J containing (a), and J' containing (b) silicon nitride ceramics.

stabilized the structure of J. However, the HREM micrograph in Figure 7 shows that the grain boundaries between two silicon nitride grains were filled by a glassy phase, and that the dihedral angle was higher in the material with J phase than that with J'. A very thick glassy phase at this J'/ Si_3N_4 interface of about 2 nm was also shown in

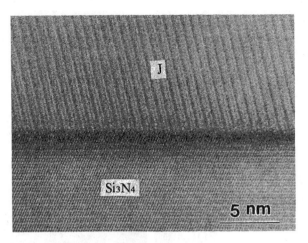

Figure 8. HREM micrograph of J/Si₃N₄ interface.

Figure 7(b). EDX analysis in the interface shows that the alumina content was less than the detection limit and was not concentrated in this residual glassy phase. In some J/Si₃N₄ phase boundaries, the absence or presence of a very thin intergranular phase if any has been observed, as shown in Figure 8. The larger amount and possibly lower viscosity of the glassy phase in the J' containing materials might be responsible for the lower creep resistance. Crystallization of refractory oxide, such as lanthanide disilicate, has been developed to improve high temperature strength, creep resistance and oxidation resistance (7,8). HREM observations have shown that the residual glassy phase is always present, not only in two Si₃N₄ grain boundaries, but also in the phase boundaries between Si₃N₄ and crystallized grains. The results suggest that further optimization might be possible by controlling nanostructures at grain and phase boundaries.

Incomplete crystallization might be due to the residual stress at the interface during cooling, due to the large differences in the mechanical and thermal properties between Si₃N₄ and grain boundary phase. Although large stresses are expected at the phase boundary between the β-Si₃N₄ and the J phase because of large differences in the thermal expansion coefficient of $3 \times 10^{-6}/°C$ and $10 \times 10^{-6}/°C$, respectively, good nanometric contact was realized, as shown in Figure 8. The network structure for the monoclinic J structure is composed of connecting Si₂(O,N)₇ double tetrahedra (17,18). The close similarity in the tetrahedra stacking of Si(O,N)₄, for example along the (201) of J phase and (110) of β-Si₃N₄, suggests that the topotactic growth of the J phase on the β-Si₃N₄ is possible in specific orientations. The presence of Al₂O₃ in liquid phase retards this overgrowth and forms thick glassy phase at phase boundaries. The differences in the nanostructures did not influence the slow crack growth behavior and the resultant high temperature strength. However, creep resistance depends largely on the nanostructure of the phase boundaries.

CONCLUSION

Y-α-sialon grains in as-sintered materials were completely surrounded by a glassy phase, which corresponded to the YAG in composition. After annealing at low temperatures, the grain boundary glass crystallized into isolated YAG grains. Most of the α/α grain boundaries were free from the glassy phase. A thick glassy phase was observed between α and large YAG phase boundaries, although no glassy film was detected between α and small YAG boundaries.

The crystallization of a J ($Yb_4Si_2O_7N_2$) phase was observed during the sintering of silicon nitride. The topotactic relation in some of the J/β-Si_3N_4 boundaries might contribute to good high temperature strength and creep resistance. The addition of a small amount of alumina retarded the formation of this type of nanostructure and resulted in lower high temperature properties.

Present work suggests that a glass-free interface is essential to improve high temperature mechanical properties.

ACKNOWLEDGEMENT

The authors are grateful to Prof. T.Sakuma, Prof. Y.Ikuhara and Dr. H.Yoshida of University of Tokyo, and Dr. H.Gu of Shanghai Institute of Ceramics, for their helpful discussion.

REFERENCES

1. G. E. Gazza, "Effect of Yttria Additions on Hot-Pressed Si_3N_4," Am. Ceram. Soc. Bull. 54[9], 778-780 (1973).
2. W. A. Sanders and D. M. Mieskowski, "Strength and Microstructure of Sintered Si_3N_4 with Rare-Earth-Oxide Additions," Am. Ceram. Soc. Bull. 64[2], 304-9 (1985).
3. A. Tsuge and K. Nishida, "High Strength Hot-Pressed Si_3N_4 with Concurrent Y_2O_3 and Al_2O_3 Additions," Am. Ceram. Soc. Bull. 57[4], 424-431 (1978).
4. K. Rajan and P. Sajgalik, "Local Chemistry Changes in Si_3N_4 Based Ceramics During Hot-Pressing and Subsequent Annealing," J. Eur. Ceram. Soc. 19, 2027-32 (1999).
5. C. T. Bodur, D. V. Szabo and K. Kromp, "Effects of Heat Treatments on the microstructure of a Yttria/Alumina-Doped Hot-Pressed Si_3N_4 Ceramic," J. Mater. Sci. 28, 2089-96 (1993).

6. D. R. Clarke and G. Thomas, "Microstructure of Y_2O_3 Fluxed Hot-Pressed Silicon Nitride," J. Am. Ceram. Soc. 61[3-4], 114-18 (1978).

7. M. K. Cinibulk, G. Thomas and S. M. Johnson, "Fabrication and Secondary-Phase Crystallization of Rare-Earth Disilicate-Silicon Nitride Ceramics," J. Am. Ceram. Soc. 75[8], 2037-43 (1992).

8. M. K. Cinibulk, G. Thomas and S. M. Johnson, "Strength and Creep Behavior of Rare-Earth Disilicate-Silicon Nitride Ceramics," J. Am. Ceram. Soc. 75[8], 2050-55 (1992).

9. Y. Bando, M. Mitomo and K. Kurashima, "An Inhomogeneous Grain Boundary Composition in Silicon Nitride Ceramics as Revealed by 300 kV Field Emission Analytical Electron Microscopy," J. Mater. Syn. Process. 6, 359-65 (1998).

10. K. H. Jack, "Sialon Ceramics: Retrospect and Prospect," pp. 15-27 in Silicon Nitride Ceramics, edited by I. W. Chen, P. F. Becher, M. Mitomo, G. Petzow and T. S. Yen. Materials Research Society, Pittsburgh, USA, 1993.

11. M. Mitomo and A. Ishida, "Stability of α-Sialons in Low Temperature Annealing ," J. Eur. Ceram. Soc. 19, 7-15 (1999).

12. T. Nishimura, M. Mitomo, A. Ishida and H. Gu, "Improvement of High Temperature Strength and Creep of α-Sialon by Grain Boundary Crystallization," pp. 171-4 in Creep and Fracture of Engineering Materials and Structures, edited by T. Sakuma and K. Yagi, Trans Tech Pub., Zurich, Switzerland, 1999.

13. T. Nishimura, M. Mitomo and H. Suematsu, "High Temperature Strength of Silicon Nitride Ceramics with Ytterbium Silicon Oxynitride," J. Mater. Res. 12, 203-9 (1997).

14. H. Yoshida, Y. Ikuhara, T. Sakuma, T. Nishimura and M. Mitomo, "High-Temperature Creep Resistance in Yb_2O_3-Fluxed Si_3N_4," pp. 653-62 in Creep and Fracture of Engineering Materials and Structures, edited by J. C. Earthman and F. A. Mohamed, The Minerals, Metals & Materials Society, USA, 1997.

15. T. Nishimura, M. Mitomo, A. Ishida, H. Yoshida, Y. Ikuhara and T. Sakuma, "Heat Resistant Silicon Nitride with Ytterbium Silicon Oxynitride," pp. 632-37 in 6th International Symposium on Ceramic Materials and Components for Engines, edited by K. Niihara, S. Hirano, S. Kanzaki, K. Komeya and K. Morinaga, Technoplaza Co., Tokyo, Japan, 1997.

16. A. Rendtel, H. Huebner and C. Schubert, "Improvement in the Creep Strength of HPSN by Optimization of Additive Content and Thermal Treatment," Key Eng. Mater. 89-91, 593-8 (1994).

17. D. P. Thompson, "The Crystal Chemistry of Nitrogen Ceramics," Mater. Sci. Forum 47, 21-42 (1989).

18. K. Liddell, D. P. Thompson, P. L. Wang, W. Y. Sun, L. Gao and D. S. Yan, "J-Phase Solid Solutions Series in the Dy-Si-Al-O-N System," J. Eur. Ceram. Soc. 18, 1479-92 (1998).

Grain Boundary Engineering in Ceramics

THE INFLUENCE OF GRAIN BOUNDARIES AND INTERPHASE BOUNDARIES ON THE CREEP RESPONSE OF SILICON NITRIDE

David S. Wilkinson
McMaster University
Department of Materials Science and Engineering
Hamilton, Ontario L8S 4L7, CANADA

ABSTRACT

The creep response of single-phase ceramic materials is often controlled by grain boundary diffusion creep processes. In multiphase ceramics such as silicon nitride however, the effect of grain boundaries and interfaces on creep is more complex, but nonetheless pervasive. In this paper I will summarize this behaviour with reference primarily to our own work in recent years.

INTRODUCTION

The creep behaviour of multiphase ceramic materials depends largely on two important factors. The first of these is the difference in intrinsic creep resistance of the individual phases that make up the material. At one limit we find materials in which both phases (in a two-phase material) contribute significantly to the overall creep response of the composite. Such is the case with Al_2O_3 / ZrO_2 composites [1] for example. In the other limit the creep resistance of on phase is so much greater than that of the other that it may be considered as purely elastic. This is the situation that will be addressed in this paper, since it applies to the creep of silicon nitride ceramics which contain a continuous amorphous phase. The second issue of concern is the degree of interconnectivity between these phases, which the microstructure exhibits. This can be characterized by considering the nature of percolation of the phases. In a previous paper [2] I identified two important thresholds. The first of can be labeled "point-contact percolation". This represents the classic form of percolation threshold in which a continuous network of contacts develops for the second phase across the microstructure. The onset of this behaviour is

To the extent authorized under the laws of the United States of America, all copyright interests in this publication are the property of The American Ceramic Society. Any duplication, reproduction, or republication of this publication or any part thereof, without the express written consent of The American Ceramic Society or fee paid to the Copyright Clearance Center, is prohibited.

found at around 16 vol% if both phases are of the same scale and are equiaxed. The percolation threshold falls considerably however when this is not the case (e.g. to a few percent for high aspect ratio whiskers). The other form of percolation can be labeled "facet-contact percolation", for which silicon nitride ceramics are a prime example. This occurs when a network develops in which the particles contact one another over a full facet. The percolation threshold in this case has not been analyzed in detail. It no doubt depends on processing conditions, but is thought to occur at around 70 vol%.

CREEP RESISTANCE OF SILICON NITRIDE

In the case of silicon nitride the crystalline phases (this is primarily crystalline Si_3N_4 but may also include secondary crystalline phases) are actually embedded in an amorphous material that forms the continuous phase of the composite. From a creep perspective this is critical and governs all of the high temperature behaviour of the material. The amorphous phase wets all except a small number of special grain boundaries. In materials processed under applied external pressure these grain boundary films develop a characteristic thickness which is typically $1 - 2$ nm [3,4,5,6]. The volume fraction of glass in most commercial silicon nitrides is at least several percent. However, from a creep perspective the effective volume fraction is just the grain boundary film thickness normalized by the grain size (typically about $0.2 - 0.5\%$) [7,8].

We have analyzed the creep response of a wide range of silicon nitride ceramics, both commercial and high purity materials [2]. The behaviour exhibited can be divided into two main categories. These are "exhaustion creep", in which an initially rapid creep rate falls off dramatically after a critical strain is reached, and "steady-state creep", in which a more-or-less constant creep rate is observed. Some materials exhibit a combination of these behaviours in which, after a critical strain, the creep rate falls to a lower but now steady-state value. We have attributed the exhaustion creep process to deformation dominated by viscous flow of the grain boundary glass. This process has been extensively modeled [7,9,10,11]. However, the essential result is that creep can only occur until a new equilibrium thickness is established amongst the grain boundary facets after a strain of order h/L, where h is the initial grain boundary film thickness and L is the grain size. This is generally close in value to the critical strain measured for exhaustion creep.

GRAIN FILM THICKNESS CHANGES DURING CREEP

We have now performed a series of detailed measurements on the thickness of grain boundary films both before and after creep deformation in silicon nitride. Film thicknesses have been measured by both lattice resolution imaging [5] and by Fresnel fringe imaging [12]. The first of these techniques gives the highest resolution with an accuracy of about ±0.1 nm. Fresnel fringe imaging has a slightly lower resolution of about ±0.15 nm. It has the advantage however of being easier to use, thus making it more amenable to gathering statistical information on film thickness for a range of grain boundaries in a single sample. We have therefore used both techniques. Both high purity [13,5,6] and commercial grades [14] of silicon nitride have been studied with similar overall conclusions. A typical distribution for a high purity silicon nitride is shown in Fig. 1.

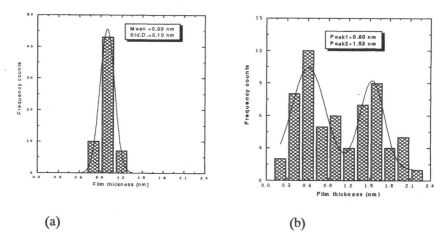

(a)   (b)

Figure 1: The distribution of grain boundary glass in a high purity silicon nitride (undoped) (a) before creep and (b) after 20h of creep at 1400°C under a stress of 200 MPa (total strain of 1.1%).

It is clear that prior to creep the distribution of film thicknesses is very narrow. In fact the standard deviation of the measurement is equal to that of the error associated with the measurement. After creep the distribution is clearly bimodal with a range of grain boundaries exhibiting film thicknesses of about 0.60 nm and a second peak at about 1.5 nm. The difference between these two stable film thicknesses is a measure of the redistribution that can occur via viscous flow of the glass between adjacent facets that are subjected to different loading conditions.   Figure 2 shows a typical creep curve for this material, along with the theoretical prediction of a model for viscous flow[11], using two different values of the initial grain boundary film thickness. The model for a thickness of 1 nm (close to that measured experimentally shown in Fig. 1a) underestimates the amount of creep that occurs prior to exhaustion, while a film thickness of 1.5 nm gives a considerably better fit.   This discrepancy could occur for several reasons. First of all the film thickness is measured at room temperature while we are really interested in this value at 1400°C, the creep temperature.   Secondly, the model used for this prediction assumes Newtonian behaviour of the glass, while the observation suggest slightly non-Newtonian behaviour with a power-law exponent of about 1.7.

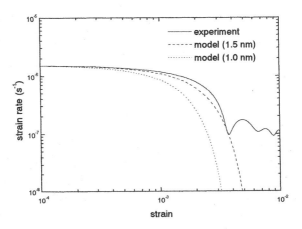

Figure 2: A plot of creep rate vs. strain (log-log scale) for a high purity silicon nitride, along with the predictions of a model for two values of the initial grain boundary film thickness.

A second check on the model can be obtained by predicting the effective viscosity of the composite from that of the grain boundary films, which, in this high purity silicon nitride, is essentially pure silica.   The experimentally-measured viscosity of the composite η ranges between $2 \cdot 10^{14}$ and $2.5 \cdot 10^{15}$ Pa-s at 1400°C, depending on the applied stress. The model suggests that this is related to that of the glass itself η_o through

Grain Boundary Engineering in Ceramics

$$\eta = \frac{18}{5} \frac{\eta_o}{f^3}$$

where f is the normalized grain boundary film thickness (see [10]). Using data for the viscosity of silica at this temperature we arrive at creep viscosity of between $1.1 \cdot 10^{15}$ and $3.6 \cdot 10^{15}$ Pa-s which agrees very well with the measurements.

One interesting question is whether or not, as the model predicts, all of the redistribution of glass occurs during the early stage creep. This is particularly important for materials which exhibit both exhaustion and steady-state creep behaviour. We have observed such behaviour in a high purity silicon nitride doped with 0.08 wt% Ba (see Fig. 3). Measurements taken after 7 hours of creep (just after the completion of exhaustion creep) and after 200 hours of creep show the same distribution of grain boundary glass. The only added feature is the appearance of a few very wide grain boundaries in the range of 3-4 μm thick. Further electron microscopy has revealed that these are all associated with cavities at adjacent triple junctions.

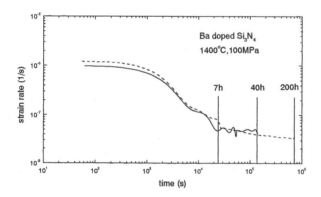

Figure 3: Creep behaviour of Ba-doped silicon nitride, exhibiting exhaustion creep followed by a new steady-state at a much lower strain rate.

We have recently studied this same phenomenon in a commercial silicon nitride, Norton NT-154 [15,14]. This material is doped with 4 wt% Y_2O_3 and densified by hot isostatic pressing. As a result it has a more complex microstructure than the high purity material, including secondary crystalline phases such as $Y_2Si_2O_7$ and $Y_5(SiO_4)_3N$ (Y,N apatite). In the as-fired (uncrept) material the boundaries between adjacent Si_3N_4 grains maintain amorphous films with a uniform thickness of 0.72 ± 0.13 nm (measured using the Fresnel fringe imaging technique). The thicknesses of the interphase boundaries are considerably larger and more variable, in the range $3.5 - 4.5$ nm. Despite this complexity the overall conclusions are unaltered. Thus if we consider the Si_3N_4 grain boundaries only following creep these once again become bimodal with peaks at 0.52 and 1.33 nm. We therefore conclude that this mechanism is not restricted to high purity materials but happens generally in all silicon nitrides that exhibit an initial period of exhaustion creep.

CONCLUSIONS

The amorphous nature of grain boundaries is of paramount importance to the creep resistance of silicon nitride ceramics. Over the past few years we have endeavoured to study this phenomenon in detail through high-resolution transmission electron microscopy. It is now possible to make statistically valid, quantitative measurements of the grain boundary film thickness. Our results show unequivocally that during creep the amorphous films are redistributed. This occurs during the first stage of creep (typically up to strains of about 0.3%). Following this the creep rate either drops to negligible levels or else a new mechanism (such cavitation or dissolution/reprecipitation creep) takes over.

ACKNOWLEDGEMENTS

I am most grateful to the Japan Society for the Promotion of Science for the provision of a visiting fellowship which allowed me to present this paper. This work has benefited from discussion with many colleagues and students over many year. Chief amongst these are Prof. G. C. Weatherly, and Dr. Q. Jin (McMaster University), Prof. J. R. Dryden (University of Western Ontario) and Dr. S. Wiederhorn, and Dr. W. Luecke (National Institute for Standards and Technology, U.S.A.)

REFERENCES

1. French, J.D., J. Zhao, M.P. Harmer, H.M. Chan, and G.A. Miller. "Creep of duplex microstructures." *Journal of the American Ceramic Society* 77(1994):2857-2865.

2. Wilkinson, D.S. "Creep mechanisms in multi-phase ceramic materials." *Journal of the American Ceramic Society* 81(1998):275-299.

Grain Boundary Engineering in Ceramics

3. Kleebe, H.J., M.K. Cinibulk, I. Tanaka, J. Bruley, R.M. Cannon, D.R. Clarke, M.J. Hoffman, and M. Rühle. High-resolution electron microscopy observations of grain-boundary films in silicon nitride ceramics. In: *Silicon Nitride Ceramics: Scientific and Technological Advances*. Edited by Chen, I.W. Pittsburgh, PA: Materials Research Society, 1993, p.65.

4. Tanaka, I., H.J. Kleebe, M.K. Cinibulk, J. Bruley, D.R. Clarke, and M. Rühle. "Calcium concentration dependence of the intergranular film thickness in silicon nitride." *Journal of the American Ceramic Society* 76(1993):911-915.

5. Jin, Q., D.S. Wilkinson, and G.C. Weatherly. "High resolution electron microscopy investigation of viscous flow creep in a high-purity silicon nitride." *Journal of the American Ceramic Society* 82(1999):1492-1496.

6. Jin, Q., D.S. Wilkinson, and G.C. Weatherly. "Redistribution of a grain boundary glass phase during creep of Si_3N_4 ceramics." *Journal of the American Ceramic Society* (1996):211-214.

7. Chadwick, M.M., D.S. Wilkinson, and J.R. Dryden. "Creep due to a non-Newtonian grain boundary phase." *Journal of the American Ceramic Society* 75(1992):2327-2334.

8. Dryden, J.R., D. Kucerovsky, D.S. Wilkinson, and D.F. Watt. "Creep deformation due to a viscous grain boundary phase." *Acta Metallurgica* 37(1989):2007-2015.

9. Chadwick, M.M., D.S. Wilkinson, and J.R. Dryden. "Creep due to a non-Newtonian grain boundary phase." *Journal of the American Ceramic Society* 75(1992):2327-2334.

10. Dryden, J.R., D. Kucerovsky, D.S. Wilkinson, and D.F. Watt. "Creep deformation due to a viscous grain boundary phase." *Acta Metallurgica* 37(1989):2007-2015.

11. Dryden, J.R. and D.S. Wilkinson. "Three dimensional analysis of the creep due to a viscous grain boundary phase." *Acta Metallurgica et Materialia* 45(1997):1259-1273.

12. Jin, Q., D.S. Wilkinson, and G.C. Weatherly. "Determination of grain-boundary film thickness by the Fresnel fringe imaging technique." *Journal of the European Ceramic Society* 18(1998):2281-2286.

13. Jin, Q., X.G. Ning, D.S. Wilkinson, and G.C. Weatherly. "Compositional dependence of creep behaviour of silicon nitride ceramics." *Journal of the Canadaian Ceramic Society* 65(1996):211-214.

14. Jin, Q., D.S. Wilkinson, G.C. Weatherly, W. Luecke, and M. Wiederhorn. "Viscous flow of grain boundary amorphous films during creep of a multiphase silicon nitride ceramic." *Journal of the American Ceramic Society* in press(2000)

15. Jin, Q., D.S. Wilkinson, and G.C. Weatherly. "The contribution of viscous flow to creep deformation of silicon nitride." *Materials Science and Engineering* A271(2000):257-265.

AN APPROACH TO
GRAIN BOUNDARY DESIGN USING CERAMIC BICRYSTALS

Y.Ikuhara, T.Watanabe, T.Yamamoto, H.Yoshida* and T.Sakuma*

Engineering Research Institute, School of Engineering, The University of Tokyo, 2-11-16, Yayoi, Bunkyo-ku, Tokyo 113-8656, Japan, ikuhara@sogo.t.u-tokyo.ac.jp
*Department of Advanced Materials Science, Graduate School of Frontier Sciences, The University of Tokyo, 7-3-1, Hongo, Bunkyo-ku, Tokyo 113-8656, Japan

ABSTRACT

Several kinds of alumina bicrystals were fabricated by a hot joining technique at 1500°C. The stability of grain boundary structure was evaluated by DV-Xα molecular orbital method, calculating the bond overlap population (BOP) and net charge (NC) as a function of the interplanar distance between two adjacent grains across the grain boundary. The most stable structure of symmetrical $\Sigma7[0\bar{1}11]/180°$ grain boundary was determined to be the structure that two oxygen-terminated $(01\bar{1}2)$ planes are joined each other at the interplanar distance of about 0.13nm. This result agrees well with the grain boundary atomic structure experimentally obtained from high resolution electron microscopy (HREM) analyses. In the [0001] symmetric tilt grain boundary, small angle grain boundaries were consisted of an array of partial dislocation with Burgers vector of $1/3[1\bar{1}00]$ to form the stacking faults between the dislocations. The behavior of grain boundary sliding was also investigated for typical grain boundaries by high-temperature creep test at 1400°C. As the result, the occurrence of grain boundary sliding was found to depend on the grain boundary characters.

INTRODUCTION

High-temperature mechanical properties of structural ceramics strongly depend on their grain boundary structures [1,2]. Grain boundary structure is also expected to be sensitive to the grain boundary characters, which has commonly been observed in metals [3]. Therefore an investigating the relationship between grain boundary structure and its characters is important to understand the effect of grain boundary structure on the mechanical properties in ceramics. For this purpose, bicrystal-experiments have an

To the extent authorized under the laws of the United States of America, all copyright interests in this publication are the property of The American Ceramic Society. Any duplication, reproduction, or republication of this publication or any part thereof, without the express written consent of The American Ceramic Society or fee paid to the Copyright Clearance Center, is prohibited.

advantage to be performed because grain boundary characters can be controlled and the fabricated boundaries are easily treated for the subsequent characterization.

In this study, several kinds of alumina bicrystals were fabricated, and the respective grain boundary structures were characterized by high-resolution electron microscopy (HREM). Furthermore, a first principles molecular orbital (MO) calculation was made using discrete-variational (DV)-Xα method [4] for a model cluster to understand the grain boundary stability, and the behavior of grain boundary sliding was investigated for typical grain boundaries by the compression creep test.

EXPERIMENTAL PROCEDURE

Single crystals of alumina (sapphire) with high purity of 99.99% were used to obtain the bicrystals of the [0001] symmetrical tilt grain boundaries including small angle boundaries, CSL boundaries and high angle boundaries [5]. In addition, the bicrystal with $\Sigma7[0\bar{1}11]/180°$ grain boundary was fabricated in order to calculate the chemical bonding state. In this case, the boundary planes were set to overlap their respective $(01\bar{1}2)$ planes [6,7], and one crystal is rotated by 180° around the $[0\bar{1}11]$ axis with respect to the other. These single crystals were cut by a diamond saw with the size of $10 \times 10 \times 5mm^3$ so as to have the surface corresponding to the grain boundary plane in the respective bicrystals. The surfaces of the blocks were then mechano-chemically polished using colloidal silica to a mirror state, and subsequently joined together so as to keep the desirable orientations at 1500°C for 10h in an air to obtain the bicrystals with the size about $10 \times 10 \times 10mm^3$. These bicrystals were cut to the several pieces for the subsequent characterization.

TEM specimens were prepared using a standard technique involving mechanical grinding to a thickness of 0.1 mm, dimpling to a thickness of 20μm and ion beam milling to electron transparency at about 4kV. HREM observations were performed using a Topcon 002B microscope with a point to point resolution of 0.18 nm. Image simulations were performed using a standard multi-slice program, and molecular orbital (MO) calculation was made by the discrete-variational (DV)-Xα method developed by Adachi [4]. By using a $(Al_2O_{10})^{14-}$ cluster model, surrounded by Mardellung potential, bond overlap population (BOP) and net charge (NC) at the grain boundary were calculated as a function of the interplanar spacing between two adjacent grains across the grain boundary.

Grain boundary sliding behavior was investigated for the [0001] 42° (high angle) grain boundary and $\Sigma7$ grain boundary by compression creep test under the constant stress of 16MPa at 1400°C in an air. The specimens for creep test were cut into the size of

Grain Boundary Engineering in Ceramics

5x5x7mm³, and the compression stress was applied to the surface of 5x5mm². In the specimen, grain boundary planes were set to incline by 45° with respect to the compression axis so that maximum shear stress can be applied to the grain boundary planes.

RESULTS AND DISCUSSION

In order to construct the atomic model of $\Sigma7[0\bar{1}11]/180°$ grain boundary (GB), the relative displacement between two crystals is needed to be determined. The displacement parallel to the GB can be fixed by considering a simple geometry between two crystals, but the displacement vertical to the GB must be determined by evaluating the bonding state as a function of the displacement. That is, BOP and NC are required to be obtained as a function of the displacement perpendicular to the GB. In this procedure, as a first approximation, lattice relaxation was not taken into account.

For setting up the optimum displacement parallel to the GB, we assume that crystal A and crystal B are connected by the Al-O bonding across the GB. Figure 1 is an atomic projection observed perpendicular to the GB, and the following procedure is required to construct the atomic model.

　　1. The distance between Al in crystal A and O in crystal B (d_1 in Fig.1) is set so as to be equal to that between O in crystal A and Al in crystal B (d_2 in Fig.1). This procedure is needed to keep force balance along the x-axis and minimize the Al-O bond length across the GB to form the strongest bonding at the boundary.

　　2. The distance between O in two crystals is set so as to have the same interatomic distance (d_3 and d_4 in Fig.1). This procedure is needed to keep force balance along the y-axis and maximize the O-O bond length across the GB to minimize the repulsive force between atoms. Thus, the relative displacement parallel to the GB along x-axis can be determined by taking into account the condition 1, and the relative displacement along the y-axis can be determined by the condition 2.

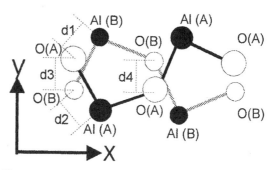

Figure 1   Atomic arrangement in the vicinity of the $\Sigma7$ grain boundary, which is project along the normal to the boundary plane. A and B represent the different two crystals across the grain boundary.

As a next step, the relative displacement perpendicular to the GB should be determined, keeping the relative displacement parallel to the GB obtained in the above. Here, d represents the interplanar distance between the oxygen-terminated surface in crystal A and B. The displacement perpendicular to the GB was determined by calculating BOP and NC at every 0.01nm as a function of d. BOP used in the present study was an average BOP between Al in crystal A and O in crystal B, and between Al in crystal B and O in crystal A.

Figure 2(a) shows the Al-O BOP as a function of d, indicating that the BOP is maximized at around d=0.13nm. Since BOP corresponds to the degree of covalency, the degree of covalency is considered to be maximum at around d=0.13nm. Fig. 2(b) shows the Coulomb's force of Al-O bonding perpendicular to the GB as a function of d, which was obtained by calculating the values of NC. In this calculation, the charges of Al and O were assumed to be the respective NC. This result indicates that the ionicity also maximizes at around d=0.13nm. Consequently, the most stable atomic structure in $\Sigma 7[0\bar{1}11]/180°$ grain boundary will be such that the two oxygen-terminated $(01\bar{1}2)$ plane are joined each other with the distance of about 0.13nm, keeping the geometric relationship mentioned above.

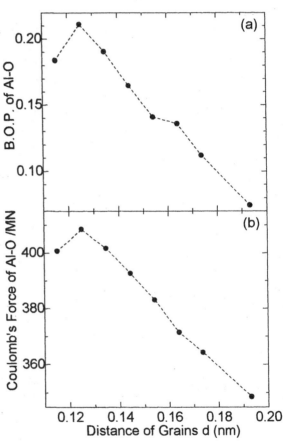

Figure 2 (a) Bond overlap population of Al-O bonding and (b) Al-O Coulomb's force perpendicular to the grain boundary as a function of distance of two adjacent grains.

Figure 3 is a high resolution electron micrograph of the Σ7 [0$\bar{1}$11]/180° grain boundary. The observed direction is parallel to the [2$\bar{1}$$\bar{1}$0] for both adjacent grains. The set of (0001) and (0$\bar{1}$11) planes can be seen in each grain, and the grain boundary plane is identified to be the (01$\bar{1}$2) plane. As seen in the micrograph, two crystals are joined together with an atomic scale. The atomic model obtained by the DV-Xα method is shown in Figure 4(a) where the oxygen atoms correspond to small circles and the aluminum atoms large ones. The parameters of the super cell for the HREM image simulations were 3.683× 3.589× 0.475nm with 742 atoms. The calculations were actually performed over a range of thickness from 2 to 20nm, and a range of defocus from 20 to -60nm. The best matching was obtained at a thickness of 11nm and defocus of -12nm as shown in Fig.4(b). This simulated image agrees well with the experimental image in Fig.4(c), which suggests that the actual structure of the grain boundary is similar to that in Fig.4(a). In conclusion, the DV-Xα calculation is considered to be effective to understand the actual atomic structure in grain boundaries in ceramics

Figure 5 shows (a) high-resolution electron micrograph of the [0001] 10° small angle tilt gain boundary. Dislocation contrast can be observed periodically along the grain boundary as indicated by the arrows. Fig. 5(b) is a Burgers circuit around a dislocation shown in (a). From this circuit, Burgers

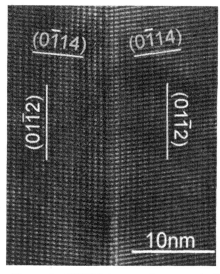

Figure 3   HREM image of symmetrical Σ7 [0$\bar{1}$11]/180° grain boundary in alumina bicrystal.

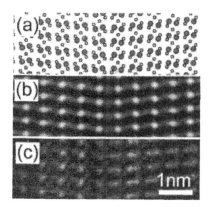

Figure 4   (a) A model of the Σ7 [0$\bar{1}$11]/180° symmetrical grain boundary which is constructed from the present study and (b) computer-simulated and (c) experimental HREM image obtained at the sample thickness of 11nm and defocus of -12nm.

vector of the dislocation can be identified to be $1/3[1\bar{1}0n]$ since the circuit is just a projected circuit along the [0001] direction. As candidate Burgers vectors, $b_1=1/3[1\bar{1}02]$, $b_2=1/3[1\bar{1}01]$ and $b_3=1/3[1\bar{1}00]$ can be considered, in which b_1 and b_2 are perfect dislocations and b_3 is partial dislocation. The sizes of Burgers vectors are $|b_1|=0.908\text{nm}$, $|b_2|=0.512\text{nm}$ and $|b_3|=0.274\text{nm}$, respectively, and, thus, b_3 has the smallest vector among them. In addition, if b_1 and b_2 dislocations are formed along the grain boundary, large twist component is introduced at the boundary. Consequently, b_3 dislocation should be formed periodically at an interval spacing of about 2nm to compensate the misorientation angle of 10° in this case. This can be reasonably explained by Frank's formula [8].

Since the dislocation of $1/3[1\bar{1}00]$ is a partial dislocation, stacking faults are considered to be introduced on the boundary plane between two dislocations for the small angle grain boundaries. Here, grain boundary energy for dislocation boundary can be calculated by the next equation [8].

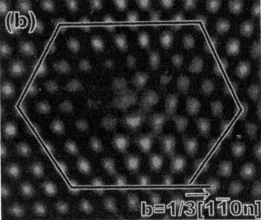

Figure 5   HREM images of (a) the [0001] 10° tilt grain boundary and (b) Burgers circuit around the dislocation in (a).

$$\gamma_{gb} = \frac{Gb\theta}{4\pi(1-v)}\left[\ln\left(\frac{b}{2\pi r_0\theta}\right)+1\right] \qquad (1)$$

Grain Boundary Engineering in Ceramics

where G, ν and r_0 are shear modulus, Poisson's ratio and core radius, respectively. In the case of small angle grain boundary in alumina in this study, grain boundary is consisted of partial dislocations with the Burgers vector of $1/3[1\bar{1}00]$ and the stacking fault on the $(1\bar{1}00)$ planes. Since the perfect dislocation with Burgers vector of $[1\bar{1}00]$ is dissociated to three partial dislocations with the Burgers vector of $1/3[1\bar{1}00]$, the stacking faults are formed on the 2/3 area of the whole grain boundary plane. Grain boundary energy of small angle grain boundary can be, therefore, expressed as the following equation.

$$\gamma_{gb} = \frac{Gb_p\theta}{4\pi(1-v)}\left(\ln\left(\frac{b_p}{2\pi r_0\theta}\right)+1\right)+\frac{2}{3}\gamma_{sf} \qquad (2)$$

where b_p is a Burgers vector of a partial dislocation and γ_{sf} is stacking fault energy. The theoretical curve obtained from this equation is inset as the solid line together with the experimental data in Figure 6 , where 130MPa, 0.24, 0.274nm, $1.5b_p$ and 0.1J/m^2 are used for G, ν, b_π, r_0 and γ_{sf} [9]. As seen in Fig. 6, the calculated curve agrees well with the experimentally measured dependence. The dotted curve is the theoretical curve due to the equation (1), which is not consistent with the experimental data.

As reported in the paper [5] in this volume, grain boundary energy strongly depends on the grain boundary characters. HREM observations revealed that low Σ CSL boundaries have coherent boundaries and high angle boundaries have a relaxed layer with the thickness less than 1nm [5]. This indicates that grain boundary atomic structure is very different between CSL boundary and high angle boundary. The behavior of grain boundary sliding was investigated for such typical grain boundaries, i.e., 38.2° (Σ7) grain boundary and 42° (high angle) grain boundary. Figure 7 shows optical micrographs of the specimens after creep test for (a) Σ7 grain boundary and (b) 42° (high angle) grain boundary.

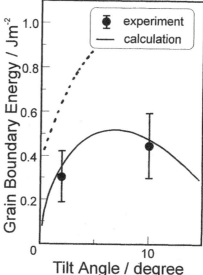

Figure 6   Grain boundary energy as a function of tilt angle for the [0001] symmetrical tilt boundaries. Solid and dotted lines represent the curves calculated by equation (2) and (1), respectively.

Σ7 grain boundary showed fracture along the boundary during creep test, but 42° grain boundary clearly showed the traces of grain boundary sliding. Judging from the trace on the surface, the distance of grain boundary sliding is 17μm which corresponds to 0.24% as strain along the grain boundary. This indicates that grain boundary sliding easily takes place in high angle grain boundaries, but not in CSL grain boundaries.

Figure 8 shows a high-resolution electron micrograph of the 42° (high angle) grain boundary after creep test for about 350ks under the stress of 16MPa at 1400°C. It can been seen that the shape of the grain boundary is wavy, and grain boundary planes move so as to be parallel to $(\bar{2}110)$ or $(1\bar{2}10)$ plane in the two crystals. This would be an evidence that grain boundary sliding accompanies grain boundary

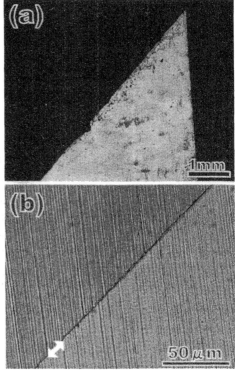

Figure 7 Optical micrographs of the samples after creep test for (a) Σ7 and (b) 42° (high angle) grain boundaries.

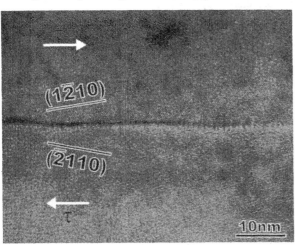

Figure 8 HREM image of the 42° (high angle) grain boundary after creep test under the stress of 16MPa at 1400°C.

Grain Boundary Engineering in Ceramics

migration. Assuming that atomic steps are existed on the grain boundary plane, atom diffusion would occur to accommodate the lattice strain at the step during grain boundary sliding [10]. This diffusion is supposed to make the boundary shape to be wavy so as that the boundary plane moves to the low index planes in the adjacent crystals. The detailed mechanism is under consideration.

CONCLUSIONS

Several kinds of alumina bicrystals were fabricated by hot-joining method. A first principles molecular orbital (MO) calculation, HREM analyses and creep tests were applied for the bicrystals in order to understand the chemical bonding state, atomic structure and mechanical properties of the respective bicrystals. The following results were obtained.

1. The displacement parallel to the GB between two crystals can be determined by a simple geometry, and the displacement perpendicular to the GB can be determined by calculating the BOP and NC between Al-O bonding across the GB.

2. The atomic structure of $\Sigma 7[0\bar{1}11]/180°$ grain boundary is formed so that the two oxygen-terminated ($01\bar{1}2$) planes are joined each other at a interplanar distance of about 0.13nm. This result agrees well with the experimental high resolution image of the grain boundary.

3. Small angle grain boundaries are consisted of a periodic array of partial dislocations with Burgers vector of $1/3[01\bar{1}0]$. Theoretical grain boundary energy, which was calculated by combining dislocation strain energy and stacking fault energy, agrees well with the experimentally obtained results.

4. Grain boundary sliding is hard to occur in low Σ CSL grain boundary, but not in high angle grain boundary at the present experimental condition. The occurrence of grain boundary sliding depends on the grain boundary characters.

ACKNOWLEDGMENTS

This work was supported by the Grant-in-Aid for Scientific Research (11450260) from the Ministry of Education, Science and Culture, Japan. A part of this work was also supposed by Toray Science Foundation.

REFERENCE

[1] S. Primdahl, A. Tholen and T. G. Langdon, "Microstructural examination of a superplastic yttria-stabilized zirconia: Implications for the superplasticity mechanism", *Acta Mater.*, **43** 1211-18 (1995).

[2] Y. Ikuhara, P. Thavorniti and T. Sakuma, "Solute segregation at grain boundaries in superplastic SiO$_2$-doped TAP", *Acta Mater.*, **45**, 5275-84 (1997).

[3] T.Watanabe, "Past, present and future of grain boundary engineering", *Ceram. Trans.*, in this volume (2000).

[4] H. Adachi, M. Tsukada and C. Satoko, "Discrete variational Xα cluster calculations I: application to metal clusters", *J.Phy.Soc.Jpn.*, **45** 875-83 (1978).

[5] T.Watanabe, H.Yoshida, T.Yamamoto, Y.Ikuhara and T.Sakuma, "Atomic structures and properties of systematic [0001] tilt grain boundaries in alumina", *Ceram. Trans.* in this Volume (2000).

[6] F.R.Chen, C.C.Chu, J.Y.Wang and L.Chang, "Atomic Structure of $\Sigma7(01\bar{1}2)$ symmetrical tilt grain boundaries in α-Al$_2$O$_3$", *Phil.Mag.A.*, **72** 529-44 (1995).

[7] T.Watanabe, Y.Ikuhara and T.Sakuma, "Evaluation of atomic grain boundary structure in alumina by molecular orbital method", *J.Ceram.Soc.Jpn.*, **106** 888-92 (1998).

[8] A.P.Sutton and R.W.Balluffi, "Interfaces in Crystalline Materials", Oxford Science Pub. (1995).

[9] K. D. P. Lagerlof, T. E. Mitchell, A. H. Heuer, J. P. Riviere, J. Cadoz, J. Castaing and D. S.Phillips, "Stacking Fault Energy in Saphire (α-Al$_2$O$_3$)", *Acta Mater.*, **32**, 99-105 (1984).

[10] M. F. Ashby, "Boundary Defects, and Atomistic Aspects of Boundary Sliding and Diffusional Creep", *Surface Science* **31** 498-542 (1972).

MECHANICAL PROPERTIES AND THERMAL STABILITY OF OXIDE EUTECTIC COMPOSITES AT HIGH TEMPERATURES

Yoshiharu Waku, Shin-ichi Sakata, Atsuyuki Mitani and Kazutoshi Shimizu
Japan Ultra-high Temperature Materials Research Institute
573-3 OKIUBE UBE City Yamaguchi, 755-0001

ABSTRACT

A novel eutectic Al_2O_3/YAG composite has recently been fabricated by accurately controlling the unidirectional solidification. The eutectic composite has a unique microstructure, in which single crystals of Al_2O_3 and single crystals of YAG are three-dimensionally and continuously connected and finely entangled without grain boundaries. A dislocation structure is observed in both Al_2O_3 and YAG phases of compressively crept specimens, indicating that the plastic deformation occurred by dislocation motion. The eutectic composite fabricated is thermally stable and has the following properties: the flexural strength at room temperature can be maintained up to just below the melting point (about 1830 ℃); the compression creep strength at 1600 ℃ and the strain rate of 10^{-4}/sec is about 13 times higher than that of sintered composites of the same composition; and no grain growth occurs, even upon heating at 1700 ℃ in an air atmosphere for 1000 hours.

INTRODUCTION

D. Viechnicki et al[1] conducted microstructural studies on an Al_2O_3/YAG system grown by the Bridgman method, and showed that the microstructure of the composite could be controlled by unidirectional solidification. In addition, it has recently been reported that a unidirectionally solidified Al_2O_3/YAG eutectic

To the extent authorized under the laws of the United States of America, all copyright interests in this publication are the property of The American Ceramic Society. Any duplication, reproduction, or republication of this publication or any part thereof, without the express written consent of The American Ceramic Society or fee paid to the Copyright Clearance Center, is prohibited.

composite has superior flexural strength, thermal stability and creep resistance at high temperatures[2-4], and therefore is a candidate for high-temperature structural materials. However, since the composite consists of many eutectic colonies, a fairly strong influence of colony boundaries on these properties may be predicted.

On the other hand, Waku et al[5-7] have recently fabricated a eutectic composite consisting of a single crystal Al_2O_3 and a single crystal YAG with neither colonies nor grain boundaries, using a unidirectional solidification method. In this paper, the high temperature strength, compressive creep strength and thermally stability, and microstructure of these Al_2O_3/YAG eutectic composites are discussed.

EXPERIMENTAL PROCEDURE

Manufacturing of Raw powder

Using commercially available Al_2O_3 powder (AKP-30, produced by Sumitomo Chemical Co., Ltd.) and Y_2O_3 powder (Y_2O_3-RU, submicron-type, produced by Shin-Etsu Chemical Co., Ltd.), wet ball milling of Al_2O_3/Y_2O_3=82/18 mol ratio was undertaken to obtain a uniform composite powder slurry. After removing the ethanol and drying the slurry using a rotary evaporator, preliminary melting to obtain ingots was performed using high-frequency induction heating in a Mo crucible.

Unidirectional Solidification

All experiments were performed using the Bridgman-type equipment at the Japan Ultra-high Temperature Materials Research Center. Ingots obtained by preliminary melting were inserted into the Mo crucible (50 mm in outside diameter by 200 mm in length by 5 mm in thickness) which were placed in a vacuum chamber, where a graphite susceptor was heated by high-frequency induction coils. After sustaining the melting temperature at 1950 ℃ (about 120 ℃ above melting point) for 30 minutes, the Mo crucible was lowered at $5mmh^{-1}$ to induce the unidirectional solidification.

Evaluation Process

The specimens used for three-point flexural and creep testing were selected so that their axial direction was parallel to the direction of the unidirectional solidification. The dimensions of the flexural test specimen were 3x4x36 mm with a 30-mm span. The three-point flexural strength was measured from room temperature to 1800 °C in an argon atmosphere at a cross head speed of 0.5 mm/min. Compressive strength tests under strain rate control were carried out on the 3x4x6 mm specimens. Specimens were heated by a graphite susceptor using high-frequency induction heating. The test temperatures were 1500 °C, 1600 °C, and 1700 °C. The strain rates were 10^{-4}/sec, 10^{-5}/sec, and 10^{-6}/sec. All creep tests were conducted in an argon atmosphere. The thermal stability of the microstructure of the composites was evaluated from microstructural changes after heat treatments in air at 1700 °C for up to 1000 hours in a furnace.

Phase identification was performed using the RAD-RB-type X-ray diffraction equipment. High-resolution transmission microscopy (HRTEM) of the composite was carried out using a JEM-2010 microscope, while the electron probe micro analysis (EPMA) was conducted with a JMX-8621MX.

RESULTS AND DISCUSSION

Microstructure of Unidirectionally Solidified Eutectic Composites

The microstructures of the upper, middle and lower planes perpendicular to the solidification direction of the Al_2O_3/YAG eutectic composite, and those of the hot-pressed plane of the sintered composite with the same composition, consisted of Al_2O_3 and YAG phases; these were determined from X-ray diffraction patterns.[6] Homogeneous microstructures with no pores or colonies are observed in the eutectic composite. This composite consists of <110> single-crystal Al_2O_3 with a hexagonal structure and <743> single-crystal YAG with a garnet structure.[7] In contrast, for the sintered composite, diffraction peaks from various planes are observed, characteristic of a polycrystalline ceramic composite with random crystal orientations.[7] Figure 1 shows a SEM micrograph illustrating the three-dimensional configuration of the single-crystal YAG in the Al_2O_3/YAG eutectic composite, from which Al_2O_3 phases had been removed by heating in graphite powder at 1650 °C for 2 hr. The configuration of the single-crystal YAG is three-dimensionally connected, extremely complex, and has a hieroglyphic-like configuration.[8] We therefore conclude that the present Al$_2$O$_3$/YAG eutectic composite has a microstructure consisting of three-dimensionally continuous and complexly entangled single-crystal Al$_2$O$_3$ and single-crystal YAG phases.[8]

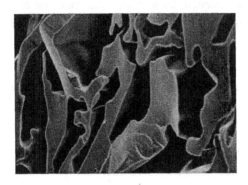

Figure 1 SEM observation of the three-dimensional configuration of the YAG phase.

Temperature Dependence of the Flexural Strength

Figure 2 shows the temperature dependence of the flexural strength of the unidirectionally solidified eutectic composite from room temperature to 1800 °C, in comparison with that of the sintered composite of the same composition. The eutectic composite maintains its room temperature strength up to 1800 °C (just below its melting point of about 1830 °C), with a flexural strength of around 350 MPa. The sintered composite, on the other hand, has a higher flexural strength at room temperature, but its strength falls precipitously above 800 °C.[7]

Sintered composites show intergranular fracture at room temperature and at 1400 °C and evidence for grain growth is clear. On the other hand, the unidirectionally solidified eutectic composites show no grain growth up to the very high temperature of 1700 °C, and the fracture is transgranular.[7] Moreover, when the test temperature reaches 1800 °C, fracture of the interface between the Al$_2$O$_3$ and YAG phases and mixed fracture of intergranular and transgranular is observed.[7]

Generally, if an interface or a grain boundary contains an amorphous phase, high-temperature strength is reduced.[9,10] It is found from HRTEM observation that the interface and grain boundary in the sintered composite contain an amorphous phase, while, the interface between the Al_2O_3 and YAG phases in the eutectic composite contains no amorphous phase.[7]

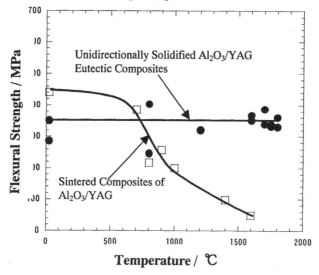

Figure 2   Temperature dependence of flexural strength of unidirectionally solidified Al_2O_3/YAG eutectic composites compared with sintered composites.

Creep Characteristics

Figure 3 shows the relationship between compressive flow stress and strain rate in the unidirectionally solidified eutectic composite and the sintered composite at test temperatures of 1500 ℃, 1600 ℃, and 1700 ℃.[8]   While the unidirectionally solidified eutectic composite and the sintered composite shared the same chemical composition and phases, their creep characteristics were markedly different. That is, at the same strain rate of 10^{-4}/sec and test temperature of 1600 ℃, the sintered composite showed flow stress of 33 MPa, while the unidirectionally solidified eutectic composite's flow stress was approximately 13 times higher at 433 MPa. Moreover, as can be seen from the diagram, the unidirectionally solidified eutectic composite has creep characteristics that surpass those of a-axis sapphire fibers.[9,11]

Figure 4 shows the bright field TEM images of dislocation structure observed in the specimens plastically deformed around 14% in the compressive test at an initial strain rate of 10^{-5}/sec and test temperature of 1600 ℃ for a unidirectionally solidified eutectic composite and a sintered composite. Dislocation structure is observed in both the Al_2O_3 phase and YAG phase for the unidirectionally solidified eutectic composite, showing that the plastic deformation occurred by dislocation motion.

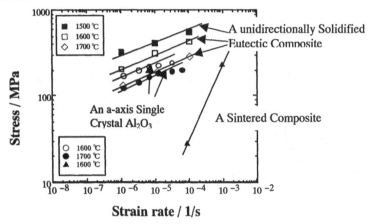

Figure 3 Comparison of compressive strength of the unidirectionally solidified eutectic composite and a sintered composite as a function of strain rate and temperature.

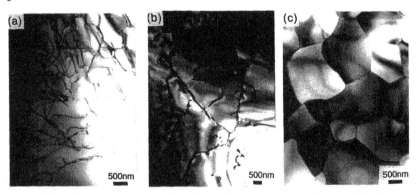

Figure 4 TEM images showing the dislocation structure of (a) Al_2O_3 phases and (b) YAG phases in the unidirectionally solidified composite, and (c) the microstructure of Al_2O_3 and YAG phases in the sintered composites, of compressively crept specimens at 1600 ℃ and strain rate of 10^{-5}/s.

From the above result, we can presume that the plastic deformation mechanism of the present eutectic composite is essentially different from that of the sintered composite similar to the micrograin superplasticity of ceramics resulting from a grain-boundary sliding or a liquid phase present at grain boundary at a high temperature.[12]

The steady state creep rate $\dot{\varepsilon}$, can be usually shown by the following equation:

$$\dot{\varepsilon} = A \sigma^n \exp(-Q/RT) \tag{1}$$

Here, A, n are dimensionless coefficients, σ is the creep stress, Q is the activation energy for the creep, T is the absolute temperature, while R is the gas constant.[13] In Figure 3, the value of n is around 1~2 for sintered composites, and 5~6 for unidirectionally solidified eutectic composites. In sintered composites, it can be assumed that the creep deformation mechanism follows the Nabarro-Herring or Coble creep models, while in unidirectionally solidified eutectic composites, the creep deformation mechanism can be assumed to follow the dislocation creep models corresponding to dislocation structure observed.[14] The activation energy Q is estimated to be about 730 kJ/mol from an Arrhenius plot,[15] which is similar to the values estimated from the high temperature creep in Al_2O_3 single crystal (compression axis is [110]) and YAG single crystal (compression axis is [110])[16,17].

Thermal Stability of Microstructure

Figure 5 shows SEM images of the microstructure after 500, 750 and 1000 hours of heat treatment at 1700 °C in an air atmosphere.[7] Even after 1000 hours of heat treatment no grain growth of microstructure was observed. The unidirectionally solidified eutectic composite was shown to be very stable during lengthy exposure at high temperature of 1700 °C in an air atmosphere. This stability resulted from the thermodynamic stability at that temperature of the constituent phases of the single-crystal Al_2O_3 and the single-crystal YAG, and the thermodynamic stability of the interface.[7]

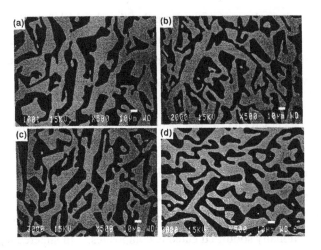

Figure 4   Thermal stability of the microstructures of a unidirectionally solidified eutectic composite at 1700 ℃ in an air atmosphere. (a) as-received and after heat treatment (b) for 500 hours, (c) 750 hours, and (d) 1000 hours.

CONCLUSION

Using Bridgman-type equipment and the unidirectional solidification method, an Al₂O₃/YAG eutectic composite was produced. The eutectic composite consisted of a single-crystal Al₂O₃ phase and a single-crystal YAG phase without grain boundaries and colony boundaries.   The eutectic composite had superior high-temperature strength characteristics with flexural strength being independent of temperature between room temperature up to 1800 ℃. This eutectic composite also has superior creep resistance and its plastic deformation mechanism is based on dislocation motion, which is different from that of the sintered composite. This eutectic composite has a very thermally stable microstructure with no grain growth in evidence even after lengthy heat treatments at 1700 ℃ in an air atmosphere.

Acknowledgment

This research has been promoted by the New SunShine Program of Agency of the Industrial Science and Technology, MITI. We also thank N. Nakagawa, T.

Wakamoto and H. Ohtsubo of Ube Research Laboratory, UBE Industries Ltd. for their assistance in experiments.

References

1. D. Viechnicki and F. Schmid, "Eutectic Solidification in the System $Al_2O_3/Y_3Al_5O_{12}$," *J. Materials Sci.*, **4**, 84-88 (1969).

2. T. Mah and T.A. Parthasarathy, "Processing and Mechanical Properties of $Al_2O_3/Y_3Al_5O_{12}$(YAG) Eutectic Composite," *Ceram. Eng. Sci. Proc.*, **11**, 1617-1627 (1990).

3. T.A. Parthasarathy, T. Mah and L.E. Matson, "Creep Behavior of an $Al_2O_3/Y_3Al_5O_{12}$ Eutectic Composite," *Ceram. Eng. Soc. Proc.*, **11**, 1628-1638 (1990).

4. T.A. Parthasarathy, Tai- ll Mah and L.E. Matson, "Deformation Behavior of an $Al_2O_3/Y_3Al_5O_{12}$ Eutectic Composite in Comparison with Sapphire and YAG," *J.Am.Ceram. Sci.*, **76**, 29-32 (1993).

5. Y. Waku, N. Nakagawa, H. Ohtsubo, Y. Ohsora and Y. Kohtoku, "High Temperature Properties of Unidirectionally Solidified Al_2O_3/YAG Composites," *J. Japan Inst. Metals*, **59**, 71-78 (1995).

6. Y. Waku, H. Otsubo, N. Nakagawa and Y. Kohtoku, Y. "Sapphire Matrix Composites Reinforced with Single Crystal YAG Phases," *J. Materials Sci.* **31**, 4663-4670 (1996).

7. Y. Waku, N.Nakagawa, T. Wakamoto, H. Ohtsubo, K. Shimizu and Y. Kohtoku, "High-Temperature Strength and Thermal Stability of a Unidirectionally Solidified Al_2O_3/YAG Eutectic Composite," *J. Materials Sci.*, **33**, 1217-1225 (1998).

8. Y. Waku, N. Nakagawa, T. Wakamoto, H. Ohtsubo, K. Shimizu and Y. Kohtoku,"The Creep and Thermal Stability Characteristics of a Unidirectionally Solidified Al_2O_3/YAG Eutectic Composite," *J. Materials Sci.*, **33**, 4943-4951 (1998).

9. D. R. Clarke, "High-Resolution Techniques and Application to Nonoxide Ceramics," *J. Am. Ceram. Soc.*, **62**, 236-246 (1979).

10. J. Echigoya, S. Hayashi, K. Sasaki and H. Suto, "Microstructure of Directionally Solidified MgO-ZrO$_2$ Eutectic," *J. Japan Inst. Metals,* **48,** 430-434 (1984).

11. D. M. Kotchick and R. E. Tressler, "Deformation Behavior of Sapphire Via the Prismatic Slip System," *J. Am. Ceram. Soc.,* **63,** 429-434 (1980).

12.F. Wakai, Y. Kodama, S. Sakaguchi, N. Murayama, K. Izeki and K. Niihara, "A Superplastic Covalent Crystal Composite," *Nature,* **344,** 421-23 (1990).

13. W. R. Cannon and T. G. Lagdon, "Creep of Ceramics," *J. Mater. Sci.,* **18,** 1-50 (1983).

14. Y. Waku and T. Sakuma, "Dislocation Machanism of Deformation and Strength of Al$_2$O$_3$-YAG Single Crystal Composite at High Temperature above 1700 K," *Journal of the European Ceramic Socity,* in press.

15. H. Yoshida, K. Shimura, S. Suginohara, Y. Ikuhara and T. Sakuma N. Nakagawa and Y. Waku, "High Temperature Deformation in Unidirectionally Solidified Eutectic Al$_2$O$_3$-YAG Single Cryatal," Proceeding of The 8th International Conference on Creep and Fracture of Engineering Materials and Structures, November 1-5, Tsukuba, 1999, **171-174,** 855-862 (2000).

16. S. Karato, Z. Wang and K. Fujino, "High-temperature Creep of Yttrium-Aluminum Garnet Single Crystals," *J. Materials. Sci.,* **29,** 6458-6462 (1994).

17. C. S. Corman, "Creep of Yttrium a Aluminum Garnet Single Crystals," *J. Materials Sci.,* **12,** 379-382 (1993).

MICROMECHANICS OF VISCOUS SLIP ALONG CERAMIC GRAIN BOUNDARIES

G. PEZZOTTI
Department of Materials, Kyoto Institute of Technology, Sakyo-ku, Matsugasaki, 606-8585 Kyoto, Japan

ABSTRACT

Internal friction characterization has been used to quantitatively assess the inherent viscosity of residual SiO_2 glass segregated to grain boundaries of polycrystalline Si_3N_4 and SiC ceramics. The intergranular SiO_2 glass was intentionally doped with foreign anions to systematically alter its inherent viscosity. The anelastic relaxation peaks of internal friction, arising from viscous slip along grain boundaries wetted by glass, was collected and analyzed with respect to its shift upon changing the oscillation frequency. As a result of this characterization, Arrhenius plots of the microscopic intergranular glass viscosity could be obtained. Creep rates were also measured in torsional geometry as a function of temperature, and the macroscopic viscosity of the polycrystals evaluated. A plot of macroscopic (polycrystal) viscosity vs microscopic (intergranular phase) viscosity revealed that the viscous flow of intergranular glass dominates the overall deformation behavior of the polycrystal only when the glass viscosity is relatively low (*i.e.*, $\log\eta_i < 6.5$). At higher intergranular glass viscosities, the creep behavior of the polycrystal obeys a different law and it is dominated by diffusive phenomena.

INTRODUCTION

Amorphous phases may segregate to grain boundaries of polycrystalline ceramics wetting the crystalline grains (*e.g.*, silicate glasses in Si_3N_4 or SiC ceramics[1,2]), thus resulting in a continuous nanometer-sized film which encompasses the ceramic grains after densification[3,4]. Despite that only a minor fraction of residual glass may be incorporated in the microstructure, the structural properties of polycrystalline ceramics are greatly influenced by such an intergranular amorphous film[5]. The grain-boundary film can actually be regarded as a "glue" which bonds neighboring grains. Advanced studies in high-resolution microscopy[6] have shown that the thickness of the intergranular film remains constant throughout the polycrystalline structure, almost independent of the reciprocal crystal orientation of neighboring grains. Regardless of the incorporation of relatively high fractions of glass, ceramic microstructures show intergranular films of nearly invariable thickness. In other words, the additional glass fraction mainly localizes at (and, thus, enlarges the size of) multigrain pockets[7]. Clarifying the geometrical and atomic structures of the intergranular glass and tailoring its intrinsic properties certainly represents a step forward to actual structural applications of engineering ceramics.

An important degree of freedom in designing the grain-boundary structure can be provided by doping the (residual) intergranular glass with foreign aliovalent anions. The role of such anions within the intergranular glass network

To the extent authorized under the laws of the United States of America, all copyright interests in this publication are the property of The American Ceramic Society. Any duplication, reproduction, or republication of this publication or any part thereof, without the express written consent of The American Ceramic Society or fee paid to the Copyright Clearance Center, is prohibited.

is presently of much interest because, depending on the anion valence, the intrinsic viscous response of the intergranular "glue" can be systematically altered. Therefore, upon controlling the inherent intergranular film viscosity, the macroscopic deformation behavior of polycrystalline ceramics can be tailored. In this paper, we shall clarify for some aspects the role of foreign anions at the grain boundary and put forward the physical link between microscopic (grain-boundary) viscosity and macroscopic viscosity of polycrystalline Si_3N_4 and SiC ceramics, whose grains are encompassed by a continuous amorphous-SiO_2 film. Besides the "traditional" creep test, which quantifies the macroscopic viscoelastic response of the polycrystal, new microscopic physical insight is provided by high-temperature internal friction experiments. In particular, the systematic detection and analysis of a relaxation peak arising from grain-boundary sliding enables the quantitativeevaluation of the intrinsic viscosity of the glass phase segregated to grain boundaries.

EXPERIMENTAL PROCEDURE

Materials
Dense polycrystalline bodies were prepared by glass-encapsulated hot isostatic pressing from high-purity Si_3N_4 or SiC starting powders, according to a schedule previously shown[8]. All the materials investigated contained relatively low amounts of SiO_2 glass ($\approx$2.5-5.0 vol%). The SiO_2 phase was either arising from surface contamination (due to exposure to atmosphere) or intentionally added to the starting powder. Addition to the starting Si_3N_4 powder of pulverized polytetrafluoroethylene or hexachloroethane, followed by a pre-firing cycle in vacuum of 2 h at 1200°C, produced grain-boundary contamination by either F⁻ or Cl⁻ ions, respectively. Note that both the selected dopants nominally involve no contamination by H^+ (which, either as OH⁻ or H_3O^+, is a SiO_2-network modifier), or further $(SiO_4)^{4-}$. After densification, the precise F and Cl amounts in the polycrystalline bodies were determined by ion selective electrode (ISE) and ion chromatography (IC) analyses, respectively. It was found to be experimentally difficult to incorporate in the dense Si_3N_4 polycrystal large chlorine amounts because, to eliminate excess carbon by pre-firing via volatilization of carbon monoxide (CO) gas, chlorine also partly volatilized as C_2Cl_6 gas. Addition of small amounts of carbon black to the starting Si_3N_4 powder was also attempted. In addition, one undoped Si_3N_4 sample was also prepared for comparison.

Two dense SiC specimens were obtained from as-received powder adding $\approx$3.0 vol% high-purity SiO_2 (but without intentional addition of anion dopants) or after oxidizing the starting SiC powder to eliminate contamination by free-carbon via volatilization of carbon monoxide (CO) gas. Annealing of dense SiC bodies in nitrogen atmosphere (3 h at 1900°C) was also attempted to produce incorporation of N anions through the glassy grain boundaries.

Segregation of anions with a smaller valence (*e.g.,* F or Cl ions) as compared to oxygen, is expected to result in a reduced number of atomic bonds per unit volume in the intergranular SiO_2 network, thus leading to a degradation of the intergranular glass viscosity. On the other hand, C and N ions substituting oxygen in the SiO^{4-} tetrahedral network, offer the possibility of 4-coordinated and 3-coordinated anion sites, respectively, which results in a higher bonding density

314

per unit volume of the SiO_2 network and increases the intergranular glass viscosity.

Transmission electron microscopy and high resolution electron microscopy characterizations of the grain-boundary film structure, shown in detail in previous reports[9-13], revealed that the incorporation of additional anions produces slight changes of the SiO_2 film thickness, the larger the nominal ionic radius of the added anion the larger the thickness (cf. Table I). The actual presence of anion dopants within the grain-boundary glass film was also confirmed by electron energy-loss spectroscopy analyses whose details are described in the references given above.

Rheological testing

A torsional pendulum device was utilized for measuring internal friction, Q^{-1}, according to either the free-decay or the forced-vibration methods. The device was enclosed in a vacuum-tight system, in which a controlled N or Ar flowing rate was maintained throughout the experiments. Q^{-1} data were automatically recorded at intervals of ≈ 5 °C. Internal friction measurements were conducted at frequencies in the range of 0.1-13 Hz, on rectangular bars $2 \times 3 \times 50$ mm in dimensions. Very high temperature (>2000 °C) was generated by a carbon resistance heater placed around the specimen and measured either by a thermocouple and/or by an infrared thermo-analyzer. Torsional creep experiments were also conducted in the torsional pendulum apparatus, by applying a constant torque. Further details about the torsional pendulum device and the experimental procedure for evaluating creep behavior have been reported elsewhere[9,14,15].

RESULTS

Internal friction behavior

Internal friction experiments were conducted on Si_3N_4 and SiC polycrystals doped with various anions at different frequencies of torsional oscillation (grain-boundary glass composition and damping frequencies are shown in Table I). Internal-friction peak components were determined from the respective experimental internal friction curve by subtracting a fitted exponential-like background component. The relaxation peaks found in various Si_3N_4 and SiC polycrystals are shown in Figs. 1(A)-(D) and Figs. 2(A)-(C), respectively. Peaks shifted as a function of the frequency of the measurement, the higher the frequency the higher the shift towards higher-temperatures. For the sake of clarity, only the (maximum) peak shift, *i.e.*, that corresponding to the minimum and maximum frequencies tested, are shown in Figs. 1 and 2. The peaks of Si_3N_4 and SiC materials added with C and/or N dopants were systematically located at higher temperatures as compared to that of the respective undoped materials. On the other hand, F and Cl anions significantly lowered the peak temperature as compared to the undoped polycrystals. The peaks of Si_3N_4 and SiC materials showed different morphologies and intensities, which are related to the different activation energy for the grain-boundary viscous flow[16], to microstructural differences[17] and to the amount of grain-boundary phase. The temperature shift (indicated in Figs. 1 and 2) upon frequency change enables one to obtain the temperature dependence of the intergranular film viscosity, as explained in the discussion section. Based on the present internal friction data, a quantitative assessment of the viscosity of the intergranular glasses can be attempted.

Torsional creep behavior

Torsional creep rates of various Si_3N_4 and SiC materials are shown in Fig. 3 as a function of temperature. Creep experiments were conducted under a constant shear stress, $\tau = 50$ MPa, a stress comparable with the maximum shear stress amplitude applied during internal friction measurements. As seen, the creep rate strongly depends on the particular anion structure of the intergranular glass film. In particular, monovalent anions (like as F^- and Cl^-) significantly degrade the creep resistance of the polycrystal, the Cl^- anion involving the most detrimental effect. On the other hand, trivalent N and/or tetravalent C anions strongly improve the creep resistance of both Si_3N_4 and SiC ceramics. The slope of the creep data indicates an apparent activation energy for the creep process. It is interesting to note that the slope decreases when F^- or Cl^- are present, while it increases when oxycarbonitride glasses are segregated at grain boundary, independent of the bulk crystalline phase. This may suggest that an enhanced diffusivity (*i.e.*, a lowered viscosity) of the intergranular SiO_2 phase, which is due to the presence of monovalent anion dopants, dominates the creep behavior of the polycrystals.

DISCUSSION

Grain-boundary film viscosity

Internal grain boundaries in polycrystalline solids are a source of internal energy dissipation which may arise from their viscous slip at elevated temperatures[18]. The sliding mechanism is schematically shown in Fig. 4. With increasing temperature, the grain-boundary film becomes viscous and it cannot sustain a shear stress. Thus, local slip occurs until the stress at grain boundaries has been locally released. For applied stress sufficiently low, no microcracking or cavitation damages occur at the grain boundary. Therefore, upon releasing the applied stress, the elastic deformation of the neighboring ceramic

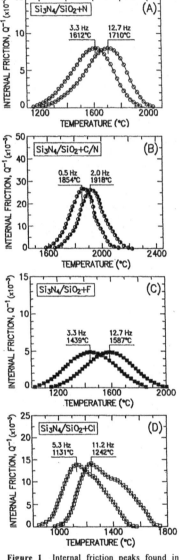

Figure 1 Internal friction peaks found in Si_3N_4 polycrystals containing different anions at SiO_2 grain boundaries. Peak shifts as a function of the damping frequency are also shown

Grain Boundary Engineering in Ceramics

grains will be recovered and grain boundaries pushed back to their original position. The polycrystalline structure will be then restored to its original undeformed state. Both the elastic deformation of the bulk ceramic grains and the size of glass pools at triple-grain junctions dictate the magnitude of the (constrained) grain-boundary slip which, according to the inherent viscosity of the intergranular glass, must occur within a finite time.

Let us consider a macroscopic shear stress (*i.e.,* a torque) which is applied to the polycrystal as a sinusoidal pulse wave of maximum stress, τ, and frequency, f. Upon monitoring, for example, the free-decay of the polycrystalline specimen oscillation, the occurrence of recoverable grain-boundary sliding leads to the presence of a peak in the internal friction ($Q^{-1}(T)$) curve. The initial stress pulse activates the microscopic displacement of the ceramic grains and viscous grain-boundary sliding eventually takes place obeying the particular applied frequency (cf. Fig. 4). In absence of external frictions, the amplitude of the free oscillation must reduce due to viscous energy dissipation. Two parameters mainly enhance the rate of internal energy dissipation, Q^{-1}: (1) the length Δu over which the reciprocal sliding between grains occurs, and, 2) the intrinsic viscosity, η_{gb}, of the grain-boundary phase. Due to the competing roles of these two parameters, a relaxation peak, located at $T=T_p$, is found in the $Q^{-1}(T)$ curve, as temperature increases. A detailed description about the origin of the internal friction peak in polycrystalline ceramics has been given in previous papers[9-14].

In principle, the elastic displacement allowed by the neighboring grains is proportional to the average grain size, d, and to the elastic shear strain, $\gamma=\tau(1-\nu)/G$, where ν and G are the Poisson's ratio and the shear modulus of the bulk ceramic crystallites, respectively. However, the actual sliding grain-boundary length may markedly differ from that calculated from the elastic displacement of neighboring grains. There are two main (competing) factors which may alter the magnitude of the maximum sliding displacement: (i) the presence of glass pools at triple-points, which also involves the presence of smoothed grain edges allowing grain rotation during sliding (cf. Fig. 4); and, (ii) a non-uniform shear stress distribution within the torsion specimen (*i.e.* zero and maximum shear stress at center and surface, respectively), which makes the total relaxation of the damping bar size-dependent and, in particular, smaller than that predictable according to the

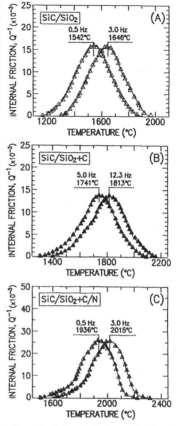

Figure 2 Internal friction peaks found in SiC polycrystals undoped or containing different anions at SiO_2 grain boundaries. Peak shift upon increasing damping frequency is also shown.

maximum (surface) strain. Since the effect of the two above factors is a priori unknown, we shall assume the following relation to represent the sliding displacement at the peak-top temperature:

$$\Delta u_{max} = \alpha d\tau (1-v)/G_U \qquad (1)$$

where G has been taken equal to the macroscopic shear modulus of the unrelaxed polycrystal, G_U, and α is an adimensional constant which takes into consideration the factors (i) and (ii), cited above. The viscosity of the glass phase filling the grain boundary channel, η_{gb}, at the internal friction peak temperature, T_p, is given by[19]:

$$\eta_{gb} = \tau \delta (t^*/\Delta u_{max}) = G_U \delta / 2\pi f \alpha (1-v) d \qquad (2)$$

where δ is the grain-boundary thickness. Note that, according to eq.(2), an increase in oscillation frequency, f, will reduce the grain-boundary viscosity at which the peak maximum manifests, thus shifting the peak toward higher temperatures.

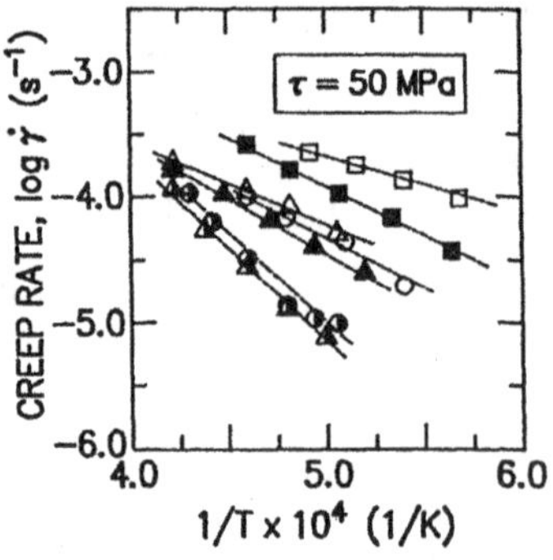

Figure 3 Torsional creep rates measured on various Si_3N_4 and SiC polycrystals are plotted as a function of temperature. Creep experiments were conducted under a constant shear stress, τ=50 MPa. Symbols represent materials as listed in Table 1.

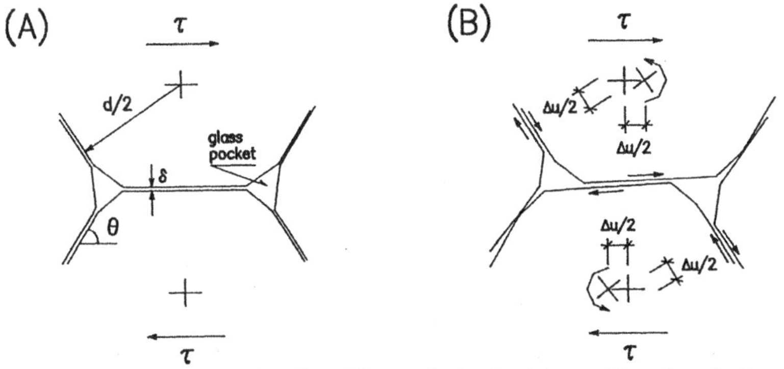

Figure 4 Schematic of grain-boundary sliding mechanism involving partial rotation of adjacent grains: (A) unstressed (or recovered) configuration; (B) configuration under shear stress.

The relaxation factor, α, can be expressed in terms of experimentally accessible parameters, according to the following considerations. The total macroscopic shear strain, γ, of the polycrystal loaded by τ at the temperature T_p consists of the sum of an elastic component and of an anelastic component arising from the elastic deformation of the bulk grains and the additional viscous displacement (rotation) at grain boundaries, respectively:

$$\gamma = (\tau/G_U) + (\Delta u_{max}/d) = \tau/G_R \qquad (3)$$

where G_R is the macroscopic shear modulus of the polycrystal after relaxation by grain-boundary sliding at the temperature T_p. Substituting from eq. (1) and

318   Grain Boundary Engineering in Ceramics

rearranging, the relaxation factor α can be expressed as:

$$\alpha = [(G_U/G_R)-1]/(1-v) \qquad (4)$$

where the relaxation modulus ratio, G_U/G_R, at the peak-top temperature is now an experimentally accessible parameter. In addition, it can be shown that, at $T=T_p$, the modulus ratio, G_U/G_R, is related to the height of internal friction peak, Q^{-1}_{max}, through the following equation[20]:

$$G_U/G_R = 1/[1-(Q^{-1}_{max}/2)] \qquad (5)$$

Substituting for α from eqs.(4) and (5), one obtains:

$$\eta_{gb} = (G_U\delta/2\pi fd)[1-(Q_{max}/2)]/(Q_{max}/2) \qquad (6)$$

Equation (6) is valid only at $T=T_p$ and represents the basic relationship which links the measured internal friction peak characteristics with both the intrinsic grain-boundary viscosity and the microstructural parameters of the ceramic polycrystal. However, it should be noted that the quantitative use of eq.(5) is still not well established, because the internal friction peak usually arises from a relaxation spectrum rather than from a single-relaxation phenomenon[16]. In addition, eq.(5) is liable to an error related to the choice of the background curve. Although this error will not significantly affect the order of magnitude of the calculated grain-boundary film viscosity, it is preferable to obtain the relaxation parameter, α, directly from a relaxation test (i.e., via eq.(4)), as shown in a

POLYCRYSTAL COMPOSITION	AMOUNT OF GLASS MODIFIERS AT G.B. (at.%)*	GRAIN SIZE d (μm)	G.-B. FILM THICKNESS δ (nm)	RELAXATION FACTOR α	DAMPING FREQUENCY f (Hz)
○ Si₃N₄/SiO₂+N	6±2	1.2	1.0±0.1	0.054	3.3 – 12.7
◑ Si₃N₄/SiO₂+N+C	(6±2)N+0.02C	1.0	1.0±0.1	0.025	0.5 – 2.0
■ Si₃N₄/SiO₂+F	9.6±0.5	1.4	1.1±0.1	0.068	3.3 – 12.7
□ Si₃N₄/SiO₂+Cl	0.6±0.2	1.0	1.3±0.1	0.070	5.3 – 11.2
△ SiC/SiO₂ (oxid.)	———	0.4	1.0±0.1	0.085	0.5 – 3.0
▲ SiC/SiO₂+C	15±4	0.8	0.9±0.1	0.070	5.0 – 12.3
▲ SiC/SiO₂+C+N	(15±5)C+5.2N	0.8	1.2±0.1	0.025	0.5 – 3.0

*with respect to A⁻/O+A⁻ ratio

Table I   Grain-boundary glass compositions, damping frequencies and other microstructural parameters involved in the calculation of η_i according to eqs.(2) and (4). The listed symbols were used throughout all the figures for identifying the investigated materials.

previous report[9]. Nevertheless, the important implication in using eqs.(2) or (6) is that a variation of oscillation frequency should produce a well defined temperature shift of the internal friction peak. This phenomenon can be used to obtain an Arrhenius plot of the η_{gb} parameter and to calculate from its slope an activation energy (peak-shift method).

Figure 5 shows the temperature dependence of intergranular film viscosity of the investigated Si_3N_4 and SiC polycrystals, as calculated according to eqs.(2) and (4), and compared with viscosity data for bulk high-purity SiO_2 glass[21]. The parameters involved in the calculation are listed in Table I[9-14]. The effect of anion dopants on the grain-boundary film viscosity is clarified by this plot, the higher the valence of the added anion the higher the viscosity. SiC prepared from pre-oxidized powder, thus containing SiO_2 glass virtually free from any foreign anion,

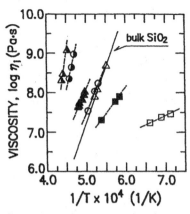

Figure 5 Temperature dependence of intergranular film viscosity of the investigated Si_3N_4 and SiC polycrystals, as calculated according to eqs.(2) and (4). Comparison is carried out with viscosity data for bulk high-purity SiO_2 glass from Ref. 21. Symbols represent materials as listed in Table I.

showed a grain-boundary viscosity almost coincident with that of bulk SiO_2 glass. It is also important to note that the effect of C and N anions on the glass phase viscosity seem to be somewhat additive, giving rise to extremely refractory silicon oxycarbonitride glasses at grain boundaries of either Si_3N_4 or SiC ceramics.

Macroscopic and microscopic viscosity

The availability of microscopic data of grain-boundary viscosity, obtained according to the internal friction characterization, offers the unique possibility of experimentally relating the viscoelastic behavior of internal grain boundaries to the macroscopic deformation behavior of the polycrystalline material. An example of the experimentally obtained relation (at 1850°C) between the macroscopic viscosity of the polycrystal, η_v, and the microscopic viscosity of the intergranular phase, η_i, is given in Fig. 6. η_v was simply calculated by dividing the applied (constant) shear stress by the torsional creep rate (*i.e.*, data from Fig. 3). For comparison, also a theoretical plot is given in Fig. 6 which was calculated according to the Roscoe equation[22], as follows:

$$\eta_v = \eta_i \{1-[(1-V_{fg})/V_f^*]\}^{-2.5} \qquad (7)$$

where V_{fg} is the volume fraction of glass and V_f^* is a critical volume fraction for maximum packing of the crystalline grains. V_f^* was determined by considering an array of regular hexagonal grains bonded without any layer of intergranular glass but with smoothed corners to accommodate triple points shaped as regular triangles. The Roscoe equation describes the flow of a viscous fluid containing a high fraction of solid particles in suspension. It is remarkable that eq.(7) can predict with accuracy the behavior of the present polycrystal, but only when

Grain Boundary Engineering in Ceramics

microscopic viscosities of relatively low magnitude are involved, namely when the intergranular phase actually flows as a liquid. Therefore, we experi-mentally find that, for $\log\eta_i < 6.5$ Pa x s, the viscous flow of the secondary glass phase controls creep, provided that relatively low stresses are applied to the present polycrystalline materials. On the other hand, for glass-phase viscosities, $\log\eta_i > 6.5$ Pa x s, the macroscopic viscosity of the polycrystalline matter is markedly lower than that predicted according to a fluid mechanics model. This is because, unless the mechanistic model behavior described by eq.(7), diffusion contribution dominates the creep behavior at temperatures below the melting point of the intergranular glass.

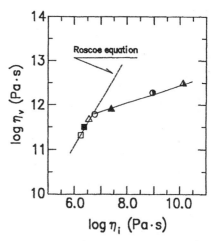

Figure 6 Experimentally obtained relation (at 1850°C) between the macroscopic viscosity of the polycrystal, η_v, and the microscopic viscosity of the intergranular phase, η_i. A theoretical plot is also shown calculated according to the Roscoe equation[22] (eq.(7)). Symbols represent materials as listed in Table I.

CONCLUSION

It has been experimentally shown that the anion structure of SiO_2 glass segregated to grain boundaries of Si_3N_4 and SiC ceramics strongly affects the macroscopic viscous behavior of the polycrystals. Glasses which contain N and C, substituting oxygen in the $(SiO)^{4-}$ tetrahedral network, offer the possibility of 3-coordinated and 4-coordinated anion sites, respectively. This could result in a higher bonding density per unit volume of the glass, thus increasing the intergranular glass viscosity. The effect of C and N ions appears to be additive, and intergranular glass viscosities improved by several orders of magnitude as compared with that of pure SiO_2 glass were found. The detrimental effect of monovalent anions was also clarified and the decrease of intergranular glass viscosity quantified. A phenomenological analysis relating the microscopic viscosity of the intergranular phase to the macroscopic creep behavior of the polycrystal suggested that, below a critical value of intergranular phase viscosity, the deformation behavior of the polycrystal is governed by the mechanics of a fluid filled by solid particles.

Acknowledgements: The author sincerely thanks Dr. K. Ota, Mr. K. Okamoto and Mr. H. Nishimura for helping with the experimental procedures.

References:

1) W. E. Luecke and S. M. Wiederhorn, "A New Model for Tensile Creep of Silicon Nitride," *J. Am. Ceram. Soc.*, **82** [10] 2769-78 (1999).
2) S. M. Wiederhorn, D. E. Roberts, T.-J. Chuang, and L. Chuck, "Damage-Enhanced Creep in a Siliconized Silicon Carbide: Phenomenology," *J. Am. Ceram. Soc.*, **71** [2] 602-08 (1988).

3) F. F. Lange, "Non-Elastic Deformation of Polycrystals with a Liquid Boundary Phase," pp. 361-81 in *Deformation of Ceramic Materials*. Edited by R. C. Bradt and R. E. Tressler, Plenum Press, New York, 1976.

4) D. R. Clarke, "On the Equilibrium Thickness of Intergranular Glass Phases in Ceramic Materials," *J. Am. Ceram. Soc.*, **70** [1] 15-22 (1987).

5) J. R. Dryden, D. Kucerovsky, D. S. Wilkinson, and D. F. Watt, "Creep Deformation Due to a Viscous Grain Boundary Phase," *Acta Metall.*, **37** [7] 2007-15 (1989).

6) H.-J. Kleebe, M. J. Hoffmann, and M. Rühle, "Influence of Secondary Phase Chemistry on Grain-Boundary Film Thickness in Silicon Nitride," *Z. Metallkd.*, **83** [8] 610-17 (1992).

7) J. Zheng, "Sinthesis, Sintering and Characterization of High-Temperature/Performance Silicon Nitride Based Ceramics," Doctoral Thesis, Institute of Scientific and Industrial Research, Osaka University (1992).

8) G. Pezzotti, "Si_3N_4/SiC-Platelet Composite Without Sintering Aids: a Candidate for Gas Turbine Engines," *J. Am. Ceram. Soc.*, **76** [5] 1313-20 (1993).

9) G. Pezzotti, K. Ota, H.-J. Kleebe, "Grain-Boundary Relaxation in High-Purity Silicon Nitride," *J. Am. Ceram. Soc.*, **79** [9] 2237-46 (1996).

10) G. Pezzotti and K. Ota, "Grain-Boundary Sliding in Fluorine-Doped Silicon Nitride,"*J. Am. Ceram. Soc.*, **80** [3] 599-603 (1997).

11) G. Pezzotti, K. Ota, and H.-J. Kleebe, "Viscous Slip along Grain Boundaries in Chlorine-Doped Silicon Nitride," *J. Am. Ceram. Soc.*, **80** [9] 2341-48 (1997).

12) G. Pezzotti, H.-J. Kleebe, and K. Ota, "Grain-Boundary Viscosity of Polycrystalline Silicon Carbides," *J. Am. Ceram. Soc.*, **81** [12] 3293-99 (1998).

13) G. Pezzotti, T. Wakasugi, T. Nishida, R. Ota, H.-J. Kleebe, K. Ota, "Chemistry and Inherent Viscosity of Glasses Segregated at Grain Boundaries of Silicon Nitride and Silicon Carbide Ceramics," *J. Non-Cryst. Solids*, in press.

14) G. Pezzotti, H.-J. Kleebe, K. Ota, and T. Nishida, "Quantitative Measurement of Grain-Boundary Viscosity in Glassy Bonded Ceramic Polycrystals," *Ceram. Transactions*, **99**, 235-44 (1998).

15) G. Pezzotti and K. Ota, "Mechanical Damping Arising from Dislocation Motion in Sapphire and Ruby Crystals," *J. Am. Ceram. Soc.*, **80** [9] 2205-12 (1997).

16) G. Pezzotti and K. Ota, "On the Width of Internal Friction Peak in Ceramic Polycrystals with Intergranular Glassy Film," *Scripta mater.*, **36** [4] 481-87 (1997).

17) T.-S. Kê, "Experimental Evidence of the Viscous Behavior of Grain Boundaries in Metals," *Phys. Rev.*, **71A** [8] 533-46 (1947).

18) C. Zener, "Theory of the Elasticity of Polycrystals with Viscous Grain Boundaries," *Phys. Rev.*, **60A** [12] 906-8 (1941).

19) D. R. Mosher and R. Raj, "Use of the Internal Friction Technique to Measure Rates of Grain Boundary Sliding," *Acta Metall.*, **22** [12] 1469-74 (1974).

20) A. S. Nowick, "Internal Friction in Metals,"*Progress in Metal Physics 4*. Edited by B. Chalmers. Pergamon Press Ltd., London, 1953.

21) K. Eguchi, "Industrial Glasses"; p. 98 in *Glass Handbook*. Edited by S. Sakka, T. Sakaino, and K. Takahashi. Asakura Shoten Publ., Tokyo, 1977.

22) R. Roscoe, "The Viscosity of Suspensions of Rigid Spheres," *Br. J. Appl. Phys.*, **3**, 267-69 (1952).

SUPERPLASTICITY AND MICROSTRUCTURAL EVOLUTION OF
YBa $_2$Cu $_3$O $_{7-x}$/25 vol%Ag COMPOSITES

J. M. Albuquerque[†], M. P. Harmer, Y.T. Chou[‡]
Department of Materials Science and Engineering and Materials Research Center
Lehigh University, Bethlehem, PA 18015.

ABSTRACT: Tensile superplasticity of YBa $_2$Cu $_3$O $_{7-x}$/ 25 vol%Ag composites was investigated. Tensile elongations close to 100% were achieved for fine grained (0.9 μm) microstructures in the temperature range of 800-850°C, and low strain rates ($\approx$10 $^{-5}$/sec). The deformation parameters determined, namely a stress exponent of n=2.3±0.5 and an activation energy of Q $_{sp}$= 456±100kJ/mol., were found to be very similar to those of the monolithic material, suggesting that the same superplastic deformation mechanism, grain boundary sliding accommodated and controlled by interface reaction, is operative. The presence of silver was shown to enhance ductility and the composite's flow behavior is predicted by Chen's soft inclusion model. Superplastic deformation was accompanied by grain growth, limited cavitation and particle agglomeration. These microstructural features and the role of strain were studied.

INTRODUCTION

Soon after YBa $_2$Cu $_3$O $_{7-x}$ was synthesized, it became apparent that the addition of silver to the polycrystalline superconductor improved not only its sinterability but the overall mechanical properties, including fracture toughness and strength. Research emphasis was placed on the effect of silver on superconducting properties (T $_c$ and J $_c$) as well, and this metal has been used since as a sheath for wires and tapes, electric contacts, as a dopant, or as a second phase in bulk YBa $_2$Cu $_3$O $_{7-x}$ composites. In recent years, as a result of creep investigations and the suitable microstructures that can be produced with the second phase addition, superplasticity has also been studied and considered as an alternative to conventional hot processing and forming. Elongations over 85% have been achieved in polycrystalline YBa $_2$Cu $_3$O $_{7-x}$ at high temperatures and low strain rates, in either compression or tension [1]. As much as 110% elongation under compression has been reported for the YBa $_2$Cu $_3$O $_{7-x}$/Ag composites [2].

Superplastic flow in ceramic materials is in general a diffusion controlled process and, under conditions of steady-state deformation, the mechanical characteristics can be represented by a equation of the form:

$$\dot{\varepsilon} = A \frac{\sigma^n}{d^p} \exp\left(\frac{-Q}{RT}\right) \qquad (1)$$

†Present Address: Departamento de Engenharia de Materiais, Instituto Superior Técnico, Lisboa, Portugal,‡Department of Chemical, Biochemical and Materials Science, University of California, Irvine, CA.

To the extent authorized under the laws of the United States of America, all copyright interests in this publication are the property of The American Ceramic Society. Any duplication, reproduction, or republication of this publication or any part thereof, without the express written consent of The American Ceramic Society or fee paid to the Copyright Clearance Center, is prohibited.

where $\dot{\varepsilon}$ is the strain rate, A is the proportional constant, σ is the flow stress, d is the grain size, n and p are constants termed, respectively, the stress exponent and the inverse grain size exponent, Q is the activation energy of deformation, and RT has the usual meaning. The parameters n, p, and Q are known to vary according to the microstructure, the specific flow/diffusion law, as well as the impurity content of the ceramic, and can be associated with various deformation mechanisms. Hence, the interpretation of experimental data usually relies on the precise determination of these variables. This can be appreciated by reviewing the data compiled in ref.[3].

The degree of ductility depends on the ability of a material to prevent flow localization, increasing as the stress exponent decreases. For superplasticity, it is generally recognized that the n-value is less than 3 and the p-value ranges from 1 to 3, depending on the rate-controlling process (e.g. grain boundary diffusion, or lattice diffusion). Superplastic ceramics are usually developed by refining the grain size to a submicron scale, since they are particularly sensitive to the grain-boundary tensile-fracture stress. For a given deformation rate, flow stress decreases as the grain size is decreased. Moreover, fine grain sizes enhance deformation through grain boundary sliding, which is considered to be the dominant superplastic deformation mechanism [3]. From the constitutive equation it is clear that high strain rates (usual in superplasticity) can be promoted by retaining a fine grained and stable microstructure at forming temperatures; by enhancing diffusivity along the controlling paths or, alternatively, by creating faster diffusion paths (e.g. adding a liquid phase). Because of the fineness of the grains, however, the ceramic microstructures are not thermally stable, and dynamic grain growth occurs during superplastic deformation (as well as in metals) [4]. The presence of a uniformly distributed second phase is expected to inhibit grain coarsening.

In 1990 Routbort et al. [5] studied the creep behavior of coarse grained ($\approx$15 μm) $YBa_2Cu_3O_{7-x}$/Ag composite with 15-30 vol% addition of silver. Testing between the temperature range of 830°C and 900°C, they found a stress exponent of n=1.2±0.2 and an activation energy of 800±160 kJ/mol. As these rheological parameters agreed with those determined for the monolithic material, they therefore suggested that the same diffusional mechanism controlled deformation. In addition, they reported a 20 fold decrease in the flow stress for samples compressed at constant strain rate (10^{-5}/sec) relative to the monolithic material. A lower decrease in flow stress, 3 to 10 fold, was reported by Hendrix et al.[6] for a 20 vol% Ag composite tested at similar strain rates but with a coarser grain size matrix, 15-25 μm . No evidence of superplasticity was found. However, this became possible by testing in compression composites with 2.5, 10 and 25 vol% silver additions to 1μm grain size matrices [2]. Compressive strains of 110% were attained and the deformation parameters were determined for the 25 vol%Ag composite tested between 800°C and 850°C. In this temperature range the stress exponent was n=2±0.1 and the grain size exponent, p=2.5±0.7. The activation energy was 760±100 kJ/mol, slightly higher than for the pure material but within the accuracy range [2,7].

In this paper we report the results of tensile testing of $YBa_2Cu_3O_{7-x}$/25 vol% Ag specimens. Constitutive relations at high temperatures are established, providing additional information on the dominant deformation mechanism of the composite, and suggesting a possible application of superplasticity on net shape forming of bulk superconductors.

EXPERIMENTAL

Commercial fine-grained $YBa_2Cu_3O_{7-x}$ superconductor (Y123 Superamic - Rhône Poulenc Inc) and high purity silver powder (99.9% - Johnson Matthey) were used. The two materials were mechanically mixed in ethanol media (200 proof) with zirconia balls, for 48 hours.

The slurry was homogenized using an ultrasonic probe for 15 minutes and was dried on a hot plate under magnetic stirring to prevent differential settling. The dried cake was crushed with a mortar and pestle, and the powder was then calcined in air at 500°C for 24 hours in a box furnace. The specimens were uniaxially pressed in a (46 mm diameter) cylindrical metallic die at 5 MPa, followed by isostatic pressing at 34 MPa. Typical batch size was 38 grams per disc. These discs were placed in an alumina crucible packed with loose powder and sintered at 875°C for 6 hours in air. A microstructure of 0.93 ± 0.15 μm with a 85% of relative density was obtained. The sintered discs were sent to a commercial machine shop (Bomas Machine Specialties, Inc.) to fabricate dog-bone tensile specimens (see Fig. 1). After machining the samples were soaked in acetone, and placed in a vacuum desiccator for aproximatly 20 minutes. Samples were then ultrasonicated for periods of 20 minutes. These precautions were taken to eliminate contamination from wax and glues used in the machining process. After drying, all surfaces were carefully hand- polished with SiC paper to a 1 μm finnish, in order to eliminate surface scratches and subsurface damage associated with machining.

A pin and clevis apparatus design was used in the tensile testing, with SiC pull rods and SiC pins applying the stress. A clam-shell type furnace was used, attached to an Instron Servohydraulic Universal Machine. The alignment of the system depended mainly on the accurate machining of the samples. Tests were performed at a constant temperatures and both constant and varying strain rates.

Figure 1. Photograph of tensile specimens of the $YBa_2Cu_3O_{7-x}$/25vol% Ag composite superplastically elongated to 95% at 850°C and 10^{-5}/sec constant strain rate.

RESULTS

The $YBa_2Cu_3O_{7-x}$/Ag composite showed superplastic elongations above 95% in uniaxial tension, as illustrated in Fig. 1 for a sample deformed at 850°C and at a strain rate of 10^{-5}/sec. No necking was observed for all the specimens. Typical microstructures from the composite before and after deformation are shown in Fig. 2. The silver phase as well as the porosity are above the percolation limit, hence forming an interpenetrated microstructure. Tests were performed on several samples both at constant and at differential strain rates in order to calculate the stress exponent, n. In Fig. 3 the stress versus strain rate at temperatures of 800°C, 825°C and 850°C are shown with their respective lines of best fitting. Both the individual constant strain-rate tests and strain-rate jump tests lead to a stress exponent of 2.3 ± 0.5, corroborating the stress exponent found in a previous study for the monolithic material [1]. These results are plotted together as a function of inverse temperature in an Arrhenius plot in order to estimate the apparent activation energy of deformation (Fig. 4). An activation energy of 456 ± 75 kJ/mol was obtained, also in close agreement with the value obtained for the monolithic material [1].

(a)   (b)

Figure 2. Typical microstructures of $YBa_2Cu_3O_{7-x}/Ag$. (a) as sintered and (b) deformed to 82% at 850°C and at 10^{-5}/sec strain rate. Light regions are the silver phase.

During deformation at 850°C, the grain size of the superconductor matrix was found to increase. Grain sizes were measured by the linear intercept method. Figure 5 shows the isothermal grain growth data at 850°C and compares the static grain growth to the grain growth during superplastic flow. The plot shows that the grain growth rate during deformation is greater than in the absence of strain. The solid curves through the data points are from a polynomial fit and have no physical meaning. In addition, microstructural examination by SEM of as sintered, annealed and deformed samples revealed a marked coarsening of the second phase during deformation. Also, the growth of the pre-existing cavities - irregularly shaped pores at the grain boundaries and triple junctions - developed preferentially in the axis perpendicular to the applied stress.

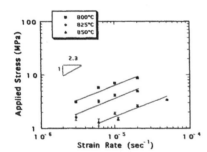

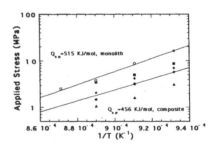

Figure 3. Plot of stress vs strain rate for $YBa_2Cu_3O_{7-x}/Ag$ at 850°C for calculation of n.

Figure 4. Arrhenius plot of applied stress vs. 1/T for the estimation of the apparent activation energy.

Density measurements (Archimedes method) were made on the samples deformed at constant strain rate (10^{-5}/sec). The plot in Fig. 6 compares data from the gauge section of the tensile samples with the annealed (but not deformed) grip region. In the departing microstructure, the 15 vol% porosity is above the percolation limit, forming an interconnected path of pores. While the density was foreseen to increase with the annealing (or testing) time due mainly to grain growth, the fact that strain did not contribute to a decrease in density (by nucleation and growth of voids) is unexpected. It is known that for polycrystalline materials, creep and cavitation occur concomitantly during high temperature deformation.

Grain Boundary Engineering in Ceramics

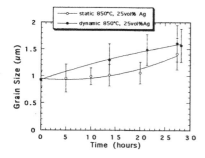

Figure 5. Grain size versus time at 850°C during static and dynamic annealing.

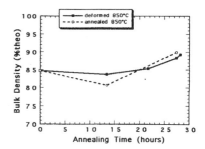

Figure 6. Relative densities of annealed and deformed samples of $YBa_2Cu_3O_{7-x}/Ag$ composite as a function of testing time.

DISCUSSION

The deformation parameters determined for $YBa_2Cu_3O_{7-x}/25$ vol% Ag composite were found to be very similar to those of the monolithic material [1]. The stress exponent determined experimentally, n=2.3±0.5, in conjunction with the calculated apparent activation energy for deformation, 456±75 kJ/mol, strongly suggests that the superplastic flow of the composite is controlled by the deformation mechanism of the more refractory (and dominant) $YBa_2Cu_3O_{7-x}$ phase. In previous testing of the monolith alone, under a similar set of conditions, the stress exponent was measured to be n=2±0.4, the inverse grain size exponent p= 1.5±0.12, and $Q_{sp}=$ 515±104kJ/mol [1]. In view of these data the proposed mechanism for the superplasticity of $YBa_2Cu_3O_{7-x}$ was grain boundary sliding controlled by interface reaction [1,8], and can be also assumed for the composite.

The presence of silver was shown to enhance ductility (elongations to failure were up to 95% true strain, Fig.1). This can be due to a decrease in stress caused by the addition of a soft phase, or to the influence of the second phase on grain size stabilization, which prevents strain hardening. Both will be analyzed separately.

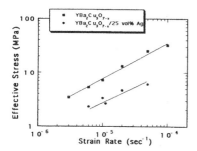

Figure 7. Comparison of effective stress dependence on strain rate from data obtained for $YBa_2Cu_3O_{7-x}$ and the composite, at 850°C.

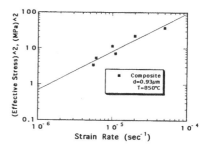

Figure 8. Application of Chen's model for soft inclusion composites to $YBa_2Cu_3O_{7-x}/25$ vol%Ag. Experimental results at 850°C are superimposed on the model prediction.

Effect of silver on stress and strain rate

Superplasticity (and grain boundary sliding) seems to be facilitated by the presence of the silver phase. In Fig.7 stress is depicted as a function of strain rate for monolithic and composite materials of comparable grain sizes. A significant increase in strain rate is observed for the composite at constant stress values. For a stress of 5 MPa, for example, the composite flows at a strain rate five times higher than that of the monolith. In order to normalize the effect of porosity the effective stress was considered. Its value is significantly higher than applied stress and it is computed with $\sigma_{eff}=\sigma_a \exp (3,7P)$, where σ_a is the applied stress and P the porosity. The multiplication factor was experimentally determined from a porosity-elastic modulus relationship [8,9].

Plastic deformation of composites depends on the volume fraction of the two phases and phase distribution, and on the relative deformation strengths. For some superplastic ceramic composites, such as cermets, the matrix phase is more refractory than the ductile second phase. This behavior was modeled by Chen [10] as non-Newtonian flow of soft incompressible spherical inclusions dispersed in a hard power law matrix, so that:

$$\dot{\varepsilon} = \left(1-V_f\right)^{-\frac{(53\,n+7)}{36}} \alpha_r \left(\frac{\sigma}{\sigma_r}\right)^n \qquad (2)$$

where n is the stress exponent, V_f is the volume fraction of inclusions, and σ_r and α_r are respectively, reference stress and strain rate, i.e. stress and strain rate of the material with no second phase and same microstructure. After data from the monolithic material were substituted into the equation for a constant silver volume fraction of 0.25 [8] the results show a good agreement with the prediction of the model (Fig. 8). This had already been observed in compression for lower volume fractions of silver [7]. Periclase-spinel $MgO-MgCr_2O_4$, (where periclase is the softer phase) was shown to follow this equation [10].

Effect of silver on the microstructure

Concerning the coarsening of the microstructure, a companion experimental study on grain growth kinetics was carried out (extended to 875°C) [8] relying on the determination of the variables involved in the grain growth law:

$$d^s - d_0^s = Kt \qquad (3)$$

where d_0 is the initial grain size, t is time and K is the rate dependent constant [11]. In that study both static and dynamic (strain induced) grain growth were shown to follow a quadratic grain growth law (s=4) in contrast to a cubic (s=3) dependency calculated for the monolithic material [8]. In both cases the time dependency of the grain growth kinetics is not affected by the application of stress. Moreover, it indicates a decrease in time dependency, resulting from a grain growth inhibition promoted by the silver second phase.

Observations of the morphological evolution of the silver phase with strain and temperature revealed that the coarsening of the departing very fine phase occured mainly for the deformed samples, the particle size remaining almost constant for the (strain-free) annealed samples. The high "mobility" of the silver phase and particularly its interaction with porosity are revealed by Fig 6. Even after large tensile strains the density of the composite was shown to be virtually unchanged as the filling up of the voids seems to retard cavitation.

Holm et al.[12], pointed out that grain growth in the presence of grain boundary particles during superplastic deformation, reflects the microstructural instability which the material undergoes. Grain growth is enhanced by particle agglomeration, resulting as a geometrical consequence of superplastic flow micromechanics [12]. The average increase of particle size has

Grain Boundary Engineering in Ceramics

been observed in superplastic deformation of metallic systems, as well as for ceramic composites [13,14]. Original models developed for duplex microstructures are based on the same form of empirical equations used for single phase materials. For duplex materials it is postulated [12,15,16] that grain growth is assisted by the enhanced particle agglomeration and that the mechanism for particle coarsening is due to collisions by stress assisted grain switching, as was originally proposed by Ashby and Verral (see inset in Fig. 9) [12]. Particles are assumed to rest on triple points. Neighbor switching events occur at a fixed frequency per unit strain, and account for all the strain increment. Following this sequence, Campenni and Cáceres [16] developed a rationale that predicts an exponential dependence of particle (grain) growth on strain:

$$D = D_0 \exp(f_s \varepsilon) \tag{4}$$

where f_s is the fraction of switches that lead to coalescence and has the same form as expressions found for single phase materials [12,16] and D_0 refers to the original particle size. When taking into account the contribution from static particle growth an expression for the instantaneous particle size is obtained:

$$D = \left[\left(\left(f_s (D_0)^s + K' \right) \exp\left(s \varepsilon f_s \right) - K' \right) / f_s \right]^{\frac{1}{s}} \tag{5}$$

where s is the grain size exponent from Eq.(3), f_s is the relative number of successful switching events, and K' is defined as $K'=K/s\dot{\varepsilon}$, with K from Eq.(3) [16].

The results comparing the experimental data for silver particle size to the predicted by the model, as a function of strain are shown in Fig. 9. A good agreement between the model prediction and measured data is shown for the case where only one over one hundred switching events ($f_s=0.01$) successfully participates in the coarsening. The result can be related to the interaction of the silver phase with porosity at low strain rates. A predicted trend of saturation of phase growth is also observed.

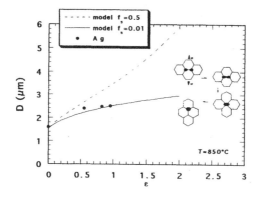

Figure 9. Strain dependence of silver phase size. Dashed and solid lines are computed solutions based on the Campenni and Cáceres model (Eq.4) for parameter $f_s=0.5$ and $f_s=0.01$, respectively. The inset shows schematically the particle aglomeration during grain switching.

When two phases are coarsenning together, it is assumed that a coupled grain growth occurs, since the growth or disappearance of one phase affects the other due to mutual topological constraints. Microstructural evolution should in this case only involve a change in scale and the relative grain size ratio between matrix grains and second phase particles should remain constant at any given time. This does not occur in the $YBa_2Cu_3O_{7-x}$/Ag composite because porosity is present in addition to the two phases. Silver strain-dependent interaction with the porosity phase (filling the voids) alters this equilibrium. Moreover, a direct correlation exists between this interaction and the enhanced strain to failure observed in the composite.

CONCLUSION

Tensile superplasticity of $YBa_2Cu_3O_{7-x}$/ 25 vol%Ag composite was studied, and elongations above 95% were recorded. The temperature and stress dependence of superplasticity followed the constitutive equation of the dominant superconductor phase. During deformation several phenomena occurred, among which grain growth and silver phase coarsening. The microstructures were studied and related to strain. By comparison with data obtained for the monolithic material silver was shown to increase the ductility by ensuring microstructural stability, preventing cavitation and slowing down the grain growth of the matrix phase.

REFERENCES
[1] J. M. Albuquerque, M. P. Harmer, and Y. T. Chou, "Tensile Superplastic Deformation of $YBa_2Cu_3O_{7-x}$ High T_c Superconductors," to be submited to *Acta Materialia* (2000).
[2] J. Yun, M. P. Harmer, Y. T. Chou, and O. P. Arora, "Observation of Superplastic Flow in $YBa_2Cu_3O_{7-x}$ Superconductors Containing Silver," pp 275-80, in *Superplasticity in Advanced Materials* ed. S. Hori, M. Tokizane and N. Furushiro, The Japan. Soc. Res. Superplasticity (1991).
[3] T. G. Nieh, J. Wadsworth and O. D. Sherby; p.111 in *Superplasticity in Metals and Ceramics*, Cambridge University Press, Cambridge, UK, 1998.
[4] J. R. Seidensticker and M. J. Mayo,"Dynamic and Static Grain Growth During the Superplastic Deformation of 3Y-TZP," *Scripta Mat.*, 38 [7] 1091-100 (1998).
[5] J. L. Routbort, K. C. Goretta, and J. P. Singh, "High Temperature Deformation of $YBa_2Cu_3O_{7-x}$ With Ag Additions," Mat. Res. Soc. Symp. Proc., 169. Mat. Res. Soc. (1990).
[6] B. C. Hendrix, T. Abe, J. C. Borofka, P. Wang, and J. K. Tien, "Hot Deformation of $YBa_2Cu_3O_{7-x}$ and Composite $YBa_2Cu_3O_{7-x}$/Ag," *J. Am. Ceram. Soc.*, 76, [4] 1008-10 (1993).
[7] J. Yun, PhD Thesis, Lehigh University, 1994.
[8] J. M. Albuquerque, PhD Thesis, Lehigh University, 1998.
[9] R. W. Rice," Comparison of Physical Property-Porosity Behavior With Minimum Solid Area Model," *J. Mater. Sci.*, 31 [6] 1509-28 (1996).
[10] I-Wei Chen," Superplastic Flow of Two-Phase Alloys," pp 5.1-5.20 in *Superplasticity*, Ed. B. Baudelet and S. Suery, CNRS, Paris (1985).
[11] J. E. Burke, and D. Turnbull, "Recristallization and Grain Growth in Metals," *Progr. Metal. Phys.*, 3, 220-92 (1952).
[12] K. Holm, J. D.Embury, and G. R. Purdy ,"The Structure and Properties of Microduplex Zr-Nb Alloys," *Acta Metall.*, 25, 1191-200 (1977).
[13] L.A.Xue, X.Wu, and I-W. Chen," Superplastic Alumina Ceramics with Grain Growth Inhibitors," *J. Am. Ceram. Soc.*, 74 [4] 842-45 (1991).
[14] L. A. Xue and R. Raj, "Grain Growth in Superplastically Deformed Zinc Sulfide/Diamond Composites," *J. Am. Ceram. Soc.*, 74 [7] 1729-31 (1991).
[15] D. S. Wilkinson, and C. H. Caceres, "On the Mechanism of Strain Enhanced Grain Growth During Superplastic Deformation," *Acta Metall.*, 32 [9] 1335-45 (1984).
[16] V.D.Campenni and C.H. Cáceres," Strain Enhanced Grain Growth at Large Strains in A Superplastic Zn-Al Alloy," *Scripta Metall.*, 22 [3] 359-64 (1988).

Keywords: Superplasticity, $YBa_2Cu_3O_{7-x}$ /Ag composites, High T_c Superconductors

ROLE OF GRAIN BOUNDARY SEGREGATION ON HIGH-TEMPERATURE CREEP RESISTANCE IN POLYCRYSTALLINE Al$_2$O$_3$

H. Yoshida, Y. Ikuhara* and T. Sakuma
Department of Advanced Materials Science,
Graduate School of Frontier Sciences, The University of Tokyo
*Graduate School of Engineering, The University of Tokyo
7-3-1 Hongo, Bunkyo-ku, Tokyo 113-8656, Japan
E-mail: yoshida@ceramic.mm.t.u-tokyo.ac.jp

ABSTRACT

High-temperature creep resistance in polycrystalline Al$_2$O$_3$ with 0.1 mol% oxides of LuO$_{1.5}$, ZrO$_2$ or MgO has been examined by uniaxial compression creep testing at 1250℃ . The creep resistance is highly improved by the doping of Lu or Zr even in the dopant level of 0.1mol%, but is reduced by Mg doping. The dopant effect on the creep resistance cannot be explained in terms of, for example, ionic radius of the dopant cation or eutectic point in Al$_2$O$_3$ -oxide of dopant cation system. Each dopant cation was found to segregate in grain boundaries, and is likely to influence grain boundary diffusion in Al$_2$O$_3$. The change in chemical bonding state with doping was estimated by the first principle molecular orbital calculation using DV-Xα method. The ionic bonding and the covalent bonding of Al - O are lowered by the introduction of V$_O^{\cdot\cdot}$ or V$_{Al}'''$, but the values of the NC in Al and O are increased by the cations doping. The change in the NC value is correlated well with the high-temperature creep resistance in Al$_2$O$_3$ with cation doping. The result indicates that the ionicity in Al and O is an important factor to determine high-temperature creep resistance in polycrystalline Al$_2$O$_3$.

INTRODUCTION

It has been pointed out that high-temperature creep deformation or plastic flow in polycrystalline Al$_2$O$_3$ is sensitively affected by a small amount of dopant cation. An addition of 0.1wt% ZrO$_2$ in Al$_2$O$_3$ with a grain size of about 1 μm severely suppresses the creep rate by a factor of about 15 at 1250℃ under the stresses of 10 ~200MPa [1]. Cho *et al.* reported that 1000 ppm yttrium or 500 ppm lanthanum-doped Al$_2$O$_3$ has improved the creep resistance in comparison with pure Al$_2$O$_3$ at 1200~1350℃ under the stress of 50MPa [2]. Such a dopant effect of MgO in the level of 0.1 wt% was also observed in high-temperature flow stress of fine-grained Al$_2$O$_3$ in conventional tensile testing; the peak stress decreases by the doping of MgO at 1450℃ under an initial strain rate of about

To the extent authorized under the laws of the United States of America, all copyright interests in this publication are the property of The American Ceramic Society. Any duplication, reproduction, or republication of this publication or any part thereof, without the express written consent of The American Ceramic Society or fee paid to the Copyright Clearance Center, is prohibited.

1×10^{-4} s^{-1} [3]. It has been pointed out that the high-temperature creep resistance in Al_2O_3 with a grain size of 1 μm is markedly improved by a doping of lanthanoid oxide such as Y_2O_3 or Lu_2O_3 even in the level of 0.05mol% [4]. In this case, the dopant cation did not form the second phase precipitations, but segregated in the vicinity of grain boundaries to improve the creep resistance in Al_2O_3. The segregation must be an origin of the change in creep deformation, but the mechanism has not been clarified yet. The purpose of this paper is to discuss role of dopant segregation at grain boundaries in terms of vacancy effect due to dopant addition. The first-principle molecular orbital calculations were done by solving the Hartree-Fock-Slater equations self consistently using DV-Xα method developed by Adachi et al. [6].

EXPERIMENTAL PROCEDURE

The materials used in this study are undoped high-purity Al_2O_3 and Al_2O_3 with 0.1mol% of $LuO_{1.5}$, ZrO_2 or MgO. High-purity alumina powders with 99.99% purity (TM-DAR, Taimei Chemicals, Japan), the lutetium acetate (99.9%, Rare Metallic, Tokyo, Japan), colloidal zirconia particles dispersed in water (99.9%, NZS-30B, Nissan Chemical Industry, Japan) and MgO powders (99.98%, Ube Chemical, Japan) were used for starting materials. The powders were mixed, ball-milled in ethanol together with 5mm diameter high-purity(> 99.9%) alumina balls for 24h, dried and shifted through a 60 mesh sieve for granulation. The green compacts were prepared by pressing the mixed powders into bars with a cemented carbide die under a pressure of 33 MPa, and then isostatically-pressed under a pressure of 100 MPa. The green compacts were sintered at a temperature in the range 1300°C - 1400°C for 2 h in air to obtain an average grain size of about 1 μm in all samples as shown in Table 1. High-temperature creep experiments were carried out under uniaxial compression in air at a constant load using a lever-arm testing machine with a resistance-heated furnace (HCT-1000, Toshin Industry, Tokyo, Japan). The applied stress and temperature were 50 MPa and 1250°C, respectively. The test temperature was measured by a Pt-PtRh thermocouple attached to each specimen and

Table I. Sintering conditions and grain size of undoped- and $LuO_{1.5}$ or ZrO_2 or MgO-doped Al_2O_3.

Sample	Sintering conditions	Grain size (μm)	Ionic radius of dopant cation (Å)*
high-purity Al_2O_3	1300°C×2h	0.9	--
$LuO_{1.5}$ –doped Al_2O_3	1400°C×2h	0.9	0.861
ZrO_2 –doped Al_2O_3	1400°C×2h	1.0	0.720
MgO –doped Al_2O_3	1300°C×2h	0.9	0.720

*All for 6-fold coordination. Ionic radius of Al^{3+} = 0.535 Å [7]

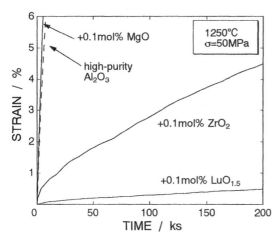

Figure 1. The creep curves in undoped, high-purity Al_2O_3 and 0.1mol% of $LuO_{1.5}$ or ZrO_2 or MgO-doped Al_2O_3 under an applied stress of 50MPa at 1250℃.

kept to within ±1℃. The size of the specimens was 6×6 mm^2 in cross-section and 8 mm in height for compression tests. Microstructures were examined with a scanning electron microscope (SEM; JSM-5200, JEOL). The grain size was measured by a linear intercept method using SEM photographs. High-resolution electron microscopy (HREM) was also performed to analyse the grain-boundary structure using a JEOL - 2010 field-emission-type electron micro-scope. In order to examine the change in chemical bonding state in alumina grain boundaries, a first principles molecular orbital calculation was made by a discrete-variational (DV) – X α method to understand the experimental EELS spectra.

RESULTS AND DISCUSSION

The creep curves at 1250℃ under an applied stress of 50MPa in the present materials are shown in Figure 1. The creep deformation in polycrystalline Al_2O_3 is very much suppressed by doping of $LuO_{1.5}$ or ZrO_2 ions [1,4,5], but is slightly accelerated by MgO doping. The doping of $LuO_{1.5}$ is much effective to improve the creep resistance in Al_2O_3 in comparison with ZrO_2. The difference of creep strain rate in the present materials is not caused by the change of grain growth rate, because the grain growth during creep deformation was negligible in the present materials at the temperature examined.

Figure 2 shows a HREM image of $LuO_{1.5}$ –doped Al_2O_3 (a) together with the data of EDS analyses taken with a probe size of 0.8 nm from a grain interior (b) and from a grain boundary (c). The HREM image is taken for the boundary at the edge-on view. No second phase or amorphous phase at the boundary was observed along the grain boundary. The EDS analysis reveals that lutetium ions are present only in the grain boundary, but not in the grain interior. Lutetium ions must segregate in grain boundaries. The microstructure in Zr^{4+} or Mg^{2+} -doped Al_2O_3 is essentially similar to that in Lu^{+3} -doped one. It is suggested that the high-temperature creep deformation is likely to take place

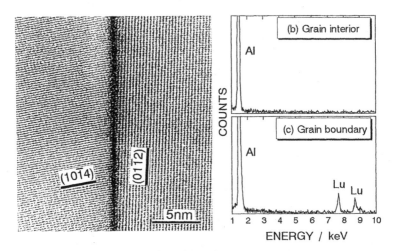

Figure 2. HREM image of as-sintered 0.1mol% $LuO_{1.5}$ –doped Al_2O_3 (a), and the EDS spectra taken with a probe size of about 1nm from (b) the grain interior and (c) the grain boundary.

mainly by Al^{3+} grain boundary sliding controlled by grain boundary diffusion [1,3-5]. The change in the creep resistance must be caused by the segregation of dopant cations at grain boundaries.

The dopant effect cannot be explained solely in terms of the ionic radius of the dopant cation; for example, Zr^{4+} and Mg^{2+} ions have similar ionic radii, which is much larger than that in Al as shown in Table 1, but the role of the dopant cations is different from each other. The eutectic point in Al_2O_3 - MgO system is about 1970℃, which is the higher than that in Al_2O_3 – ZrO_2 (~1710℃) or in Al_2O_3-Lu_2O_3 (~1800℃). But, no correlation is seen between the creep strain rate and the eutectic point. In order to find the origin of the change in the creep resistance due to the cation doping, change in chemical bonding state in the cation-doped Al_2O_3 was examined by a first-principle molecular orbital calculation.

The model clusters of undoped-Al_2O_3 and Al_2O_3 with dopant cation used are shown in Figure 3 (a) - (f). Fig. 3 (a) shows the model clusters for undoped Al_2O_3, $[Al_5O_{21}]^{27-}$, which represents a part of the corundum structure. One can estimate the interaction between an Al ion at the center of the cluster (marked as Al_C in Fig.3 (a)) and surrounding O or Al ions by using the model cluster. Fig.3 (b) and (c) show model clusters of $[Al_5O_{20}]^{25-}$ and $[Al_4O_{21}]^{30-}$, which have O^{2-} ion vacancy ($V_O^{··}$) and Al^{3+} ion vacancy (V_{Al}'''), respectively. In order to estimate the interaction between Al_C and vacancy, the vacancy is introduced at the nearest O or Al site as shown in Fig.3 (b) and (c). Fig. 3 (d) shows a model cluster of $[Al_2Lu_3O_{21}]^{27-}$, in which Lu^{3+} ions are substituted at Al sites.

Grain Boundary Engineering in Ceramics

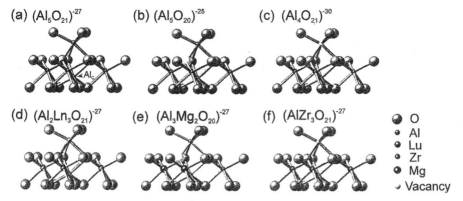

Figure 3   Atomic structure of cluster models (a) $[Al_5O_{21}]^{27-}$, (b) $[Al_5O_{20}]^{25-}$, (c) $[Al_4O_{21}]^{30-}$, (d) $[Al_2Lu_3O_{21}]^{27-}$, (e) $[Al_3Mg_2O_{20}]^{27-}$ and (f) $[AlZr_3O_{21}]^{27-}$.

Fig.3 (e) and (f) shows the $[Al_3Mg_2O_{20}]^{27-}$ and $[AlZr_3O_{21}]^{27-}$ clusters, which include the dopant cations and accompanying O^{2-} or Al^{3+} vacancies for electronic neutrality.

From the first-principle molecular orbital calculations, the ionic net charges (NC) of each atom and the bond overlap population (BOP) between atoms can be obtained as the parameters of chemical bonding state. The NC and the BOP are regarded as the effective ionic charge and the covalency between atoms, respectively [6,8-10]. Figure 4 shows the BOP of (a) Al - O ions and (b) dopant cation - O obtained from the cluster models. The horizontal broken lines denote the value of Al - O in pure-Al_2O_3 cluster. The BOP of Al-O in $V_O^{\cdot\cdot}$ or V_{Al}''' -introduced Al_2O_3 and cation-doped Al_2O_3 is smaller than that in pure-Al_2O_3. This result indicates that the covalency between Al and O ions decreases by the presence of vacancy or dopant cations. The BOP of dopant cation – O is also smaller than that of Al – O in pure Al_2O_3. This fact indicates that the covalency between dopant cation and O ions is lower than that of Al – O. The BOP seems not to be related to the change in the high-temperature creep resistance in Al_2O_3.

Figure 5 shows the NC of (a) Al and O ions and (b) dopant cations. The horizontal broken lines in the figure denote the value in pure-Al_2O_3. The absolute values of NC of Al and O ions become smaller in the V_{Al}''' or $V_O^{\cdot\cdot}$-introduced clusters. The decrease of NC is clearer in Al^{3+} vacancy-introduced cluster. This result suggests that the chemical bonding strength is reduced by the presence of vacancies and that the reduction is larger in V_{Al}''' – doped cluster. On the other hand, the value of NC in Lu-doped cluster is larger than that in Al_2O_3 cluster, and NC of Al and O in $[Al_3Mg_2O_{20}]^{27-}$ or $[AlZr_3O_{21}]^{27-}$ is larger than that in $[Al_5O_{20}]^{25-}$ or $[Al_4O_{21}]^{30-}$, respectively. One can conclude that the dopant cations have

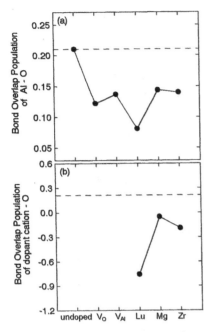

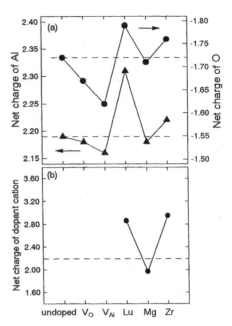

Figure 4   Bond overlap population of (a) Al - O and (b) dopant cation − O obtained from molecular orbital calculations.

Figure 5   Net charges for (a) Al and O and (b) dopant cation obtained from model clusters.

an effect to increase the ionicity between Al and O ions in Al_2O_3. The values of NC of Al, O and dopant cations in the Lu or Zr-doped cluster becomes larger, but the NC value in the Mg-doped cluster becomes smaller than that in pure-Al_2O_3. This result indicates that NC is likely to correlate to the creep resistance in cation-doped Al_2O_3.

Figure 6 shows the creep strain rate against (a) BOP of Al - O, and (b) the product of NC in Al and O ions. The effect of dopant cations on the creep resistance cannot be explained solely from the order of the BOP in cation-doped Al_2O_3; the BOP of Mg or Zr-doped cluster is nearly the same, but the dopant effect on the creep strain rate is very different. However, the absolute value of the production of NC shows a good correlation with the creep resistance in cation-doped Al_2O_3. The product of NC of Al and O must correspond to the Coulomb's attraction force between cation and anion. The present result must reflect the fact that the chemical bonding strength in Al_2O_3 is mainly associated with ionic bonding [11].

In our previous report, the change in chemical bonding state in Al_2O_3 grain boundaries with segregation of Lu ions was detected by HREM - electron energy loss spectroscopy

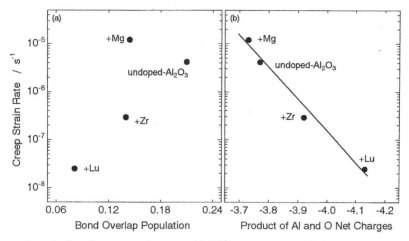

Figure 6   A plot of creep strain rate at 1250°C under the applied stress of 50 MPa in the present materials against (a) bond overlap population of Al – O and (b) the product of Al and O net charges obtained by molecular orbital calculations.

(EELS) analysis [12]. The change in the energy-loss near-edge structure (ELNES), which contains fine structure of unoccupied density of states in the conduction band (DOS) [13], can be explained by the DOS obtained from the molecular orbital calculations using Lu-doped Al_2O_3 cluster [12]. This result suggests that the chemical bonding state in Al_2O_3 is influenced by the presence of dopant cations rather than by the grain boundary atomic structure. The model clusters used in this study were very simple ones, but seem to be useful to explain the change in the chemical bonding state in Al_2O_3 grain boundaries due to the segregation of dopant cations. The configuration of dopant ions at grain boundaries in Al_2O_3 would be different from those of Fig. 3, but, we can assumed the simple clusters for the grain boundary models with the segregation in order to estimate the change in the atomic bonding state on introducing vacancy or dopant cations into Al_2O_3.

CONCLUSIONS

The creep resistance of polycrystalline Al_2O_3 with a grain size of about 1 μm is highly improved by the doping of $LuO_{1.5}$ or ZrO_2 even in the level of about 0.1mol%, but is reduced slightly by MgO doping at temperature of 1250°C under an applied stress of 50MPa. The high-temperature creep deformation is likely to take place mainly by grain boundary sliding accommodated by grain boundary diffusion, and the change in the creep resistance is likely to be caused by the change in the grain boundary diffusivity due to the segregation of cations at grain boundaries. The ionic bonding and the covalent bonding of Al - O are lowered by the introduction of $V_O^{\cdot\cdot}$ or $V_{Al}^{\prime\prime\prime}$, but the values of NC in Al and

O are increased by the cations doping. The change in NC value is correlated well with the high-temperature creep resistance in Al_2O_3 with cation doping. It is suggested that the ionicity in Al and O is an important factor to determine high-temperature creep resistance in polycrystalline Al_2O_3.

ACKNOWLEDGEMENT

The authors wish to express their gratitude to the Ministry of Education, Science and Culture, Japan, for the financial support by a Grant-in-Aid for Developmental Scientific Research (2)-1045 0254 for Fundamental Scientific Research. We also wish to express our thanks to Research Fellowships of the Japan Society for the Promotion of Science for Young Scientists for their financial aid.

REFERENCES

1) H. Yoshida, K. Okada, Y. Ikuhara and T. Sakuma, *Phil. Mag. Lett.*, **76** (1997) 9 - 14.

2) J. Cho, M.P. Harmer, H.M. Chan, J.M. Rickman and A.M. Thompson, *J. Am. Ceram. Soc.*, **80** (1997) 1013-17.

3) Y. Takigawa, Y. Ikuhara and T. Sakuma, *J. Mater. Sci.*, **34** (1999) 1991-97.

4) H. Yoshida, Y. Ikuhara and T. Sakuma, *J. Mater. Res.*, **13** (1998) 2597-2601.

5) H. Yoshida, Y. Ikuhara and T. Sakuma, *Phil. Mag. Lett.*, **79** (1999) 249-256.

6) H. Adachi, M. Tsukada and C. Satoko, *J. Phys. Soc. Jpn.*, **45** (1978) 875-883.

7) R.D. Shannon, *Acta. Cryst.*, **A32** (1976) 751-767.

8) H. Adachi, *Ceramics*, **27** (1992) 495-501.

9) I. Tanaka, J. Kawai and H. Adachi, *Solid State Commun.*, **93** (1995) 533-36.

10) Y. Ikuhara, Y. Sugawara, I. Tanaka and P. Pirouz, *Interface Science*, **5** (1997) 5-16.

11) W.D. Kingery, H.K. Bowen, and D.R. Uhlman, *Introduction to Ceramics*, New York: John Wiley and Sons (1976) p.185.

12) H. Yoshida, Y. Ikuhara and T. Sakuma, *J. Inorganic Mater.* **1** (1999) 229-234.

13) D.B. Williams and C.B. Carter, Transmission Electron Microscopy. Plenum Press, New York, 1996.

ANISOTROPIC THERMAL CONDUCTION MECHANISM OF β-Si3N4 GRAINS AND CERAMICS

Koji WATARI#, Lionel POTTIER*, Bincheng LI*, Daniel FOURNIER*, Motohiro TORIYAMA#

#National Industrial Research Institute of Nagoya, Kita-ku, Nagoya 462-8510, Japan
*Laboratoire d'Instrumentation de l'Universite Pierre et Marie Curie, UPR A0005 du CNRS, 10 rue Vauquelin, F-75005 Paris, France

ABSTRACT

The thermal conductivity of a β-Si3N4 single crystal grain in a sintered material could be successfully measured by thermoreflectance microscopy. Anisotropic thermal conductivity is found inside the Si3N4 grain, and the conductivity along the c-axis of β-Si3N4 crystal is about three times higher than that along the a-axis. Based on the results, the anisotropic thermal conduction mechanism of β-Si3N4 grains and ceramics is discussed combining with its growth mechanism.

INTRODUCTION

Recently a significant increase in the thermal conductivity of sintered β-Si3N4 has been achieved. Hirosaki et al. successfully obtained a β-Si3N4 ceramic with a room temperature thermal conductivity of 124 $W \cdot m^{-1} \cdot K^{-1}$ by gas-pressure sintering at 2273 K with additions of Y2O3 and Nd2O3 [1]. In another work, grain-oriented Si3N4 was fabricated by tape-casting a powder slurry with rodlike β-Si3N4 single crystal particles and Y2O3 addition, and its conductivity showed a value of 120 $W \cdot m^{-1} \cdot K^{-1}$ along the casting direction [2]. Very recently, a β-Si3N4 ceramic with a conductivity of 155 $W \cdot m^{-1} \cdot K^{-1}$ was developed by an ultra-high temperature HIP process [3].

In order to further enhance the conductivity of the sintered material, the thermal conduction mechanism of β-Si3N4 single crystal and ceramic should be understood. In this case, data about the thermal conductivity of high-purity β-Si3N4 single crystal is required to clarify the relationship between the conductivity of β-Si3N4 and the type and amount of impurity, and to estimate an intrinsic thermal conductivity. It is, however, difficult to obtain cm-size β-Si3N4 single crystals for conventional thermal conductivity measurements. In this work, β-Si3N4 ceramic with a very large grain size was produced by an ultra-high temperature HIP process. The thermal conductivity of very large β-Si3N4 single crystal grains in the sintered material was measured by thermoreflectance microscopy. Using this data, the thermal conduction mechanism of β-Si3N4 grains and ceramics is discussed.

EXPERIMENTAL PROCEDURE

Fabrication of ceramic with very large grains

To the extent authorized under the laws of the United States of America, all copyright interests in this publication are the property of The American Ceramic Society. Any duplication, reproduction, or republication of this publication or any part thereof, without the express written consent of The American Ceramic Society or fee paid to the Copyright Clearance Center, is prohibited.

The sintered β-Si3N4 with very large grains was fabricated by the addition of rod-like, high-purity β-Si3N4 single crystal particles to a mixture of raw Si3N4 powders with a sintering aid, followed by selective grain growth during firing. The β-Si3N4 single crystal particles were obtained through the growth from a melt flux in the Si3N4-Y2O3-SiO2 system [4]. Their mean diameter and length are 1.3 and 5.4 μm, respectively.

High-purity α-Si3N4 powder and 5 mass % Y2O3 as a sintering aid were mixed by ball milling. The β-Si3N4 single crystal particles of 5 vol % were added to this slurry. After drying, the powders were placed into a Si3N4-BN-SiO2 crucible, and were then HIPed at 2773 K for 2 h under a N2 gas pressure of 200 MPa.

The crystalline phase of the sintered material was identified as β-Si3N4 by X-ray diffractometry using Cu$K\alpha$ radiation. The bulk density of the specimen, 3205 kgm-3, was measured by a displacement method, and is almost equal to the theoretical density of β-Si3N4.

Measurement of thermal conductivity by thermoreflectance microscopy
In order to minimize the effect of the presence of a grain-boundary phase on the thermal conductivity of β-Si3N4 grains in the sintered material, the removal of the grain-boundary phase around grains is necessary. To remove the grain-boundary phase, the surface of the specimen was polished, and it was chemically etched by a mixture of solutions of NaOH and KOH. Figure 1 shows a SEM photograph of the chemically etched surface of the specimen. An extremely large elongated grain (mean diameter : 17 μm and length: 100 μm) is observed in the material. In the present work, measurement of the thermal conductivity of this grain was performed to establish the value of the experimental thermal conductivity of β-Si3N4 single crystal.

The crystal structure of β-Si3N4 is hexagonal. Therefore, it is postulated that heat flow in β-Si3N4 is different depending on the crystal axis. Yasutomi et al. verified the crystallographic orientation of β-Si3N4 ceramic containing elongated β-Si3N4 particles, and found that the elongated grains grew remarkably in the [0001] direction [5], indicating that the favored axis for grain growth is the c-axis of β-Si3N4. In the present work, the longitudinal axis of the very large grain as seen in Fig. 1 coincided with the crystal c-axis. On the other hand, all directions normal to the c-axis corresponds to an a-axis.

The experimental set-up of thermoreflectance microscopy has been reported in detail elsewhere [6-8], but is briefly described here. Figure 2 shows a schematic diagram of the thermoreflectance microscope. An intensity-modulated Ar+ laser beam focused by an optical microscope onto the surface of the very large β-Si3N4 grain (Fig. 1) excites a thermal wave. The resulting distribution of the surface temperature modulation is read by a second laser beam via thermoreflectance. The amplitude and phase of the thermoreflectance signal are extracted by lock-in detection.

RESULTS

The two-dimensional phase map of the thermoreflectance signals obtained on the very large grain of Fig. 1 is shown in Figure 3. The horizontal direction in Fig. 3 corresponds to the crystal c-axis of the grain (vertical direction of Fig. 1). It is found that heat flow is more marked along the

Grain Boundary Engineering in Ceramics

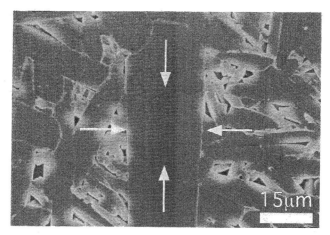

Figure 1. SEM photograph of the polished and chemically etched surface of the specimen. An extremely large elongated grain formed by addition of β-Si3N4 single crystal particles and high-temperature firing, is observed on the polished surface. The c-axis of β-Si3N4 is parallel to the edges of this very long rod. The four white arrows point to the center of the region of the measurements.

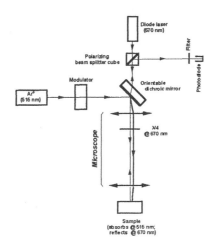

Figure 2. Schematic diagram of the thermoreflectance microscope.

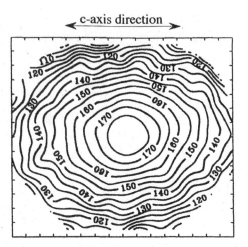

Figure 3. Contour lines of the phase (in degrees) of the thermoreflectance signal obtained on the surface of the very large Si3N4 grain (Fig. 1). The heating beam's modulation frequency is 300 kHz. At this frequency the range of the thermal wave is short enough for the wave to remain inside the grain. XY tick spacing: 1 μm. The center of the pattern is the location of the center of the (fixed) probe spot. For a given position [X,Y] (in μm) of the center of the heating spot, the map gives the corresponding phase $\phi(X, Y)$.

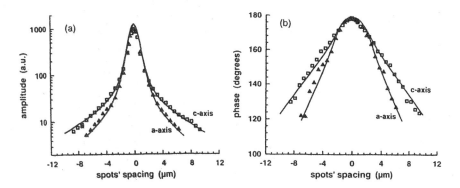

Figure 4. Dependence of the amplitude (a) and the phase (b) on the spacing between the heating and probe spots, along the a-axis or c-axis, for a β-Si3N4 single crystal grain. Modulation frequency is 300 kHz.

Grain Boundary Engineering in Ceramics

c-axis, indicating that anisotropy in thermal conductivities is obvious in the β-Si3N4 single crystal grain. To determine thermal diffusivity and conductivity of the very large grain, the amplitude and phase of the thermoreflectance signal were also recorded as functions of the separation of the two spots along either of the two principal axes in the surface plane (i.e. c-axis and a-axis), at a modulation frequency of 0.3 MHz, and are then indicated in Figure 4. The thermal diffusivity was extracted from the slope $d\phi/d\chi$ of the phase ϕ of the transverse deflection of the probe beam as a function of the probe beam offset χ using the 'phase method' [9, 10]. The principal thermal diffusivities obtained from the results of Fig. 4 are 0.84 and 0.32 cm2s-1, respectively, along the c-axis and a-axis. The corresponding thermal conductivity κ was calculated using the following relation,

$$\kappa = \rho \, C \, D \qquad\qquad (1)$$

where ρ is the density of β-Si3N4 (3200 kgm-3), C the experimental specific heat (670 JK-1kg-1 [11]) and D the thermal diffusivity. The calculated thermal conductivity is 180 W · m-1 · K-1 along the c-axis, and 69 W · m-1 · K-1 along the a-axis.

DISCUSSION

There are no reports, previous to the present work, about thermal conductivity of β-Si3N4 single crystals. In this paper, it could be clarified from the measurement of the conductivity of grains in the sintered materials by thermoreflectance microscopy that β-Si3N4 single crystal has a thermal conductivity of 180 W · m-1 · K-1 at room temperature. This value is the highest experimental value for Si3N4. Slack et al. reported that diamond, BN, SiC, BeO, BP, AlN, Si, GaN, and GaP possess intrinsic thermal conductivities higher than 100 W · m-1 · K-1, and mentioned that these non-metallic solids are high-thermal-conductivity materials [12]. The obtained conductivity of β-Si3N4 single crystal is higher than the theoretical thermal conductivities of Si (150 W · m-1 · K-1), GaN(130 W · m-1 · K-1), and GaP (100 W · m-1 · K-1). Therefore, β-Si3N4 is one of the high-thermal-conductivity insulating materials.

Anisotropic thermal conductivity in the β-Si3N4 single crystal was found in the present work. The conductivity along the c-axis is about three times higher than that along the a-axis. It is assumed that the anisotropic thermal conductivity is closely associated with the crystal structure (hexagonal) of β-Si3N4 or with the distribution of crystal defects. The relationship between crystal structure and thermal conductivity of β-Si3N4 has not been clarified yet, but it can be discussed from results of thermal expansion relating to crystal structure and atomic packing density in a crystal. It has been reported that the thermal expansion coefficient of β-Si3N4 for the c-axis is slightly higher than that for the a-axis [13]. On the other hand, it is generally accepted that the direction with lower thermal expansion coefficient shows higher thermal conductivity [14]. According to this thermal expansion rule, the conductivity along the a-axis of β-Si3N4 crystal should be higher than that along the c-axis. It is not agree with experimental result concerning measurement of the conductivities of β-Si3N4 single crystal. Hence, the anisotropic thermal conductivities of β-Si3N4 single crystal grain can not be explained by the thermal expansion behavior relating to crystal structure and atomic packing density in a crystal.

In a previous work, experimental observations and theoretical calculation of phonon mean free path showed that thermal conductivity of sintered β-Si3N4 at room temperature is independent of grain size, but is controlled by the internal defect structure of the grains such as point defects and dislocations [15]. Therefore, the anisotropic conductivity may be ascribable to diverse distribution of crystal defects depending on the crystal axis. Watari et al. reported that removal of the crystal defects and purification into grains concurrently proceeds with grain growth [3]. Furthermore, Yasutomi et al. found that elongated grains grew remarkably in the c-axis direction, compared to the a-axis direction [5]. These results indicate that selective removal of crystal defects in the c-axis direction of β-Si3N4 single crystal grain is possible. Consequently, it is thought that anisotropic thermal conductivity appears in the β-Si3N4 single crystal grain.

CONCLUSION

The thermal conductivity at room temperature of a β-Si3N4 single crystal grain is determined as 180 W $\cdot$ m-1 $\cdot$ K-1 along the c-axis, and 69 W $\cdot$ m-1 $\cdot$ K-1 along the a-axis. The anisotropic thermal conductivity is considered to be attributed to active grain growth in the c-axis direction to remove the crystal defects and purify into the β-Si3N4 single crystal.

REFERENCES
[1] N.Hirosaki, Y.Okamoto, M.Ando, F.Munakata, and Y.Akimune, "Thermal Conductivity of Gas-Pressure-Sintered Silicon Nitride,"*J. Ceram. Soc. Jpn.*, **104**, 49-53 (1996).
[2] K.Hirao, K.Watari, M.E.Brito, M.Toriyama, and S.Kanzaki, "High Thermal Conductivity in Silicon Nitride with Anisotropic Microstructure," *J. Am. Ceram. Soc.*, **79**, 2485-88 (1996).
[3] K.Watari, K.Hirao,M.E.Brito, M.Toriyama, and S.Kanzaki, "Hot-Isostatic-Pressing to Increase Thermal Conductivity of β-Si3N4 Ceramics," *J. Mater. Res.*, **14**, 1538-42 (1999)
[4] K.Hirao, T.Nagaoka, M.E.Brito, and S.Kanzaki, "Microstructure Control of Silicon Nitride by Seeding with Rodlike β-Silicon Nitride Particles" *J. Ceram. Soc. Japan*, **101**, 1071-73 (1993).
[5] Y.Yasutomi, Y.Sakaida, N.Hirosaki, and Y.Ikuhara, "Analysis of Crystallographic Orientation of Elongated β-Si3N4 Particles in In Situ Si3N4 Composite by Electron Back Scattered Diffraction Method," *J. Ceram. Soc. Jpn.*, **106**, 980-83 (1998).
[6] B-C. Li, L.Pottier, J.P.Roger, D.Fournier, K.Watari and K.Hirao, "Measuring the Anisotropic Thermal Diffusivity of Silicon Nitride Grains by Thermoreflectance Microscopy," *J. Euro. Ceram Soc.*, **19**, 1631-39 (1999).
[7] L.Pottier, "Micrometer Scale Visualization of Thermal Waves by Photoreflectance Microscopy," Appl. Phys. Lett., **64**, 1618-19 (1994).
[8] L.Fabbri, D.Fournier, L.Pottier, and L.Esposito, "Analysis of Local Heat Transfer Properties of tape-cast AlN Ceramics using Photothermal Reflectance Microscopy," *J. Mater. Sci.*, **31**, 5429-36 (1996).
[9] A.Salazar, A. Sanchez-Lavege, A.Ocariz, J.Guitonny, G.Pandey, D.Fournier, and A.C. Boccara, "Thermal Diffusivity of Anisotropic Materials by Photothermal Methods," *J. Appl. Phys.*, **79**, 3984-93 (1996).
[10] M.Berolotti, R. Li Voti, G.Liakhou, and C. Sibilia, "On the Photodeflection Method Applied to Low Thermal Diffusivity Measurements," *Rev. Sci. Instrum.*, **64**, 1576-83 (1993).

[11] K.Watari, Y.Seki, and K.Ishizaki, "Thermal Coefficients of HIPed Si3N4 Ceramics," *J. Ceram. Soc. Jpn.*, **97**, 174-81 (1989).

[12] G.A.Slack, R.A.Tanzilli, R.O.Pohl and J.W.Vandersande, "The Intrinsic Thermal Conductivity of AlN," *J. Phys. Chem. Solids*, **48**, 641-47 (1987).

[13] C.M.B.Henderson and D.Tayloe, "Thermal Expansion of the Nitride and Oxynitride of Silicon in Relation to Their Structures," *Trans. Br. Ceram. Soc.*, **74**, 49-53 (1975).

[14] W.D.Kingery, "The Thermal Conductivity of Ceramic Dielectrics"; pp.182-235 in *Progress in Ceramic Science*, Vol. 2. Edited by J.E.Burje. Pergamon Press, New York, 1962.

[15] K.Watari, H.Hirao, M.Toriyama, and K.Ishizaki, "Effect of Grain Size on the Thermal Conductivity of Si3N4," *J. Am. Ceram. Soc.*, **82**, 777-79 (1999).

DISLOCATIONS IN Al$_2$O$_3$-20wt%ZrO$_2$ (3Y) CERAMICS

S. D. De la Torre *,[a] and D. Rios-J [a]. [a] *Advanced Mat. Res. Center CIMAV. M. de Cervantes 120. CP.31109. Chihuahua, Mexico.* H. Kume [b], Y. Nishikawa [b], S. Inamura [b], A. Kakitsuji [b], H. Miyamoto [b], K. Miyamoto [b] and L. Gao [b]. [b] *Tech. Research Institute of Osaka Pref. 2-7-1 Ayumino, Izumi 594-1157 Osaka, Japan.* H. Tsuda [c] and K. Morii [c]. [c] *Met. and Mat. Sci. Osaka Pref. University, Gakuen-Cho, Sakai 599-8531. Osaka, Japan.*

ABSTRACT

Zirconia toughened alumina (ZTA) ceramics have been fabricated for ball bearing applications. The lattice plane resolution and microanalysis of Al$_2$O$_3$–20wt%(3Y) ZrO$_2$ composites have been conducted by high-resolution transmission electron microscopy (HREM). It is shown that the material microstructure accumulates a large percentage of microstrain originating from evident dislocation-defects created during the t $\rightarrow$ m-ZrO$_2$ toughening transformation.

INTRODUCTION

The positive strengthening effect that ZrO$_2$ plays when incorporated into the microstructure of several ceramic composites has been extensively studied [1-3]. However, the advent of more precise characterization equipment makes it possible to observe the microstructure on a nearly atomic-level, which can verify what has been theoretically established. Thus, microanalysis using high-resolution transmission electron microscopy (HREM) can contribute in understanding the strengthening mechanism taking place, specifically at the surfaces or interfaces of advanced materials. Previous phase contrast microscopy work has shown interesting lattice images of Al$_2$O$_3$-ZrO$_2$ composites, revealing not only extremely fine structure defects but also interfaces matching [4-6]. It is known that the atomic microstructure of matter mostly defines the ultimate mechanical properties of the bulk material after suitable sintering. In our work, the flexural strength of pure Al$_2$O$_3$ was improved (from σ=300 to 1150 MPa) by dispersing 20 wt % (3Y)-ZrO$_2$ into the fine Al$_2$O$_3$-matrix [7]. Analyzing this material designed for ball bearings, we came to the conclusion that the strengthening mechanism operating in the sintered material seems to be a function of a series of cooperative or contributing factors [8-9], including:

*Invited researcher at TRI-Osaka.

To the extent authorized under the laws of the United States of America, all copyright interests in this publication are the property of The American Ceramic Society. Any duplication, reproduction, or republication of this publication or any part thereof, without the express written consent of The American Ceramic Society or fee paid to the Copyright Clearance Center, is prohibited.

- Both fine (nano-scale) grain size and chemical composition of final product,
- A surface free of defects,
- Martensitic transformation (retention of some percentage of t-ZrO$_2$),
- Strong atomic bonding between Al$_2$O$_3$//ZrO$_2$ interfaces,
    - Semi-coherent (metastable) interfaces,
    - Sort of pseudo-plastic interfaces, and
    - Microstrain accumulation at heterogeneous interfaces.

Note that neither the order nor the extent in which these determinant factors contribute to the strengthening mechanism of the material at that ZrO$_2$ concentration are well understood. As an attempt to better understand the effect of these factors, we briefly report some dislocation structures of incoherent twin boundaries found in our most strengthened Al$_2$O$_3$-ZrO$_2$ ceramics.

EXPERIMENTAL PROCEDURE

The Al$_2$O$_3$-ZrO$_2$ raw material and composites preparation procedures are described elsewhere [7]. Using conventional HREM procedures for sample preparation [10] thin discs of 3 mm diameter about 10 μm in thickness were prepared, followed by ion-thinning operations (Gatan-691, setting 3~5eV and 4~3°). The final sample thickness was less than 10 nm. During HREM operation it was found that unless an extremely thin (~1 nm) carbon layer could be deposited on the sample surface, the resolved microstructure will not be stable. Stable images are particularly required when working at more than 1500x magnification. Specimens were examined at 200 kV in a Hitachi-HF2000 electron microscope, using its maximum high-resolution capability (0.5 nm) and a computer-operated EDAX system attached to the microscope.

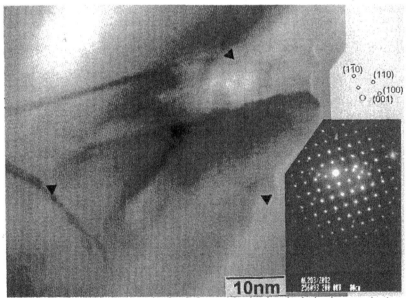

Figure 1. Transmission micrograph of a ZrO$_2$ particle showing the lattice image and micro twinning. Arrows indicate twinning. Insert is the corresponding diffraction pattern.

RESULTS AND DISCUSSION

The typical microstructure, fracture mode and the sort of interfaces occurring in this material have been recently described [8]. Twining of the microstructure was evident during the martensitic transformation (t → m-ZrO₂). Figure 1 shows a HREM photo of a ZrO₂ particle. A group of micro twins is indicated with arrows, with the corresponding diffraction pattern (DP). The pattern suggests simultaneous presence of the t- and monoclinic forms of zirconia, whereas apart from the direct beam optical axis, other big diameter spots are mostly about the existence of twins. The diffraction conditions in Fig.1 shown that the principal reflections operating for t-ZrO₂ were (1 $\bar{1}$ 0) and (001). Such indexing indicates lattice parameters a = b = 3.64 Å and c = 5.27 Å. α = β = γ = 90° (P42/nmc). On the other hand, (110) and (100) were the main reflections for m-ZrO₂ as indicated in the diffraction pattern draw. For the latter case a = 5.1505 Å, b = 5.2116 Å and c = 5.3173 Å. α = β = 90° and γ = 99.230°. Note the dark contrast originated during micro twinning. This is associated with microstrain accumulation at those places. A clear evidence of much thicker and faulted twins also occurring in this material is the photo in Figure 2. Insert lines in Figure 2 are to indicate the lattice orientation.

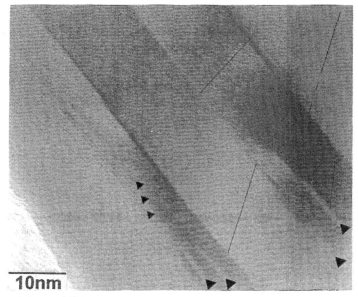

Figure 2. Group of thick twins typically found in the Al₂O₃-20wt%ZrO₂ material. In the lower part of the picture it is seen how the twins become faulted, similarly occurring on the left side of the left twin (faults indicated by arrows). The lattice plane direction is denoted with lines.

Figure 3 shows a typical lattice image of dislocations visible within ZrO₂ particles. Figure 4 is the raw tracing sketch of an enlarged area in Figure 3 to show details of the defects. To discuss Figure 4 we first refer to the right side of the sketch, i.e. to include the area between point N and W. That is a free defects zone, which according to the DP's analyzed in different zones but free of dislocations they exclusively include t-ZrO₂. However, as soon as the monoclinic form of ZrO₂ tends to appear the lattice swells as seen from V-W. Such swelling creates the topographical conditions required to create

lattice defects or dislocations. Those dislocations imaged as in U show a single extra half
111 plane, which are apparently of edge-type. The sort of "dislocation-cloud" or diffuse
effect observed in photo of Figure 3, from M-N, O-P and Q-R, always takes place at the
dislocations boundary. Such diffuse zone is the result of dislocations stacking where
microstrain is largely accumulated. If the photo in Figure 3 or sketch in Figure 4 are
viewed at a glancing angle looking either along or against the lines, the transformation
strain around the dislocations becomes readily apparent.

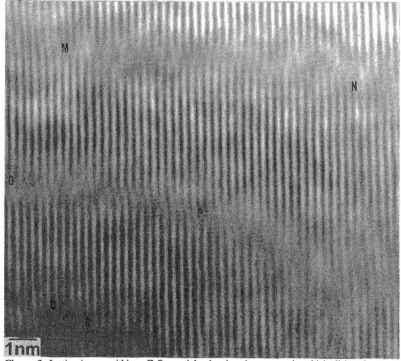

Figure 3. Lattice image within a ZrO$_2$ particle showing the manner in which dislocations are
terminated to form long range and "horizontal strain fields" (black and white contrast).

The strain created by those dislocations is relatively long range, roughly extending some
10 nm from M-N, for example. We assume that the strain accumulated in the material
microstructure is not uniform since the black and white contrast was not found
topographically uniform. On the other hand, it can also be true that the thickness of the
analyzed foil was not constant. However, the latter assumption might be disregarded
since the fringes would disappear in such a case. In practice, the dislocation-defects
analysis was found to be quite complex since different reflections can operate as a
function of the analyzed film (tilt) orientation. The defects seen in Figure 4 are of the
positive and negative types, which makes a complex Burger's circuit. Moreover, some of
these positive edge dislocations (as in U) are found distant from the negative ones (as in
M), whereas dislocations S and T seem to be coplanar. For the sake of comparison, there
is a separation of nearly 8.5 nm between the dislocation marked M with a counterpart
found between the points Q and R, and just one or two planes separating dislocations S
and T.

Grain Boundary Engineering in Ceramics

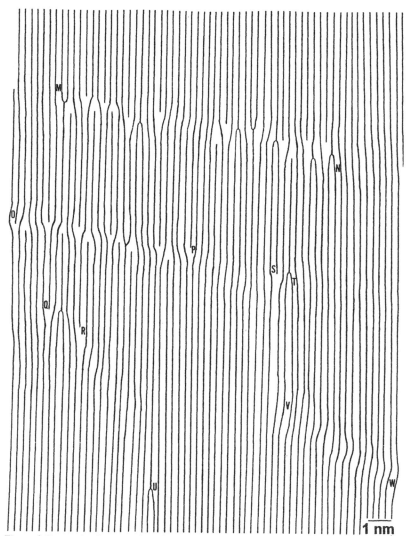

Figure 4. Raw tracing from an enlarged area of sample in Figure 3 to illustrate defect dislocations more clearly.

Although the extent by which strain accumulation contributes to the strengthening mechanism of Al_2O_3-20wt%(3Y)ZrO_2 ceramics was not quantified, a large strain content, as deduced from the large number of dislocations found and the diffraction contrast, has been found heterogeneously distributed in the entire composites microstructure. Strain was mainly accumulated at the interfaces of the ZrO_2 and Al_2O_3 particles.

CONCLUSIONS

Edge type-like dislocations in strengthened Al_2O_3-20wt%(3Y)ZrO_2 ceramics were visible by normal diffraction contrast. A direct correlation between the strained regions and the diffraction contrast was evident. The regions of dark diffraction contrast appear as the t- to m-ZrO_2 transformation takes place. Lattice images within ZrO_2 particles show the manner in which dislocations are terminated to form long range and "horizontal strain fields". The strain created by those dislocations is relatively long range. We assume that the strain accumulated in the material microstructure is not uniform since the black and white contrast was not found topographically uniform.

Acknowledgements
SDT gratefully acknowledges the financial support from NEDO-Japan and SNI-Conacyt, Mexico. Authors acknowledge to Dr. Koji Tanaka of Osaka Industrial University (Ikeda City) his assistance for the TEM samples preparation.

REFERENCES

1. A.H.Heuer, F.F.Lange, M.V.Swain and A.G.Evans, "Transformation Toughening: An Overview", *J. Am. Ceram. Soc.* **69** [3], p.i-iv and 169-298 (1986).
2. D.J.Green, R.H.J.Hannink and M.V.Swain, "Transformation Toughening of Ceramics", *CRC Press* Florida (1989).
3. S.P.S.Bawdal, J.Drennan, R.H.J.Hannink and A.E.Hughnes, "Influence of Interfaces on the Properties and Performance of Zirconia Based Advanced Ceramics", p.109-126 in *Key Eng.Materials*, Trans Tech Pub. Vols.111-112 (1995).
4. S.P.K.Lanteri, T.E.Mitchel and A.H.Heuer, "Structure of Incoherent ZrO_2/Al_2O_3 Interfaces", *J. Am. Ceram. Soc.* **69** [3] p.256-58 (1986).
5. B.T.Lee, K.Hiraga, D.Shindo and A.Nishiyama, "Microstructure of Pressureless-sintered Al_2O_3-24vol%ZrO_2 Composite Studied by HREM", *J. of Mat. Sci.* **29** p.959-964 (1994).
6. A.H.Heuer, R.Chaim and V.Lanteri, "Review: Phase Transformations and Microstructural Characterization of Alloys in the System Y_2O_3-ZrO_2", p.3-20 *in Advances in Ceramics*, Vol.**24**: Sci. and Tech. of Zirconia III (1988).
7. H.Kume, Y.Nishikawa, S.Inamura, H.Miyamoto, K.Yamabe and T.Maeda, "Strengthening of Al_2O_3-ZrO_2 Composites by HIP Sintering", *Rev. High Pressure Sci. Tech.* **7** p.1087-89 (1998), and "Aqueous Precipitation and Attrition Milling – Combined Processes to Fabricate Strengthened Al_2O_3-5wt%ZrO_2 Composites", *Advances in Ceramic Matrix Comp.* Ceramic Transactions **103** (2000) in press.
8. S.D.De la Torre, H.Kume, Y.Nishikawa, S.Inamura, A.Kakitsuji, H.Miyamoto, K.Miyamoto, D.Rios-J., H.Tsuda and K.Morii, "Interfaces in Alumina-Zirconia Ceramics for Ball-bearing Applications", *Mat. Sci. Forum.* Vols. **315-317** (2000) in press.
9. S.D.De la Torre, H.Kume, S.Inamura, A.Kakitsuji, Y.Nishikawa, K.Miyamoto, H.Miyamoto, J.Hong and L.Gao, "Assessing the Zirconia-Alumina Interface of a Metastable Solid Solution", p.295-300 in *Proc. of Int. Symp. on Designing, Proc. and Prop. of Advanced Eng. Mat.* Toyohasi Japan, JSPS AEM 156 Committee. Eds.T.Kobayashi, M.Umemoto and M.Morinaga (1977).
10. W.E.Lee and W.M.Rainforth, *"Ceramic Microstructure. Properties Control by Processing"*. Chapman & Hall 1994 (First Edition).

ENHANCED FRACTURE RESISTANCE OF HIGHLY ANISOTROPICIZED POROUS SILICON NITRIDE

Yoshiaki Inagaki,[+] Yasuhiro Shigegaki,[++] Naoki Kondo,[+++] Yoshikazu Suzuki,[+++] Tatsuya Miyajima,[*+++] Tatsuki Ohji,[*+++]

[+]Synergy Ceramics Laboratory, Fine Ceramic Research Association, Nagoya, 462-8510 Japan

[++]Ishikawajima-Harima Heavy Industries Co., Ltd., Tokyo, 135-8733 Japan

[+++]National Industrial Research Institute of Nagoya, Nagoya, 462-8510 Japan

INTRODUCTION

A number of approaches have been applied to improve the fracture toughness of ceramics. As for silicon nitride, there have been many studies about toughening by controlling microstructures, such as grain size and morphology, grain alignment, and boundary chemistry to enhance a variety of crack wake toughening mechanisms including grain bridging and grain pull-out.[1-7] Aligning fibrous silicon nitride grains using a tape-casting technique with seed particles, Hirao et al.[8] obtained high fracture toughness, ~11 MPa·m$^{1/2}$ as well as high fracture strength, ~1100 MPa when a stress was applied parallel to, or a crack extended normal to, the grain alignment. In this case, a greater number of fibrous grains are involved with the crack wake toughening mechanisms, and then the toughness can rise steeply in a very short crack extension.[9] Moreover, Kondo et al.[10] succeeded in better alignment of fibrous grains through a superplastic deformation in plane-strain compression, realizing higher fracture toughness, ~13 MPa·m$^{1/2}$, and higher fracture strength, ~1650 MPa.

In brittle ceramic materials, strains which are undesirably generated cannot be relaxed by plastic deformation. This often leads to the abrupt fracture even with a small strain, particularly for high-elastic-modulus materials. Then, if the elastic modulus can be lowered while retaining strength, strain-to-failure can become larger, ensuring the structural reliability. Based on such a concept, Shigegaki et al.[11] developed a porous silicon nitride with a porosity of 14.4 %,

To the extent authorized under the laws of the United States of America, all copyright interests in this publication are the property of The American Ceramic Society. Any duplication, reproduction, or republication of this publication or any part thereof, without the express written consent of The American Ceramic Society or fee paid to the Copyright Clearance Center, is prohibited.

where the fibrous silicon nitride grains were uniaxially aligned; this material has a high fracture strength, ~1080 MPa, as well as a low elastic modulus, 246 GPa. Since the pores exist around the aligned fibrous grains, enhanced operations of the grain bridging and pull-out can be anticipated when the crack propagates perpendicularly to the direction of grain alignment. Furthermore, the presence of pores surrounding the silicon nitride grains causes cracks to tilt or twist, namely crack deflections.[12, 13] Keeping this in mind, this study is aimed at investigating fracture resistance behaviors of this porous silicon nitride. For this purpose, we employed the chevron-notched-beam (CNB) technique, since the crack extends in a stable manner and the crack resistance can be clearly related to the crack extension distance. This technique also can give macroscopically the Mode I crack extension due to the ligament configuration; otherwise a crack tends to deflect along the grain alignment.

EXPERIMENTAL PROCEDURES

Materials

The fabrication of the porous silicon nitride (hereafter PSN) has been reported elsewhere,[11, 14] but is briefly described here. For starting powders, β-silicon nitride whiskers were mixed with 5 wt% yttria and 2 wt % alumina. The green sheets which were formed by tapecasting were stacked and bonded under pressure. Sintering was performed at 1850°C under a nitrogen pressure of 1 MPa. Figure 1 shows the microstructure of the material; the fibrous grains of silicon nitride are well aligned toward the casting direction, and the pores exist among the grains. For comparison, the fracture resistance behavior of a reference material (hereafter RSN), which was fabricated using α-silicon nitride powders with sintering additives 5 wt% yttria and 2 wt% alumina,[11] was characterized by the same testing procedures.

Fracture Testing

Flexural test specimens, whose nominal dimensions are 4.0 mm in height, 3.0 mm in width, and 35 mm in length, were cut from the sintered billets so that the tensile axis and tensile surface were parallel to the casting direction and the stacking plane, respectively. Details of the employed chevron-notched specimen geometry and testing procedures were the same as those that had been used for a previous study.[15] The shape of the ligament was a regular triangle with an edge length of 3 mm, and the initial crack length, a_0, was 1.4 mm. The width of the chevron-notch was 100 μ m, and the fracture testing was carried out at room temperature when three-point bending with a lower span of 30 mm at a cross-head speed of 0.01 mm/min. A bending fixture of silicon carbide was used to realize high rigidity of the testing system, which is needed for stable crack growth. The true load-displacement (*L-D*) curve was determined by subtracting the compliance of the testing machine and the fixture, which was obtained in advance by an independent calibration, from the experimentally observed curve.

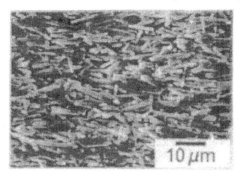

Fig 1. SEM micrograph of polished and plasma-etched top surface of the PSN.

RESULTS AND DISCUSSION

Five specimens were used for each characterization, and stable crack growth until the completion of the test was obtained for a few specimens of both the PSN and RSN with sufficient reproducibility of the *L-D* curves; the other measurements resulted in undesirable catastrophic failure during testing. Successful examples of the *L-D* curves for the two materials are shown in Fig. 2.

While the RSN showed a smooth *L-D* curve which is normally observed for chevron-notched beam tests of monolithic ceramic materials, the PSN demonstrated serrations in the almost entire region, suggesting repetition of crack initiations and arrests. The curve of the PSN in Fig. 2 is that after leveling the serration. The effective fracture energy, γ_{eff}, is defined as follows:

$$\gamma_{eff} = W_{WOF} / 2A \tag{1}$$

where W_{WOF} is the energy under *L-D* curve, and A is the area of the specimen web portion. By substituting W_{WOF} calculated from Fig. 2 into Eq. (1), one can obtain 492.7 J/m^2 and 70.2 J/m^2 as γ_{eff} of the PSN and RSN, respectively. Note that γ_{eff} of the PSN is about 7 times larger than that of the RSN. It is also enormously high compared with those of other toughened silicon nitrides, such as 96 J/m² of a silicon-carbide-whisker-reinforced dense silicon nitride,[15] and 246 J/m² of a highly anisotropic silicon nitride produced by superplastic deformation.[16] (These values were measured using exactly the same CNB test techniques.)

The fracture toughness is defined as follows:

$$K_{IC} = (2E' \gamma_{eff})^{1/2} \tag{2}$$

where E' is the Young's modulus for plane strain condition, which is given by $E' = E / (1- n^2)$ with n being the Poisson's ratio. Substituting 246 GPa and 325 GPa as the Young's moduli of the PSN and RSN, respectively[11] and 0.25 as the Poisson's ratio into Eq.(2), K_{IC} becomes 16.1 MPa·m$^{1/2}$ for the PSN and 7.0 MPa·m$^{1/2}$ for the RSN.

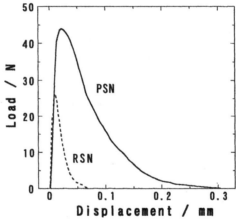

Fig 2.   Load-deflection diagrams of the chevron-notched bend tests for the PSN and RSN.

Moreover, a stress intensity factor, K_I, can be estimated from the CNB tests using Bluhm's synthesized numerical slice model, as follows;[17, 180]

$$K_I = [(E' / 2) (dx/dA)]^{1/2}P \tag{3}$$

where x is the compliance, A is the crack area, and P is the load. The crack length corresponding to a load can also be calculated from the change in specimen compliance, and the variation of fracture resistance with crack extension thus can be obtained (however, the bridging traction in the crack wake as shown later would influence the specimen compliance, raising some concern about the accuracy of this technique). The fracture resistance curves (R-curves) of the PSN and RSN are shown in Fig. 3, where a and W represent the crack length and the height of a specimen, respectively. The PSN showed a stronger rising R-curve than the RSN. The resistance increases almost lineally with the crack extension and reaches about 20 MPa·m$^{1/2}$ around the a/W of 0.9. This fracture toughness is also extremely high compared with those of other toughened silicon nitrides.

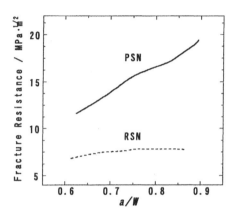

Fig 3. R-curve behaviors in the chevron-notched bend tests for the PSN and RSN.

This strong rising R-curve behaviors is attributable to processes or mechanisms which occur in a wake behind a crack tip, so called crack wake toughening.[19-21] The crack wake toughening mechanisms of silicon nitride are classified into two categories: (1) bridging of the crack by unbroken fibrous reinforcing grains that is partially debonded from the matrix (grain bridging), (2) frictional pull-out of fibrous grains (grain pull-out). For these mechanisms to operate, debonding of the interface between the matrix and the fibrous grains is a necessary condition. In dense silicon nitrides where interfaces are relatively strong, such debonding is restricted and then the toughening mechanisms are not fully operated. In the PSN, however, the pores existing around the grains would enhance the debonding, leading to the activated crack-wake toughening mechanisms. It has been known that a steep rise of fracture resistance in a short crack extension (typically several microns to several ten microns) is primarily attributable to grain bridging, while grain pull-out gives rise to an increased resistance after a long crack extension (typically several hundred microns).[9, 22] The R-curve of the PSN in Fig. 3 is very likely caused by the latter. Figure 4 shows the SEM image of the fracture surface of the test specimen where the fibrous grains have been pulled out, supporting this speculation.

Fig 4.   SEM micrograph of the ligament area of fracture surface for the PSN.

As stated in the introduction, the PSN has a fracture strength, ~1080 MPa, which is very high for a silicon nitride with a porosity of 14.4%. It has been known, however, that an increased resistance after a long crack extension such as the R-curve of the PSN contributes little to an increase of fracture strength.[9] Thus, it can be presumed that the toughening with a very short crack extension, such as the grain bridging and crack deflection, also should work actively.

SUMMARY

The fracture resistance behavior of a porous silicon nitride with aligned fibrous grains was investigated by using a chevron-notched-beam technique when a crack propagated perpendicularly to the grain alignment. The fracture energy was determined to be ~ 500 J/m², which was about 7 times larger than that of a dense silicon nitride used as a reference material. The very strong rising R-curve was observed and the plateau fracture toughness resulted in ~ 20 MPa·m$^{1/2}$. The large fracture energy and strong rising R-curve behavior were primarily attributable to the pull-out effect of the aligned fibrous grains, which was enhanced by pores surrounding the grains.

ACKNOWLEDGEMENTS

This work has been promoted by AIST, MITI, Japan as a part of the Synergy Ceramics Project under the Industrial Science and Technology Frontier (ISTF) Program. Under this program, part of the work has been supported by NEDO. The authors are members of the Joint Research Consortium of Synergy Ceramics.

REFERENCES

[1]E. Tani, S. Umebayashi, K. Kishi, K. Kobayashi, and M. Nishijima, "Gas-Pressure Sintering of Si_3N_4 with Concurrent Addition of Al_2O_3 and 5 wt% Rare Earth Oxide : High Fracture Toughness Si_3N_4 Si_3N_4 with Fiber-like Structure," *Am. Ceram. Soc. Bull*, 65 [9] 1311-15 (1995).

[2]T. Kawashima, H. Okamoto, H. Yamamoto, and A. Kitamura, "Grain Size Dependence of the Fracture Toughness of Silicon Nitride," *J. Ceram. Soc. Jpn.*, 98 [3] 235-42 (1990) ; (in Japanese)

[3]M. Mitomo, and S. Uenosono, "Microstructural Development During Gas-Pressure Sintering of a-Silicon Nitride," *J. Am. Ceram. Soc.*, 75 [1] 103-108 (1992).

[4]Y. Goto, and A. Tsuge, "Mechanical Properties of Unidirectionally Oriented SiC-Whisker Reinforced Si_3N_4 Fabricated by Extrusion and Hot-Pressing," *J. Am. Ceram. Soc.*, 76 [1] 1420-24 (1993).

[5]N. Hirosaki, Y. Akimune, and M. Mitomo, "Effect of Grain Growth of β-Silicon Nitride on Strength, Weibull Modulus and Fracture Toughness," *J. Am. Ceram. Soc.*, 78 [6] 1687-90 (1995).

[6]P. F. Becher, H. T. Lin, S. L. Hwang, M. J. Hoffmann, and I-W. Chen, "The Influence of Microstructure on the Mechanical Behavior of Silicon Nitride Ceramics;" pp. 147-58 in *Silicon Nitride Ceramics*. Edited by I-W. Chen, P. F. Becher, M. Mitomo, G. Petzow, and T. S. Yen. Material Research Society, Pittsburgh, PA, 1993.

[7]K. Hirao, T. Nagaoka, M. E. Brito, and S. Kanzaki, "Microstructure Control of Silicon Nitride by Seeding with Rodlike b-Silicon Nitride Particles," *J. Am. Ceram. Soc.*, 78 [7] 1857-62 (1995).

[8]K. Hirao, M. Ohashi, M. E. Brito, and S. Kanzaki, "Processing Strategy for Producing Highly Anisotropic Silicon Nitride," *J. Am. Ceram. Soc.*, 78 [6] 1687-90 (1995).

[9]T. Ohji, K. Hirao, and S. Kanzaki, "Fracture Resistance Behavior of Highly Anisotropic Silicon Nitride," *J. Am. Ceram. Soc.*, 78 [11] 3125-28 (1995).

[10]N. Kondo, Y. Suzuki, and T. Ohji, "Superplastic Sinter-Forging of

Silicon Nitride with Anisotropic Microstructure Formation," *J. Am. Ceram. Soc.*, 82 [4] 1067-69 (1999).

[11]Y. Shigegaki, M. E. Brito, K. Hirao, M. Toriyama, and S. Kanzaki, "Strain Tolerant Porous Silicon Nitride," *J. Am. Ceram. Soc.*, 80 [2] 495-98 (1997).

[12]K. T. Faber, and A.G.Evans, "Crack Deflection Processes I. Theory," *Acta. Metall.*, 31 [4] 565-76 (1983).

[13]K. T. Faber, and A.G.Evans, "Crack Deflection Processes II. Experimental," *Acta. Metall.*, 31 [4] 577-84 (1983).

[14]Y. Shigegaki, M. E. Brito, K. Hirao, M. Toriyama, and S. Kanzaki, "Processing of a Novel Multilayered Silicon Nitride," *J. Am. Ceram. Soc.*, 79 [8] 2197-200 (1996).

[15]T. Ohji, Y. Goto, and A. Tsuge, "High Temperature Toughness and Tensile Strength of Whisker Reinforced Silicon Nitride," *J. Am. Ceram. Soc.*,74 [4] 739-45 (1991).

[16]N. Kondo, Y. Inagaki, Y. Suzuki, and T. Ohji, "Fracture Resistance Behavior of Highly Anisotropic Silicon Nitride Produced by Superplastic Deformation," *J. Ceram. Soc. Jpn.* (in review).

[17]J. I. Bluhm, "Slice Synthesis of a Three Dimensional Work of Fracture Specimen," *Eng. Fract. Mech.*, 7 [3] 593-604 (1975).

[18]M. Sakai, K. Yamasaki, "Numerical Fracture Analysis of Chevron-Notched Specimens: I, Shear Correction Factor, k," *J. Am. Ceram. Soc.*,66 [5] 371-75 (1983).

[19]P. F. Becher, S. L. Hwang, H. T. Lin, and T. N. Tiegs, "Microstructural Contribution to the Fracture Resistance of Silicon Nitride Ceramics;" pp. 87-100 in *Tailoring of Mechanical Properties of Si₃N₄ Ceramics*. Edited by M. J. Hoffmann and G. Petzow. Kluwer Academic Publishers, the Netherlands, 1994

[20]P. F. Becher, "Microstructural Design of Toughened Ceramics," *J. Am. Ceram. Soc.*, 74 [2] 255-69 (1991)

[21]R.W. Steinbrech, "Toughening Mechanisms for Ceramic Materials," *J. Eur. Ceram. Soc.*, 10 [3] 131-42 (1992)

[22]T. Ohji, N. Kondo, Y. Suzuki, and K. Hirao, "Grain Bridging of Highly Anisotropic Silicon Nitride," *Mat. Let.*, 40, 5-10 (1999).

... *Tailoring of Mechanical Properties of Si_3N_4 Ceramics*. ...

PROCESSING AND GRAIN BOUNDARY STRUCTURE OF 3Y-TZP/BaFe$_{12}$O$_{19}$ AND 3Y-TZP/NaAl$_{11}$O$_{17}$ COMPOSITES

Yoshikazu Suzuki, Masanobu Awano, Naoki Kondo and Tatsuki Ohji
National Industrial Research Institute of Nagoya, Hirate-cho, Kita-ku, Nagoya
462-8510, Japan

ABSTRACT

Fine-grained 3 mol%-Y$_2$O$_3$-doped ZrO$_2$ (3Y-TZP) matrix composites with magnetoplumbite- or β-alumina-structured second phase, such as 3Y-TZP/BaFe$_{12}$O$_{19}$ and 3Y-TZP/NaAl$_{11}$O$_{17}$, were fabricated and their microstructure, mechanical and functional properties were evaluated. Effects of grain boundary structure on these properties will be discussed.

INTRODUCTION

Ceramic materials with magnetoplumbite (M-type) or related β-alumina (β-type) structure generally possess hexagonal platelike or needlelike shapes, and they show attractive magnetic and electrical properties due to their unique crystal structures. Barium hexaferrite, BaFe$_{12}$O$_{19}$, is a typical M-type compound with platelike shape and has the easy-magnetization axis perpendicular to the basal plane. It is widely used as economical permanent magnets. While, sodium β-alumina, "NaAl$_{11}$O$_{17}$" with ideal composition, has high ionic conductivity due to mobile alkali cations located in a conduction plane between spinel-structured alumina blocks. Incorporating such ceramic materials as a second-phase in structural ceramics may bring some magnetic and electrical functions to them, as well as some improved mechanical properties. Recently, the authors have investigated the synthesis and properties of 3Y-TZP/BaFe$_{12}$O$_{19}$[1-5] and 3Y-TZP/NaAl$_{11}$O$_{17}$[6] composites.

In this paper, we briefly introduce our recent development in these two composite systems and compare them to each other. The grain-boundary structure of the 3Y-TZP/BaFe$_{12}$O$_{19}$ *in situ* composite will be discussed.

To the extent authorized under the laws of the United States of America, all copyright interests in this publication are the property of The American Ceramic Society. Any duplication, reproduction, or republication of this publication or any part thereof, without the express written consent of The American Ceramic Society or fee paid to the Copyright Clearance Center, is prohibited.

CHARACTERISTICS OF M-TYPE AND β-TYPE STRUCTURES

The similarity and the difference between M-type ($M^{2+}XO_3[X_{11}O_{16}]$) and β-type ($M^+O[X_{11}O_{16}]$ or $M^{3+}O_2[X_{11}O_{16}]$) crystal structures have been widely discussed.[7-9] Since the M-type structure has the close-packed interspinel layer, it gives higher leach resistance and thermal stability against volatilization compared to the β-type.[9] Many M-type ferrites possess excellent magnetic properties. On the other hand, since the β-type structure has the defective interspinel layer, it allows the ion-exchange and the ionic conductivity.

3Y-TZP/BaFe$_{12}$O$_{19}$ *IN SITU* COMPOSITE [3-5]

A fine-grained 3Y-TZP/BaFe$_{12}$O$_{19}$ *in situ* composite (3Y-TZP grain size of ~100-300 nm and BaFe$_{12}$O$_{19}$ particle size of ~200-500 nm) has been fabricated successfully via the pressureless reactive sintering of 3Y-TZP, BaCO$_3$, and γ-Fe$_2$O$_3$ powders at 1300°C for 2 h.[3] High magnetization and coercivity values were obtained for the composite.[3] Its microstructure is shown in Fig. 1

Figure 1   SEM micrograph of the 3Y-TZP/20 wt% BaFe$_{12}$O$_{19}$ *in situ* composite
sintered at 1300°C for 2 h.

Figure 2 shows a TEM photograph of the 3Y-TZP/20 wt% $BaFe_{12}O_{19}$ *in situ* composite sintered at 1300°C. In this figure, some microcracks were observed along with the interface between 3Y-TZP matrix and $BaFe_{12}O_{19}$. This phenomenon is attributable to (1) the thermal expansion mismatch between isotropic t-ZrO_2 grains and strongly anisotropic $BaFe_{12}O_{19}$ ones, and (2) the volume decrement due to the CO_2 evaporation. Well-crystallized hexagonal $BaFe_{12}O_{19}$ particles were surrounded by nanoparticles, which indicates the progress of the *in situ* reactions between $BaCO_3$ and Fe_2O_3. To illuminate the nanoparticles at the interphase boundary, a high-resolution TEM micrograph is given in Fig. 3.

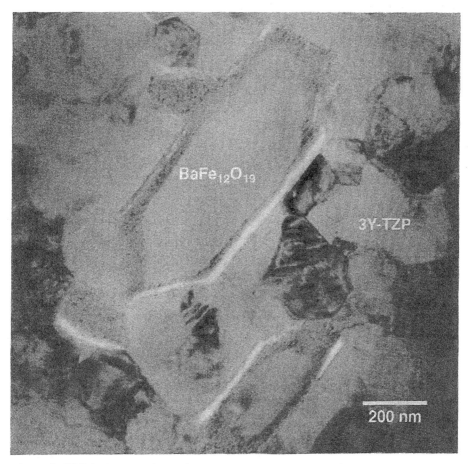

Figure 2 TEM photograph of the 3Y-TZP/20 wt% $BaFe_{12}O_{19}$ *in situ* composite sintered at 1300°C for 2h.

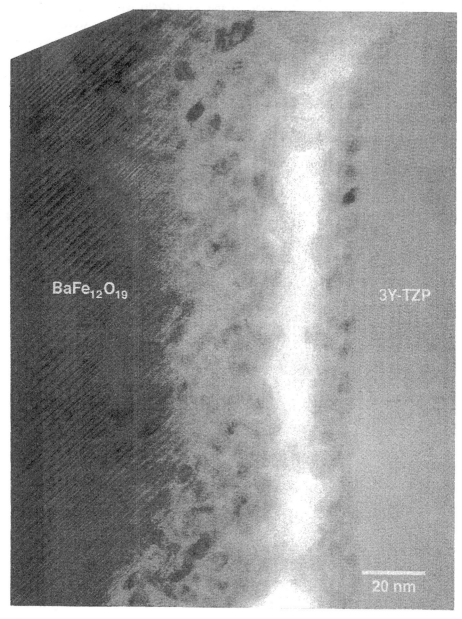

Figure 3 High-resolution TEM micrograph of the 3Y-TZP/20 wt% $BaFe_{12}O_{19}$ *in situ* composite sintered at 1300°C for 2h.

Grain Boundary Engineering in Ceramics

3Y-TZP/NaAl$_{11}$O$_{17}$ COMPOSITES [6]

Fine-grained 3Y-TZP/β-Al$_2$O$_3$ (~ NaAl$_{11}$O$_{17}$) composites have been prepared by pressureless sintering of planetary-ball-milled 3Y-TZP and β-Al$_2$O$_3$ powders. In the composites, β-Al$_2$O$_3$ particles with about 0.2-3 μm in size were homogeneously dispersed in the TZP matrix with the grain size of ~ 200 nm. It is obvious that long-term planetary ball-milling was effective to produce the fine-grained microstructure. The microstructure of the 3YTZP/20 vol% β-Al$_2$O$_3$ composite is shown in Fig. 4. X-ray diffraction analysis revealed that the composite consisted of only tetragonal ZrO$_2$ (t-ZrO$_2$) and β-Al$_2$O$_3$, indicating good phase compatibility of these two phases.

There was no dramatic change in mechanical properties by the addition of fine β-Al$_2$O$_3$, but a certain improvement was observed for the hardness and strength. Fracture toughness of the composite was almost the same as that of the pure TZP, indicating that a rather strong bonding may exist between the matrix and fine β-Al$_2$O$_3$ particles. Details of the mechanical properties are described elsewhere.[6]

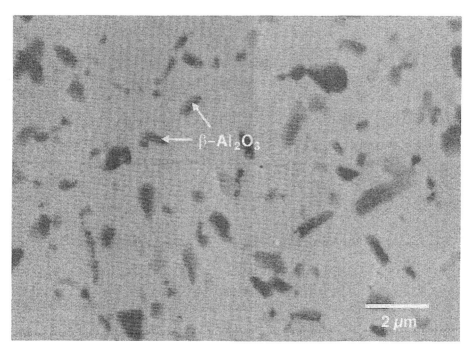

Figure 4 SEM photograph of the 3Y-TZP/20 vol% β-Al$_2$O$_3$ composite sintered at 1400°C for 2 h.

SUMMARY

In this paper, we have briefly described the fine-grained TZP matrix composites that include magnetoplumbite or β-alumina structured second phases. Good mechanical and functional properties can be harmonized in one component via such composite designs.

ACKNOWLEDGEMENT

A part of this work was supported by STA, Japan, under the Fluidity Promoting Research System and the Encouragement of Basic Research at National Research Institutes (the Special Coordination Funds for Promoting Science and Technology).

REFERENCES

1. Y. Suzuki, M. Awano, N. Kondo and T. Ohji, "Effects of Plastic Deformation on Microstructure and Magnetic Properties of 3Y-TZP/Ba-M Type Ferrite Composite," *J. Jpn. Soc. Powder & Powder Metall.*, **46** [6] 604-609 (1999).

2. Y. Suzuki, M. Awano, N. Kondo and T. Ohji, "Enhanced Magnetization of 3Y-ZrO_2/Ba-Hexaferrite Composite by Post-Plastic Deformation," *J. Am. Ceram. Soc.*, in press.

3. Y. Suzuki, M. Awano, N. Kondo and T. Ohji, "Preparation and Characterization of Fine-Grained 3Y-TZP/$BaFe_{12}O_{19}$ *In Situ* Composites," *J. Am. Ceram. Soc.*, **82** [9] 2557-59 (1999).

4. Y. Suzuki, M. Awano, N. Kondo and T. Ohji, "Synthesis and Properties of Pressureless-Sintered 3Y-TZP/Ba-M Type Ferrite *In Situ* Composites," *J. Jpn. Soc. Powder & Powder Metall.*, **47** [2] 208-212 (2000).

5. Y. Suzuki, H. A. Calderón, N. Kondo and T. Ohji, "*In Situ* Formation of Hexaferrite Magnets within 3Y-TZP Matrix: La_2O_3-ZnO-Fe_2O_3 and BaO-Fe_2O_3 Systems," *J. Am. Ceram. Soc.*, in press.

6. Y. Suzuki, M. Awano, N. Kondo and T. Ohji, "Fabrication of Fine-Grained 3Y-TZP/β-Al_2O_3 Composites," *J. Jpn. Soc. Powder & Powder Metall.*, **46** [12] 1292-1296 (1999).

7. V. Adelsköld, "X-Ray Studies of Magnetoplumbite, PbO•6Fe_2O_3, and Other Substances Resembling Beta-Alumina, Na_2O•11Al_2O_3," *Ark. Kemi. Minerali. Geol.*, **12A** [29] 1-9 (1938).

8. J. M. P. J. Verstegen and A. L. N. Stevels, "The Relation between Crystal Structure and Luminescence in β-Alumina and Magnetoplumbite Phases," *J. Lumin.*, **9** [5] 406-14 (1974).

9. P. E. D. Morgan and E. H. Cirlin, "The Magnetoplumbite Crystal Structure as a Radwaste Host," *J. Am. Ceram. Soc.*, **65** [7] C114-15 (1982).

ATOMIC STRUCTURES AND PROPERTIES OF SYSTEMATIC [0001] TILT GRAIN BOUNDARIES IN ALUMINA

Tsuyoshi Watanabe, Hidehiro Yoshida, Takahisa Yamamoto, Yuichi Ikuhara and Taketo Sakuma
Department of Materials Science, Faculty of Engineering,
The University of Tokyo,7-3-1 Hongo, Bunkyo-ku, Tokyo 113-8656, Japan

ABSTRACT

Several kinds of alumina bicrystals were systematically fabricated by joining two single crystals at high temperature. The alumina bicrystals have the [0001] symmetrical tilt grain boundaries which include low sigma boundaries ($\Sigma 3$, $\Sigma 7$), high sigma boundaries ($\Sigma 19$ etc.), small angle boundaries and high angle boundaries. Grain boundary structures were mainly observed by high-resolution electron microscopy (HREM). It was found that the grain boundary with incoherent atomic structure took high energy and the boundary with coherent atomic structure took low energy. The structure of $32.0°$ incoherent tilt boundary is disordered, and took high energy of about $0.6 J/m^2$. The atomic structure of the $\Sigma 3$ boundary was completely coherent as well as $\Sigma 1$ grain boundary. This boundary took extremely low energy of about $0.05 J/m^2$. The $2.0°$ tilt boundary was consisted of an array of periodic dislocations. This boundary with coherent and incoherent atomic structure takes the middle energy of about $0.3 J/m^2$.

To the extent authorized under the laws of the United States of America, all copyright interests in this publication are the property of The American Ceramic Society. Any duplication, reproduction, or republication of this publication or any part thereof, without the express written consent of The American Ceramic Society or fee paid to the Copyright Clearance Center, is prohibited.

INTRODUCTION

Mechanical properties in structural ceramics are dependent on the grain boundary structures. Therefore, an investigation of the grain boundary structures are needed to understand the nature of mechanical properties, however, the grain boundary structures are also influenced by grain boundary characters. One of the most effective method to clarify the relationship between grain boundary structures and characters is using bicrystals because they have single grain boundaries. The advantage of fabricating bicrystals is the controllability of grain boundary characters, for example rotation axis, rotation angle and grain boundary plane. Thus, the bicrystal-experiments enable us to study the effect of grain boundary characters. In this study, we systematically investigate the grain boundaries in alumina ceramics using several kinds of bicrystals.

So far there have been a couple of reports on the bicrystal experiments in alumina. For example, the relationship between relative energies and tilt angles of grain boundary was reported to be determined by using bicrystals containing $[\bar{1}100]$ tilt boundaries. In their experiments, they insisted that the energy of small angle boundary is a function of tilt angle according to dislocation core model, however, the energy of high angle grain boundary does not depend on the tilt angle [1]. However, we found that coincidence site lattice (CSL) boundaries in ceramics take lower energies in the same as ones in metals [2], [3].

In this study, grain boundary energies in alumina were examined as a function of [0001] tilt angles using several kinds of bicrystals which included small angle boundaries, large angle boundaries and CSL boundaries. In addition, atomic structure of the respective grain boundaries were observed by high-resolution electron microscopy (HREM) in order to understand the origin of grain boundary energies.

EXPERIMENTAL PROCEDURE

Alumina bicrystals was fabricated by joining two single crystals with purity of 99.99%. Figure 1 is an illustration of the bicrystals prepared in this study. As seen in the figure, the $[11\bar{2}0]$ axes of two crystals in the bicrystals are tilted around the [0001] axis by θ from boundary plane. The bicrystals have, thus, the [0001] symmetric tilt grain boundaries with the rotation angle of 2θ, respectively. Table 1 shows tilt angles 2θ of the bicrystals. The coincidence orientations are obtained at 21.8°, 27.8°, 38.2°, 46.8° and 60.0°, and these angles correspond to Σ21, 13, 7, 19 and 3, respectively. The tilt angles were selected to include these values. In Table 1, Σ values and boundary planes of bicrystals are inset.

In order to fabricate bicrystals, single crystals were cut into $10.0 \times 5.0 \times 10.0 \text{mm}^3$ with the surface planes parallel to the corresponding boundary planes. The surfaces of two single crystals were mechano-chemically polished to the mirror state with colloidal silica solution and joined together by annealing in air at 1773K for 36ks. Perfectly joined bicrystals was obtained by this procedure.

The grain boundary energies were measured by thermal grooving technique. The

Grain Boundary Engineering in Ceramics

Table 1. Tilt angles 2θ, Σ values and corresponding grain boundary planes for the bicrystals used in the present study.

2θ	Σ	G. B. plane
2.0°	~1	~$\{1\bar{1}00\}$
10.0°	~1	~$\{1\bar{1}00\}$
21.8°	21	$\{4\bar{5}10\}$
27.8°	13	$\{3\bar{4}10\}$
32.0°		
38.2°	7	$\{2\bar{3}10\}$
42.0°		
46.8°	19	$\{3\bar{5}20\}$
60.0°	3	$\{1\bar{2}10\}$

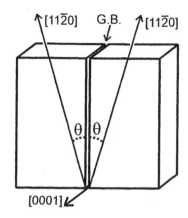

Figure 1. A schematic of bicrystals which have the [0001] symmetrical tilt grain boundary. θ corresponds to a half of a tilt angle, and is selected as shown in Table 1.

specimens were annealed in air at 1673K for 7.2ks to thermally groove the grain boundaries. It is known that the relationship among the dihedral angle of the groove, the grain boundary energy and surface energy can be expressed as the following equation,

$$\gamma_{gb}=2\gamma_s\cos(\psi/2) \qquad (1)$$

In this equation, ψ, γ_{gb} and γ_s are dihedral angle, grain boundary energy and surface energy, respectively. Thus, dihedral angle and surface energy are needed in order to calculate the grain boundary energy. The dihedral angles were measured from groove profiles obtained by atomic force microscopy (AFM). On the other hand, surface energy in alumina was substituted with the data reported by Nikolopouos [4]. The surface energy was estimated to be 1.25J/m² at the temperature of about 1673K.

The groove profiles were analyzed directly from the surface images obtained by AFM (SHIMADZU, SPM-9500E). Scanned area was a square of which length of side is 1.0μm. The measurements of dihedral angles were made for about 20 different points in all specimens, and their average values was used to calculate the grain boundary energy.

Atomic structures of grain boundaries were observed by high-resolution electron microscopy (HREM). TEM specimens were prepared using a standard technique involving mechanical grind to a thickness of 0.1mm, dimpling to a thickness of 20mm and ion beam milling to electron transparency at about 4kV. HREM observations were performed using the Topcon 002B microscope operating at 200kV with a point to point resolution of 0.18nm. The beam directions were set parallel to the [0001] direction.

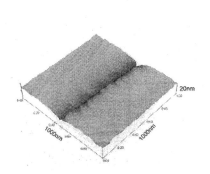

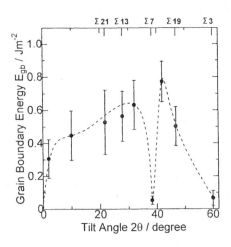

Figure 2. AFM image of a grain boundary in alumina annealed in air at 1673K for 7.2ks. In addition to the grain boundary grooving, ledges are formed on the surface because the surface plane is not exactly parallel to the (0001) plane.

Figure 3. The relationship between grain boundary energy and tilt angle. Low-energy cusps are observed at low Σ boundary.

RESULTS AND DISCUSSIONS

Surface shapes of alumina bicrystals annealed at the above condition was observed using AFM. Figure 2 shows one of the AFM images, which was taken from the 46.8° tilt boundary. Surface shapes of other boundaries were similar except Σ3 and Σ7 boundaries. Many ledges were observed on the surface of bicrystals because the (0001) planes of two crystals in bicrystals was not exactly parallel to the surface. However, the heights of ledges were negligible small to measure grooving profiles in the present study. On the other hand, grooves of Σ3 and Σ7 boundaries were extremely smaller than others.

Figure 3 shows the relationship of grain boundary energy γ_{gb} and tilt angle 2θ. As seen in the figure, grain boundary energies except Σ3, Σ7 and small angle boundaries were almost the constant at about 0.6-0.7J/m². CSL boundaries having low sigma values such as Σ3 and Σ7 take extremely lower energies of 0.05J/m² than the other boundaries. The energies of small angle boundaries are lower than high angle boundaries but higher than CSL boundaries with low sigma values. The 2.0° tilt boundary takes the energy of approximately 0.3J/m². This value was a little lower than high angle boundaries, and several times higher than low sigma boundaries. The 10.0° tilt boundary takes the energy of approximately 0.4J/m², which was a little bit higher than the 2.0° tilt boundary.

In order to consider the relationship between grain boundary energy and grain boundary structure, three typical type of grain boundaries were observed by HREM. Figure 4 shows the HREM image of (a) 32.0° high angle boundary, (b) 2.0° small angle boundary and (c) 60.0° Σ3 boundary. It was found that the atomic structure of the 32.0° high angle boundary is distorted, and forms a relaxed layer with a thickness of about 1nm. The relaxed layer is similar structure to "an extended grain boundary" [5], and considered to be formed by reducing high grain boundary energy. On the other hand, 2.0 small angle

Grain Boundary Engineering in Ceramics

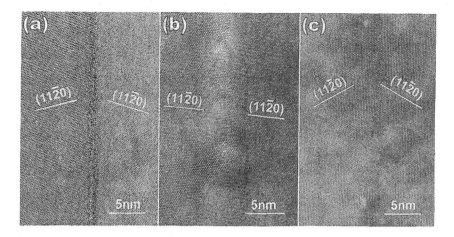

Figure 4. HREM image of the grain boundary with the tilt angle of (a) 32.0°, (b) 2.0° and (c) 60.0° (Σ3) in the respective bicrystals.

grain boundary is consisted of periodic array of dislocations. The atomic structure of the boundary between dislocations is coherent. Fig.4 (c) shows that the structure of the 60.0° Σ3 boundary is completely coherent and its image is the same as the single crystal.

These results indicate that the boundary with incoherent atomic structure takes high energy and the boundary with coherent atomic structure takes low energy. The 32.0° tilt boundary takes high energy of about $0.6J/m^2$, and Σ3 coherent boundary takes extremely low energy of about $0.05J/m^2$. The 2.0° tilt boundary consists of periodic dislocations. This boundary has the incoherent atomic structure around the dislocations however it generally has the coherent one. This boundary with coherent and incoherent atomic structure takes the middle energy of about $0.3J/m^2$.

The Σ3 boundary is {11$\bar{2}$0} twin boundary. Fig.5 (a) shows the atomic structure model of alumina projected from the [0001] direction. Large and small circles represent oxygen ions and aluminum ions, respectively. In the crystal structure, the oxygen sublattice takes the hexagonal closed-packed structure and alumina ions occupy 2/3 of the octahedral position which is composed of the oxygen ions. The ions in Fig.5 are classified by the component of the [0001] direction. The large black circles represent oxygen ions with the z-values of 3/12, 7/12 and 11/12, and the large white circles represent oxygen ions with the z-values of 1/12, 5/12 and 9/12, where z-value is coordinate component along the [0001] direction. The small circles with A, B, and C represent the aluminum ions with the z-values of 2/12, 4/12, 8/12 and 10/12, the z-values of 0, 2/12, 6/12 and 8/12, and the z-values of 0, 4/12, 6/12 and 10/12, respectively. Fig.5 (b) is the atomic structure rotated by 60° around the [0001] axis from one shown in Fig.5 (a). When the atomic structure in Fig.5 (b) is translated to [0001] direction by 6/12, the oxygen ions represented by the black and white circles exchange, but the aluminum ions do not exchange as shown in Fig.5 (c). Comparing Fig.5 (a) with Fig.5 (c), it is obvious that the difference of them is only a sequence of aluminum ions. The Σ3 boundary in the present study can be

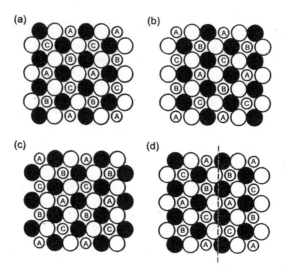

Figure 5. The atomic structure models of (a) an alumina single crystal, (b) the single crystal rotated around [0001] axis by 60°, (c) the single crystal rotated around the [0001] axis by 60° and translated to the [0001] direction by half of c-axis and (d) Σ3 grain boundary. (In the schematics, large and small circles represent oxygen and aluminum ions, respectively.)

obtained by joining the two structures in (a) and (c) as shown in Fig.5 (d). The boundary takes thus the completely coherent manner, and only the sequence of aluminum ions is different, comparing to the structure of perfect single crystal. That is the reason why the Σ3 boundary looks like single crystal in HREM image and takes extremely low energy.

CONCLUSION

Alumina bicrystals with [0001] symmetrical tilt grain boundaries including small angle boundaries and CSL boundaries were fabricated by hot-joining method. The grain boundary energies were estimated from groove profile measured by AFM and the grain boundary atomic structures were observed by HREM. It was found that the grain boundary energies are dependent on the grain boundary characteristics. The boundaries having low sigma indices take extremely lower energies and the 2.0° tilt boundary takes a little bit lower energy than others. The atomic structures of the boundaries with high energies are incoherent and the atomic structures of the boundaries with low energies are coherent.

ACKNOWLEDGEMENTS

This work was supported by the Grant in Aid for Scientific Research (11450260) from the Ministry of Education. The authors aould like to thank Mr. T. Saito of JFCC for technical assistance of HREM observations.

REFERENCE

[1] J. F. Shackelford and W. D. Scott, "Relative energies of [$\bar{1}$100] tilt boundaries in aluminum oxide," J. Am. Ceram., vol.51, pp688-692, (1968)

[2] , "Inter facial energies of tilt boundaries in aluminium experimental and theoretical determination," Scr. metall., vol.5, pp889-894, (1971)

[3] , "Grain boundary diffusion mechanisms in metals" Metall. Trans. A, vol.13A, pp2069-2095, (1982)

[4] P. Nikolopoulos, "Surface, grain-boundary and interfacial energies in Al_2O_3 and Al_2O_3-Sn, Al_2O_3-Co systems," J. Mater. Sci., vol.20, pp3993-4000, (1985)

[5] Y. Ikuhara, H. Kurishita and H. Yoshinaga, "Interface in polymer, ceramic and metal matrix composites," edited by H. Ishida, pp673, Elsevier Sci. Pub. (1988)

CRYSTALLOGRAPHIC ORIENTATION ANALYSIS AROUND STABLE CRACKS IN MgO

Yorinobu TAKIGAWA and Yoshiyuki YASUTOMI
Research and Development Laboratory
Japan Fine Ceramics Center
2-4-1 Mutsuno, Atsura-ku, Nagoya
456-8587, Japan

ABSTRACT

Crystallographic orientation analysis by electron back scattered diffraction (EBSD) is conducted around stable cracks in MgO, which is achieved by single edge-notched beam (SENB) method with crack stabilizers. It is found that the most of the cracks propagate along grain boundaries under stable fracture, while almost all the cracks propagate through the grain interior under unstable fracture in the present study. Analysis of the grain boundary characteristics around stable cracks by EBSD reveals that all the boundaries where cracks propagate are random boundaries. By using the coincidence of reciprocal lattice points (CRLP) method, it is clarified that the cracks preferably propagate along grain boundaries whose matching is relatively low. Since such boundaries are expected to have higher grain boundary energy, it is concluded that the stable cracks tend to propagate along high-energy boundaries.

INTRODUCTION

The mechanical properties and fracture behavior of polycrystalline ceramics greatly depend not only on the composition but also on the microstructure [1, 2]. Thus, it is important to know what kind of crystallographic planes or boundaries cause cracks to propagate. In the present study, crystallographic orientation analysis by electron back scattered diffraction (EBSD) was conducted around stable cracks in polycrystalline MgO in order to understand the effect of microstructural factors on microcrack propagation.

To the extent authorized under the laws of the United States of America, all copyright interests in this publication are the property of The American Ceramic Society. Any duplication, reproduction, or republication of this publication or any part thereof, without the express written consent of The American Ceramic Society or fee paid to the Copyright Clearance Center, is prohibited.

EXPERIMENTAL PROCEDURE

The samples used in this study are high purity MgO polycrystals. *In-situ* observation of crack extension was conducted for single edge-notched beam (SENB) specimens equipped in a three-point bend fixture having crack stabilizers. The details of the experiment have been reported previously [3, 4]. In order to understand the effect of microstructural factors on microcrack propagation, crystallographic orientation analysis is conducted by EBSD method on the specimen with microcracks [5, 6]. The orientations are analyzed on 120 grains around the microcrack. Details of the EDSD used in this study has already reported in a literature [6].

RESULTS AND DISCUSSION

Figure 1 shows a load-displacement curve of SENB specimen in MgO obtained by a three-point bending test with crack stabilizers. The load drops sharply after exhibiting the peak, which is the typical trend of unstable fracture. After the sharp drop, the load gradually decreases. Stable fracture must occur in this region.

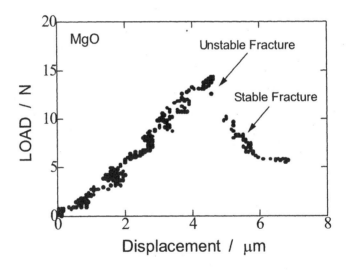

Figure. 1.   Load-displacement curve of a single-edge notched beam specimen in MgO tested by three point bending with crack stabilizer.

Crystallographic orientation analysis is conducted around the crack in both unstable and stable fracture regions. In the unstable fracture region, the same crystallographic orientations are observed across the crack, and it is clarified that the crack mainly propagates through the grain interior. In contrast, crystallographic orientation is different between two points across the crack at the most part in stable fracture region. These results indicates that intergranular fracture is dominant in the stable fracture region.

It is important to know what kind of crystallographic planes or boundaries cause preferential crack propagation. In the present study, paying attention to the stable fracture region, the characteristics of crack propagating grain boundaries is analyzed. Figure 2 shows the SEM micrograph of a stable crack propagating along the grain boundary. An example of grain boundary characteristics is also shown in the micrograph. The two crystals forming the grain boundary have a relationship 47 degrees rotated around the [5 14 22] axis, and the grain boundary is shown to be a random one. More than 20 crack extending boundaries are analyzed in this study, and all boundaries are found to be random boundaries, rather than special boundaries such as low-angle boundaries or low sigma coincidence boundaries.

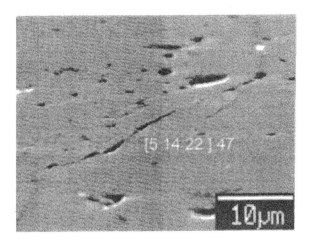

Figure 2.   SEM micrograph of a stable crack propagating along the grain boundary. Grain boundary character is also shown in the micrograph.

In order to evaluate the matching of such random boundaries, coincidence of reciprocal lattice points (CRLP) method is used in this study [7]. The sum of overlapped volumes of reciprocal lattice points in the two crystals, $V(\theta)$, are estimated in each grain boundary where the crack propagates. Figure 3 shows a result of the estimation. The value of $V(\theta)$ greatly depends on the rotation angle, θ. For example in the case of the boundary shown in Fig. 2, the value of $V(\theta)$ exhibits relatively lower one.

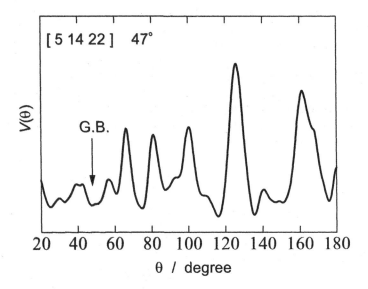

Figure 3.   The change in sum of overlapped volumes of reciprocal lattice points, $V(\theta)$, with rotation angle, θ, of a MgO lattice about the [5 14 22] axis of another similar lattice.

Figure 4 shows the comparison of the distribution of $V(\theta)$ between in total boundaries and in crack extending boundaries. The value of $V(\theta)$ in the total boundaries are estimated in more than 100 boundaries. In the total boundaries, $V(\theta)$ has a peak around 0.3-0.4, and there are some boundaries where the value of $V(\theta)$ is high. In contrast, in the crack extending boundaries, boundaries having high value of $V(\theta)$ are not exist, and the peak value is lower than that in total boundaries. Since the higher the value of $V(\theta)$ the better the matching of the two

Grain Boundary Engineering in Ceramics

grains, this result indicates that the cracks preferably propagate along grain boundaries whose matching is relatively low. It is generally accepted that the boundary exhibiting lower matching have higher grain boundary energy. From these results, it is experimentally shown that the stable cracks preferably propagate along high-energy boundaries.

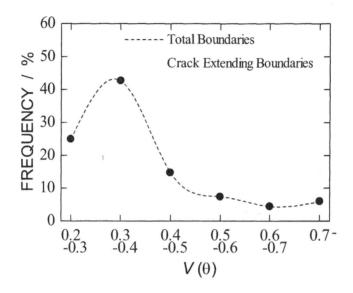

Figure 4.   Comparison of the distribution of $V(\theta)$ between total boundaries and crack extending boundaries.

CONCLUSION
     Crystallographic orientation analysis by electron back scattered diffraction (EBSD) was conducted around stable cracks in MgO, which is achieved by SENB method with crack stabilizers. The following conclusions are obtained.
1.   Most of the cracks propagate along grain boundaries under stable fracture, while almost all the cracks propagate through the grain interior under unstable fracture.
2.   Cracks preferably propagate along grain boundaries whose matching is relatively low. Since such boundaries are expected to have higher grain boundary energy, it is concluded that the stable cracks tend to propagate along high-energy

boundaries.

ACKNOWLEDGEMENTS

The authors would like to thank Mr. K. Hiramatsu, Mr. S. Ogawa and Mr. T. Kuribayashi for technical assistance. This work is supported by NEDO as part of the Synergy Ceramics Project under the Industrial Science and Technology Frontier (ISTF) Program promoted by AIST, MITI, Japan.

REFERENCES
1. V. D. Krstic, "Fracture of Brittle Solids in the Presence of Termoelastic Stress," *J. Am. Ceram. Soc.*, **67** [9] 589-593 (1984).
2. H. P. Kirchner and R. M. Gruver, "Strength-Anisotropy-Grain Size Relations in Ceramic Oxides," *J. Am. Ceram. Soc.*, **53** [5] 232-236 (1970).
3. D. J. Dingley and V. Randle, "Review Microstructure Determination by Electron Back-scatter Diffraction," *J. Mater. Sci.*, **27** [17] 4545-66 (1992).
4. Y. Yasutomi, Y. Sakaida, N. Hirosaki and Y. Ikuhara, "Analysis of Crystallographic Orientation of Elongated β-Si3N4 Particles in In Situ Si_3N_4 Composite by Electron Back Scattered Diffraction Method," *J. Ceram. Soc. Jpn.*, **106** [10] 980-83 (1998).
5. A. Okada, K. Hiramatsu and H. Usami, "R-curve Evaluation techniques for Toughened Ceramics," pp.770-75, in *Proceedings of 6th International Symposium on Ceramic Materials and Components for Engines*, Ed. by K. Nihara, S. Hirano, S. Kanzaki, K. Komeya and K. Morinaga (1997).
6. K. Hiramatsu, A. Okada and H. Usami, "*in situ* Observation of Stable Crack Propagation in SENB Specimens for R-Curve Evaluation," pp.335-40, in *Proceedings of the 22nd Annual Conference on Composites, Advanced Ceramics, Materials and Structures*, Ed. by D. Bray, The American Ceramic Society (1998).
7. Y. Ikuhara and P. Pirouz, "Orientation Relationship in Large Mismatched Bicrystals and Coincidence of Reciprocal Lattice Points (CRLP)," *Mater. Sci. Forum*, **207-209**, 121-24 (1996).

Grain Boundary Engineering in Ceramics

VISCO-ELASTIC ANALYSIS OF INTERNAL FRICTION ON ENGINEERING CERAMICS

Shuji Sakaguchi
National Industrial Research Institute of Nagoya
2268-1, Shimo-Shidami, Moriyama-ku, Nagoya 463-8687 JAPAN

ABSTRACT

Internal friction of alumina and silicon nitride was measured using a torsion pendulum, and was analyzed with the four-element visco-elastic model. The grain size dependence of internal friction of alumina was also discussed. Alumina showed no peaks in temperature dependence of internal friction, and it increased exponentially with ascending temperature. The calculated activation energy was relatively close to the value for creep deformation, but the calculated strain rate was much larger than the experimental data. Silicon nitride showed a peak in temperature dependence of internal friction, and this temperature dependence was well expressed with the four-element model. However, the calculated activation energy was smaller than the experimental value. The calculated relaxation time was quite short, and it does not explain the transient creep behavior of silicon nitride. Grain size exponents for internal friction of alumina between 0.55 and 0.87 were obtained. This grain size exponent shall become 1, from the discussion of the diffusion creep model.

INTRODUCTION

Engineering ceramics are expected to be used for high temperature applications, and the grain boundary phenomena of the engineering ceramics are focused, because the mechanical properties of the ceramics at elevated temperatures are strongly related to the grain boundary properties. Internal friction expresses the damping phenomenon of the mechanical vibration of the material [1-9]. The temperature dependence of the internal friction is related to the visco-elastic properties of the material. In this study, the temperature dependence of internal friction of alumina and silicon nitride ceramics was measured, and analyzed with the four-element visco-elastic model. The analyzed data were compared to the experimental data on creep deformation. The grain size dependence of the internal friction on alumina ceramics is also discussed, for understanding the effect of the grain boundary on the internal friction.

ANALYSIS

The four-element visco-elastic model in Fig. 1 is used for the analysis. Equation (1) is obtained for expressing the total vibration of this model, from the equations for each element.

To the extent authorized under the laws of the United States of America, all copyright interests in this publication are the property of The American Ceramic Society. Any duplication, reproduction, or republication of this publication or any part thereof, without the express written consent of The American Ceramic Society or fee paid to the Copyright Clearance Center, is prohibited.

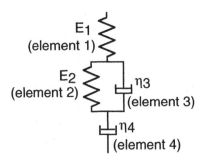

E_1 (element 1)

E_2 (element 2) η_3 (element 3)

η_4 (element 4)

Figure 1   Four-element visco-elastic model

$$\frac{E_2}{\eta_4}\sigma + [1 + \frac{E_2}{E_1} + \frac{\eta_3}{\eta_4}]\dot{\sigma} + \frac{\eta_3}{E_1}\ddot{\sigma} = E_2\varepsilon + \eta_3\dot{\varepsilon} \tag{1}$$

where, σ and ε are total stress and strain of this model, E_1 and E_2 are elastic moduli of elements 1 and 2, η_3 and η_4 are viscosity of elements 3 and 4.

If we put a fixed stress, σ_0, to this model, time dependence of the strain is obtained in equation (2).

$$\varepsilon = \frac{\sigma_0}{E_1} + \frac{\sigma_0}{E_2}[1 - \exp(-\frac{E_2}{\eta_3}t)] + \frac{\sigma_0}{\eta_4}t \tag{2}$$

Internal friction can be expressed as follows. If we put a cyclic strain of

$$\varepsilon(t) = \varepsilon_0 \sin \omega t \tag{3}$$

to this model, the stress appears as,

$$\sigma(t) = A \sin \omega t + B \cos \omega t \tag{4}$$

and the internal friction, Q^{-1}, can be obtained as,

$$Q^{-1} = B / A \tag{5}$$

From these relations, internal friction can be expressed, using the parameters of each element,

$$Q^{-1} = [\frac{E_2^2}{\eta_4} + \eta_3(1 + \frac{\eta_4}{\eta_3})\omega^2] / [E_2(1 + \frac{E_2}{E_1})\omega + \frac{\eta_3^2\omega^3}{E_1}] \tag{6}$$

EXPERIMENTS

Tested samples were sintered alumina ceramics of 99.9 % purity, and silicon nitride ceramics with alumina and yttria as sintering additives. Alumina single crystal was also tested for the reference.

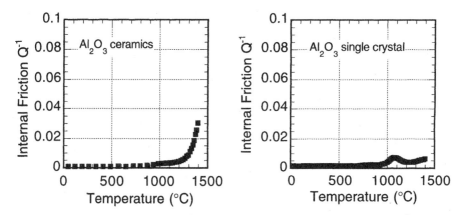

Figure 2   Temperature dependence of internal friction for alumina ceramics and alumina single crystal.

Internal friction was measured using torsion pendulum equipment, supplied by Rhesca Co. (Type: MR-2001C) [4]. Test pieces were rectangular bars of 80 x 5 x 1 mm. Temperature dependence was measured in the range from room temperature to 1400 °C, with heating rate of 5 °C/min, in air.

Grain size dependence of internal friction was examined, using annealed alumina. Five kinds of specimens with different average grain size, 1.1 μm, 1.8 μm, 3.8 μm, 4.4 μm and 5.7 μm, were prepared by annealing in air under different conditions.

RESULTS

Alumina

Figure 2 shows the temperature dependence of internal friction of alumina ceramics and alumina single crystal (sapphire). The internal friction is completely different, due to the existence of the grain boundary phase. In the data on alumina ceramics, the typical peak of the internal friction was not observed, and the internal friction increased exponentially with an increase in temperature. This result was analyzed, assuming that $1/E_2 = 0$, as no peak was observed. We could obtain the parameters;

$$E_1 = 162.3 \times (1 - 9.36 \times 10^{-5} \, T) \tag{7}$$

$$\eta_4 = 9.8 \times 10^{-7} \times \exp(3.37 \times 10^4 / T) \tag{8}$$

where T is temperature in K. Using these parameters, the calculated temperature dependence of the internal friction could be obtained. It is shown in Fig. 3.

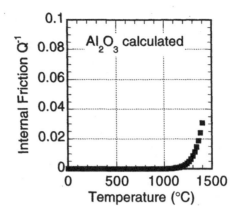

Figure 3   Calculated internal friction, using the obtained parameters for alumina.

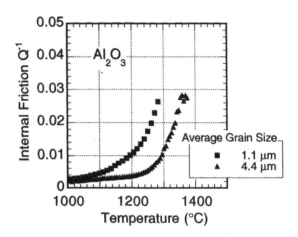

Figure 4   Temperature dependence of internal friction of alumina, having different average grain sizes.

Figure 4 shows the temperature dependence of internal friction for alumina ceramics with an average grain size of 1.1 μm and 4.4 μm. The increase of internal friction begins at a lower temperature for the specimen having smaller grain size. The grain size dependence exponent for the internal friction was obtained.

Silicon Nitride

The temperature dependence of internal friction for silicon nitride is shown in Fig. 5. This result was used for the four-element analysis, and the following parameters were obtained;

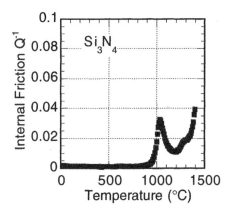

Figure 5 Temperature dependence of the internal friction for silicon nitride.

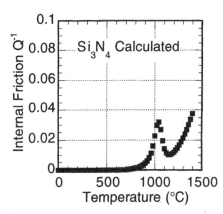

Figure 6 Calculated internal friction, using the obtained parameters for silicon nitride.

$$E_1 = 122 \ (\text{GPa}) \tag{9}$$

$$E_2 = 30 \ (\text{GPa}) \tag{10}$$

$$\eta_3 = 2.0 \times 10^{-15} \times \exp (5.0 \times 10^4 / T) \tag{11}$$

$$\eta_4 = 2.0 \times 10^{-3} \times \exp (1.5 \times 10^4 / T) \tag{12}$$

With these parameters, the calculated temperature dependence of internal friction could be obtained, as shown in Fig. 6.

DISCUSSION

Deformation Parameters on Alumina

With the parameter η_4, we could obtain the activation energy which corresponds to that for creep deformation. The activation energy was calculated as 280 kJ/mol, and this value was relatively close to the value obtained in creep deformation on alumina ceramics [10].

We could calculate the strain rate of the steady state creep, with this parameter. Figure 7 shows the calculated and the experimental strain rate from 1350°C to 1450°C [4]. It shows the calculated values are more than 100 times larger than the experimental data. This is caused by that the results of internal friction expresses the very beginning of the deformation. Therefore, the activation energy for generating the source of the diffusion is similar, but the total diffusion rate is much smaller on creep deformation, compare to the case of internal friction test.

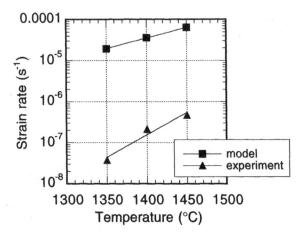

Figure 7   Experimental and calculated strain rate of alumina between 1350°C and 1450°C.

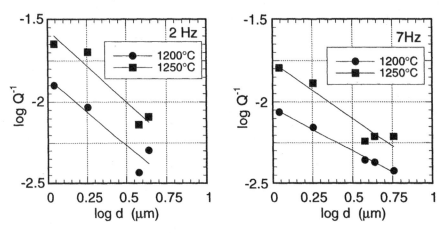

Figure 8   Grain size dependence of internal friction for alumina at 1200°C and 1250°C.
The gradient indicates the grain size exponent.

Grain Size Exponent of Alumina

Figure 8 shows the grain size dependence of the internal friction at 1200°C and 1250°C. The gradient of the line indicates the grain size exponent of the internal friction for the following equation;

$$Q^{-1} = C \cdot d^{-n} \tag{13}$$

Grain Boundary Engineering in Ceramics

Table I   Grain size exponent for internal friction at each test condition.

Temperature	Frequency 2 Hz	7 Hz
1200 °C	0.81	0.55
1250 °C	0.87	0.74

The obtained grain size exponent for each test condition is shown in Table I. In the case of creep deformation of ceramics, it is known that the grain size exponent of 2 is obtained for fine grain ceramics. Alumina with less than 10 μm is typical. However, the grain size exponent for the internal friction is much smaller than 2, relatively close to 1.

We shall discuss the meaning of the grain size exponent of 2, in Nabarro-Herring model [11]. This exponent corresponds to: (1) the source of the ionic diffusion is formed at the grain boundary, then the amount of the diffusion source is proportional to the amount of grain boundary, and, (2) the diffusion path length is proportional to the average grain size. The grain size exponent becomes 2, as both of these phenomena work. We can estimate that, in the case of the internal friction test, only the former phenomenon affects the grain size dependence, as the internal friction test does not require the macroscopic deformation and it does not require the diffusion passing through one grain. So, it can be reasonable that the grain size exponent for internal friction is smaller than that for creep deformation, with a value relatively close to 1.

Deformation Parameters on Silicon Nitride

For silicon nitride, the activation energy of steady state creep, calculated from the obtained parameter, was 130 kJ/mol. It is much smaller than the experimental activation energy for creep deformation of silicon nitride (500 to 600 kJ/mol) [10]. We could obtain the relaxation time from the parameters. The calculated relaxation time is very short. For example, the relaxation time at 1300°C was 0.0043 sec. Therefore, we could not explain the transient creep deformation of silicon nitride, using these parameters.

We assumed these phenomena were also caused by the difference between the internal friction test and creep deformation test. It is known that the creep deformation on silicon nitride is caused both by grain boundary sliding and cavity formation. However, in the internal friction test, we have no cavity formation, as the applied stress was smaller. The calculated maximum stress was about 5 MPa on this internal friction test. The mechanical damping caused by the grain boundary phase is more clearly indicated by the internal friction test than the creep deformation test.

CONCLUSIONS

The temperature dependence of internal friction for alumina and silicon nitride was measured and analyzed using the four-element visco-elastic model. The grain size dependence of internal friction for alumina was also measured, and the grain size exponent was discussed.

(1) Alumina showed no characteristic peak in internal friction, and the calculated activation energy was relatively close to the value for creep deformation. However, the calculated strain rate was much larger than the experimental value. This is caused by that the internal friction indicates the beginning part of the deformation process on ceramic materials.

(2) Silicon nitride showed a peak in internal friction, and the temperature dependence was well expressed by the four-element model. However, the calculated activation energy was smaller than that of the experimental data. The calculated relaxation time was very short, so we could not express the transient creep behavior with the obtained parameters. The creep deformation of silicon nitride is mainly caused by the grain boundary sliding and the cavity formation, but the internal friction test does not make the cavity.

(3) The grain size dependence of alumina was measured, and the grain size exponent for internal friction was obtained in the range between 0.55 and 0.87. It is smaller than the grain size exponent for creep deformation, which is typically 2. The grain size exponent for internal friction shall be 1, from considering the diffusion creep model.

REFERENCES

1. K. Matsushita, S. Kuratani, T. Okamoto, M. Shimada, "Young's Modulus and Internal Friction in Alumina Subjected to Thermal Shock," *J. Mater. Sci. Lett.*, **3**, [4], 345-48, (1984).

2. S. Sakaguchi, N Murayama, F. Wakai, "Internal Friction of Si3N4 at Elevated Temperatures," *J. Ceram. Soc. Jpn.*, **95**, [12], 1219-22, (1987) (in Japanese).

3. I. Tanaka, G. Pezzotti, K. Matsushita, Y. Miyamoto, T. Okamoto, "Impurity-enhanced Intergranular Cavity Formation in Silicon Nitride at High Temperatures," *J. Am. Ceram. Soc.*, **74**, [4], 752-59, (1991)

4. S. Sakaguchi, N. Murayama, Y. Kodama, F. Wakai, "The Relation between Internal Friction and Tensile Creep Deformation on Alumina Ceramics," *J. Alloys Compd.*, **211/212**, 361-64, (1994).

5. A. Lakki, R. Schaller, "Anelastic Relaxation Associated with the Intergranular Phase in Silicon Nitride and Zirconia Ceramics," *J. Alloys Compd.*, **211/212**, 365-68, (1994).

6. K. Matsushita, "Internal Friction in Ceria Ceramics Doped with Alkali Earth Metal Oxides," *J. Alloys Compd.*, **211/212**, 374-77, (1994).

7. K. Watari, S. Sakaguchi, S. Kanzaki, T. Hamasaki, K. Ishizaki, "Quantity and Character of Grain Boundary Phase in Mixed α'/β' Sialon Ceramics," *J. Mater. Res.*, **9**, [11], 2741-44, (1994).

8. K. Ota, G. Pezzotti, "Internal Friction Study of Sialon Ceramics," *Phil. Mag. A*, **73**, [1], 223-35, (1996).

9. G. Roebben, L. Donzel, S. Sakaguchi, M. Steen, R. Schaller, O. Van der Biest, "Internal Friction under Low-amplitude Torsional and High-amplitude Uniaxual Load in Silicon Nitride," *Scripta Mater.*, **36**, [2], 165-171, (1997).

10. W. R. Cannon, T. G. Langdon, "Creep of Ceramics. I. Mechanical Characteristics," *J. Mater. Sci.*, **18**, [1], 1-50, (1983).

11. F. R. N. Nabarro, "Steady-state Diffusional Creep," *Phil. Mag.*, **16**, 231-37, (1967).

TEM STUDIES OF REACTION-BONDED SI3N4/SIC COMPOSITES

Kenji KANEKO and Naoki KONDO*

Engineering Research Institute, The University of Tokyo,
2-11-16 Yayoi Bunkyo-ku Tokyo 113-8656 Japan
*Synergy Ceramics Laboratory/NIRIN, Shidami Human Science Park,
2268-1, Shimo-Shidami, Moriyama-ku, Nagoya, 463-8687 Japan

ABSTRACT

Nano-structures of silicon nitride (Si_3N_4)-silicon carbide (SiC) composites synthesized using the reaction-bonding method were carefully studied by transmission electron microscopy (TEM). Various sizes of SiC particles were observed at grain boundaries, embedded on Si_3N_4 particles and included within. Interfaces between the SiC particles and Si_3N_4 matrices are carefully analyzed. The presence or absence of a glassy phase at the interfaces of either SiC/Si_3N_4 or Si_3N_4/Si_3N_4, SiC/SiC was also investigated.

INTRODUCTION

Si_3N_4 has outstanding thermal and mechanical properties at high temperature and therefore is being considered for the high-temperature mechanical components in heat engines and turbines. Incorporation of nano-size SiC particles into Si_3N_4 ceramics improves the initial properties of Si_3N_4, as increasing its hardness, creep and oxidation resistance compared to monolithic Si_3N_4 ceramics.[1] The improvement of mechanical properties should be strongly related to the change of microstructure and chemistry of the composite, such as the grain morphology of Si_3N_4, distribution of SiC particles, structures and chemistries of each Si_3N_4/Si_3N_4, Si_3N_4/SiC and SiC/SiC interface.[1]

Several methods have been used to fabricate Si_3N_4-SiC composites, particularly with nano-size SiC grains. They are fabricated from the following types of powders as;

1) amorphous Si-C-N powder method, either fabricated from CVD powder[1, 2] or

To the extent authorized under the laws of the United States of America, all copyright interests in this publication are the property of The American Ceramic Society. Any duplication, reproduction, or republication of this publication or any part thereof, without the express written consent of The American Ceramic Society or fee paid to the Copyright Clearance Center, is prohibited.

prepared from precursor powders,[3-5] 2) mechanical mixing method (mechanical mixing of Si_3N_4 and SiC powders),[6-9] 3) carbon coating method (carbon coating on Si_3N_4 powder),[10] 4) reaction-bonding method (reaction-bonding of Si and SiC or C) mixed powder,[11, 12] and so on.

Here, the present work is aimed at analyzing the formation mechanisms of the nano-sized β-SiC particles dispersed or embedded in β-Si_3N_4 matrix, sintered with reaction-bonding (RB) method, since reaction-bonding method should be the important fabrication procedures to make Si_3N_4-SiC composite.[12] We employed conventional high-resolution transmission electron microscopy (HRTEM) to focus on the structures and chemistries of each Si_3N_4-Si_3N_4, Si_3N_4-SiC and SiC-SiC interface.

SPECIMEN PREPARATION AND EXPERIMENTAL METHODS

A Si_3N_4/SiC composite specimen was prepared by reaction bonding followed by hot pressing with a composition of 73.6 wt. % Si_3N_4 - 18.4 wt. % SiC – 5.0 wt. % Y_2O_3 – 3.0 wt. % Al_2O_3 (the weight ratio of Si_3N_4 : SiC should be 8 : 2). Further information with respect to the specimen fabrication is reported by Kondo et al[12].

Standard mechanical and ion thinning methods were applied to prepare TEM specimens. HRTEM was carried out on a JEM 2010 equipped with a LaB_6 gun having a point resolution of 0.18 nm operated at 200 kV. Details of the working conditions of the TEM are described elsewhere[13].

RESULTS AND DISCUSSIONS

Nano-sized SiC particles, with grain sizes less than 100 nm, are dispersed within the β-Si_3N_4 matrix, as seen in Fig. 1. The nano-sized SiC particles are also present at the grain boundary regions, and partially within the Si_3N_4 matrix. Liquid phases consist of Y_2O_3, Al_2O_3 and possibly SiO_2 are observed at both grain boundaries and at multigrain junctions. The difference in Si_3N_4 or SiC was easily distinguished from the presence of stacking fault within the grain by HRTEM or the presence of Si-C bondings by EELS.

Dispersed SiC particles within Si_3N_4 grains (intragranularly) have sizes less than 100 nm. On the other hand, larger particles are usually located at Si_3N_4 grain boundaries (intergranularly), with sizes between 50 and 300 nm. The HRTEM micrographs show the characteristic growth morphology of different types of such grains: Intragranular nano-sized SiC particles (less than 100 nm) embedded within Si_3N_4 matrix, and intergranular SiC particles (100 – 300 nm) located at the grain boundary regions, as seen in Fig. 1. Dislocations caused by strain due to the intergranular SiC particles were clearly observed, additionally there is an intragranular phase of nano-size turbostratic carbon, less than 10 nm.

HRTEM studies show the presence of amorphous phases not only at grain boundaries or at multigrain junctions, but also at the phase boundaries formed by the Si_3N_4 matrix and the intergranular nano-sized SiC particles. The grain boundary in Fig. 2(a) has an intergranular amorphous film with a thickness of about 2.0 nm, whereas Fig. 2(b) shows a grain boundary film with a thickness of about 1.0 nm with multigrain junctions. The only exceptions were a few percent of the boundaries having a low misorientation angle and no apparent films between Si_3N_4 and SiC grains at the grain boundaries in Fig. 2(c). Therefore, unlike the case of monolithic Si_3N_4 ceramics, the film thickness is not uniform, possibly due to the non-equilibrium of the grain boundary compositions.

In Si_3N_4 matrix grains, various state of intragranular nano-sized SiC particles can be categorized into three types; 1) one with a special orientation with Si_3N_4 matrix grain that makes it difficult to observe the amorphous phase, seen in Fig. 3(a), 2) another one with random orientation with respect to the matrix, which is usually surrounded by amorphous phase, seen in Fig. 3(b), and 3) the other one included within the amorphous phase, seen in Fig. 3(c). The relationship between the orientation and the presence of amorphous films is not particle size dependent.

The shapes of intragranular nano-sized SiC particles are also variable, as seen from Fig. 1, some of them are spherical shapes and others are faceted ones. The faceted ones, usually with ~ 100 nm grain size, are clearly showing the presence or absence of amorphous phases at the interfaces. On the other hand, neither the clear interfaces nor clear stacking faults can be seen in the spherical particles of less than ~ 10 nm grain size, due to the strong lattice fringe of Si_3N_4 matrix, seen also in Fig. 1 and also seen in the left side of Fig. 3(a). The moiré patterns in the particle were occasionally observed. The presence of spherical shape, presumably the most stable form of the grain prior to the growth may have been formed by various reasons, possibly due to the dynamical grain growth relationship between the SiC and Si_3N_4, or the reaction between the Si_3N_4 matrix and carbon gas, or the reaction between Si_3N_4 matrix and the carbon precipitates. At the earliest stage of the grain growth, the grains are expected to maintain spherical shape due to no preferential direction of growth. Faceted SiC grains usually have flat interfaces with Si_3N_4 matrix, as the preferential development of interfaces.

CONCLUSION

The structures of grain boundaries and phase boundaries were carefully studied by HRTEM imaging. The thickness of amorphous films at Si_3N_4 grain boundaries varies in the present material. Amorphous films were occasionally observed at phase boundaries formed by the Si_3N_4 matrix and the SiC particles. The phase boundaries for both inter- and intra-granular nano-sized SiC particles are categorized in Table 1. If there were amorphous films present at the phase boundaries, the thickness of these films varied due to the orientation. The SiC particles may be made from the reaction between the Si_3N_4 matrix and carbon gas, or between the reaction Si_3N_4 matrix and the carbon precipitates.

Table 1
The Category of Boundaries

Category	Grain size	Amorphous phase	Grain shape
Phase Intra	less than 10 nm	nano-size turbostratic carbon Not known (strong Si_3N_4 lattice fringe)	Spherical
	up to ~ 30 nm	Yes	Spherical
	up to ~ 100 nm	Yes	Faceted
		No (special orientation)	Faceted
Phase Inter	50 ~ 300 nm	Yes	Faceted
		No (special orientation)	Faceted

Acknowledgment

We would like to acknowledge Dr. Y. Ikuhara (Univ. of Tokyo, Japan), Dr. Y. Suzuki and Dr. T. Ohji (NIRIN, Japan) for valuable discussions.

Reference
1. K. Niihara, "New Design Concept of Structural Ceramics –Ceramic Nanocomposites-," *J. Ceram. Soc. Jpn.*, **99** [10] 974-982 (1991).
2. X. Pan, J. Mayer and M. Rühle, "Silicon Nitride Based Ceramic Nanocomposites," *J. Am. Ceram. Soc.*, **79** [3] 585-590 (1996).
3. T. Hirano, K. Izaki and K. Niihara, "Microstructure and Thermal Conductivity of Si_3N_4/SiC Nanocomposites Fabricated From Amorphous Si-C-N Precursor Powders," *Nanostructures Materials*, **5** [7/8] 809-818 (1995).
4. T. Rouxel, F. Wakai, and K. Izaki, "Tensile Ductility of Superplastic Al_2O_3-Y_2O_3-Si_3N_4/SiC Composites," *J. Am. Ceram. Soc.*, **75** [9] 2363-2372 (1992)
5. J. Douza, P. Sajagalik and M. J. Reece, "Characterization of Si_3N_4+SiC Nanocomposite," *Fourth Euro Ceramics*, **4** 67-74 (1995).
6. P. Greil, G. Petzow and H. Tanaka, "Sintering and HIPping of Silicon Nitride-Silicon Carbide Composite Materials," *Ceramics International*, **13** [1] 19-25 (1987).
7. K. Ramoul-Badache and M. Lancin, "Si_3N_4-SiC Particulate Composites: Devitrification of the Intergranular Phase and its Effect on Creep," *J. Euro. Ceram. Soc.*, **10** [4] 369-379 (1992).
8. T. Hirano, T. Ohji, and K. Niihara, "Effect of Matrix Grain Size on the Mechanical Properties of Si_3N_4/SiC Nanocomposites Densified with Y_2O_3," *Material Letters*, **27** [5] 53-58 (1996).
9. J.–F. Yang, Y.-H. Choa, J. P. Singh and K. Niihara, "Fabrication and Mechanical Properties of Si_3N_4/SiC Nanocomposite with Pressureless Sintering and Sinter-Post-HIPing," *J. Ceram. Soc. Jpn.*, **106** [10] 951-957 (1998).
10. T. Yanai, T. Hamasaki, and K. Ishizaki, "High Temperature Mechanical Strength of Si_3N_4/SiC Nanocomposite Prepared from Carbon Coated Si_3N_4 Powder," *J. Ceram. Soc. Jpn.*, **103** [10] 1017-1021 (1995).
11. S.-Y. Lee, "Fabrication of Si_3N_4/SiC Composite by Reaction-Bonding and Gas-Pressure Sintering," *J. Am. Ceram. Soc.*, **81** [5] 1262-1268 (1998).
12. N. Kondo, K. Kaneko, Y. Suzuki and T. Ohji, "Middle Stage Heat Treatment for Microstructure Control of Reaction-Bonded Silicon Nitride-Silicon Carbide Composite," *J. Ceram. Soc. Jpn.*, **submitted**.
13. K. Kaneko, M. Yoshiya, I. Tanaka and S. Tsurekawa, "Chemical Bonding of Oxygen in Intergranular Amorphous Layers in High-Purity β-SiC Ceramics," *Acta Materialia*, **47** [4] 1281-1287 (1999).

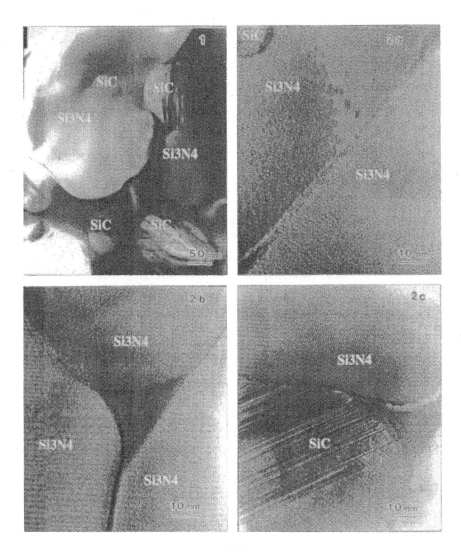

Figure 1. Both inter- and intragranular nano-sized SiC particles are dispersed within the β-Si3N4 matrix. The grain boundary has an intergranular amorphous film with a thickness of about 2 nm, Fig. 2(a), and with a thickness of about 1 nm, Fig. 2(b), also with no apparent amorphous films, Fig. 2(c).

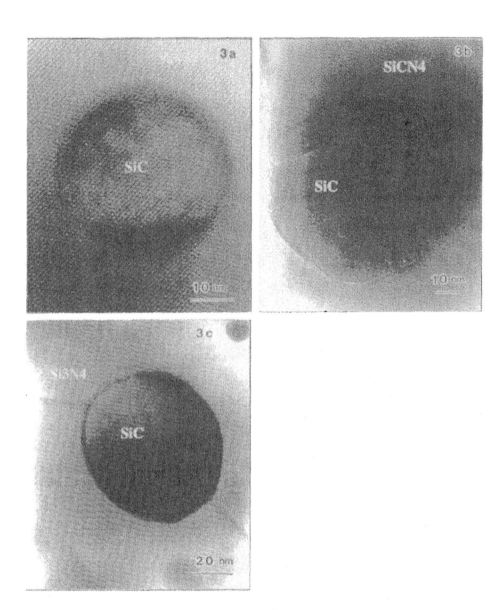

Figure 3(a) shows an intragranular nano-sized SiC particle with a special orientation, so that difficult to observe the amorphous phase. Fig. 3(b) shows the SiC particle with random orientation and surrounded by an amorphous phase. Fig. 3(c) shows the SiC particle surrounded by thick amorphous phase.

Intergranular Film

THE EFFECTS OF INTERGRANULAR FILMS ON THE MECHANICAL BEHAVIOR OF SELF-REINFORCED CERAMICS

Paul F. Becher, Gayle S. Painter, and Hua-Tay Lin
Metals and Ceramics Division
Oak Ridge National Laboratory
Oak Ridge, TN 37831-6068

ABSTRACT

Self-reinforced silicon nitride ceramics exhibiting high toughness and strength are obtained by tailoring the size and number of the bridging grains. In each material, bridging mechanisms rely on debonding of the reinforcing grains from the matrix to increase toughness. Interfacial debonding can be influenced by the nature of the intergranular films, whose composition and properties are a function of the sintering aids. In silicon nitride ceramics that use alumina as one of the sintering aids, the amorphous intergranular films surround β-Si_3N_4 grains on which epitaxial β- SiAlON layers form. In this case, interfacial debonding is dependent on the composition of the SiAlON layer, becoming more difficult as the Al (and O) content of the SiAlON increases. This appears to be a result of the formation of strong Si-O, Si-N and Al-O bonds across the glass-crystalline interface. Observations by others also reveal that the interfacial debonding conditions required in the toughening process can be altered in the presence of amorphous Si-O-N intergranular films and in "special" grain boundaries.

In addition to its influence on fracture, the amorphous intergranular films containing Si, Al, O, N and RE (i.e., rare earth elements) will also influence the temperature dependent viscosity of the amorphous phase. Measurements of softening temperature (T_g) reveal that T_g decreases with increase in oxygen to nitrogen ratio and with the substitution of larger rare earth ions in the film. These reductions in the T_g of the amorphous intergranular films lower the creep resistance of the ceramic and limit its life at elevated temperatures. It is apparent that sintering additives used to promote the densification of self-reinforced ceramics can play an important role in improving their fracture and creep resistances. However, greater understanding at the basic level is needed for further insight into how their chemistry controls the structure/bonding and properties of the amorphous films and interfaces in self-reinforced polycrystalline ceramics.

BACKGROUND

Additives to promote densification are commonly employed to produce monolithic ceramic components. In a number of cases, these additives result in the formation of amorphous intergranular films; a feature that is further promoted when lower purity raw materials are used to reduce costs. These intergranular films can degrade the mechanical behavior of ceramics both at room and elevated temperatures. Thus, considerable effort has been made to produce ceramics where such films are eliminated by use of more sophisticated processing and/or high purity raw materials, adding to cost. In addition, one can modify the additives in an attempt to crystallize the amorphous films after densifying the body. Typically, the material in triple points can be

To the extent authorized under the laws of the United States of America, all copyright interests in this publication are the property of The American Ceramic Society. Any duplication, reproduction, or republication of this publication or any part thereof, without the express written consent of The American Ceramic Society or fee paid to the Copyright Clearance Center, is prohibited.

crystallized; however, crystallizing that between grain faces is far more difficult, often impossible, to achieve. As an alternative, one should consider whether or not the properties of these amorphous intergranular films can be tailored to actually enhance both the fracture and creep resistances of the ceramics and, if so, to understand how the film composition can be used to achieve this.

Such issues abound in silicon nitride ceramics where sintering additives are required to promote densification and also influence the development of a self-reinforced microstructure wherein elongated grains can serve to toughen the ceramic. Densification proceeds in the presence of an oxynitride glass that serves as a liquid sintering agent in which growth of the elongated grains also occurs. The result is a microstructure of larger elongated grains in an often finer grained matrix, where all grains are surrounded by the amorphous intergranular film when the energy of the unwetted "special" boundary is greater than the sum of the energies of the two interfaces formed when the boundary is wetted. As summarized by Ernst et al., the film thickness between two grains can reach an equilibrium thickness, which is a function of the specific constituents of the film in silicon nitride ceramics.[1,2] However, as these authors point out, this film thickness does not appear to depend on the amount of a specific composition amorphous phase. Numerous observations are reported which demonstrate that silicon nitrides contain continuous amorphous intergranular films.

The growth of elongated grains in silicon nitride ceramics provides a means to increase the fracture toughness, as observed by a number of researchers. The toughening mechanism involves the formation of elongated grains that bridge the crack in the region behind the crack tip, which imposes a closure stress on the crack. Bridging grains can be partially debonded but still intact close to the crack tip and, then with continued advance of the crack tip, pull out of the matrix on one side of the crack plane. Both types resist opening of the crack and dissipate strain energy by frictional processes such that the toughening contribution can be enhanced by the presence of larger diameter elongated grains.[3,4] A distinctly bimodal microstructure consisting of larger (1-2 μm diameter) elongated grains dispersed in a matrix with a submicron grain size was one approach to achieve both high strength and toughness, Table I. Additional studies indicate that similar toughening could be obtained by introducing a strong degree of alignment of the reinforcing grains where the matrix and reinforcing grain sizes were similar.[5]

However, the process of forming bridging grains requires that the crack tip be diverted around a potential bridging grain rather than cutting through it. This can be accomplished if the crack tip is deflected when it reaches the interface between the intergranular film and the elongated grain and the grain separates from the film (i.e., debonds).[6] Earlier evidence indicated that the composition of this intergranular phase may be a factor in the interfacial debonding/toughness process where it was found that transgranular fracture and low toughness were associated with an amorphous SiO_xN_y intergranular film.[7] Additional analytical studies have indicated that interface strength can be weakened when larger lanthanide ions are present in the intergranular films.[8] These observations and the fact that a variety of sintering aids have been used to develop toughened silicon nitrides with mixed results stimulated studies of the effects the intergranular film composition could have on the interfacial properties and the mechanical behavior of self-reinforced ceramics.

Table I. Effect of Microstructure on the Mechanical Properties of Silicon Nitride Containing 6.25 wt. % Y_2O_3 plus 1 wt. % Al_2O_3.[4]

Material Description	Average Grain Diameter	Fracture Toughness MPa√m	Fracture Strength MPa
Large elongated grains in fine grain matrix	0.3 μm / 2 μm Distinctly Bimodal	10.6	1144 ± 126
Medium sized reinforcing grains and matrix grains	~ 1 μm Broad Monomodal	6.3	850 ± 75
Small elongated grains in fine grained matrix	~0.2 μm/~0.5 μm Distinctly Bimodal	4.4	925 ± 75
Fine equiaxed grains	~0.4 μm Monomodal	3.5	165

EXPERIMENTAL METHODS

The influence of the intergranular film composition on interfacial debonding has been examined using β-Si_3N_4 whiskers embedded in oxynitride glasses. This allows one to control the glass composition and alter the radial compressive thermal expansion mismatch stresses imposed on the interfaces to assess its role on debonding. The experimental technique involves the generation of a crack in the glass by indentation that intercepts the prism plane of the whisker.[9] The angle between the crack and prism planes can be varied allowing one to characterize the energy required to debond the interface. Basically, the minimum angle at which the crack tip begins to deflect up along the interface will increase, as the interface becomes weaker. This minimum angle is known as the critical angle for debonding and large critical angle values are sought to promote interface debonding in the ceramics.

Three silicon nitrides with the same bimodal microstructures were also prepared using different ratios of the yttria and alumina sintering aids to selectively reduce the alumina content of the epitaxial SiAlON layers.[10] Quantitative analysis confirmed that the same microstructural characteristics were maintained in each composition: average diameter of matrix grain 0.3 μm, average diameter of reinforcing grains - 2 μm, distribution of grain diameters maintained, and area fraction of intergranular glass phase same - ~ 0.04. The fracture strengths were determined in four point flexure using six 4 mm wide by 3 mm thick bars with 400 grit surface finishes and beveled edges for each composition, which were loaded at 70 MPa/s. The fracture toughness was determined using the precracked applied moment double cantilever beam method.[6,10]

Si-RE-Al oxynitride glasses were prepared by heating pellets in a covered BN-coated BN or Graphite crucible. In the case of the oxynitride glasses, this crucible was then placed on a BN powder bed inside a larger graphite crucible and then place in a gas-pressure sintering furnace and heated to 1000°C under vacumm (~ 0.1 torr). At this point, nitrogen was then introduced to raise the gas overpressure to 5 MPa nitrogen gas overpressure for melting the oxynitride glasses or either a vacuum (0.1 torr) or 5 MPa argon gas overpressure was used for oxide glasses. The softening points were determined by heating the samples at a rate of 2°C/m in a dual rod dilatometer versus a NIST sapphire standard to temperatures of < 1100°C with the softening point defined by the onset of rapid permanent deformation of the glass sample.

RESULTS AND DISCUSSION

Interfacial Debonding Associated with Intergranular Films

As seen in Table II, the largest critical debond angles are associated with the absence of the formation of an epitaxial β-SiAlON layer on the β-Si$_3$N$_4$ whiskers. In glasses containing aluminum, this occurs when the nitrogen content of the glass is low, but can also be retarded by significantly lowering the aluminum content of the glass. By decreasing the aluminum content, the viscosity of the glass increases, which lowers the rate of SiAlON formation. The current results also indicate that residual stresses acting on the interface have a more limited effect as compared to that due to the formation of the epitaxial SiAlON layer. One then wonders if the amount of aluminum could not be lowered sufficiently to promote debonding while still having a sufficient amount to effectively promote densification.

As shown in Table III, the aluminum (and oxygen) content of the epitaxial SiAlON layers on the β-Si$_3$N$_4$ grains does decrease as yttria was substituted for alumina in the sintering aids. This is accompanied by an increase in critical debond angle and, more importantly, by an increase in the toughness for the same reinforcing microstructural characteristics. One can, then, by tailoring the sintering aids, enhance the interfacial debonding required to cause crack deflection and the formation of bridging grains, which are critical to the toughening process.

Recent first principles cluster calculations have addressed the strength of the interface between the amorphous intergranular film and the grains by characterizing the strength within and between Si- and Al-based tetrahedral clusters centered along the interface.[12,13] Such tetrahedral clusters are common to both the crystalline and amorphous phases. The calculations reveal that the binding energies of the SiN$_4$ tetrahedra increase as oxygen replaces the nitrogen with SiO$_4$ exhibiting very high binding energies. In the case of the formation of Si$_{6-z}$Al$_z$O$_z$N$_{8-z}$, where Al replaces some of Si, the binding energies of the Al-based tetrahedra exhibit the same increase with oxygen substitution for nitrogen. In fact, the AlO$_4$ tetrahedron has only slightly lower binding energy than does the SiO$_4$ tetrahedra. The combination of Si- and Al-based tetrahedra on either side of the interface will cause the SiAlON-glass interface to be stronger than a Si$_3$N$_4$-glass interface. In the presence of an epitaxial layer, the calculations reveal that the interface strength will also increase with the Al and O content of the SiAlON layer, consistent with the observations in Tables I and II.

Table II. Debonding Conditions for β-Si$_3$N$_4$ Whiskers Embedded in Oxynitride Glass [9,11]

Glass Composition, Eq.%	Compressive Stress on Interface, MPa [a]	z-Value of Epitaxial Beta Si$_{6-z}$Al$_z$O$_z$N$_{8-z}$ Layer	Critical Debond Angle, Degrees
55Si25Al20Y10N90O	~350	0	~70
55Si10Al35Y10N90O	~515	0	~70
57Si43La20N80O	~590	0	~68
55Si10Al35Y20N80O	~530	0	~72
55Si10Al35Y20N80O After annealing	~530	0.15	~50
55Si25Al20Y20N80O	~320	1.0	~55
41Si30Al29Yb23N77O	~455	1.6 – 2.0	~55
46Si27Al27La27N73O	~555	1.6 – 2.0	~50

Table III. Characteristics of Self-Reinforced β-Si$_3$N$_4$ with an Amorphous Intergranular Film. [10]

Sintering Additive Composition, wt. %	z-Value of Epitaxial Beta Si$_{6-z}$Al$_z$O$_z$N$_{8-z}$ Layer	Critical Debond Angle, Degrees	Areal Fraction Glass, %	Areal Fraction Large Grains, %	Steady-State Fracture Toughness, MPa√m
6.25Y$_2$O$_3$+1Al$_2$O$_3$	0.01	~75	4.7	32	10.5 - 11.0
5Y$_2$O$_3$+2Al$_2$O$_3$	0.03	~70	3.4	29	8.5 - 9.0
4Y$_2$O$_3$+2.8Al$_2$O$_3$	0.06	~60	3.7	29	7 - 7.5

---

[a] Calculated stress based on measured values of thermal expansion coefficients, softening temperatures, and Young's modulus of the glasses and silicon nitride (e.g., ref. 9,10).

Viscous flow of amorphous intergranular phase

The dilatometric softening temperature(T_g) of SiREAl oxynitride glasses decreases significantly when nitrogen in the glass network is replaced by oxygen. This occurs due to loss of strength of network bonding as three-fold coordination of nitrogen is replaced by the two-fold coordination of oxygen. This would result in shifting the viscosity-temperature curve of the intergranular oxynitride glass phase to lower temperatures as nitrogen is replaced by oxygen. This is reflected in the lowering of the temperature to reach a specific viscosity (the softening point) as the nitrogen content of the glass decreases, Figure 1. While the viscosity of an amorphous film under the constraint of neighboring rigid grains may well be higher than a bulk glass of the same composition, the *trends* in viscosity with compositional changes will be quite similar.

One can sense that in oxidizing environments the glass phase and hence the ceramic can exhibit creep rates that increase with time due to an increase in the oxygen content of the intergranular glass phase. In fact, it is possible with some intergranular glass compositions for the viscosity to be diminished sufficiently in this manner for creep to occur at temperatures below 1000°C. This effect would be localized to the sample region and expand into the interior as oxygen was able to diffuse inward along the intergranular glass phase. As the viscosity of the glass decreases, the diffusion rates at a given temperature within the glass will also increase, raising the rate of oxygen penetration. After creep testing at 1370°C, observations reveal the development of a surface/subsurface layer where the intergranular phase is enriched in oxygen and the promotion of intergranular cavitation within this region consistent with a lowering of the amorphous phase viscosity.[14] Additional support for losses in the glass viscosity in silicon nitride ceramics due to oxygen enrichment are indicated failure that originate in the surface/sub-surface region due to viscous flow of the glass phase observed as ligaments in the fracture origin region after tensile testing in air at temperatures as low as 850°C .[15]

These processes can, then, markedly influence the lifetime of ceramics. However, the viscosity (as well as other properties) of intergranular amorphous phases can be controlled by the judicious selection of the sintering additives (e.g., rare earth oxides, aluminum oxide, etc.). For example, substituting yttrium for aluminum in the glass phase substantially increases the softening temperature over the range of nitrogen contents, Fig. 1. As one recalls from the debonding and toughening results, substitution of yttrium for aluminum in the intergranular glass produced a substantial increase in the fracture resistance of the self-reinforced silicon nitrides. Thus, procedures are available to significantly improve the fracture and creep resistance of silicon nitride by tailoring compositions. Finally, in glasses with fixed cation ratios, replacing yttrium by rare earth ions in the lanthanide series raises the glass softening temperature, especially with the substitution of smaller ions such as lutetium, Fig. 1. If a trend similar to that seen with yttrium follows, substitution of lutetium for aluminum should raise the softening point further. Thus, careful selection of rare earth additives, in combination with Si:Al ratios, offers the potential for further improvements in creep resistance in sinterable self-reinforced silicon nitride ceramics.

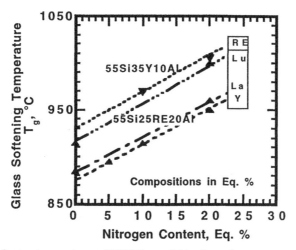

Figure 1. The softening temperatures of SiREAl oxynitride glasses increase as oxygen is replaced by nitrogen, as aluminum is replaced by yttrium, and as the yttrium is replaced by smaller sized lanthanide series ions.

CONCLUSIONS

It is clear that intergranular films can influence the toughening response and the creep resistance of self-reinforced silicon nitrides. To promote densification, additives such as alumina and yttria are employed, resulting in a fluid amorphous phase that promotes sintering processes and grain growth. The presence of aluminum (and oxygen from the silicon nitride powder) results in the formation of an epitaxial SiAlON layer on the silicon nitride grains. The initial studies revealed that the presence of an epitaxial SiAlON layer was detrimental to the interfacial debonding process, which is required to obtain toughening by crack deflection and grain bridging. However, interfacial debonding is promoted when the aluminum (and oxygen) content of the SiAlON layer is diminished. Thus, one can make compromises between enhancing the densification and increasing the toughening effects by tailoring the Al:Rare Earth ratio in the additives.

The intergranular glass properties are also a function of the densification additive composition. The glass softening temperature (hence viscosity) can be increased by raising its nitrogen content, substituting smaller lanthanide series ions for yttrium, increasing the silicon level, or by increasing the RE: Al ratio. By raising the glass softening temperature, one increases the viscosity at maximum use temperature. This has two effects. First, the rates of diffusion of species into and out of the ceramic will be lowered as the viscosity increases. This will limit the penetration of any reaction zones due to environmental effects. Second, limiting flow of the intergranular glass phase will increase the creep resistance of the ceramics, which is obviously desirable for elevated temperature applications. The results suggest that by understanding the influence of additive chemistry one can tailor self-reinforced ceramics with both high strength and toughness at room temperature and creep resistance and prolonged life at elevated temperatures.

Still other issues regarding the tailoring of grain boundaries and amorphous intergranular films in these systems require our attention. As noted earlier, transgranular fracture occurs in silicon nitrides containing Si-O-N intergranular films.[7] However, molecular orbital calculations{ref,13} indicate the strengths of the interfaces associated with this system may exhibit a maximum at an intermediate O:N ratio and be lowest when oxygen is absent. Recent observations indicate that debonding occurs accompanied by toughening effects in self-reinforced silicon nitrides composed of large elongated grains having a highly fibrous texture.[16] This material is produced with sintering additives of 5 wt.% yttria with 2 wt.% alumina and the elongated grains exhibit a thick SiAlON epitaxial layer. In this case however, initial analysis of this material suggests that the grain boundaries between aligned prism surfaces are "special" boundaries with no amorphous film present. Furthermore, Pezzotti and Kleebe have begun to address the important issue of the effects of the stresses created by crystallized triple point phases, which cannot be neglected in the interfacial debonding process.[17] Therefore, these studies suggest that additional questions remain as to the role of the structure and composition of the grain boundary on the ability to obtain fracture resistance (and creep resistance).

ACKNOWLEDGEMENTS

The research was sponsored by the Division of Materials Sciences and Engineering , U. S. Department of Energy under contract DE-AC05-96OR22464 with Lockheed Martin Energy Research Corp.

REFERENCES

1. F. Ernst, O. Kienzle, and M. Rühle, "Structure and Composition of Grain Boundaries in Ceramics," *J. Europ. Ceram. Soc.* **19**[6-7] 665-73 (1999).
2. H.- J. Kleebe, "Structure and Chemistry of Interfaces in Si₃N₄ Ceramics Studied by Transmission Electron Microscopy,"*J. Ceram. Soc. Jpn* **105**[6] 453-75 (1997).
3. C. W. Li and J. Yamanis, Ceram. Eng. Sci. Proc.,"Super-Tough Silicon Nitride with R-Curve Behavior," 1[7-8] 632-45 (1989). T. Kawashima, H. Okamoto, H. Yamamoto, and A. Kitamura,, "Grain Size Dependence of the Fracture Toughness of Silicon Nitride Ceramics," *J. Ceram. Soc. Jpn*, **99**: 1-4, (1991). M. Mitomo, "Toughening of Silicon Nitride Ceramics by Microstructural Control," pp. 101-07 in *Proc. Science of Engineering Ceramics*, S. Kimura and K. Niihara, eds., Ceram. Soc. Jpn, Tokyo, 1991.
4. P. F. Becher, E. Y. Sun, K. P. Plucknett, K. B. Alexander, C-H Hsueh, H-T Lin, S. B. Waters, C. G. Westmoreland, E-S Kang, K. Hirao, and M. Brito, "Microstructural Design of Silicon Nitride with Improved Fracture Toughness, Part I: Effects of Grain Shape and Size," *J. Am. Ceram. Soc.*, **81**[11] 2821-30 (1998). P. F. Becher, "Microstructural Design of Toughened Ceramics," *J. Am. Ceram. Soc.*, **74**[2] 255-69 (1991).

Grain Boundary Engineering in Ceramics

5. K. Hirao, H. Imamura, K. Watari, M. E. Brito, M. Toriyama, and S. Kanzaki, "Seeded Silicon Nitride: Microstructure and Performance," pp. 469-74 in *Science of Engineering Ceramics II*, K. Niihara, T. Sekino, E. Yasuda, and T. Sasa (eds.), Ceramic Society of Japan, Tokyo (1999).

6. P. F. Becher, C. H. Hsueh, K. B. Alexander, and E. Y. Sun, "The Influence of Reinforcement Content and Diameter on the R-Curve Response in SiC Whisker-Reinforced Alumina," *J. Am. Ceram. Soc.*, 79[2] 298-304 (1996).

7. G. Pezzotti, I. Tanaka, and T. Nishida, "Intrinsic Fracture Energy of Polycrystalline Silicon Nitride," *Phil. Mag. Lttrs*, 67[2] 95-100 (1993).

8. I. Tanaka, H. Adachi, T. Nakayasu, and T. Yamada, "Local Chemical Bonding at Grain Boundaries of Si$_3$N$_4$ Ceramics," pp. 23-34 in *Ceramic Microstructure: Control at the Atomic Level*, A. P. Tomsia and A. Glaeser (eds.), Plenum Press, New York, 1998.

9. P. F. Becher, E. Y. Sun, C. H. Hsueh, K. B. Alexander, S. L. Hwang, S. B. Waters, and C. G. Westmoreland, "Debonding of Interfaces Between Beta-Silicon Nitride Whiskers and Si–Al–Y Oxynitride Glasses," *Acta Metall.*, 44[10] 3881-93 (1996).

10. E. Y. Sun, P. F. Becher, C-H Hsueh, S. B. Waters, K. P. Plucknett, K. Hirao, and M. Brito, "Microstructural Design of Silicon Nitride with Improved Fracture Toughness, Part II: Effects of Additives," *J. Am. Ceram. Soc.*, 81[11] 2831-40 (1998).

11. E. Y. Sun, P. F. Becher, C. H. Hsueh, G. S. Painter, S. B. Waters, S. L. Hwang, and M. J. Hoffmann, "Debonding Behavior Between β-Si$_3$N$_4$ Whiskers and Oxynitride Glasses With And Without An Epitaxial β-SiAlON Interfacial Layer," *Acta Mater.*, 47[9] 2777-85 (1999).

12. P. F. Becher, G. S. Painter, E. Y. Sun, C. H. Hsueh, and M. J. Lance, "The Importance of Amorphous Intergranular Films in Self-Reinforced Ceramics," submitted to *Acta Mater.*

13. G. S. Painter, P. F. Becher, and E. Y. Sun, to be published.

14. A. A. Wereszczak, M. K. Ferber, T. P. Kirkland, K. L. More, M. R. Foley, and R. L. Yeckley "Evolution of Stress Failure Resulting from High-Temperature Stress-Corrosion Cracking in a Hot Isostatically Pressed Silicon Nitride," *J. Am. Ceram. Soc.*, 78[8] 23129-40 (1995).

15. A. A. Wereszczak, T. P. Kirkland, H. T. Lin, and S. K. Lee, "High Temperature Inert Strength and Dynamic Fatigue of Candidate Silicon Nitrides for Diesel Exhaust Valves," *Ceram. Engr. and Sci. Proc.*, in press.

16. Brito, K. Watari, K. Hirao, and M. Toriyama, "Special Boundaries in Highly Anisotropic Silicon Nitride," Acta Materialia Int'l Mater. Conf. On Ceramic And Bimaterial Interfaces, Seville, Sept. 20-23, 1999, to be published.

17. G. Pezzotti and H.-J. Kleebe, "Effect of Residual Microstresses at Crystalline Multigrain Junctions on the Toughness of Silicon Nitride," *J. Europ. Ceram. Soc.*, 19[4] 451-55 (1999).

GLASS IN AND ON CERAMIC OXIDES

C. Barry Carter, N. Ravishankar and Svetlana Yanina
Dept of Chem. Eng. & Materials Science, University of Minnesota,
421 Washington Ave., SE, Minneapolis, MN 55455 USA

ABSTRACT

A brief review is presented of some recent advances in the systematic study of reactions and interactions involving glass in crystalline ceramic oxides. The systematic approach uses and extends earlier studies using bicrystals to investigate clean grain boundaries. The special features of this work are that the crystallography and morphology of the adjoining crystalline grains are then known and that the thickness and chemistry of the glass layer at the interface can be controlled. Examples are illustrated for alumina and magnesia. Observations of migrating grain boundaries and triple junctions is reported.

INTRODUCTION

Either intentionally or because of unintended impurites, glass is often present during the processing or high-temperature use of ceramic oxides.[1] At elevated temperatures, the glass will become liquid or highly viscous. There can be many consequences of this phase change. For example, a solid which contains glass at grain boundaries may lose its mechanical strength at higher temperatures. Liquid may be intentionally included during processing in order to enhance densification at lower temperatures. The liquid-phase sintering (LPS) of alumina generally involves the formation of a silicate liquid at the sintering temperature. The liquid may relocate to the surface or to internal and external interfaces. Mass transport may take place at the solid/liquid, solid/vapor and liquid/solid interfaces.[2]

A method for systematically examining the different factors involved in these processes has been developed based on the well-known bicrystal approach to the study of nominally clean grain boundaries. Several different types of studies have been made on different ceramic oxides.

The influence of the silicate phase on the evaporation from a ceramic (sapphire) substrate can be examined by systematically heating the single-layer sample.[3,4] The same examination gives new insight into how a glass liquid dewets a surface,

To the extent authorized under the laws of the United States of America, all copyright interests in this publication are the property of The American Ceramic Society. Any duplication, reproduction, or republication of this publication or any part thereof, without the express written consent of The American Ceramic Society or fee paid to the Copyright Clearance Center, is prohibited.

the role of surface faceting, and the interplay between these two phenomena.

Grain boundaries can be prepared which include a thin glassy intergranular film. These samples allow systematic studies of how glass moves in and out of grain boundaries, and how grain boundaries which contain such glass layers move.

EXPERIMENTAL

The thin glass films are prepared using pulsed-laser deposition (PLD). Films have been deposited on different single-crystals of various crystallographic orientations and on polycrystalline material. Most of the studies have used single crystals of alumina, magnesia or spinel. Examples using thin films, approximately 10 nm to 100 nm thick, of such silicates as anorthite ($CaAl_2Si_2O_8$) and celsian ($BaAl_2Si_2O_8$) will be illustrated below.

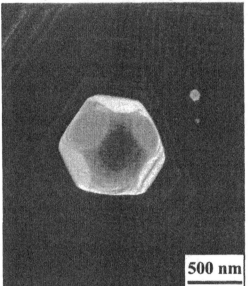

500 nm

Figure 1 An SE image from a sapphire surface containing a celsian film which has been heat treated at 1770°C for 10 min. The celsian crystallizes on the sapphire surface. The polyhedral morphology of the crystallite suggests that the hexacelsian phase has formed in this case.

The samples are heat treated at temperatures of up to 1850°C in both air and in vacuum either before, during or after processing. After sectioning the samples for viewing in the SEM, samples are coated with a thin film of Pt (~1 nm) to avoid charging in the microscope. The surface of the heat-treated samples is characterized using a field-emission scanning electron microscope (FEG-SEM) (Hitachi S900) operated at 5 kV in the secondary electron (SE) or backscattered electron (BSE) imaging modes. Other samples were examined using atomic force microscopy (AFM). Select samples were also examined by transmission electron microscopy (TEM) using samples prepared in the usual manner [5].

Tricrystals were fabricated by bonding the polished face of a bicrystal to a single-crystal substrate which had itself been coated with glass. The surface perpendicular to the bicrystal boundary was first polished with diamond lapping films using the tripod-polishing technique. This yields a T-shaped tricrystal with

Grain Boundary Engineering in Ceramics

three intersecting grain boundaries. The bicrytals and tricrystals were annealed at 1650°C for different times.

OBSERVATIONS

Surface evaporation

The presence of the silicate phase affects the evaporation from the sapphire substrate. In the example shown in Fig. 1, the celsian has crystallized and is present as polyhedral particles on the substrate. This silicate phase has crystallized in the form of hexacelsian because of the nucleation advantage when compared with the stable, monoclinic form. Evaporation is prevented from the regions beneath the particle. The steps on the surface advance due to evaporation and are 'pinned' by the particle to give a raised plateau on the surface. An SE image from a sample annealed at 1850°C for 2 h is shown in Fig. 2. The near-hexagonal region seen in the middle is the initial position of the celsian particle which has fully evaporated in this case. The evaporation from the basal surface of the sapphire takes place by the removal of atoms from the step sites causing the steps to move. When the steps encounter an obstacle, they are pinned and have to 'bow' around the obstacle[6]. A significant bending of steps around the crystallized celsian particles gives the morphology of steps that is observed in the present case.

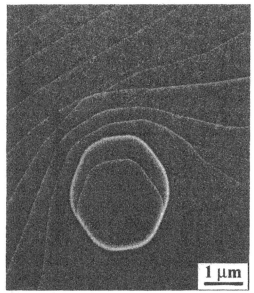

1 μm

Figure 2 An SE image from a sample annealed at 1850°C for 2 h. In this case, the celsian has fully evaporated from the surface. The near-hexagonal shape at the center indicates the position of the particle before evaporation. Steps on the sapphire surface bend around the particle.

Surface dewetting

Figure 3 is an SE image from a glass/{10$\bar{1}$0} alumina sample which has been heat treated at 1700°C for 7 h. The glass film was initially continuous but now appears as droplets distributed across the sapphire surface. The shape of the glass islands is similar to that expected for an insoluble liquid droplet in contact with a flat surface and can be used to derive values for the interfacial tensions from the familiar Young's equation.[7] There are, however, two special features: (i) the surface of the alumina was flat before the heat treatment, but is no longer so afterwards, and (ii) the islands appear D-shaped in projection.

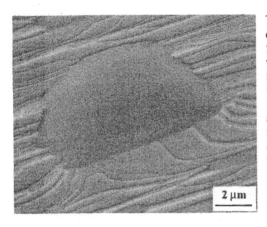

The alumina surface is faceted in essentially the same way as is found in the absence of glass or when the glass present as an intergranular film.[8-10] The surface facets parallel to two new planes: $\{10\bar{1}1\}$ and $\{\bar{1}012\}$ which intersect along a $<11\bar{2}0>$ direction. The straight lines seen in this image and in the VLM image correspond to the $<11\bar{2}0>$ direction in sapphire. The direction of the facets changes in the vicinity of the glass island. This change in direction is more pronounced on the side of the glass island which appears to be flattened.[11] The crystallography of the facets and of the adjoining surfaces can be determined using electron-beam backscattered diffraction (EBSD) patterns.

Figure 3 A secondary-electron image of a dewet anorthite droplet on the m-plane of sapphire. The morphology of the particle is affected by the facetting of the unstable sapphire surface.

Thermal treatment of the dewet droplet on MgO is shown in Fig. 4. The sample has been annealed at 1690°C for 900 s in three 300-s increments. A depression occurs where the droplet had been located. In contrast, the area of the substrate outside, but close to, the droplet is elevated above the surface level. The remnant of the silicate phase adheres to the wall and protrudes out of the depression. The shape of the depression in the cross section is asymmetric and a ridge ~30 nm in height circles the boundary of the original droplet. Narrow ridges on the inside of the well were produced when the droplet stopped moving between heat treatments. Acid treatment of the dewet droplet on the surface revealed a similar depression underneath, and a plateau around, the droplet.[12] Considering the fact that the original substrate was produced by cleaving the MgO

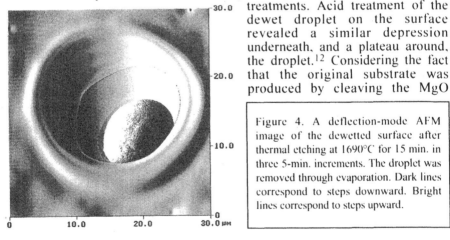

Figure 4. A deflection-mode AFM image of the dewetted surface after thermal etching at 1690°C for 15 min. in three 5-min. increments. The droplet was removed through evaporation. Dark lines correspond to steps downward. Bright lines correspond to steps upward.

Grain Boundary Engineering in Ceramics

crystal along the (001) plane, such a depression/elevation structure cannot be related topographically to the cleavage geometry. Therefore, it must have been produced in the course of annealing/quenching through transfer of substrate material (MgO) from the glass to the air/ceramic/silicate triple junction, i.e., from within the droplet to the droplet edge.

Moving grain boundaries in a bicrystal

The migration of grain boundaries in polycrystalline materials can occur under a variety of driving forces. Grain growth in a single-phase material and Ostwald ripening of a second phase are two common processes involving boundary migration. It is invariably difficult to study this process because it occurs at high temperatures and is necessarily internal to the sample. During solid-state sintering, mass transport across grains controls the densification process. During LPS, a liquid film is presumed to be present at the grain boundaries which results in an enhanced mass transport from one grain to its neighbor grain, or to more distant grains, leading to faster densification. Again, it has previously not been possible to study the mass transport across a boundary containing a liquid film. The use of bicrystals and tricrystals with glass layers in the boundary can provide a controlled geometry by which to study these phenomena.

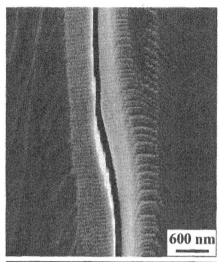

600 nm

Figure 5 An SE image showing boundary migration in a bicrystal bounded by the m-plane and the c-plane of sapphire. The original position of the boundary to the right is seen from the boundary groove.

A secondary-electron (SE) image from a bicrystal that has been annealed at 1650°C for 2 h and subsequently cooled to room temperature is shown in Fig. 5. Part of the grain boundary, which was initially straight, has migrated to the left. The initial position of the boundary is evident on the right side of the boundary because there is a remnant of the grain boundary groove. In order to investigate whether the migrated grain boundary extends along the depth of the sample, focused-ion beam thinning (FIB) was used. Figure 6 is an image acquired in the FIB from a sample which has been sectioned perpendicular to the migrated grain boundary. The depth of the section is about 6 μm. It is seen that the migrated boundary does indeed extend into the interior of the crystal.

The Tricrystal

The SE image of migration of the boundary in a tricrystal is shown in Fig. 7. The original tricrystal boundary is T-shaped. After annealing at 1650°C, the initially horizontal portion of the boundary migrates while the vertical segment of the boundary does not move. The bounding planes of the vertical segment of the boundary are both of basal orientation while the horizontal segments have different bounding planes on either side of the boundary. Thus, the difference in the bounding plane orientation can be considered to be the primary requirement for boundary migration.

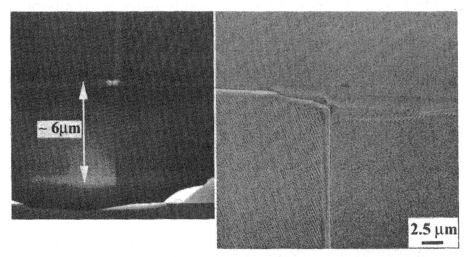

Figure 6 An FIB image of a section perpendicular to the migrated boundary. The boundary migration extends along the depth of the sample.

Figure 7 Boundary migration in a tricrystal. The horizontal segment is bounded by unlike planes and moves while the vertical segment which is bounded by basal planes does not migrate.

CONCLUSION

The use of bicrystals and tricrystals, which have been manufactured to contain glass whose initial composition is known, provide a means for systematically studying processes which occur during liquid-phase sintering or in subsequent heating of glass-containing oxide ceramics. The role of crystallography, environment and changes in glass composition can each be studied. The crystallization and wetting behavior of the liquid depend on the crystallography. The presence of a liquid phase causes grain boundaries to migrate. Such controlled experiments with bicrystals and tricrystals shed light on the mechanism of densification in the presence of a liquid phase.

ACKNOWLEDGEMENTS

This research has been supported by the U.S. Department of Energy through grant DE-FG02-92ER45465-A004. The authors would like to acknowledge Prof. Stan Erlandsen for access to the FESEM, Chris Frethem for technical assistance with the instrument and Prof. H. Schmalzried for discussions.

REFERENCES

1. German, R.M. *Sintering Theory and Practice*; John Wiley and Sons, Inc.: New York, p 226, 1996.

2. Ramamurthy, S., Carter, C.B. and Schmalzried, H., "Interaction of Silicate Liquid with Sapphire Surfaces," *Phil. Mag*, **80** (in press) (2000).

3. Ramamurthy, S., Hebert, B.C., Carter, C.B. and Schmalzried, H., "Interaction of Silicate Liquid with Sapphire Surfaces," *Mat. Res. Soc. Symp. Proc.*, **398** 295-300 (1996).

4. Ramamurthy, S. *Amorphous-Crystalline Interfaces in Oxide-Ceramics*, Ph.D., U. of Minnesota 1996.

5. Ravishankar, N., Johnson, M.T. and Carter, C.B., "Migrating Interfaces In Sapphire Bicrystals And Tricrystals," *Microscopy & Microanalysis*, **6** [suppl 2] in press (2000).

6. Morrissey, K.J. and Carter, C.B., "Surface Steps on α-Alumina Films" *Mat. Res. Soc. Symp. Proc.*, **41** 137–142 (1985).

7. Young, T., "An Essay on the Cohesion of Fluids" *Phil. Trans. Roy. Soc. London*, **95** 65-87 (1805).

8. Susnitzky, D.W. and Carter, C.B., "Structure of Alumina Grain Boundaries Prepared with and without a Thin Amorphous Intergranular Film" *J. Am. Ceram. Soc.*, **73** [8] 2485-93 (1990).

9. Mallamaci, M.P. *Interfaces Between Alumina and Silicate-Glass Films*, Ph. D. Thesis, Cornell University 1995.

10. Mallamaci, M.P. and Carter, C.B., "Faceting of the Interface between Al_2O_3 and Anorthite Glass," *Acta Materialia*, **46**[8] 2895-2907 (1997).

11. Ramamurthy, S., Hebert, B.C. and Carter, C.B., "Dewetting of Glass-Coated {1010} α-Al_2O_3 Surface," *Phil. Mag. Lett.*, **72** [5] 269-75 (1995).

12. Yanina, S.V. and Carter, C.B., "On Dewetting of Reactive Silicate Glass Films on Single-Crystal Ceramic Substrates," *Mat. Res. Soc. Symp. Proc.*, **586** (2000).

STABILIZATION OF SURFACE FILMS IN CERAMICS

Jian Luo and Yet-Ming Chiang
Department of Materials Science and Engineering
Massachusetts Institute of Technology
Cambridge, MA 02139, USA

ABSTRACT

Disordered films of 1-2 nm thickness have been observed on the surfaces of binary oxide systems. The systems Bi_2O_3-ZnO, Bi_2O_3-Fe_2O_3, WO_3-TiO_2, MoO_3-Al_2O_3 and Bi_2O_3-SiO_2 have been examined. Sufficient data now exists to interpret these films as an equilibrium state. The thermodynamics of film formation, and the criteria for selecting systems that will exhibit stable surface films, are discussed.

INTRODUCTION

Recently, a unique class of wetting behavior has been observed at grain boundaries and heterophase interfaces in ceramics, wherein a thin amorphous intergranular film is stabilized at an "equilibrium" thickness that is approximately 1 nanometer. It was initially proposed by Clarke [1] that the stability of the film was due to a balance between a long-range attractive van der Waals (dispersion) force and short-range repulsive forces. Intergranular amorphous films have also been found to form at solid-state equilibrium [2-3]. Motivated by the possibility of a free-surface counterpart to equilibrium intergranular films, we have conducted experiments in five oxide systems.

EXPERIMENTAL PROCEDURE

Doped oxide powders were fired at elevated temperatures in closed containers to achieve thermal equilibrium, and air-quenched. The powder surfaces were then characterized by high resolution electron microscopy (HREM) using a JEOL JEM 2010 microscope and a Topcon / Akashi EM-002B microscope. Compositional distributions were mapped using a Fisons / Vacuum Generators HB603 analytical scanning transmission electron microscope (STEM) equipped with an X-ray analyzer.

To the extent authorized under the laws of the United States of America, all copyright interests in this publication are the property of The American Ceramic Society. Any duplication, reproduction, or republication of this publication or any part thereof, without the express written consent of The American Ceramic Society or fee paid to the Copyright Clearance Center, is prohibited.

In single crystal experiments, a small amount of Bi_2O_3 powder was dispersed on the surfaces of ZnO crystals. The samples were then annealed and quenched. A JEOL Model 6320 scanning electron microscope (SEM) equipped with a field emission gun and an optical microscope were used to examine the crystal surfaces. Atomic force microscopy (AFM) characterization of surface facets was carried out using the tapping mode of a Digital Instruments Nanoscope IIIa scanning probe microscope. Surface composition was measured using a Physical Electronics Model 660 scanning Auger microprobe.

RESULTS

Summary of Observations

Studies of five binary oxide systems revealed the widespread existence of nanometer-thick amorphous films. These films were observed to form on the surfaces of binary oxides (Bi_2O_3 on ZnO, Bi_2O_3 on Fe_2O_3, WO_3 on TiO_2) and intermediate compounds (Al-Mo-O and Bi-Si-O), as well as for attractive (Bi_2O_3 on ZnO) and repulsive (Bi_2O_3 on Fe_2O_3, WO_3 on TiO_2) dispersion interaction systems. Typical HREM images are shown in Fig.1, and the experimental observations are summarized in Table 1.

Table 1. Average value and standard deviation of thickness for surface films observed in this study.

System	Samples	Thickness $\pm$ standard deviation
Bi_2O_3 on ZnO {11$\bar{2}$0} facets ($T_{Eutectic} = 740°C$)	Bi_2O_3-saturated, 780°C	1.54 ± 0.28 nm
	Bi_2O_3-saturated, 700°C	1.30 ± 0.31 nm
	Bi_2O_3-saturated, 650°C	1.14 ± 0.21 nm
	0.023 mole % Bi_2O_3 (unsaturated), 780°C	1.41 ± 0.29 nm
	0.023 mole % Bi_2O_3 (unsaturated), 700°C	0.92 ± 0.31 nm
	0.023 mole % Bi_2O_3 (unsaturated), 650°C	0.80 ± 0.15 nm
Bi_2O_3 on ZnO {10$\bar{1}$0} facets	saturated & unsaturated 650°C - 780°C	~ 0 (no film)
Bi_2O_3 on Fe_2O_3 ($T_{Eutectic} = 960°C$)	Bi_2O_3-saturated, 800°C	0.97 ± 0.28 nm
WO_3 on TiO_2 ($T_{Eutectic} = 1233°C$)	WO_3-saturated, 1100°C	0.66 ± 0.15 nm
MoO_3 on Al_2O_3	MoO_3-saturated, 450°C - 650°C	~ 0 (no film)
Films on Al-Mo-O	Al_2O_3-saturated, 650°C, 12 hours	~ 2-5 nm
Bi_2O_3 on glassy SiO_2	Bi_2O_3-saturated, 750°C - 800°C	~ 0 (no film)
Films on Al-Si-O	SiO_2-saturated, 750°C	~ 0 (no film)
	SiO_2-saturated, 800°C	2.06 ± 0.46 nm

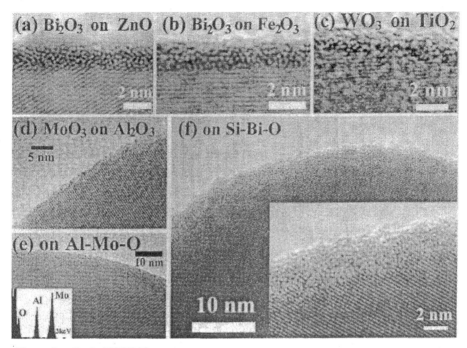

Figure 1. Typical HREM images:

(a) Bi_2O_3 on ZnO (0.023 mole % Bi_2O_3, 780°C, unsaturated).

(b) Bi_2O_3 on Fe_2O_3 (Bi_2O_3-saturated, equilibrated at 800°C, $T_{eutectic}$=960°C).

(c) WO_3 on TiO_2 (WO_3-saturated, equilibrated at 1100°C, $T_{eutectic}$=1233°C).

(d) Surface of MoO_3-saturated Al_2O_3 particle (450-650°C), showing absence of a distinguishable surface film.

(e) Surface amorphous film on an Al-Mo-O phase particle. The inset is an EDX spectrum of the particle, indicating it is an intermediate Al-Mo-O compound.

(f) Surface amorphous film on an Si-Bi-O phase particle. The inset is the same film at a higher magnification.

The largest number of experiments were conducted for Bi_2O_3 on ZnO, and the results are summarized as follows:

Bi_2O_3 on ZnO: Powder Experiments

Nanometer-thick amorphous films were observed on more than 150 particle surfaces. Eighty-five such surfaces whose orientation could be clearly characterized by electron diffraction were identified to be $\{11\bar{2}0\}$ facets (Fig. 2(a)), while no distinguishable films formed on the $\{10\bar{1}0\}$ surfaces. X-ray compositional mapping using a ~ 1 nm electron probe showed that these films are bismuth-enriched (Fig. 2(b)).

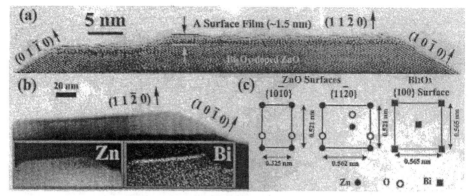

Figure 2. (a) HREM image of Bi_2O_3-saturated ZnO surface (T > $T_{eutectic}$), viewed along the [0001] zone axis. The {11$\bar{2}$0} facets exhibit films, while the {10$\bar{1}$0} facets are devoid of films.
(b) STEM bright field image, Bi and Zn maps. Surface films are Bi-enriched.
(c) Schematic of the terminating atomic planes, showing the epitaxy possible between Bi_2O_3 and the ZnO {11$\bar{2}$0} surface.

The thickness of films formed on {11$\bar{2}$0} surfaces is nearly constant along the surface (Fig. 2(a)). For example, 97 films were measured in Bi_2O_3-saturated ZnO annealed at 780°C, and this population yielded a mean value for the film thickness of 1.54 nm with a standard deviation of 0.28 nm. Furthermore, variations in equilibration time and the amount of excess Bi_2O_3 had no discernible influence on the average value of and variability in film thickness (Fig. 3). Therefore, in saturated samples these films appear to have an equilibrium thickness at a fixed temperature.

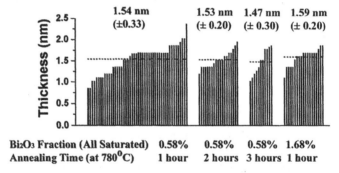

Figure 3. Film thickness vs. equilibration time and Bi_2O_3 excess for Bi_2O_3 on ZnO at 780°C. Each bar represents one measured thickness. Doping levels of 0.58 mole % and 1.68 mole % both represent saturated samples in which the chemical potentials are the same while the secondary phase fractions vary. The results show that the thickness is not supply-controlled.

Grain Boundary Engineering in Ceramics

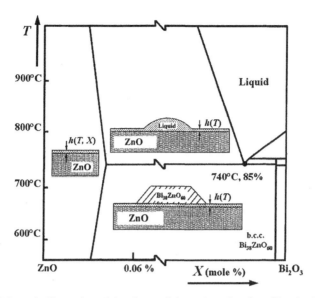

Figure 4. Schematic illustration of the observed formation of surface films in the phase fields of the ZnO - Bi$_2$O$_3$ binary phase diagram.

Films of this character were observed in Bi$_2$O$_3$-saturated samples equilibrated both above and below the bulk eutectic temperature, as well as those containing a lesser amount of Bi$_2$O$_3$ within the solubility limit. The observations are schematically illustrated in Fig. 4. The film thickness decreases slightly with annealing temperature, and is slighter lower in unsaturated samples (Fig. 5). Films equilibrated at subsolidus temperatures also appeared to be partially or completely disordered (Fig. 1(a)-(c)).

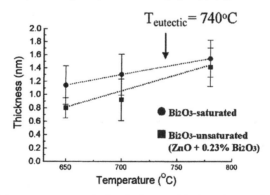

Figure 5. A slight temperature and chemical potential dependence of film thickness was evident, but no sharp discontinuity is evident at the eutectic temperature for bulk phases. Bars represent +/- one standard deviation.

Bi₂O₃ on ZnO: Single Crystal Experiments

Observations of quenched droplets showed that the bulk liquid did not completely wet the ZnO (11 $\bar{2}$ 0) surface (Fig. 6), providing additional evidence for the co-existence of bulk nonwetting liquid with a "liquid-like" surface film of an equilibrium thickness.

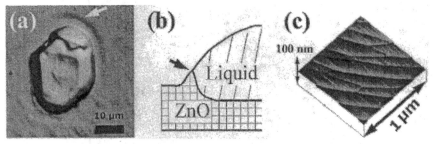

Figure 6. (a) Optical micrograph of a quenched Bi-enriched droplet on the ZnO (11 $\bar{2}$ 0) surface. The liquid did not completely wet the surface, and retracted further upon cooling, as seen by the trace of a ridge formed at the firing temperature.
(b) Schematics of the ridging configuration at the firing temperature.
(c) AFM image showing (11 $\bar{2}$ 0) faceting.

A few facetted micro-crystals of δ-Bi₂O₃ were also observed to form on the ZnO surface in both air-quenched and slowly-cooled samples. In Figure 7(a), there is a common orientation to all the micro-crystals. The crystal directions (labeled) clearly indicate an epitaxial relation between the precipitate and the ZnO substrate.

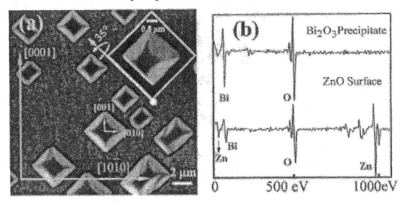

Figure 7. (a) Field emission SEM image of oriented δ-Bi₂O₃ precipitates on the ZnO (11 $\bar{2}$ 0) surface. The facets on the microcrystals are of {111} orientation. Inset shows the {111} facet after tilting 35° around the <110> axis.
(b) Auger spectrum from typical facetted micro-crystals and the ZnO surface.

These surface films are Bi₂O₃-enriched (Fig. 2(b)), but Auger analysis (Fig.7(b)) showed that they not are Bi₂O₃-rich on an absolute basis. The average measured molar percentages of Bi_2O_3 in the film is 18%, being distinctly Zn-rich compared to the near-eutectic equilibrium bulk liquid ($\sim$83 mole % Bi_2O_3).

MODELING AND INTERPRETATION

The surface amorphous films observed in this study should not be viewed simply as a confined bulk liquid slightly modified in composition and structure by the adjacent crystal. The films are true surface phases. In the two-component systems studied, the surface films have been observed to co-exist with one or two bulk phases, being thermodynamically allowed due to the additional degree of freedom provided by having a surface. Correspondingly the film thickness becomes "thermodynamically-determined," and the composition and structure of the "liquid-like" surface films are distinct from that of the bulk liquid phase.

A thermodynamic model can be found in reference 4 and 5. The essential physical concepts of this model are briefly described as follows:

Equilibrium-Thickness and Anisotropy

The existence of an "equilibrium" thickness firstly suggests that the film formation lowers the surface energy, and secondly requires an attractive long-range interaction preventing the film from thickening infinitely. Specifically for Bi_2O_3 on ZnO, the dispersion interaction is attractive (the equivalent pressure is $\sim$2.5 MPa for a 1.5 nm thick film). The polycrystalline ZnO surface energy ($\gamma_c \cong 700$ mJ$\cdot$m^{-2}) is higher than that of the eutectic liquid ($\gamma_l \cong 200$ mJ$\cdot$m^{-2}) [6]. Therefore, if the crystal-liquid interface energy, γ_{cl}, is less than 500 mJ$\cdot$m^{-2}, a short-range repulsion is implied and the surface film can reach an equilibrium thickness in the same manner as an intergranular liquid film.

Preferential film formation at the $\{11\bar{2}0\}$ facets of ZnO can be understood as the existence of a low interfacial energy, γ_{cl}, due to epitaxy or ordering of the bismuth-enriched film at this surface. Figure 2(c) shows that a good dimensional match exists between the ZnO $\{11\bar{2}0\}$ surface and the δ-Bi_2O_3 $\{100\}$ plane. Observation of oriented growth of δ-Bi_2O_3 on the $\{11\bar{2}0\}$ facets (Fig.7(a)) supported the existence of epitaxy.

Stability of Subsolidus Amorphous Films

Analogous to the subsolidus intergranular films [3] and surface melting of pure solids [7], surface amorphous (liquid) films are stable below the solidus temperature if the increased volume energy for amorphization ($\Delta G_{amorph}\cdot h$, h being the film thickness) is more than compensated by reduction of the surface energy ($\Delta \gamma \cong \gamma_c$-($\gamma_{cl}$+$\gamma_l$)). Specifically for Bi_2O_3 on ZnO, the requirement is:

$$\Delta G_{amorph}\cdot h \; < \; \Delta \gamma = 500 mJ \cdot m^{-2} - \gamma_{cl} \cdot \qquad (1)$$

Neglecting the mixing energies for simplicity, the amorphization term can be estimated from the fusion entropies (ΔS_{fusion}) of film-forming oxides:

$$\Delta G_{amorph} = \Delta S_{fusion} \cdot \Delta T \cdot \qquad\qquad (2)$$

The energies to form a 1nm amorphous film at an undercooling of $\Delta T = 100°K$ range from 29 mJ·m^{-2} for Bi$_2$O$_3$ to 167 mJ·m^{-2} for ZnO, easily being more than compensated by $\Delta \gamma$ for an epitaxial low energy interface (Eq. 1).

Based on the magnitudes of individual terms, we proposed that dispersion forces are relatively unimportant well below the solidus where the amorphization energy is a dominant consideration preventing unconstrained film thickening [5]. The existence of subsolidus films for both attractive and repulsive dispersion force systems supports this point of view. The observed film thicknesses decrease in the order h(Bi$_2$O$_3$ on ZnO, Fe$_2$O$_3$) > h(WO$_3$ on TiO$_2$) > h(MoO$_3$ on Al$_2$O$_3$), scaling with increasing free energy of fusion for the film-forming additive, further supporting this model.

CONCLUSIONS

Experimental observations have been made of surface amorphous films in several binary oxide systems. A model for the stability of such films has been proposed and examined. The existence of stable surface films always implies the film formation lowers the surface energy. Near and above the solidus temperature, the presence of attractive dispersion interactions are critical for the film to have an equilibrium thickness. Well below the solidus temperature, however, the volume free energy of forming the disordered film is believed to be a dominant consideration, and oxides with low entropies of fusion are likely to be constituents of energetically stable surface amorphous films.

Acknowledgements – Funding from U.S. Department of Energy, Office of Basic Energy Sciences, grant No. DE-FG02-87ER45307; Shared Experimental Facilities at MIT supported by NSF; J. –R. Lee for providing the Bi$_2$O$_3$ – ZnO powders; R.H. French for assistance in calculating Hamaker constants; M. Frongillo, A. J. Garratt-Reed, E. L. Shaw, and A. N. Soukhojak for experimental assistance.

REFERENCES

1. D. R. Clarke, "On the Equilibrium Thickness of Intergranular Glass Phases In Ceramic Materials," *J. Am. Ceram. Soc.*, **70** [1], 15-22 (1987).
2. H. D. Ackler and Y. -M. Chiang, "Model Experiment on Thermodynamic Stability of Retained Intergranular Amorphous Films," *J. Am. Ceram. Soc.*, **80** [7] 1893-96 (1997).
3. H. Wang and Y. -M. Chiang, "Thermodynamic Stability of Intergranular Amorphous Films in Bi-Doped ZnO," *J. Am. Ceram. Soc.*, **81** [1], 89-96 (1998).
4. J. Luo, and Y. –M. Chiang, "Existence and Stability of Nanometer-Thick Amorphous Films on Oxide Surfaces," *Acta Materialia*, in press (2000).
5. J. Luo, and Y. -M. Chiang, ""Equilibrium-thickness Amorphous Films on {1 1 $\bar{2}$ 0} Surfaces of Bi$_2$O$_3$-Doped ZnO," *Journal of European Ceramic Society*, **19**, 997-701 (1999).
6. J. -G. Li, "Some Observations on Wetting in Bi$_2$O$_3$ – ZnO System," *J. Mater. Sci. Eng. Lett.*, **13**, 400-403 (1994).
7. J. G. Dash, "Surface Melting," *Contemporary Physics*, **30** [2] 89-100 (1989).

Adsorption and Wetting Mechanisms at Ceramic Grain Boundaries

R. M. Cannon,* M. Rühle,[#] M. J. Hoffmann,[@] R. H. French,[+] H. Gu,[&] A. P. Tomsia* and E. Saiz*

*MSD, Lawrence Berkeley National Laboratory, Berkeley CA 94720, USA
[#]Max-Planck-Institut für Metallforschung, Seestrasse 92, D-70174 Stuttgart, Germany
[@]Institut für Keramik im Maschinenbau, Univ. Karlsruhe, D-76131 Karlsruhe, GE
[+]CR&D, E.I. du Pont de Nemours & Co., Wilmington, DE 19880, USA
[&]Shanghai Institute of Ceramics, Shanghai, 200050, China

ABSTRACT

Various models of impurity segregation at grain boundaries are discussed. Also, some general differences are identified for effects of partial wetting plus adsorption versus complete wetting on key aspects of microstructures, including the impurity partition, for polycrystals containing a liquid phase. Behavior for some oxide and silicon-based ceramics and selected metals is compared. For some, evidence exists that submonolayer adsorption occurs. However, in other ceramics, nanometer thick, disordered layers are seen via TEM. The latter have been attributed to wetting by a liquid phase, but this is overly simple and can lead to erroneous expectations. For example, in a growing list of instances, compositions of boundary films are found to differ from those of the second phase with which they coexist. Thus, considering these in terms of multilayer adsorption or prewetting layers is useful, even if models are only qualitatively applicable to ceramic materials.

INTRODUCTION

The properties of ceramics depend upon the characteristics of the internal interfaces and any associated intergranular phases plus the morphologies of the bulk phases. In recent decades, the effects of grain boundary impurity adsorption on microstructural evolution or on properties have typically been conceived in terms of (sub)monolayer levels, e.g., for MgO inhibiting grain growth [1 and refs. therein] or Y_2O_3 inhibiting creep [2,3] in Al_2O_3. Often, however, liquid-phase sintering is used and the resultant presence of intergranular films and the specific impurities in them profoundly alter macroscopic properties, often beneficially. For example, intergranular films strongly influence mechanical properties, e.g., degrading hot strength [4-6] whilst enhancing fracture toughness [7-11] of Si_3N_4 or SiC based materials, or the electrical properties in thick-film resistors [12], varistors [13,14], or high T_c superconductors [15]. Recently, it has been

To the extent authorized under the laws of the United States of America, all copyright interests in this publication are the property of The American Ceramic Society. Any duplication, reproduction, or republication of this publication or any part thereof, without the express written consent of The American Ceramic Society or fee paid to the Copyright Clearance Center, is prohibited.

shown that nanometer thick, amorphous grain boundaries form in Bi_2O_3 doped ZnO even when sintered below the eutectic temperature; such structures may be generally associated with the enhanced sintering rates often imparted to materials in the solid state regime by additives that specifically also form stable liquids, i.e., cause deep eutectics [16]. It must be realized that the actual composition and bonding of the interfacial films can differ from those of any associated bulk glass phase [12,17-22,11]. Providing a context to understand these films is a goal of this paper.

Originally thin intergranular films (IGFs) were held to be kinetically limited remnants of processing. However, such films exist after extensive grain growth which entails diffusional reconstruction of the entire microstructure and should, thereby, yield near-equilibrium interfaces. In Si_3N_4-based materials, 1-2 nm thick, amorphous oxynitride films usually form between particles during liquid phase sintering [23,24]. Extensive study of the Si_3N_4 system has shown that the boundary thickness is nearly constant within a given material (despite processing history) and is invariant over a wide range of liquid fraction, although it adjusts depending upon the exact liquid composition [25-28,5] implying that the thickness reflects an equilibrium condition [29,30]. However, in ZnO/Bi_2O_3 under normal sintering conditions, the capillary forces (that drive densification) do reversibly influence the grain boundary composition that, therefore, derives from a balance among nonlocal plus local forces acting across a boundary [21,31]. Thus, the films at room temperature often represent a high temperature, equilibrium configuration only altered by cooling. Presumably the hot films include the same excess elements per unit area found at room temperature as diffusion along the boundary into triple junctions may be negligible during cooling, especially in siliceous films [18]. However, it is very possible that before cooling, the films were less ordered and even wider owing to incorporation of material from the adjoining grains.

THEORY

Impurity effects on interfaces in ceramics have often been viewed either in terms of segregation/adsorption mechanisms that involve monolayer levels or less, or else as complete wetting by a liquid often leaving retained glassy layers. Although often not appreciated, mounting evidence reveals that a wide spectrum of interesting behavior exists between these extremes. A rich literature exists concerning adsorption, multilayer adsorption and prewetting films for vapor species attaching to solid surfaces, with many variations actually seen [32] and perhaps more predicted by theory [33-35]. For internal interfaces, adding layers of adsorbed material requires separating condensed phases from both sides of the interface, and so larger disjoining forces are needed. Consequently, we first describe adsorption theories and then consider the stability of thin liquid-like films in terms of a force balance across them. Finally, implications regarding the impurity partition and how the morphology of liquid pockets and solid grains depend upon the degree of wetting for internal interfaces are outlined.

Adsorption Theories

The Langmuir-McLean (L-M) model assumes that a fixed density of interface sites exists for impurity adsorption, and that a single adsorption energy, G^{ad}, pertains at all coverages [32,36]. The adsorption statistics relate the fractional occupancy of the special sites, θ, and X, the atom fraction of impurity in solution. When $-G^{ad}/kT$ is large, adsorption obtains at dilute solution, and the Gibbsian adsorption level, Γ, the number of excess impurity ions per unit area and the areal density of surface sites, Γ°, relate directly as $\theta = \Gamma/\Gamma^{\circ}$, so that:

$$\Gamma/(\Gamma^{\circ} - \Gamma) \approx XK \approx aK', \quad X \ll 1 \qquad (1)$$

where $K = \exp\{-G^{ad}/kT\}$ and a is the activity of the impurity.

It is important whether or not adsorption becomes significant at solute levels below the solubility limit. This attribute, dictated by the relative sizes of the energies for precipitation and for adsorption, can be described in terms of the activity a^*, or concentration, X^*, at the solubility limit versus that for adsorption, taken at $\theta = 0.5$ and denoted $a_{0.5}$. The choice to characterize the activity for adsorption as that to give $\theta = 0.5$ arises as adsorption typically does not occur abruptly at a certain activity, but may increase gradually, Eq. (1), in contrast with a precipitation reaction.

Although simple, the L-M model provides a basis to visualize more realistic ones. In this spirit, the adsorption at internal interfaces in solids has often been contemplated as relaxing the atom size misfit [36,37 and refs therein]. This model can be generalized to allow interactions among adsorbate atoms. The Fowler model invokes a mean field approximation to obtain an analytical model wherein $G^{ad} = G^{\circ} + Z\theta U$, with Z, the coordination number among adsorbate atoms and U, an adsorbate-adsorbate interaction energy [38]. A qualitative change ensues when $-ZU/kT > 4$; an abrupt jump in Γ occurs at a critical activity as for a first order transition. For ionic materials, a space charge region also evolves as a secondary response. Primary adsorption of charged species (point defect or solute) causes electrostatic interactions that can be compensated by a nearby double layer region in which adsorption or suppression of other charged species happen in response [37,40-42]. This leads to more complex temperature dependencies because the induced electrostatic fields must counterbalance the tendencies for preferential adsorption of any species and the need for the remote lattice to be neutral despite the presence of aliovalent impurities.

Multilayer adsorption of gaseous species on solids, including H_2O on silicates, occurs widely (often initiating at $\sim a^*/4$) and many phenomenological models for it have emerged [32], e.g., as layer by layer extensions of the Langmuir model [39]. Subsequent, more realistic lattice gas type models reveal additional types of behavior [33-35]. Gross trends depend upon whether the adsorbate-substrate interaction is strong or intermediate relative to the adsorbate-adsorbate interaction. For either, adsorption initiates at adsorbate chemical potentials, μ, below that for condensation, μ^*, in some temperature range, with the extent increasing as the vapor pressure rises to that for condensation (coexistence) [33]. Often, extensive multilayer adsorption ensues above a fraction of the critical temperature. If the fluid is fully wetting, $\varphi = 0$, then the adsorption level increases without bound as $a \rightarrow a^*$; in contrast, Γ remains finite if discrete droplets form with a finite layer between them, meaning $\varphi > 0$, Fig. 1a (with φ, the contact angle). Wetting often occurs only for $T > T_w$. A rise in adsorption level, Γ, may occur with and promote interfacial roughening transitions [33]. For strong substrate interactions, adsorption is extensive at lower temperature [33], as for the classical Langmuir behavior. Extensions to fluid/fluid systems allows intermixing of substrate atoms into the film to be treated readily; results are found to depend upon long range London dispersion forces [35]. For two condensed fluids, at least one will wet the other above a critical wetting temperature, T_w; even in the single liquid phase region of the other, a thick, but graded surface region containing the component that imparts low surface energy will often be stable [43]. The excess energy of that entire layer is essentially the surface tension, which was shown to obey the Gibbs adsorption isotherm.

Although very idealized, a lattice gas model for a grain boundary exhibits both impurity adsorption at low temperature (albeit in highly supersaturated material) and increasing adsorption above a grain boundary disorder temperature with complete wetting achieved upon approaching the melting point [44,45], as depicted in Fig. 1b. More generally with solid substrates, epitaxy can enhance the initial adsorption energy; however, the associated misfit strain can limit the thickness of solid adsorbate layers prior to droplet or island formation as observed by surface analysis [46,47]. Analogous effects may actually pertain for liquid-like layers, as inferred from recent molecular dynamics simulations that show epitaxial transitions for IGFs in ceramics [48,49].

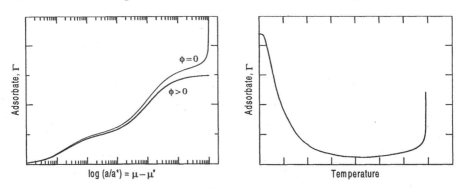

Fig. 1. *Schematic showing plausible increases in adsorbate level with adsorbate activity, with examples of Γ being finite ($\varphi > 0$) and infinite ($\varphi = 0$) at the coexistence condition, a^*, in (a) and variation of excess adsorbate versus temperature from lattice gas, grain boundary model (with constant solute level) after Kikuchi & Cahn [44] in (b).*

Force Balance

Alternately, to explain the equilibrium, liquid-like films at interfaces in ceramics, Clarke et al. invoked a force balance, similar to those used to describe low temperature colloidal interactions [29,30]. This can be contemplated as the interaction between two large, flat solid surfaces across a layer of liquid expressed in terms of local attractions and repulsions plus any applied force, F_{ap}, induced by capillary and external loads:

$$F = F_a + F_r - F_{ap} \tag{2}$$

The signs are such that tension, $F_{ap} > 0$, must be applied to separate attractive solids.

The London dispersion force was regarded to be the dominant attraction [29,30]. It can be expressed as [50]:

$$F_a = H/(24\pi h^3) \tag{3}$$

where h is the layer thickness, and F_a scales with the Hamaker constant, H. In principle, H can be computed from the dielectric functions of the associated materials, if the inner layer and outer solids are homogenous of known composition [51-54]. If the two solids are the same, the force is attractive and tends to increase with the difference in refractive indices, n_o, of the solid and liquid, i.e., $H \propto (\Delta n_o)^2$ [50].

Clarke first proposed that the repulsion derives from a steric force in the rather polymeric silicate or similar liquids [29], which was modeled using diffuse interface theory [55]. It was also realized that for partially ionic systems, some charged species from the liquid could preferentially adsorb onto the solid faces, inducing an electrostatic repulsion [30]. Each of these forces can be approximated as [56]:

$$F_r = -A\exp(-h/\xi) \tag{4}$$

The correlation length ξ should scale with a molecular bond or chain length for a steric repulsion or with the Debye length for electrostatic repulsion. The magnitude, A, is hard to quantify. For a steric repulsion, it may scale with the difference in degree of order or epitaxy induced in the normally disordered molecules by virtue of bordering each of the solids [29,57]. An electrostatic repulsion increases with the adsorbed surface charge. With $H/(24\pi A\xi^3)$ below a critical value, an energy minimum ($F_a + F_r = 0$) exists at a finite separation, $h_e \sim (3\text{-}20)\xi$; otherwise the two solids are attractive at all h.

This formulation describes trends, but needs expansion. For the thin disordered layers of interest, 0.5-2 nm, the residual Coulombic attraction may not be negligible versus dispersion forces, and to describe an actual minimum energy with liquid excluded, a Born type (orbital overlap) repulsion is needed. More critical are modifications that treat how much of the atomic species from the liquid would remain during a forced approach of the two grains. For the present theories [29,30,57-59], once a film has been compressed below a critical thickness, it is unfavorable to have any ordered liquid adjoining the solid faces; all of it would be abruptly expelled, an unlikely situation for most real grain boundaries. An explicit energy is needed for the last desorption, or equivalently for the initial adsorption expected at a clean interface, which could be potent enough to cause spontaneous separation of a clean boundary to admit liquid. Using diffuse interface theories, several order parameters and gradient terms will be needed to account for epitaxy and gradients in orientation, composition and structure [60].

Although this argument focused on interactions across a uniform gap, the forces given by Eqs. (3 and 4) can be modified to describe the interactions between spherical particles [50,56]. More generally, it is also necessary to anticipate changes in particle shape that would ensue if kinetics permit it.

Junction Condition

Situations of particles dispersed in liquid or grain arrays with intergranular liquid can be described in terms of interfacial thermodynamics. The excess free energy associated with all the interfaces is:

$$G^x = \Sigma(\gamma_{gb}A_{gb} + \gamma_{sl}A_{sl}) \tag{5}$$

where γ_{sl} is the solid-liquid surface tension, and γ_{gb}, the grain boundary tension (for solid/solid contacts). Minimizing this locally, in the isotropic limit, requires that the dihedral angle, ψ, for liquid at a liquid/grain boundary junction, Fig. 2, is [61]:

$$2\gamma_{sl} \cos(\psi/2) = \gamma_{gb} \tag{6}$$

This applies for a boundary that is fully equilibrated with the liquid regarding adsorbate, which may markedly lower the interfacial energy from that of the pure boundary, $\gamma_{gb}°$.

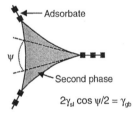

Fig. 2. Impurity atoms in a solid/liquid system are partitioned between being adsorbate at interfaces and residing within the second phase at triple or quadra junctions. The equilibrium wetting behavior and ψ are dictated by the energy of the boundary having equilibrium adsorption at an activity appropriate to the two phase coexistence condition, i.e., at $a = a^*$.

If any attraction exists between very widely spaced solids leading to an equilibrium spacing, h_e, the excess energy G^x, per unit area of the two parallel interfaces and the intervening liquid layer, is less than $2\gamma_{sl}$ by an interaction energy that obtains from integrating the intermediate range interaction, Eqs. (3 and 4) including other relevant terms; that is $\Delta\gamma = \int(F_a + F_r)dh$. Thus, the excess energy of the entire disordered region between the two solids is the grain boundary energy $\gamma_{gb} = 2\gamma_{sl} + \Delta\gamma$ [62], and:

Grain Boundary Engineering in Ceramics

$$\cos(\psi/2) = 1 + \Delta\gamma/2\gamma_{sl} \qquad (7)$$

So $\psi > 0$ for any attractive interaction, any $\Delta\gamma < 0$. Only the fully repulsive case coincides with complete *wetting* of a boundary by a liquid, i.e., $\psi \rightarrow 0$ [62]. Note, this can only obtain from Eqs. (2-4) if $H \rightarrow 0$ [56]. For incomplete wetting, the extra solute atoms in the film constitute Gibbsian adsorbate, Γ. This may be essentially multilayer adsorption, as has been quantified in cases exhibiting nanometer thick, equilibrium films of silicate: between Si_3N_4 grains, where $\Gamma \approx 30$ nm^{-2} of oxygen [17,18] or Pb-ruthenate grains, where $\Gamma \approx 20$ nm^{-2} of silicon [12]. For these situations having a thick but not infinitely thick disordered layer at equilibrium in the presence of bulk liquid, the interfaces have been denoted as *moist* [63,62]; the liquid is not fully wetting. It must be anticipated that such interfaces have gradients across them and nowhere exactly match the bulk liquid in either composition or structure. The term *dry* then pertains when interfaces are clean or have submonolayer adsorption [63,62].

Network Morphologies

Insights regarding the partition of impurity, between the boundaries and the bulk liquid phase, as well as the morphology of grains and liquid pockets can derive from quasi-equilibrium network models. In excess liquid, if any interparticle attraction occurs, some type of particle network should be (meta)stable. This would entail adjustments in particle shapes, although kinetics may limit the rates. In fact, a direct parallel exists between sintering of a particle compact in a vapor phase and the evolution driven by colloidal interactions of a particle network submersed in excess liquid [62]. Then, it follows that the arrays may evolve toward the quasi-equilibrated microstructures for polycrystals containing a second phase, described by Smith [61] and others [64,65], that depend sensitively on ψ. We consider a periodic model of grains with intergranular liquid, and then distinguish whether or not excess liquid is present.

The driving force to densify (consolidate) a particle network and exclude liquid can be quantified based on minimizing the total interfacial energy, G^x, Eq. (5), as described in [66,67]. This force can be expressed in terms of the difference in the average chemical potentials for solid atoms located at the surface of liquid pockets, μ_p, and at grain boundaries, i.e., the contact flats, μ_{gb}. Assuming each pocket has maintained the equilibrium dihedral angle, this is:

$$\mu_p - \mu_{gb} = -2\gamma_{gb}\Omega/d - \kappa_p\gamma_{sl}\Omega - (p_{ex} - p_{in})\Omega \qquad (8)$$

where d is the particle size, Ω, the atom volume and γ_{gb} satisfies Eq. (7). The last term describes the effects of a pressure in the surrounding fluid acting on the array, p_{ex}, which may differ from that in the fluid in the pockets, p_{in}. The first term always favors densification, i.e., reduction in internal liquid volume and increase in grain boundary area; it effectively arises from the interparticle attractions. The curvature of a pocket wall, $\kappa_p(N, \psi)$, is defined as positive if the radius of curvature lies within the pocket and is negative if the opposite, the case depicted in Fig. 2. It can also be visualized from this sketch that the smaller a pocket volume, v_p, the higher $|\kappa_p|$, and that lower ψ gives greater $|\kappa_p|$. The curvature of pocket surfaces depends upon the grain coordination number, N, around the pocket and ψ as elaborated by Kingery and Francois [68]. For a low angle, κ_p is negative, which, in turn, provides a driving force for a pocket to actually grow. Under certain conditions a metastable equilibrium situation would develop with $\mu_p = \mu_{gb}$ and the pockets having constant curvature shapes [61,64,65].

Even with a fully submersed particle array and no liquid/vapor interfaces to pull particles together, the intermediate range particle attractions (acting across any film) provide a force to densify the solid network. When ψ is high enough that $\kappa_p > 0$, the first two terms in Eq. (8) both drive densification. Then, liquid may become entrapped within isolated pockets. Such liquid would be under hydrostatic pressure and so have a driving force to diffuse completely out into the liquid reservoir, if possible. In contrast, when ψ is small, or N is large, leading to negative κ_p, a specific pore size does exist for which the first two terms in Eq. (8) balance, leading to a stable equilibrium pocket size and shape for which

$$(v_p)^{1/3}/d \propto -1/\kappa_p d = 1/[4\cos(\psi/2)] - (p_{ex} - p_{in})/2\gamma_{gb}(-\kappa_p) \qquad (9)$$

A smaller pocket should grow; a larger one should shrink. It can be shown that for any $\psi > 0$, some grain boundary formation, i.e., contact flattening, is energetically favorable, but the amount is less the lower ψ and the higher N [66]. If $60° > \psi > 0$, two grain junctions should form via neck growth at interparticle contacts, but the liquid should wet along all triple lines, allowing $p_{ex} = p_{in}$. With $\psi > 60°$, the liquid should be isolated into pockets at particle intersticies, for which $p_{ex} \neq p_{in}$ without (slow) exchange with the liquid reservoir. The four grain junctions will have stable liquid for $\psi < 109°$.

Thus, for a narrow particle size distribution, a network with a stable liquid content, $v_f{}^*$, that simply depends upon the value of ψ may evolve, as $\kappa_p d = f(\psi)$ dictates this volume fraction if $p_{ex} = p_{in}$. Such periodic structures have been computed for the monodispersed grain size limit [64,65]. More generally, a grain array with a narrow size distribution and isotropic interface energies may evolve into a network similar to those for soap froths [61]. Such a structures may be susceptible to coarsening, but may remain morphologically stable if the kinetics favor normal grain growth. However, complete disruption may result from nonlinear, anisotropic growth kinetics [62].

When only a small amount of liquid is present, further capillary forces increase the size of the contact flat and also alter h (via F_{ap}) and thereby γ_{gb}. A driving force to increase the density of a solid/liquid compact arises from any extant liquid/vapor meniscus [70]. With all the internal porosity eliminated, the meniscus at the edge of the compact dictates the pressure within the (adjoining) liquid, i.e., $(p_{ex} - p_{in}) = -\kappa_l \gamma_{lv}$ [69]. If the liquid is reentrant between particles, it will be under tension (if $p_{ex} = 0$). For smaller levels of liquid, $v_f < v_f^*$, yet with a low enough dihedral angle for the liquid to be interconnected, it will all experience internal tension ($-\kappa_l \gamma_{lv}$), which will be balanced by a compression acting across each grain boundary, even if a film. This will induce both the area of the contact flat to increase and any film to thin, yielding $h < h_e$. The net effect on the contact area follows from Eqs. (9) (although there is another small term that depends on the size of the piece [66] and a further adjustment if Ω differs for either the atoms in the solid or bulk liquid and the film [31]). For isolated pockets, the sign and magnitude of p_{in} depend on $\kappa_p(N, \psi)$ with the magnitude rising as v_f decreases. However, as a film thickness changes, γ_{gb} will change, as dictated by the Gibbs adsorption isotherm. In particular, if a film thins, then γ_{gb} will increase leading to a reduction in ψ and then to an adjustment in curvature of the internal pockets.

The force balance model, Eqs. (2-4), pertains obviously at coexistence, where material can exchange readily between a film and a plentiful bulk liquid. Although if the compositions of the film or of material being exchanged differ from that of the liquid, refinements are needed. For situations in polycrystalline arrays with the activity of liquid forming materials below standard coexistence conditions, a^*, films can still be stable, albeit with $h < h_e$. The material exchange to adjust a film can be envisioned between pockets stabilized by negative curvature or with the lattice at lower activities. Then, the concentrations of films and even pockets may depend the upon pocket sizes, as illustrated in a diffuse interface model by Bobeth et al. [59]. Any liquid at triple and quadra grain junctions that is stabilized by $\kappa_p < 0$, with $\mu < \mu^*$ [71], formally corresponds to capillary condensation.

The prior discussions emphasize that for particles dispersed in excess liquid, the condition of complete wetting of all grain boundaries ($\psi \to 0$) corresponds to complete deflocculation in a colloidal sense, i.e., fully repulsive particles at all h. Full wetting means that in the presence of excess liquid, particles may exhibit their equilibrium shapes without contact flats. In contrast, with attractive particles, $\psi > 0$ and necks form at particle contacts in response. This favors some contact flattening, whether the resultant grain boundaries are moist or dry. Realizing this yields a basis to interpret experiments that involve immersing the solid phase (either powder or initially dense polycrystal) in a massive amount of liquid at high temperature [56,72,63,62]. These can probe particle interactions and roles of interfacial composition as systems evolve without capillary forces, induced by liquid/vapor menisci, that can cause contact flats even if $\psi = 0$.

ASSESSMENT

Grain boundary segregation can sometimes be considered in terms of boundaries being well ordered to within an atom spacing of the core and the adsorbate atoms located in specific sites that typically exist having a more propitious size or bonding environment. Such atomic structures can be found using high resolution transmission electron microscopy (TEM) for simple tilt boundaries, e.g., in [74-76], although rarely has an adsorbate level been quantified and related directly to the boundary structure in a ceramic. Nonetheless, adsorption at ceramic boundaries has been widely seen [e.g., 37,77,2,3 and refs. therein], sometimes associated with ion size [78] or adsorption in space charge regions [79,80], often in boundaries giving sharp TEM images, and often correlated with properties. Only more recently were actual adsorption levels quantified and shown to be submonolayer in some situations [80]; this specific study also found that an isoelectric condition existed and that the adsorbed species associated with the space charge changed on either side of this condition. In another instance, quantification revealed levels that appear to actually exceed a monolayer, even with additions just below the solubility limit to form bulk liquid phase [81].

Evidence about thicker, disordered interfaces from TEM or spectroscopy can be related to microstructural attributes and trends from theory. Microstructural features more obviously reflect equilibrium when they are consistent in several situations, namely, normal polycrystals, colloidal samples having particles dilutely suspended in liquid, and dense polycrystals exposed to excess liquid which can penetrate the grain boundaries. (Key attributes, including grain shapes, are presumably retained on cooling especially with silicate based liquids.) Disruptive penetration has been seen for several oxides [72,56,73,82,83] and Si_3N_4 clusters and polycrystals [62]. The occurrences of extensive deflocculation would seem to imply full wetting ($\psi \to 0$) inconsistent with the formation of equilibrium boundary films. Instead these disruptions may derive from transient, nonequilibrium wetting, perhaps analogous to that on free surfaces [84,85], or other nonequilibrium factors [62] and require cautious interpretation. In support of this, the degree of disruption was greater with Bi_2O_3 penetrating ZnO when the liquid was not first chemically equilibrated with the compact [73].

Table I. Computed Hamaker Constants with Various Interlayers

Solid	n_o	SiO_2 Film H, zJ	Pb-Al-SiO_x Film, H, zJ	Specific Film	H, zJ
MgO	1.70	8 (1)	3 (1)		
Al_2O_3	1.75	19 (1)	10 (1)		
ZnO	1.94	9 (2)	3 (2)	Bi_2O_3-ZnO eut.	4 (3)
c-ZrO_2+Y_2O_3	2.04-1.92	27-23 (1)	15-13 (1)		
Si_3N_4	2.02	33 (4)	21 (4)	Y-Al-Si-O-N	8 (4)
$SrTiO_3$	2.3	~45 (e)	~30 (e)		
SiC	~2.4	~60 (e)	~50 (e)		
TiO_2	2.56	40 (1)	26 (1)	10TiO_2-90SiO_2	25 (5)
Si	3.44	80 (1)	63 (1)		
$Pb_2Ru_2O_7$	3.56	28 (1)	31 (1)		

1, ref. [51]; 2, ref. [54]; 3 ref. [21]; 4, ref. [52]; 5, ref. [57]; e, estimated
zJ, zeptoJoule = 10^{-21} J

The dispersion forces may be the best understood component of the force balance envisaged for IGFs, Eqs. (1-3). The Hamaker constant has been computed for several pure solids with various intervening films, using the full spectral method [51]. Examples in Table I are listed for a set of solids ranked in terms of refractive index, n_o, increasing from 1.7 to 3.6; these compare H for glass films of SiO_2 or of Pb-Al-silicate, having n_o of 1.50 and 1.62, respectively, less than for all the solids. A few more results are based on measurements of the optical properties for glasses with compositions more similar to those in equilibrium with the solid. The computed forces do tend to be lower for glasses nearer in composition to that of the solid, but do not strictly follow the approximation $H \propto (\Delta n_o)^2$. A complication is that evidence is mounting that the compositions of films are rarely the same as those of the bulk glass in the system [17-21,12,11], and there may be gradients across films requiring refinements in theory.

Based simply upon trends in dispersion forces, Table I, it has been expected that multilayer IGFs derived from silicate liquids should be stable and thicker for grains of low index oxides such as MgO or Al_2O_3 and absent for high index materials, e.g., Si [29]. This trend is, however, only roughly obeyed owing, in part, to variations caused by anisotropy, constituents from the solid dissolved into the films, and perhaps temperature. No amorphous boundary film could be found in Si containing SiO_2, the example with the highest predicted dispersion force, even in polycrystals with different histories [86]. Silicate based films have been found or inferred in most other materials in Table I including those with relatively high index and computed H including several SiC materials containing oxygen [10,87-89]. For TiO_2 particles in fused SiO_2, which also have high H, Ackler and Chiang proposed that both primary and secondary minima exist [57,58]. Initially dispersed particles flocculate leaving moist boundary films; whereas liquid SiO_2 does not penetrate pure TiO_2 polycrystals. Despite many investigations of extensive interconnected glassy pockets and films in polycrystals of Al_2O_3 [90-96,92] or MgO [97,82] (including impurity effects on ψ) there has been relatively little TEM study of actual films. For either, after being extensively penetrated by silicate liquids [72,56,82,83] or dispersed as powder [56,98,62] or precipitate [99], extensive wetting is indicated but a discernable minority fraction of boundaries remain intact or are obviously not wetted. Whether these boundaries are moist, dry or crystallographically special remains to be ascertained. However, TEM studies of Al_2O_3 with small levels of glass indicate that dry boundaries are often special, e.g., twins [90,93], and that anisotropy is very important [95,96].

Polycrystalline Si_3N_4 has been the prototype system for equilibrium IGFs; it has repeatedly been found that a very high fraction of grain boundaries are an amorphous silicate film [5]. The thickness depends as much or more upon the liquid composition as on the boundary crystallography (or the material source, if pure, Fig. 3), although a few special boundaries are dry [25-27,5]. A series of experiments with Si_3N_4 particles equilibrated with a RE-Al-silicate (RE being one of Y or a rare earth) showed that small particle clusters form in excess liquid. And the mean film thickness depends upon the specific rare earth used, but it is virtually the same as in polycrystals with 5-12% liquid [27,25], Fig. 3. Highly anisotropic grain growth rates occur, indicating that important differences exist in local interfacial structure [100,101], and apparently caused disruption of large flocs of particles [62]. Thus, attractive interparticle forces exist, as expected from particle attraction models, but are of the order of other capillary forces. Analytical TEM studies show the RE/Si ratio in these films changes from <1/2 using Yb to > 2 with La when the glass has a 1:1 ratio [11]. Other films contain more N than do the pockets and can have higher concentrations of Ca [17-19].

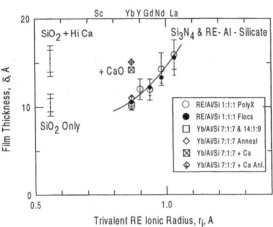

Fig. 3. Boundary thickness in Si_3N_4 versus ion radius of rare earth in RE + Al silicate from polycrystals with 12 vol% liquid and flocs in glass with 4 vol% particles (RE:Al:Si are atom ratios in residual glass) [27]. Also shown are values from polycrystals with Yb/Al >> 1, and an identical one with added Ca [25,102]. At left are thicknesses from pure Si_3N_4/SiO_2 and with 500 ppm CaO, from [26 and 18]. Films in other materials with only SiO_2 are similar [5,28,6].

Another, more isotropic model system is Pb-ruthenate particles dispersed in excess Pb-Al-glass, which yields thick-film resistors. The particles form a network with similar 1-1.5 nm thick IGFs for seemingly all boundaries, yet exhibit significant flattening at virtually all contacts [12]. The larger clusters affirm that partially densified particle networks can evolve, as proposed, for attractive situations having moderate to low values of ψ. With TiO_2 doping, the film thicknesses increase, and the Ti is preferentially located at the particle surfaces and films, rather than in the glass.

An important recent result emerged from a doped SiC having 1-2 nm thick IGFs for virtually all interfaces after sintering and grain growth above 1900°C. After a long anneal at 1200-1400°C, most boundaries are ordered (i.e., having no obvious amorphous core) but contain 1+ ML of Al, and excess O; these elements are in several lattice planes grown epitaxially on adjacent basal planes [89]. In another case with high H, extensive glassy regions found in many materials have been taken to imply the ubiquitous existence of IGFs in ZrO_2 doped with Y_2O_3 and SiO_2 [103]. However, a recent investigation found instead that actual grain boundaries were ordered yet contain significant Y and Si adsorbate [104,105]. This begs the question whether these are, in fact, also disordered

Grain Boundary Engineering in Ceramics

above a critical temperature? In both cases, it is unresolved whether the excess impurity in the ordered regions is metastable relative to being rejected to particles.

Recently, in three oxide systems, thick disordered grain boundaries have been seen in which the primary constituent in the film is not SiO_2. A combination of quantitative analytical TEM and high resolution TEM has revealed interfaces having a monolayer or so of cation adsorbate within a disordered region 1-2 nm wide and involving an equivalent or larger amount of material from the solid.

The system of ZnO-Bi_2O_3, which has been widely studied owing to its use as a varistor [13], has been carefully re-examined by Chiang and coworkers (see also earlier refs. in [21,31,73]). The boundaries can be disordered layers when formed either above or below the minimum eutectic temperature in the system; the films have Zn/Bi ratios of 1-2, which markedly exceeds that in associated bulk liquid. As the grains are largely unfaceted, it is easily seen that $\psi > 0$ under ambient pressure conditions. Long exposure of dense polycrystals to Bi_2O_3 rich liquid, pre-equilibrated with ZnO, leads to liquid being drawn into the system; yet it tends to maintain a network, albeit with more tendency to deflocculate if hotter. This behavior comports with proposed expectations for adsorbate films between attractive particles with low ψ. However, the amount of Bi adsorbate decreases with increasing capillary pressure and the Bi is reversibly driven from the boundaries into a second phase by application of high (> GPa) hydrostatic pressure implying that the force minimum is rather shallow, and that Ω for Bi is larger in the IGF than in the bulk phase [21,31].

Similarly it was found that Fe-doped $SrTiO_3$ polycrystals having excess TiO_2 exhibited similar ~1 nm disordered grain boundaries containing ~ 1 ML of excess Ti and half that of Fe in the films [106-108]. Rather similar results were recently found in Al_2O_3 doped with CaO-SiO_2. Again the boundaries were wide disordered layers, with the Ca/Si ratio being much higher than in the liquid pockets [22]. Obviously these results imply caution in interpreting earlier studies of oxides containing glassy impurities.

In addition, earlier TEM studies of quenched samples of less pure ZnO-Bi_2O_3 [13] and TiO_2-rich $SrTiO_3$ [109] revealed that each has a wetting transition for which $\psi \to 0$ above a critical temperature, about 1150°C and 1450°C, respectively. The transition temperature was markedly lower with other impurities present for $SrTiO_3$ [109]. In the former it is clear that ψ decreases from ~ 70°, with rising temperature, and that with complete wetting, the liquid layers are nearer a micron than a nanometer thick.

In all three systems, whatever SiO_2 exists in the sample is partitioned into the second phase pockets in preference to the films [21,31,108,22]. In the ZnO-Bi_2O_3 case, and likely for Al_2O_3-CaO and $SrTiO_3$, the dispersion forces across these particular disordered regions are lower than they would be for a silicate film. Nonetheless, a disjoining force that is not based upon the steric force of a polymeric oxide is needed to explain these observations. The films are more likely diffuse adsorption layers that form near but below critical wetting transitions.

In metallic systems, an extensive literature on grain boundary segregation has widely invoked L-M type models [36]. Recently, qualitative changes in properties, such as diffusion anomalies, have been found to be exhibited in several systems above some transition temperature. These have been interpreted as wetting or prewetting transitions, but often without benefit of exhaustive TEM investigation [110-113]. However, the behavior of Bi in Cu may serve as a prototype for many metallic and ceramic systems. Auger spectroscopy studies reveal that at an activity well below the solubility limit, a transition clearly occurs from near-monolayer to multilayer adsorption [114,115] at temperatures for which the equilibrium liquid seemingly does not wet the grain boundaries. This must entail a significant change in the interface structure [115,116], perhaps from ordered with site specific L-M type adsorption to disordered layers such as those seen in ceramic systems.

CONCLUDING REMARKS

A considerable literature discusses boundary adsorption and related theories for monolayer levels or less, although in only a few instances are details related to atomistic models of the grain boundaries for ceramics. Recently thick, disordered boundaries have been evaluated for several systems, some being silicate rich and others not. Increasing evidence shows these are more often equilibrium, nonwetting adsorbate films often involving multilayer impurity levels and often having different compositions than the bulk phases from which they are equilibrated. More recently, examples have been reported of nanometer thick, disordered interface layers that incorporate considerable material from the adjoining grains but contain no polymeric oxide, e.g., SiO_2.

Considering theories for interfacial wetting, interfacial films and sintering of materials with low dihedral angles has yielded a framework for comprehending the microstructures of solid-liquid systems including the impurity partition between IGFs and the bulk liquid and the attributes of complete wetting. Models based on colloid-like force balances have been invoked to explain these IGFs in which a steric disjoining force is implied to offset attractive dispersion forces. These rationalize some trends, but have little predictive power especially regarding specific impurity effects or films not based on silicates, and lack generality. Aspects of multilayer adsorption or prewetting models may yield guidance, although these have not been developed for atomic forces that pertain for ceramics.

Multilayer adsorption of gases exhibits near universal behavior, becoming stable well before coexistence. Similar behavior is duplicated in many lattice gas models, including one for a grain boundary in which adsorption increases above a disorder temperature that scales with the melting point. Such behavior may not be so general for internal interfaces, as London dispersion forces, and perhaps strain or other factors, can limit adsorption and layer thickness, and these apparently shift with choice of impurity ion. Nonetheless, evidence is accumulating to anticipate that some model of multilayer adsorption pertains that increases with temperature analogous to premelting or prewetting layers; that is Γ increases and ψ decreases with heating until a wetting transition occurs prior to or at the melting point. However, some L-M like adsorption likely also prevails at lower temperature, kinetics permitting.

In no case has the entire adsorption spectrum expected with increasing activity (from monolayer adsorption, to nonwetting film formation below coexistence, to boundaries at coexistence) been well documented for any one ceramic, as has been seen in Bi doped Cu. A caution is that with diminishing levels of additives, other, unintended impurities can have larger effects; as an example, low levels of Ca impurity can alter oxynitride films in Si_3N_4 [28].

Acknowledgement

This work was supported by the Director, Office of Energy Research, Office of Basic Energy Sciences, Materials Sciences Division of U. S. Dept. of Energy under Contract No. DE-AC03-76SF00098. The authors acknowledge insightful discussions with X.-Q. Pan of, U. Michigan, I. Tanaka and M. Yoshiya of Kyoto Univ., Y.-M. Chiang, W. C. Carter and C. A. Bishop of MIT and M. Gülgün at MPI, Stuttgart.

References

1   S. J. Bennison and M. P. Harmer, *Ceram. Trans.*, **7** 13-49 (1990).
2   J. Cho, M. P. Harmer, H. M. Chan, J. M. Rickman and A. M. Thompson, *J. Am. Ceram. Soc.*, **80** 1013-17 (1997).
3   H. Yoshida, Y. Ikuhara and T. Sakuma, *J. Mater. Res.*, **13** 2597-2601 (1998).
4   M. J. Hoffmann, in *Tailoring of Mechanical Properties of Si₃N₄ Ceramics*, ed. M. J. Hoffmann and G. Petzow, Kluwer Academic Publ, Dordrecht, The Netherlands, pp. 59-72 (1994).
5   H.-J. Kleebe, *J. Ceram. Soc. Jpn.*, **105** 453-75 (1997).
6   Q. Jin, D. S. Wilkinson and G. C. Weatherly, *J. Am. Ceram. Soc.*, **82** 1492-96 (1999).
7   P. F. Becher, E. Y. Sun, K. P. Plucknett, K. B. Alexander, C.-H. Hsueh, H-T. Lin, S. B. Waters and C. G. Westmoreland, *J. Am. Ceram. Soc.*, **81** 2821-30 (1998).
8   E. Y. Sun, P. F. Becher, K. P. Plucknett, C.-H. Hsueh, K. B. Alexander, S. B. Waters, K. Hirao and M. E. Brito, *J. Am. Ceram. Soc.*, **81** 2831-40 (1998).
9   N. P. Padture, *J. Am. Ceram. Soc.*, **77** 519-23 (1994).
10 J. J. Cao, W. J. MoberlyChan, L. C. DeJonghe, C. J. Gilbert and R. O. Ritchie, *J. Am. Ceram. Soc.*, 79 461-69 (1996).
11 M. J. Hoffmann, H. Gu and R. M. Cannon, *Mater. Res. Soc. Symp. Proc.* Interfacial Engineering for Optimized Properties II in press, (2000).
12 Y.-M. Chiang, L. A. Silverman, R. H. French and R. M. Cannon, *J. Am. Ceram. Soc.*, **77** 1143-52 (1994).
13 J. P. Gambino, W. D. Kingery, G. E. Pike and H. R. Phillip, *J. Am. Ceram. Soc.*, **72** 642-45 (1989).
14 F. Greuter, *Solid State Ionics*, **75** [1] 67-78 (1995).

15 R. Ramesh, S. M. Green and G. Thomas, in *Studies of High Temperature Superconductors. Advances in Research and Applications*, vol. 5, Nova Science Publ., Commack, NY, pp. 363-403 (1990).
16 J. Luo, H. Wang and Y.-M. Chiang, *J. Am. Ceram. Soc.*, **82** 916-20 (1999).
17 H. Gu, R. M. Cannon and M. Rühle, *J. Mater. Res.*, **13** 476-87 (1998).
18 H. Gu, X. Pan, R. M. Cannon and M. Rühle, *J. Am. Ceram. Soc.*, **81** 3125-35 (1998).
19 H. Gu, R. M. Cannon H. J. Seifert and M. J. Hoffmann, "Solubility of Si_3N_4 in Liquid SiO_2," submitted *J. Am. Ceram. Soc.*
20 I. Tanaka, H. Adachi, T. Nakayasu and T. Yamada, *Ceramic Microstructure: Control at the Atomic Level*, ed. A. P. Tomsia and A. Glaeser, Plenum Press, New York, pp.23-34 (1998).
21 Y.-M. Chiang, J.-R. Lee and H. Wang, *Ceramic Microstructure: Control at the Atomic Level*, ed. A. P. Tomsia and A. Glaeser, Plenum Press, New York, pp. 131-147 (1998).
22 R. Brydson, S.-C. Chen, F. L. Riley, S. J. Milne, X. Pan and M. Rühle, *J. Am. Ceram. Soc.*, **81** 369-79 (1998).
23 D. R. Clarke and G. Thomas, *J. Am. Ceram. Soc.*, **60** 491-95 (1977).
24 L. K. V. Lou, T. E. Mitchell and A. H. Heuer, *J. Am. Ceram. Soc.*, **61** 392-96 (1978).
25 H.-J. Kleebe, M. K. Cinibulk, R. M. Cannon and M. Rühle, *J. Am. Ceram. Soc.*, **76** 1969-77 (1993).
26 I. Tanaka, H.-J. Kleebe, M. K. Cinibulk, J. Bruley, D. R. Clarke and M. Rühle, *J. Am. Ceram. Soc.*, **77** 911-14 (1994).
27 C.-M. Wang, X. Pan, M. J. Hoffmann, R. M. Cannon and M. Rühle, *J. Am. Ceram. Soc.*, **79** 788-92 (1996).
28 X. Pan, H. Gu, R. van Weeren, S. C. Danforth, R. M. Cannon and M. Rühle, *J. Am. Ceram. Soc.*, **79** 2313-20 (1996).
29 D. R. Clarke, *J. Am. Ceram. Soc.*, **70** 15-22 (1987).
30 D. R. Clarke, T. M. Shaw, A. Philipse R. G. Horn, *J. Am. Ceram. Soc.*, **76** 1201-04 (1994).
31 H. Wang and Y.-M. Chiang, *J. Am. Ceram. Soc.*, **81** 89 -96 (1998).
32 A. W. Adamson and A. P. Gast, *Physical Chemistry of Surfaces*, 6th Ed., John Wiley & Sons, Inc., Ch. XVII (1997).
33 R. Pandit, M. Schick and M. Wortis, *Phys. Rev.*, **B26** 5112-40 (1982).
34 M. Schick, in *Liquids at Interfaces, Les Houches Summer School Lectures, Session XL VIII*, J. Charvolin, J.F. Joanny, and J. Zinn-Justin, Editors. 1989, Elsevier: Amsterdam.
35 S. Dietrich, in *Phase Transitions in Surface Films*, H.E.A. Taub, Ed., Plenum Press: New York. p. 391-423 (1991).
36 P. Lejcek and S. Hofmann, *Crit. Rev. Sol. State Mater. Sci.*, **20** 1-85 (1995).
37 W. D. Kingery, *J. Am. Ceram. Soc.*, **57** 1-8 & 74-83 (1974).
38 R. H. Fowler and E. A. Guggenheim, *Statistical Thermodynamics*, Cambridge Univ. Press, Cambridge, UK (1939).
39 S. Brunauer, P. H. Emmett and E. Teller, *J. Am. Chem. Soc.*, **60** 309-19 (1938).
40 R. B. Poeppel and J. M. Blakely, *Surf. Sci.*, **15** 507 (1969).
41 M. F. Yan, R. M. Cannon and K. K. Bowen, *J. Appl. Phys.* **54** 764-78 (1983).
42 J. Maier, *Solid State Ionics*, **32/33** 727-33 (1989).
43 J. W. Cahn, *J. Chem. Phys.*, 66 3667-72 (1977).
44 R. Kikuchi and J. W. Cahn, *Phys. Rev.*, **B36** 418 (1987).
45 R. Kikuchi and J. W. Cahn, *Phys. Rev.*, **B21** 1893-97 (1980).
46 D. J. Eaglesham and M. Cerullo, *Phys. Rev. Lett.* 64 1943-46 (1990).
47 J. A. Venables, J. S. Drucker, M. Krishnamurthy, G. Raynerd and T. Doust, *Mater. Res. Soc. Symp. Proc.* **198** 93-104 (1990).
48 S. Blonski and S. H. Garofalini, *J. Am. Ceram. Soc.* 80 1997-2004 (1997).

49 M. Yoshiya, I. Tanaka and H. Adachi, "Intergranular Glassy Film in High-Purity Si_3N_4-SiO_2 Ceramic: I Structure and Energy," submitted *Acta Mater.*

50 J. N. Israelachvili, *Intermolecular and Surface Forces*, 2nd Ed., Academic Press, NY (1992).

51 R. H. French, R. M. Cannon, L. K. DeNoyer and Y.-M. Chiang, *Solid State Ionics*, **75** 13-33 (1995).

52 R. H. French, C. Scheu, G. Duscher, H. Müllejans, M. J. Hoffmann and R. M., Cannon, *Mater. Res. Soc. Symp. Proc.*, **357** 243-58 (1995).

53 R. H. French, H. Müllejans, D. J. Jones, G. Duscher, R. M. Cannon and M. Rühle, *Acta Mater.*, **46** 2271-87 (1998).

54 R. H. French, Optical Properties and Hamaker Constant Data Base, Dupont Corp. Wilmington, DE (1998).

55 J. W. Cahn and J. E. Hilliard, *J. Chem. Phys.*, **28** 258-67 (1958).

56 T. M. Shaw and P. R. Duncombe, *J. Am. Ceram. Soc.*, **74** 2495-505 (1991).

57 H. D. Ackler and Y.-M. Chiang, *J. Am. Ceram. Soc.*, **82** 183-89 (1999).

58 H. D. Ackler and Y.-M. Chiang, *Ceramic Microstructure: Control at the Atomic Level*, ed. A. P. Tomsia and A. Glaeser, Plenum Press, New York, pp. 131-147 (1998).

59 M. Bobeth, D. R. Clarke and W. Pompe, *J. Am. Ceram. Soc.*, **82** 1537-46 (1999).

60 R. Kobayashi, J. A. Warren and W. Craig Carter, *Physica D*, **140** 141-50 (2000).

61 C. S. Smith, *Trans. AIME*, **175** 15-51 (1948); *Metal. Rev.*, **9** 1-48 (1964).

62 R. M. Cannon and L. Esposito, *Z. Metallkd.*, **90** 1002-15 (1999).

63 L. Esposito, E. Saiz, A. P. Tomsia, R. M. Cannon, in *Ceramic Microstructure: Control at the Atomic Level*, ed. A. P. Tomsia and A. Glaeser, Plenum P., New York, pp. 503-12 (1998).

64 W. Beere, *Acta Metall.*, **23** 131-38 (1975).

65 P. J. Wray, *Acta Metall.*, **24** 125-35 (1976).

66 R. M. Cannon, "On the Effects of Dihedral Angle and Pressure on the Driving Forces for Pore Growth or Shrinkage," unpublished, (1980).

67 R. Raj, *J. Am. Ceram. Soc.*, **70** C210-11 (1987).

68 W. D. Kingery and B. Francois, *Sintering and Related Phenomena*, eds. G. C. Kuczynski, N. A. Hooton and C. F. Gibbon, Gordon and Breach, New York, NY, pp. 471-96 (1967).

69 H.-H. Park, S.-J. L. Kang and D. N. Yoon, *Metall. Trans.*, **17A** 325-30 (1986).

70 R. B. Heady and J. W. Cahn, *Metall. Trans.*, **1** 185 (1970).

71 R. Raj, *J. Am. Ceram. Soc.*, **64** 245-48 (1981).

72 P. L. Flaitz and J. A. Pask, *J. Am. Ceram. Soc.*, **70** 449-55 (1987).

73 J.-R. Lee and Y.-M. Chiang, *Solid State Ionics*, **75** 79-88 (1995).

74 K. L. Merkle, J. F. Reddy and C. L. Wiley, Ultramicroscopy, **18** 281-84 (1985).

75 M. Rühle, *J. de Phys.*, *Colloque* C4, **46** C4-281-92 (1985).

76 T. Höche, P. R. Kenway, H.-J. Kleebe and M. Rühle, *J. Am. Ceram. Soc.*, **77** 339-48 (1994).

77 W. D. Kingery, in *Ceramic Microstructures '86: Role of Interfaces*, eds. J. A. Pask and A. G. Evans, Plenum P., NY, pp. 281-94 (1987).

78 C.-W. Li and W. D. Kingery, *Adv. in Ceram.* **10** 368-78 (1984).

79 Y.-M. Chiang, A. F. Henriksen, W. D. Kingery and D. Finello, *J. Am. Ceram. Soc.*, **64** 385-89 (1981).

80 J. A. S. Ikeda, Y.-M. Chiang, A. J. Garratt-Reed and J. B. Vander Sande, *J. Am. Ceram. Soc.*, **76** 2447-59 (1993).

81 A. Roshko and W. D. Kingery, *J. Am. Ceram. Soc.*, **68** C-331-33 (1985).

82 S. Ramamurthy, M. P. Mallamaci, C. M. Zimmerman, C. B. Carter, P. R. Duncombe and T. M. Shaw, *JMSA ISSN*, **2** 113-28 (1996).

83 J.-J. Kim and M. P. Harmer, *J. Am. Ceram. Soc.*, **81** 205-08 (1998).

84 P.-G. deGennes, *Rev. Mod. Phys.*, **57** 827-63 (1985).

85 R. M. Cannon, E. Saiz, A. P. Tomsia and W. C. Carter, *Mater. Res. Soc. Symp. Proc.*, **357** 279-92 (1995).

86 G. Duscher, Ph.D. Dissertation, Univ. of Stuttgart and Max-Planck Institute for Metals Research, Stuttgart, Germany, (1997).

87 H.-J. Kleebe, *J. Eur. Ceram. Soc.*, **10** 151-59 (1992).

88 H. Gu, SiC films

89 X.-F. Zhang, M. E. Sixta and L. C. DeJonghe, "Grain Boundary Evolution in Hot-Pressed ABC-SiC," *J. Am. Ceram. Soc.*, in press (2000).

90 S. C. Hansen and D. S. Phillips, *Philos. Mag. A*, **A47** 209-34 (1983).

91 Y. K. Simpson, C. B. Carter, K. J. Morrissey, P. Angelini, and J. Bentley, *J. Mater. Sci*, **21** 2689-96 (1986).

92 W. A. Kaysser, M. Sprissler, C. A. Handwerker, and J. E. Blendell, *J. Am. Ceram. Soc.*, **70** 339-43 (1987).

93 D. W. Susnitzky and C. B. Carter, *J. Am. Ceram. Soc.*, **73** 2485-93 (1990).

94 C. A. Powell-Dogan and A. H. Heuer, *J. Am. Ceram. Soc.*, **73** 3670-76 (1990).

95 D.-Y. Kim, S. M. Wiederhorn, B. J. Hockey, J. E. Blendell and C. A. Handwerker, *J. Am. Ceram. Soc.*, **77** 444-53 (1994).

96 J. E. Blendell, W. C. Carter and C. A. Handwerker, *J. Am. Ceram. Soc.*, **82** 1889-900 (1999).

97 B. Jackson, W. F. Ford and J. White, *Trans. Br. Ceram. Soc.*, **62** 577-607 (1963).

98 E. Saiz, A. P. Tomsia and R. M. Cannon, to be published.

99 A. P. Tomsia, A. M. Glaeser and J. S. Moya, *Key Eng. Mater.*, **111-112** 191-208 (1995).

100 M. Kramer, M. J. Hoffmann and G. Petzow, *J. Am. Ceram. Soc.*, **76** 2778-84 (1993).

101 M. Kramer, M. J. Hoffmann and G. Petzow, *Acta Metall. Mater.*, **41** 2939-47 (1993).

102 J. S. Vetrano. H.-J. Kleebe, E. Hampp, M. J. Hoffmann, M. Rühle and R. M. Cannon, *J. Mater. Sci.*, **28** 3529-38 (1993).

103 M. Rühle, N. Claussen and A. H. Heuer, *Adv. in Ceramics*, **12** 352-70 (1984).

104 Y. Ikuhara, P. Thavorniti and T. Sakuma, *Acta Mater.* **45** 5275-84 (1997).

105 P. Thavorniti, Y. Ikuhara and T. Takuma, *J. Am. Ceram. Soc.*, **81** 2927-32 (1998).

106 F. Ernst, O. Kienzle and M. Rühle, *J. Eur. Ceram. Soc.*, **19** 665-73 (1999).

107 O. Kienzle and M. Rühle, *Proc.of Euromat '99'*, Wiley-VCH Verlag GmbH, Weinheim, Germany. in press.

108 O. Kienzle, *Atomistische Struktur und Chemische Zusammensetzung innerer Grenzflächen von Strontiumtitanat*, Ph.D. Dissertation, Univ. of Stuttgart and Max-Planck Institute for Metals Research, Stuttgart, Germany, (1999).

109 M. Fujimoto and W. D. Kingery, *J. Am. Ceram. Soc.*, **68** 169-73 (1985).

110 E. I. Rabkin, L. S. Shvindlerman, B. B. Straumal, *Intl. J. Mod. Phys.*, **B5** 2989-3028 (1991).

111 B. B. Straumal, T. Muschik, W. Gust and B. Predel, *Acta Metall. Mater.*, **40** 939-45 (1992).

112 B. B. Straumal, O. I. Noskovich, V. N. Semenov, L. S. Shvindlerman, W. Gust and B. Predel, *Acta Metall. Mater.*, **40** 795-801 (1992).

113 E. I. Rabkin and W. Gust, in *Interfaces in Ceramics*, ed. J. Nowotny, Elsevier Publ. (1993).

114 M. Menyhard, B. Blum and C. J. McMahon, Jr., *Acta Metall.*, **37** 549 (1989).

115 L.-S. Chang, E. I. Rabkin, B. Straumal, P. Lejcek, S. Hofmann and W. Gust, *Scripta Metall. Mater.*, **37** 729-35 (1997).

116 U. Alber, H. Müllejans and M. Rühle, "Bismuth Segregation at Copper Grain Boundaries," to be published.

INTERGRANULAR GLASSY FILMS IN Si_3N_4-SiO_2 CERAMICS:
Morphology, chemistry, atomic structure and energetics

Isao TANAKA

Department of Energy Science and Technology,

Kyoto University, Sakyo, Kyoto 606-8501 Japan

Masato YOSHIYA, Shigeto R. NISHITANI and Hirohiko ADACHI

Department of Materials Science and Engineering,

Kyoto University, Sakyo, Kyoto 606-8501 Japan

ABSTRACT

High purity Si_3N_4-SiO_2 ceramics fabricated by the glass capsulation technique of hot isostatic pressing have been used as reference materials for liquid phase sintered Si_3N_4 ceramics. Experiments found that the intergranular glassy films (IGF) in these materials exhibit a constant thickness of $1.0\pm0.1nm$, and N/(N+O) ratio of 30%. First principles molecular orbital calculations were used to interpret the experimental ELNES (electron energy loss near edge structures): The high N content in the IGF was well confirmed. The presence of N can be ascribed to the topological requirements imposed by minimization of the number of broken bonds in the IGF. Atomistic modeling of the IGF was made on the basis of this experimental information. Pair-potentials calculations found that the oxygen adsorption or formation of the Si-O-N film is energetically favorable for a twist boundary of Si_3N_4. The geometric misorientational strain can be relieved by the presence of the Si-O bonds.

To the extent authorized under the laws of the United States of America, all copyright interests in this publication are the property of The American Ceramic Society. Any duplication, reproduction, or republication of this publication or any part thereof, without the express written consent of The American Ceramic Society or fee paid to the Copyright Clearance Center, is prohibited.

INTRODUCTION

Significant progress in the study of intergranular glassy films (IGF) of silicon nitride ceramics has been achieved in the last decade by combining results from high resolution electron microscopy (HREM), Angström-scale analytical microscopy and theoretical calculations. High-purity Si_3N_4 ceramics fabricated by the glass capsulation technique of hot isostatic pressing[1] played a central role in these experimental works. This material contains only a few mol% of SiO_2 that is derived from the starting powder. This serves as a good reference system for liquid-phase sintered Si_3N_4 ceramics. Experimental information on the chemistry and chemical bondings of the reference material have been used to model the IGF. Theoretical calculations were made to discuss the structure and energetics of the IGF using these models. In this paper, this recent work will be reviewed.

HREM AND ELNES OF IGF

High resolution electron microscopy (HREM) studies of high purity Si_3N_4-SiO_2 ceramics were carefully made in order to measure the film thickness of the IGF. The thickness was first reported to be 1.0 ± 0.1 nm, independent of the grain misorientation and volume fraction of the glassy phase.[2] A more detailed anlalysis[3] found the constant thickness of 1.1 ± 0.1 nm. The HREM structure at the interface between Si_3N_4 grain and the IGF is found to be abrupt.

A spatially-resolved EELS investigation of the IGF was done by Gu and coworkers.[4] They found that the N/(O+N) ratio of the glassy film is approximately 30% by the analysis of spectral intensities of O- and N-K edge ELNES (electron energy loss near edge structures). No other elements were detected from the film. This means that the film is composed of Si-oxynitride glass. They have also recorded ELNES at various edges from the film.[5] The spectra from the film and Si_3N_4 matrix are most clearly discriminated at the Si-L_{23} edge as shown in Fig. 1 (left). The difference between the two spectra is two fold: 1) the energy of the first peak in the IGF is higher than that of the Si_3N_4 grain by 0.9 eV; and 2) the first peak is broader in the IGF.

The coordination number of Si in all polytypes of SiO_2, Si_3N_4 and Si_2N_2O are four except for the high-pressure phases, stishovite-SiO_2 and spinel-Si_3N_4. The Si-O bond length in these crystals is in the range of 1.61 to 1.62 Å. It is in the range of 1.70 to 1.75 Å for the Si-N bond. It is therefore natural to assume that the Si-N and Si-O bond lengths in the IGF are in the same range. Yoshiya et al[6] made first principles calculations of SiO_2, Si_3N_4 and Si_2N_2O using model clusters to see the theoretical dependence of the peak energies on the N/(N+O) ratio, M_N. The quantity can be related to the number of N atoms included in each SiX_4 tetrahedron in average, X_N, as

$$X_N = \frac{12 M_N}{\left(3 M_N + 2(1 - M_N)\right)} \qquad (1)$$

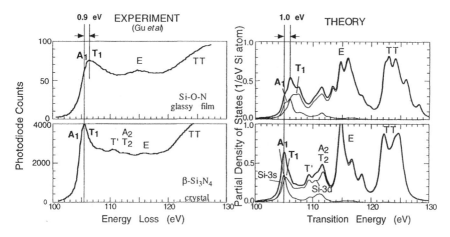

Fig. 1 (Left) Experimental Si-L_{23} edge ELNES from β-Si_3N_4 and IGF in Si_3N_4-SiO_2 ceramics by Gu et al[4].

(Right) Theoretical ELNES to compare with the experimental spectra[6].

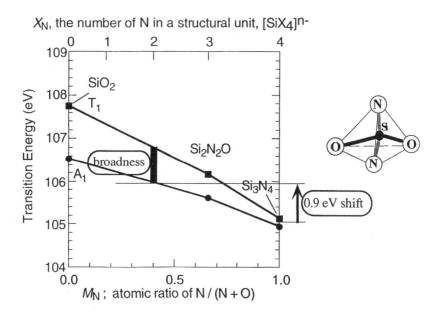

Fig. 2 Theoretical peak energies of Si-L_{23} edge ELNES obtained for three crystals.

Figure 2 shows theoretical transition energies as a function of M_N or X_N. Calculations were made by a first principles molecular orbital method using model clusters. Experimental ELNES from the IGF shows 0.9 eV-shift of the peak A_1. This approximately corresponds to $M_N = 0.4$ or $X_N = 2$ according to Fig. 2. The separation between peaks A_1/T_1 is then approximately 1 eV, which is also consistent with the broadening of the first peak as observed experimentally. In order to verify the reproduction of the experimental ELNES from the IGF, an example model cluster $[Si_7N_{13}O_7]^{25-}$ was constructed. A central tetrahedron of the cluster was taken to be $[SiN_2O_2]^{6-}$ corresponding to $X_N = 2$. Outer tetrahedra were systematically removed so that coordination numbers of O and N are kept to be 2 and 3, respectively. Bond-length between Si-O was adjusted to be 1.61 $\approx$. Theoretical Si-L_{23} edge spectrum obtained by the cluster model of the film can be compared with experimental spectrum in Fig. 1. Relative energy of the first peak as compared with Si_3N_4 and the broadening of the first peak is found to be well reproduced. The present model is confirmed to satisfy the necessary condition to reproduce the experimental spectrum of the glassy film without inclusion of further complexity or inhomogeneity. The next question is how to fit the modeled structure into the IGF.

ATOMIC STRUCTURE OF THE IGF

In sintered materials, β-Si_3N_4 grains often exhibit an elongated shape in the c-direction, and the surface area of the prismatic plane ($10\bar{1}0$) is greatest. The best choice of the model for the IGF should be a Si-O-N layer sandwiched between two prismatic planes, as schematically drawn in Fig. 3. The HREM shows that the structure at the interface between β-Si_3N_4 grain and the IGF is abrupt. The line profile across the IGF by Gu[5] found that no significant dissolution of O takes place into the neighboring β-Si_3N_4 grains. An example of the atomic structure of the film that satisfies these conditions is given in Fig. 4. In this model, the Si-O-N layer is sandwiched between two prismatic planes of the same orientation. This is chosen for simplicity although this type of specially oriented interface may not accommodate the glassy film in reality. The structure of the film is constructed on the basis of β-Si_3N_4 structure similar to the case of the model cluster used for Fig. 1. Si atoms are systematically removed in order to keep the coordination numbers of O and N to be 2 and 3, respectively. For future refinement of the atomic arrangement, periodicity of a' is kept to be the same as the lattice constant of a in β-Si_3N_4. Periodicity of c' is kept four times as c in β-Si_3N_4.

Thickness of the film in this model as measured between the first O / N layers is approximately 0.9 nm, and the ratio $M_N = 0.43$. Although whole N concentration can be varied depending upon the structure, a certain amount of N needs to be present at the interface between β-Si_3N_4 and the film when inclusion of the broken bond is not allowed. When only $[SiO_4]^{4-}$ molecules are adsorbed on the prismatic plane of β-Si_3N_4, as schematically drawn by Clarke,[7]

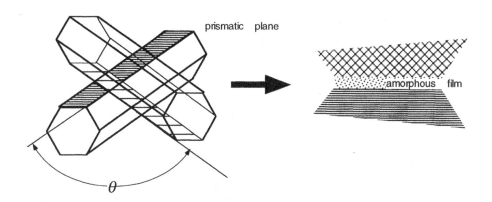

Fig. 3   Schematic picture of the IGF between two β-Si$_3$N$_4$ (prism) planes.

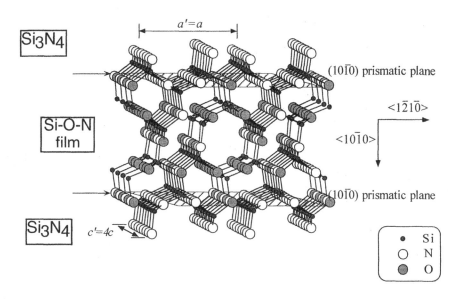

Fig. 4   An example modeled structure of the Si-O-N film
sandwiched in two pararell prismatic planes of β-Si$_3$N$_4$.
The strucure is used for the computation of energetics shown in Fig. 6.

a network structure above the $[SiO_4]^{4-}$ layer cannot be generated. In other words, the presence of N into the intergranular glassy film is a topological requirement imposed by the absence of broken bonds in a system composed only of Si, O and N. On the other hand, the presence of glass modifier-impurities in the intergranular film may significantly alter the bonding situation and the atomic arrangement. Once the topological requirement is released, there are much greater degrees of freedom for the structure, which may be intimately related to the interfacial bond strength and other properties.

ENERGETICS OF THE IGF

The phenomena of the IGF have been clarified by the combination of the HREM, ELNES and theoretical calculations. However, it is still an open question whether the oxygen adsorption at the interface of the $Si_3N_4//Si_3N_4$ is energetically favorable (Fig. 5). Firstly we have adopted a modeled IGF structure given in Fig. 4, and made structural optimization for a given geometry in order to evaluate the interface energy using interatomic potentials.[8] Calculations were done for four configurations, i.e., $Si_3N_4//Si_3N_4$ and $Si_3N_4/Si\text{-}O\text{-}N/Si_3N_4$ twist boundaries with $0°$ and $180°$ misorientation. Computational details were described elsewhere.[8] The energy penalty paid by the formation of the interface was divided by the area of the interface in the modeled structure in order to obtain the interface energy. Figure 6 shows the interface energies for four configurations plotted as a function of M_N. The interface energy of the $0°$-twist $Si_3N_4//Si_3N_4$ interface is zero by definition. When the Si-O-N film is present, the interface energy increases by 0.40 J/m^2 because of the chemical mismatch. It is natural because the Si-O-N film is very artificial, which is not expected to be present in a bulk state although the structure is optimized with respect to the given initial structure. In the "dry" boundary, the interface energy increases up to 1.08 J/m^2 when the misorientation is $180°$. The energy gain can be ascribed to the geometric misorientational strain. The strain energy is found to be relieved when the Si-O-N film is present to be 0.91 J/m^2. The magnitude of the strain energy can be correlated with the bond-length distribution in the vicinity of the interface as shown in ref. 8. It is suggested that the oxygen adsorption or the formation of the Si-O-N film is energetically favorable to decrease the geometric strain energy associated with the misorientation. The scenario was initially derived from the computation of a modeled boundary with special orientation.[8] More profound calculations on the energies associated with the IGF have been made employing molecular dynamic technique.[9] The IGF made in this way is a more realistic model for a random $Si_3N_4/Si\text{-}O\text{-}N/Si_3N_4$ twist boundary. As can be found in the companion paper in this volume,[9] the conclusion obtained by the latter calculation is the same as that by the simple model.

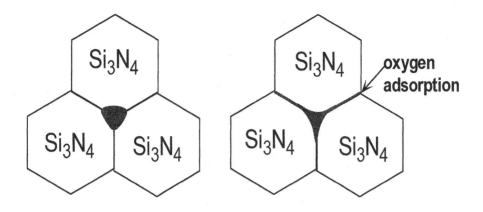

Fig. 5　Two ways to put excess oxygen in Si₃N₄ ceramics.

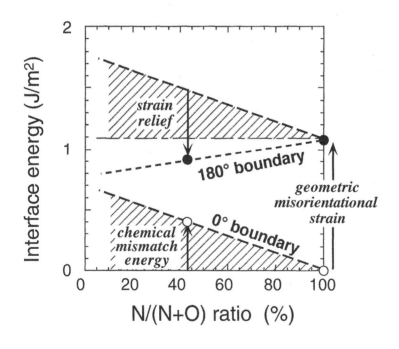

Fig. 6　Interface energies of the modeled Si₃N₄ boundaries.

Acknowledgments

This work is supported by the ACT program from JST. Helpful discussion with members of the International Group for Fundamentals of Intergranular Glassy Film (IGF)^2, especially with R.M. Cannon and H. Gu are gratefully acknowledged.

REFERENCES

1. I. Tanaka, G. Pezzotti, T. Okamoto, Y. Miyamoto and M. Koizumi: "Hot isostatic press sintering and properties of silicon nitride without additives", *J. Am. Ceram. Soc.*,72, (1989), 1656-1660.
2. I. Tanaka, H.-J. Kleebe, M.K. Cinibulk, J. Bruley, D.R. Clarke, and M. Rühle, "Calcium Concentration Dependence of the Intergranular Film Thickness in Silicon Nitride," *J. Am. Ceram. Soc.*, **7 7** [4] 911-14 (1994).
3. H. Gu, X. Pan, R. M. Cannon, and M. Rühle, "Dopant distribution in grain-boundary films in calcia-doped silicon nitride ceramics", *J. Am. Ceram. Soc.* 81, (1998) 3125-35.
4. H. Gu, M. Ceh, S. Stemmer, H. Müllejans, and M. Rühle, "A quantitative approach for spatially-resolved electron energy-loss spectroscopy of grain boundaries and planar defects on a subnanometer scale", *Ultramicroscopy*, 59 (1995) 215-227.
5. H. Gu, R.M. Cannon, and M. Rühle, "Composition and Chemical Width of Ultrathin Amorphous Films at Grain Boundaries in Silicon Nitride," *J. Mater. Res.*, **1 3** (1998) 376-387.
6. M. Yoshiya, H. Adachi and I. Tanaka, "Interpretation of Si-L$_{23}$ edge electron energy loss near edge structures (ELNES) from intergranular glassy film of Si$_3$N$_4$ ceramics", *J. Am. Ceram. Soc.*, 82 (1999) 3231-36.
7. D.R. Clarke "Grain boundaries in polyphase ceramics", *J. Phys. Paris* C4 (1985) 51-59.
8. M. Yoshiya, I. Tanaka, and H. Adachi, "Energetical role of modeled intergranular glassy film in Si$_3$N$_4$-SiO$_2$ ceramics", *Acta Mater.* in press.
9. M. Yoshiya, K. Tatsumi, I. Tanaka and H. Adachi, "Molecular dynamics study for intergranular glassy film in high purity Si$_3$N$_4$-SiO$_2$ ceramics" this volume.

MOLECULAR DYNAMICS STUDY OF INTERGRANULAR GLASSY FILM IN HIGH-PURITY Si₃N₄-SiO₂ CERAMICS

Masato Yoshiya[†], Kazuyoshi Tatsumi[†], Isao Tanaka[‡], and Hirohiko Adachi[†]
[†]Department of Materials Science and Engineering, Kyoto University,
Yoshida, Sakyo, Kyoto 606-8501, JAPAN
[‡]Department of Energy Science and Technology, Kyoto University,
Yoshida, Sakyo, Kyoto 606-8501, JAPAN

ABSTRACT

Theoretical calculations in atomic scale for grain boundaries in Si_3N_4-SiO_2 ceramics are conducted. The obtained structure shows structurally and chemically abrupt interface between intergranular glassy film (IGF) and the adjacent grain. It is found that the IGF at the grain boundary plays an important role to relieve strain energy due to misorientation of adjacent grains. The interfacial structure comes to be more stable by the presence of the IGF, which results in lower interface energy. Calculations with several initial thicknesses of the IGF reveals that a further increase in the IGF thickness (larger than 21.5 Å) does not contribute to a release in the strain energy. Estimations of the instability of the silicon oxynitride indicate that interface energy has a minimum value when the thickness is 7.0 Å.

INTRODUCTION

It is well known that mechanical properties of sintered Si_3N_4 ceramics are often determined by ubiquitous intergranular glassy film (IGF) composed of silicon oxynitride at grain boundary. Many of foreign elements contained in the sintering additives often segregate at the grain boundary and modify the nature of the IGF. Therefore, fundamental understandings of the nature of the IGF in atomic scale are needed in order to improve the mechanical properties in a rational manner.

Recent advances in field emission (FE) guns used in transmission electron microscopy (TEM) and in scanning-TEM (STEM) enable us to directly observe the IGF. Statistical analyses of IGF thicknesses done by Kleebe and coworkers[1-3] using high-resolution electron microscopy (HREM) have shown that the thickness of the IGF in high-purity Si_3N_4-SiO_2 system is almost constant along the grain boundary and is independent from the misorientation of adjacent grains except for some special boundaries. By line-profiling using electron energy loss spectroscopy (EELS) across the grain boundary, Gu and coworkers[4] have revealed the chemical thickness of the high-purity Si_3N_4 is 1.33 ± 0.25 nm and the N/(N+O) ratio is 0.30 ± 0.12. In our previous study,[5] the N/(N+O) ratio is found to be 0.43 ± 0.06 by analysis of electron energy loss near edge structure (ELNES) obtained from the IGF[6] with an

To the extent authorized under the laws of the United States of America, all copyright interests in this publication are the property of The American Ceramic Society. Any duplication, reproduction, or republication of this publication or any part thereof, without the express written consent of The American Ceramic Society or fee paid to the Copyright Clearance Center, is prohibited.

aid of first principle molecular orbital (MO) calculations. Although these ratios are obtained by thoroughly independent ways, agreement of the ratio was satisfactory. In addition, an example model structure of the IGF which satisfies the N/(N+O) ratio, density of Si, and electronic states by ELNES was proposed.

In our previous study, [7] using the model structure of the intergranular Si-O-N phase, interface energy and atomistic structure in the 180°-twist Si_3N_4/Si-O-N/Si_3N_4 grain boundary were evaluated. Although the model actually has periodicity like a crystal, it is found that the intergranular Si-O-N phase is energetically favorable when misorientation of adjacent grains is present and that O in the Si-O-N phase plays an important role to reduce the strain energy due to the misorientation. However, it is still unclear whether discussion using a fixed model is applicable to more general boundaries having amorphous structures.

In the present study, molecular dynamics (MD) calculations have been performed in order to understand these issues in atomic scale by introduction of an amorphous structure. A 90°-twist grain boundary is employed as a representative of random grain boundaries. First, we will see the "MD" heat treatment of the Si_3N_4/Si-O-N/Si_3N_4 grain boundary. Then, we discuss the effect of the interfacial structure on the interface energy. Finally, we will see the dependence of the interface energy on IGF thickness

COMPUTATIONAL PROCEDURE

In the present study, the molecular dynamics (MD) method is employed for the calculation of energy and structure of glassy phase at the grain boundary. MOLDY code [8] is used for the MD calculations. With the MD method, an amorphous structure can be obtained with the help of heat. Temperature and pressure are controlled at every 100 steps by the Nosé-Hoover thermostat scheme [9,10] and Parrinello-Rahman scheme,[11] respectively. The Busing-type pair potential set for the Si-O-N system reported by Kawamura and coworkers [12] is employed for the calculation of the IGF. It is reported that glass structures in SiO_2-Na_2O,[13] Na-Si-O-N,[12] Ca(Mg)-Al-Si-O [14] system calculated with the potential set are in good agreement with experimental results.

In the present study, structure and energy are evaluated at 0 K to exclude the kinetic energy contribution. The parameter set employed in the present study was designed to reproduce atomic motion at high temperature. Thus, it is expected that the absolute bond strength between atoms at low temperature will be estimated to be greater, compared with that when the potential set obtained by first-principles calculations [15,16] is used. However, we evaluate the relative energy, not the absolute energy, to discuss the energetical role of the IGF in a similar way as in our previous study [7]. We will see the effects of the "tight pair-potential" on the energy and the structure in the following section.

The time step, i.e., the time interval in the calculation, is set to 1 fs (10^{-3} ps) to prevent abnormal atomic motion. In order to minimize "artifacts", much attention is paid to the choice of temperature and time length for thermalization, cooling rate, size of Si_3N_4 grain region perpendicular to Si_3N_4/Si-O-N interface, and size of supercell. Although a smaller time step leads to higher "MD melting temperature", the temperature for thermalization is determined by evaluating atomic motion at various temperatures. The cooling rate and size of Si_3N_4 grain and supercell are determined in similar ways by evaluating atomic motions with various physical parameters in MD calculations.

THERMALIZATION OF 90°-TWIST Si₃N₄/Si-O-N/Si₃N₄ BOUNDARY

A model structure of the Si-O-N phase similar to the model used in the previous study[7] is employed as the starting structure. No dangling bond exists in the model. In other words, the composition of the model is stoichiometric. The anion ratio of N/(N+O) is 0.32 and the density of Si is 0.75 of bulk β-Si₃N₄. Height of the model is 13.17 Å, which is a bit larger than the thickness of the IGF observed by HREM in high-purity Si₃N₄-SiO₂ ceramics.

Procedure of thermalization is shown in Fig. 1. Initial amorphous structure of Si-O-N was obtained by calculation at high temperature with halved charge of ions, which leads to reduction of long-range electrostatic energy. The initial amorphous Si-O-N is thermalized to achieve thermal equilibrium for 100 ps. In order to reduce the statistical ambiguity of the amorphous structure, three sets of amorphous structure are obtained. The first one is obtained after initial thermalization. The second and third ones are obtained after extra thermalization for 50 ps and 100 ps, respectively. Thermalization of amorphous Si-O-N is done at $T_a = 12000$ K under constant volume condition. With these amorphous Si-O-N structures, a supercell for 90°-twist Si₃N₄/Si-O-N/Si₃N₄ boundary is constructed, as schematically illustrated in Fig. 2. In the supercell, amorphous Si-O-N is sandwiched between two Si₃N₄ grains with a spacing of 2 Å. One of two Si₃N₄ grains is rotated along <10$\bar{1}$0> direction so that a (10$\bar{1}$0) prismatic plane is shared as an interface with Si-O-N and Si₃N₄. Since MD calculations are done under a three-dimensional periodic boundary condition, two grain boundaries of Si₃N₄/Si-O-N/Si₃N₄ appear in one supercell. As shown in Fig. 1(b), the supercell is thermalized for 50 ps at $T_t = 8000$ K and is cooled down to 0 K with the cooling rate of 40 K/ps under constant pressure of 0.1 MPa. Time length and temperature for thermalization was long and high enough for atoms to reconstruct the interfacial structure. Although T_t is lower than T_a by 4000 K, the difference of temperature did not affect the results; they were chosen to save computing time for the cooling.

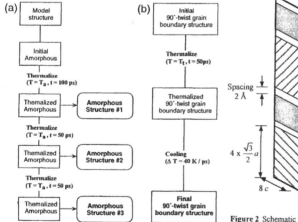

Figure 1. Procedure of heat treatment in MD calculation to obtain (a) amorphous Si-O-N glass for IGF and (b) final 90°-twist boundary structure. In the present study, $T_a = 12000$ K and $T_t = 8000$ K.

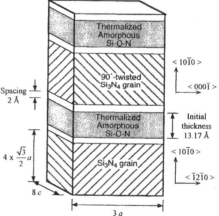

Figure 2 Schematic illustration of supercell for the 90°-twist grain boundary. a and c denote lattice constants of β-Si3N4. Since the calculation is done under the three-dimensional periodic boundary condition, two grain boundary appear in one supercell.

The total energy of the supercell is plotted as a function of time in Fig. 3. During the thermalization, thermal equilibrium was achieved at about $t = 30000$ fs (30 ps). Stepwise cooling was done with a decrement of 200 K. When the temperature was decreased by 200 K at the beginning of each step of cooling, a variation of total energy is found. However, thermal equilibrium is achieved after 5 ps at each temperature step. Thus, it is expected the final total energy is close to its local minimum with respect to structure. For comparison, three Si_3N_4/Si_3N_4 dry grain boundary structures are also calculated with T_t = 1000 K for 20 ps and the cooling rate of 20 K/ps.

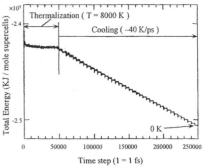

Figure 3. An example plot of total energy as a function of time step. A step corresponds to 1 fs in the present study. Step-wise cooling is done with a step, $\Delta T = -200$ K.

As a result of the heat treatment in the MD calculation, a grain boundary structure of two Si_3N_4 grains with the IGF at the grain boundary is obtained. Figure 4 shows one of three calculated structure of Si_3N_4/Si-O-N/Si_3N_4 grain boundary projected to a plane perpendicular to the grain boundary. As seen in the figure, structurally and chemically abrupt interfaces are generated at both sides of the IGF. This is consistent with HREM observation and elemental analysis done by EELS line-profiling across the grain boundary.

EFFECT OF PRESENCE OF INTERGRANULAR GLASSY FILM

In order to confirm the energetical role of the IGF, the interface energy of Si_3N_4/Si-O-N/Si_3N_4 grain boundary is compared with that of the Si_3N_4/Si_3N_4 dry grain boundary. In the present study, the interface energy for Si_3N_4/Si-O-N/Si_3N_4, γ_{SON}, is defined as,

$$\gamma_{SON} = \frac{\{E_{total} - (E_{SN} + E_{glass})\}}{2 \cdot A}, \tag{1}$$

where E_{total} is the total energy of supercell, E_{SN} is the total energy of pure β-Si_3N_4 crystal with the same number of particles, and E_{glass} is the averaged total energy for the Si-O-N glass. The numerator is divided by the number of grain boundaries in a supercell, 2, and interfacial area, A. Interface energy for Si_3N_4/Si_3N_4 dry boundary, γ_{SN}, is defined in a similar way except for a term, E_{glass}. From Eq.(1), the averaged interface energies, $\gamma_{SON} = 6.0$ J/m^2 and $\gamma_{SN} = 11.31$ J/m^2, are obtained. Although these values seems to be much larger than the interface energy in other systems, the larger interface energy is ascribed to the choice of "tight pair-potential", as mentioned before. Relative values are still meaningful.

Evaluating these interface energies, it is concluded that the presence of the IGF is energetically favorable; the IGF is found to play an important role to reduce the interface energy. Although these findings have already been reported in our previous study in the case of 180°-twist boundary,[7] it should be noted that the introduction of the amorphous structure, or dangling bonds, does not change the previous findings. These results obtained for grain boundary with large misorientation would be applicable to different grain boundaries, except for special boundaries.

Grain Boundary Engineering in Ceramics

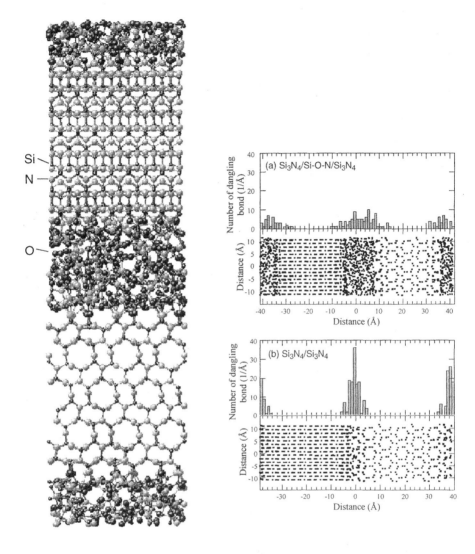

Figure 4. One of three calculated structures of the Si_3N_4/Si-O-N$/Si_3N_4$ grain boundary projected to a plane perpendicular to the grain boundary. The large white ball, large black ball, and small black ball are N, O, and Si, respectively.

Figure 5. Spatial distribution of dangling bonds (a) in Si_3N_4/Si-O-N$/Si_3N_4$ and (b) in Si_3N_4/Si_3N_4 dry grain boundary with those atomic positions projected to a plane perpendicular to the grain boundary.

Bond lengths for Si-N or Si-O bonds in Si_3N_4, Si_2N_2O, and SiO_2 crystals are around 1.61 Å or 1.73 Å, respectively. In the present study, with an analogy to the crystals, dangling bond is defined as the Si-X (X=N,O) bond whose bond length is larger than 1.80 Å. Since distribution of bond length is small due to the "tight pair-potential", the effect of the cut-off value for bond length on the results were negligible. Number of dangling bond, N_{db} is calculated as,

$$N_{db} = \sum_{i=0}^{3} \left\{ N_{Si(i)} \times (4 - i) \right\},$$ (2)

where i is the coordination number of Si and $N_{Si(i)}$ is the number of Si whose coordination number is i. $(4-i)$ denotes the number of dangling bonds of Si(i). Spatial distribution of atoms with dangling bond both in Si_3N_4/Si-O-N/Si_3N_4 grain boundary and Si_3N_4/Si_3N_4 dry grain boundary are shown Fig. 5. At the interface of the Si_3N_4/Si_3N_4 dry grain boundary, dangling bonds segregate. On the other hand, such segregation is not found at the Si_3N_4/Si-O-N interface, although a small amount of dangling bond is nearly uniformly present inside of the IGF. It is obvious that the number of dangling bonds is significantly decreased by the presence of the IGF in the Si_3N_4/Si-O-N/Si_3N_4 boundary. The significant decrease of dangling bond in Si_3N_4 grain interior in the Si_3N_4/Si-O-N/Si_3N_4 grain boundary affects the interface energy. Therefore, the Si_3N_4/Si_3N_4 dry grain boundary is energetically more expensive than the boundary with the IGF due to the presence of a higher amount of dangling bonds, which originated from the geometric strain due to misorientation of Si_3N_4 grains.

DEPENDENCE OF INTERFACE ENERGY ON IGF THICKNESS

Statistical HREM observations show thickness of the IGF in high-purity Si_3N_4-SiO_2 ceramics is ca. 10 Å with quite a small distribution. Although the thickness is determined by dynamic transportation of atoms in the system containing grains, IGF and triple pockets during sintering, the presence of the IGF with constant thickness strongly implies that there must be some static factors to keep the IGF thickness constant. In order to understand the effect of the IGF thickness on the interface energy, MD calculations with different initial thicknesses of the IGF are made.

Calculations are done with initial thickness of 0.25, 0.5, 1, 1.5, 2, 2.5, 3 in unit of height of the model of the IGF, 13.17 Å. For each initial thickness, three calculations are done with different amorphous structures, as indicated in Fig. 1(a). In the present study, IGF thickness is defined as,

$$h = \frac{\left(H_{supercell} - H_{SN/SN} \right)}{2},$$ (3)

where $H_{supercell}$ is the height of the supercell obtained under constant pressure condition and $H_{SN/SN}$ is the height of the supercell for the Si_3N_4/Si_3N_4 dry grain boundary having the same number of particles as in the grain interior in the supercell for the Si_3N_4/Si-O-N/Si_3N_4 grain boundary. The numerator is divided by the number of grain boundaries in a supercell, i.e., 2. The thickness contains the effect of the excess volume at the Si_3N_4/Si-O-N interface, although the volume of the amorphous film is not possible to be defined uniquely.

Grain Boundary Engineering in Ceramics

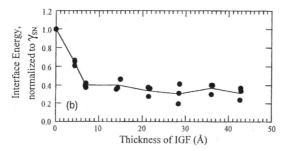

Figure 6. Plots of interface energy normalized to the interface energy for Si₃N₄/Si₃N₄ dry grain boundary, γ_{SN} as a function of IGF thickness. Solid line connects averaged values. At h ≥ 21.5 Å, the interface energy is seen to be almost constant.

Obtained interface energies normalized to γ_{SN} are plotted as a function of the IGF thickness in Fig. 6. Averaged values at each initial thickness are connected with a solid line. As seen in Fig. 6, interface energies at given thickness shows a small distribution while interface energy at $h = 0$ shows no distribution. This is ascribed to the ambiguity of the amorphous structure of the IGF. With an increase in h, the interface energy significantly decreases at $h \leq 7.0$ Å. At $h \geq 21.5$ Å, however, the interface energy is almost constant. Therefore, it is concluded that a further increase in the IGF thickness does not contribute to a release in the strain energy due to misorientation of adjacent grains.

In the present study, as defined in Eq.(1), we neglect the energy penalty of the presence of the silicon oxynitride glass at the grain boundary: N/(N+O) ratio of the IGF much exceeds the solubility limit of N in silica glass that is as small as ca. 4%. Thus the instability of the silicon oxynitride can be a counter energy factor to determine the IGF thickness. Estimations of the instability indicate that the interface energy has a minimum value at $h \sim 7.0$ Å. However, calculations for much larger systems containing Si₃N₄ grains, IGFs and triple pockets are required to evaluate the transportation of N and O accompanied by dissolution or formation of the silicon oxynitride at the grain boundary. It is beyond the scope of the present study.

CONCLUSION

In the present study, molecular dynamics calculations of the interface energy and the structure of the IGF in Si₃N₄/Si-O-N/Si₃N₄ 90°-twist grain boundary were made. It was confirmed that the presence of the IGF at the grain boundary is energetically favorable. Although some dangling bonds are present in the IGF, they are not segregated at the Si₃N₄/IGF interface, while dangling bonds segregate at the Si₃N₄/Si₃N₄ dry grain boundary. Therefore, it is concluded that the IGF plays an important role to relieve strain energy due to misorientation of adjacent grains. MD calculations with several initial IGF thicknesses indicate that a further increase in the IGF thickness (larger than 21.5 Å) does not contribute to a release in the strain energy. Estimations of the instability of the silicon oxynitride indicate the interface energy has a minimum value at $h \sim 7.0$ Å.

ACKNOWLEDGMENT

This work was supported by the ACT program from JST.

REFERENCES

1. H.-J. Kleebe, M.J. Hoffman, and M. Rühle, "Interface of Secondary Phase Chemistry on Grain-Boundary Film Thickness in Silicon Nitride," *Z. Metallkd.*, 83 [8] 610-17 (1992).
2. H.-J. Kleebe, M.K. Cinibulk, R.M. Cannon, and M. Rühle, "Statistical Analysis of the Intergranular Film Thickness in Silicon Nitride Ceramics," *J. Am. Ceram. Soc.*, 76 [8] 1969-77 (1993).
3. H.-J. Kleebe, M.K. Cinibulk, I. Tanaka, J. Bruley, J.S. Vetrano, and M. Rühle, " High-Resolution Electron Microscopy study on Silicon Nitride Ceramics"; pp. 259-74 in *Tailoring of Mechanical Properties of Si_3N_4 Ceramics*. Edited by M.J. Hoffmann and G. Petzow, Kluwer Academic Publishers, Dordrecht, The Netherlands, 1994.
4. H. Gu, R.M. Cannon, and M. Rühle, "Chemistry and Bonding of Grain Boundary and Interface in Silicon Nitride Based Materials," *J. Mater. Res.*, 13 [2] 376-387 (1998).
5. M. Yoshiya, I. Tanaka, and H. Adachi, "Interpretation of Si-$L_{2,3}$ Edge Electron Energy Loss Near Edge Structures (ELNES) from Intergranular Glassy Film of Si_3N_4 Ceramics," *J. Am. Ceram. Soc.*, 82 [11] 3231-36 (1999).
6. Gu H., Ceh M., Stemmer S., Müllejans H., and Rühle M., "A Quantitative Approach for Spatially-Resolved Electron Energy-Loss Spectroscopy of Grain Boundaries and Planar Defects on a Subnanometer Scale," *Ultramicroscopy*, 59 215-227 (1995).
7. M. Yoshiya, I. Tanaka, and H. Adachi, "Energetical Role of Modeled Intergranular Glassy Film in Si_3N_4-SiO_2 Ceramics," *Acta Mater.*, Accepted.
8. K. Refson, "Molecular Dynamics Simulation of Solid n-Butane," *Physica*, 131B 256-266 (1985).
9. S. Nosé, "A Molecular Dynamics Method for Simulations in the Canonical Ensemble," *Mol. Phys.*, 52 [2] 255-268 (1984).
10. W. G. Hoover, "Canonical Dynamics: Equilibrium Phase-Space Distribution," *Phys. Rev. A*, 31 1695-1697 (1985).
11. M. Parrinello and A. Rahman, "Polymorphic Transitions in Single Crystals: A New Molecular Dynamics Method," *J. App. Phys.*, 52 [12] 7182-7190 (1981).
12. H. Unuma, K. Kawamura, N. Sawaguchi, H. Maekawa, and T. Yokokawa, "Molecular Dynamnics Study of Na-Si-O-N Oxynitride Glass," *J. Am. Ceram. Soc.*, 76 [5] 1308-12 (1993).
13. H. Ogawa, Y. Shiraishi, K. Kawamura, and T. Yokokawa, "Molecular Dynamics Study on the Shear Viscosity of Molten $Na_2O\cdot2SiO_2$," *J. Non-Cryst. Sol.*, 119 151-158 (1990).
14. M. Okuno and K. Kawamura, "Molecular Dynamics Calculations for $Mg_3Al_2Si_3O_{12}$ (pyrope) and $Ca_3Al_2Si_3O_{12}$ (grossular) glass structures," *J. Non-Cryst. Sol.*, 191 249-259 (1995).
15. B.W.H. van Beest, G.J. Kramer, and R.A. van Santen, "Force Fields for Silicas and Aluminophosphates Based on Ab Initio Calculations," *Phys. Rev. Lett.*, 64 [16] 1955-58 (1990).
16. W.Y. Ching, Y.N. Xu, J.D. Gale, and M. Rühle, "Ab-Initio Total Energy Calculation of α- and β-Silicon Nitride and the Derivation of Effective Pair Potentials with Application to Lattice Dynamics," *J. Am. Ceram. Soc.*, 81 [12] 3189-96 (1998).

MICROSTRUCTURAL ASPECTS OF SUPERPLASTIC DEFORMATION IN A FINE-GRAINED SILICON NITRIDE DOPED WITH A SILICA-CONTAINING ADDITIVE

Rong-Jun Xie   Mamoru Mitomo   Guo-Dong Zhan
National Institute for Research in Inorganic Materials, Namiki 1-1, Tsukuba-shi, Ibaraki 305-0044, Japan

A superplastic fine-grained silicon nitride ceramic was fabricated by hot pressing a powder mixture of fine β-Si_3N_4 and cordierite at 1750°C for 5 min. The superplastic flow behavior was investigated under uniaxial compression at temperatures between 1450 and 1650°C over a range of strain rates between 1 x 10^{-5} to 1 x 10^{-3} s^{-1}. It was found that the material exhibited a Newtonian flow under these conditions. Detailed microstructural investigation of the superplastically deformed specimens by SEM and TEM revealed that (i) an as-hot-pressed monolithic ceramic evolved into a composite ceramic with elongated silicon oxynitride embedded in a fine-grained beta-silicon nitride matrix; (ii) the *in-situ* formed silicon oxynitride grains were textured through a preferred nucleation and growth mechanism and (iii) the beta-silicon nitride matrix retained its fine and equiaxed morphology, indicating a grain-boundary sliding deformation mechanism.

INTRODUCTION

A number of researchers have investigated superplasticity in silicon nitride ceramics, such as SiC/Si_3N_4,[1] α/β-sialons,[2-4] α-rich Si_3N_4[5] and fine-grained β-Si_3N_4.[6,7] Microstructural analysis of these ceramics have revealed that, apart from fine-grained β-Si_3N_4 ceramics which exhibit excellent microstructural stability, most of them undergo microstructural coarsening (concurrent grain growth) or phase changes (α to β phase transformation) during superplastic deformation, and that these changes in microstructure greatly affected the flow behaviors.

Recently, we have developed a superplastic silicon nitride-silicon oxynitride *in situ* composite,[8,9] starting with a fine-grained β-Si_3N_4 microstructure and using a transient liquid produced during the formation of Si_2N_2O grains. Remarkable

To the extent authorized under the laws of the United States of America, all copyright interests in this publication are the property of The American Ceramic Society. Any duplication, reproduction, or republication of this publication or any part thereof, without the express written consent of The American Ceramic Society or fee paid to the Copyright Clearance Center, is prohibited.

changes in the phases and microstructure were found to occur during superplastic deformation, resulting in a silicon oxynitride-reinforced silicon nitride composite analogous to "self-reinforced" silicon nitride ceramics. Moreover, texture developed in the Si_2N_2O grains during deformation. Although previous studies[8,9] have focused on the superplastic flow behavior in the materials of interest, the present investigation was undertaken to characterize the changes in grain morphology, the evolution of phases and texture development during superplastic deformation, and to evaluate the influence of these parameters on flow behavior.

EXPERIMENTAL PROCEDURE

The starting materials were 93 wt% ultrafine β-Si_3N_4 (Denki Kagaku Kogyo K.K., Japan, SN-BF97M) with an average particle size of 0.20 μm and 7 wt% cordierite (2MgO•2Al$_2$O$_3$•5SiO$_2$; Marusu-yuyaku Co., Nagano, Japan, SS600). The powder mixture was ball-milled in n-hexane for 2 h using silicon nitride balls. Densification was conducted by hot pressing at 1750°C for 5 min under a pressure of 20 MPa in nitrogen atmosphere. The sintered materials reached full density (> 98 % theoretical density).

The deformation was evaluated by uniaxial compression tests under a constant displacement rate. Specimens (approximately 2.5 x 3 x 5 mm) were diamond cut and finished so that parallelism of the two end surfaces was ensured. Experiments were performed under a nitrogen atmosphere of 0.1 MPa in a resistance furnace with a tungsten-heating element. The testing temperatures varied from 1450°C to 1650°C and the initial strain rates ranged from 1 x 10^{-5} to 1 x 10^{-3} s^{-1}.

Phase identification was determined from X-ray diffraction (XRD) analysis using CuKα radiation. The volume fraction of Si_2N_2O was calculated by using calibration lines. Scanning electron microscopy (SEM) and transmission electron microscopy (TEM) were used to characterize the microstructure both before and after deformation. The SEM samples were cut, polished, and then plasma-etched with CF$_4$ containing 7.8 wt% O$_2$. Thin foils for TEM were prepared from longitudinal slices parallel to the stress axis by the standard procedures of grinding, dimpling, and ion-beam thining, followed by carbon deposition to minimize charging under the electron beam.

RESULTS

Microstructural observations

An SEM micrograph of the microstructure of the as-sintered Si_3N_4 is illustrated in Fig. 1, which indicates a fully dense and homogeneous microstructure with equiaxed Si_3N_4 grains and a grain boundary glassy phase. Results from conventional XRD show that the as-sintered material consists primarily of β-Si_3N_4 as a crystalline phase. The mean grain diameter of the β-Si_3N_4 particle is around 0.17 μm.

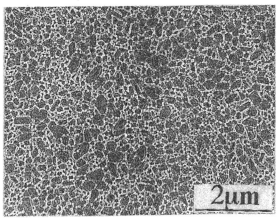

Figure 1 SEM image of the as-sintered material, showing the fine and equiaxed β-Si_3N_4 grains and a continuous grain boundary glassy phase.

A dramatic change in phase and microstructure occurs during deformation. Figure 2 shows a TEM micrograph of the material deformed at 1550°C under a strain rate of 1 x 10-4 s^{-1}. The material typically consists of rod-like Si_2N_2O grains that are prismatic in morphology, fine β-Si_3N_4 grains and a continuous grain boundary glassy phase. The elongated Si_2N_2O grains appear to be embedded in the fine-grained β-silicon nitride matrix. Some smaller, round-shaped β-Si_3N_4 particles are entrapped within the large Si_2N_2O grains and a high density of stacking fault is found in the Si_2N_2O grains.

The β-Si_3N_4 matrix grains, however, tended to retain their equiaxed shape after deformation, as shown in Fig. 3. The average grain diameter of the β-Si_3N_4 particle is about 0.19 μm, which indicates that dynamic grain growth did not take place for β-silicon nitride allowing the matrix to maintain its excellent microstructural stability. Similar observations have been made for a superplastic fine-grained β-silicon nitride ceramic,[7] which was also made from the same ultrafine β-silicon nitride powder. These findings of microstructural stability are in sharp contrast to the results reported by Wu and Chen,[2] where strain-induced grain growth of the β-silicon nitride took place even at temperatures as low as 1550°C. This discrepancy must be related to differences in the starting powders used. The present material was fabricated from an ultrafine β-silicon nitride powder instead of the widely-used-α-silicon nitride powder.

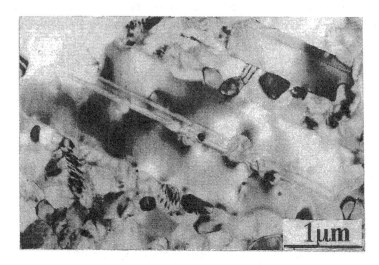

Figure 2 TEM image of a material deformed at 1550°C and 1×10^{-4} s^{-1}, showing the elongated Si$_2$N$_2$O grains embedded in a fine-grained β-Si$_3$N$_4$ matrix.

Figure 3 Bright-field TEM image of the deformed material (1550°C, 1×10^{-4} s^{-1}), showing that the matrix grains retained their equiaxed shapes and remained fine with a grain diameter of 0.19 μm.

Concurrent grain growth and texture development in the Si₂N₂O phase

Microstructural examination revealed a formation of silicon oxynitride grains during superplastic deformation, resulting in a silicon oxynitride - silicon nitride *in-situ* composite ceramic that is morphologically analogous to "self-reinforced" silicon nitride ceramics. Fig. 4(a) shows the relationship between the amount of Si_2N_2O formed and the true strain. Clearly, the amount of the Si_2N_2O phase increased with strain and reached its maximum of 22 vol% at strains over - 0.70. The figure also indicates that the amount of the Si_2N_2O phase depends on both the temperature and the duration of compression. Figure 4(b) shows the volume fraction of Si_2N_2O for the materials annealed and deformed, respectively, for various periods. The deformed material had about 8.5 vol.% Si_2N_2O more than the annealed one, clearly indicating deformation-enhanced grain growth of the Si_2N_2O grains.

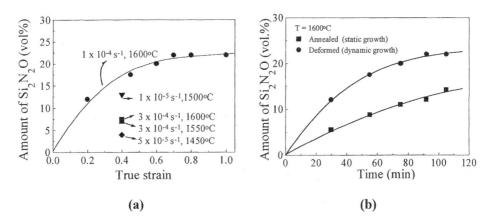

(a)   (b)

Figure 4 The volume fraction of the Si_2N_2O phase *vs* (a) strain and (b) time.

Another unique microstructure characteristic is texture development in the silicon oxynitride phase which formed *in situ*. Figure 5 shows an SEM image of the specimens deformed at 1550°C under a strain rate of 1×10^{-4} s⁻¹, clearly indicating a preferred orientation for the Si_2N_2O grains, in which the grain length is aligned normal to the direction of the compressive stress. The fact that this preferred orientation is absent in the annealed samples, even though they have a similar microstructure as the deformed samples, is evidence that the texture was developed by superplastic deformation.

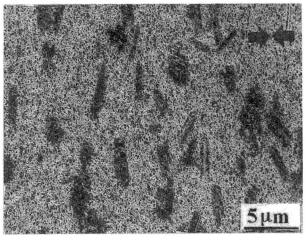

Figure 5 SEM image of the material deformed at 1550°C and 1 x 10^{-4} s^{-1} to -0.70, showing strong texture development in the Si$_2$N$_2$O grains

Deformation Behavior

As a detailed examination of the deformation behavior has been reported earlier, [8,9] only a brief summary is presented here.

Well-defined steady-state deformation clearly took place in the present material over a large range of strain rates. Strain hardening, commonly observed in some superplastic silicon nitride ceramics, is absent in the present material, even though deformation occurred at lower strain rates. This is primarily due to the observed microstructural stability in the matrix. Based on this evidence, it is reasonable to conclude that the dynamic grain growth of the silicon oxynitride grains did not result in strain hardening.

A strain rate sensitivity of $m \sim 1$ was calculated from the slope of flow stress-strain rate curve, indicating a Newtonian flow behavior. The activation energy for superplasticity, Q_{sp} can be obtained from an Arrhenius plot in which strain rate is plotted as s function of the inverse absolute temperature for a given flow stress. The calculated apparent activation energy is approximately 450 ± 43 kJ/mol at a flow stress of 100 MPa for the present material.

DISCUSSION

Microstructural stability in the fine-grained matrix during superplastic deformation

In structural ceramics, such as silicon nitride, the formation of abnormally large grains is frequently found during long-duration exposure to elevated temperatures, due to the usual anisotropic growth characteristic of the ceramics. This is clearly undesirable in superplastic ceramics because it may lead to the formation of cavities at isolated coarse grains or within clusters of coarse grains and may result in strain hardening. To date, several kinds of silicon nitride ceramics have been shown to be superplastic, and of these the α-phase powder is often used for specimen preparation.

During superplastic deformation, specimens often experience an α→β-phase transformation and grain growth of the β-silicon nitride, which has an unstable microstructure and thus results in remarkable strain hardening. In the present investigation, however, the microstructural stability of the β-silicon nitride matrix grains after deformation benefits from two important features of the material, which contribute to the steady-state deformation and the absence of strain hardening. The first is the use of an ultrafine β-silicon nitride starting powder, which has a lower surface energy than the α phase. The second feature is the absence of an α to β phase transformation, which is known to provide a large driving force for grain growth. Moreover, since the morphology of the β-silicon nitride matrix grains did not change, remaining equiaxed after deformation, grain boundary sliding is believed to be the dominant deformation mechanism.

Concurrent nucleation and grain growth of Si_2N_2O during superplastic deformation

An interesting aspect of superplasticity is that the deformation process enhances concurrent nucleation and grain growth of silicon oxynitride. Silicon oxynitride (Si_2N_2O) is formed by a chemical reaction between the $β-Si_3N_4$ and SiO_2 in the liquid. This reaction takes place through a process of solution-diffusion-precipitation. The relatively small β-silicon nitride particles dissolve into the Mg-Al-Si-O liquid, interact with the SiO_2 constituent of the liquid, and finally the Si_2N_2O nuclei precipitate onto the retained $β-Si_3N_4$ grains. Microstructural investigation indicates that the grain growth of the Si_2N_2O was enhanced by deformation, this appears to have been a function of the strain, duration of compression, and temperature. It is possible to explain this enhanced grain growth in terms of the extensive grain-boundary-sliding that occurs during superplastic deformation, the accelerated dissolution of $β-Si_3N_4$ into the liquid under the externally applied stress, and the related enhancement of grain boundary diffusion and grain boundary mobility.

Dynamic grain growth is commonly observed in superplastic ceramics and can alter the superplastic deformation behavior, being associated with increased flow resistance, strain hardening, as well as cavitation. It is interesting to note that concurrent nucleation and grain growth of Si_2N_2O did not result in strain hardening in the present material, which is in contrast with the results reported by Ohashi et al.[10] They observed that pronounced strain hardening occurred when the rod-like Si_2N_2O grains grew from the glass matrix during compressive deformation. One possible reason for this difference may be the fact that the newly formed Si_2N_2O is a minor phase, which contributes a little to the total strain, and that Si_2N_2O is textured in the present material.

Deformation-induced texture development and its influence on the flow behavior

The most important effect of deformation on microstructural development of the present material is the texture formation in the Si_2N_2O grains. Si_2N_2O grows anisotropically, resulting in a pseudo-hexagonal prism morphology, as shown in Fig. 2. Deformation promotes the growth of elongated Si_2N_2O grains in a direction normal to the compressive stress and affects the alignment of the rod-like grains in the same direction. A previous study[11] indicated that texture in the Si_2N_2O phase depended strongly on the strain and the amount of time-dependent formation of Si_2N_2O during deformation.

Chen et al.[2,3] reported similar deformation-induced texture development in β-sialon ceramics, and modeled the kinematic process for this texture formation in terms of grain rotation. In our case, however, a grain rotation mechanism could not operate, because the as-sintered ceramic does not contain Si_2N_2O, which was produced *in situ* during deformation. Thus, a different mechanism must have contributed to this development of texture.

It is possible that the applied stress provided a bias for preferential nucleation of Si_2N_2O onto those β-Si_3N_4 grains under tension, resulting in the development of the well-aligned Si_2N_2O nuclei. These well-aligned nuclei began to grow under the applied stress and finally developed into the elongated grains, resulting in the resultant texture. Therefore, preferential nucleation and anisotropic grain growth are the mechanisms for the texture development.

As pointed out earlier, the concurrent nucleation and growth of Si_2N_2O did not have a detrimental effect on the flow behavior of the material investigated. It is due primarily to its *in situ* formation and texture development during deformation. During the formation Si_2N_2O, a transient liquid was produced as a result of a reaction between the Si_3N_4 and the additive, which greatly reduced the flow resistance. Moreover, the textured Si_2N_2O grains provided a fiber-strengthening effect toughening the deformed material. It was, therefore, possible to achieve good ductility in the present material. A similar finding has been reported by Huang and Chen[3] where *in situ* formed and textured β-sialon grains contributed to the enhanced formability of the ceramics investigated.

CONCLUSION

(1) A fine-grained β-silicon nitride ceramic doped with cordierite, having an initial grain size of ~ 0.17 μm, was superplastically deformed at a temperature ranging between 1450 to 1650°C. It exhibited a well-defined steady-state deformation and a Newtonian flow behavior.

(2) Concurrent nucleation and grain growth of silicon oxynitride accompanied the superplastic deformation. The β-silicon nitride matrix, however, remained fine and equiaxed, providing excellent microstructural stability, due to the use of an ultrafine β-silicon nitride starting powder.

(3) Deformation-induced texture was developed in the silicon oxynitride grains that formed *in situ*. The textured Si_2N_2O grains did not have a negative effect on the ductility of the investigated material.

References

1. F. Wakai, Y. Kodama, S. Sakaguchi, N. Murayama, K. Izaki, and K. Niihara, "A Superplastic Covalent Crystal Composite," *Nature* (London), **344**, 421-23 (1990).
2. X. Wu and I. W. Chen, "Exaggerated Texture and Grain Growth in a Superplastic SiAlON," *J. Am. Ceram. Soc.,* **75** [10] 2733-41 (1992).
3. S. L. Hwang and I. W. Chen, "Superplastic Forming of SiAlON Ceramics," *J. Am. Ceram. Soc.,* **77** [10] 2575-85 (1994).
4. A. Rosenflanz and I. W. Chen, "Classical Superplasticity of SiAlON Ceramics," *J. Am. Ceram. Soc.,* **80** [6] 1341-52 (1997).
5. T. Rouxel, F. Rossignol, J. L. Besson, and P. Goursat, "Superplastic Forming of an α-Phase Rich Silicon Nitride," *J. Mater. Res.,* **12** [2] 480-92 (1997).
6. M. Mitomo, H. Hirotsuru, H. Suematsu, and T. Nishimura, "Fine-Grained Silicon Nitride Ceramics Prepared from β-Powder," *J. Am. Ceram. Soc.,* **78** [1] 211-14 (1995).
7. G. D. Zhan, M. Mitomo, T. Nishimura, R. J. Xie, T. Sakuma, and Y. Ikuhara, "Superplastic Behavior of Fine-Grained β-Silicon Nitride Material under Compression," *J. Am. Ceram. Soc.,* **83**(4) 841-47, 2000.
8. R.J. Xie, M. Mitomo, G.D. Zhan, and T. Emoto, "Superplastic Deformation in Silicon Oxynitride-Silicon Nitride *In situ* Composites", *J. Am. Ceram. Soc.,* in review.
9. R. J. Xie, M. Mitomo, G.D. Zhan, "Superplasticity in a Fine-Grained Beta-Silicon Nitride Ceramic Containing a Transient Liquid," *Acta. Mater.,* **48**(9) 2049-2058, 2000.
10. M. Ohashi, Y. Iida, S. Hampshire, "Nucleation Control for Hot-Working of Silicon Oxynitride based Ceramics", *J. Mater. Res.,* **14** [1] 170-77 (1999).
11. R.J. Xie, M. Mitomo, Y.J. Kim, and Y.W. Kim, "Texture Development in a Silicon Oxynitride-Silicon Nitride *In situ* Composite via Superplastic Deformation," *J. Am. Ceram. Soc.,* in review.

TEM CHARACTERISTICS OF GRAIN BOUNDARIES IN SUPERPLASTIC SILICON NITRIDE CERAMICS

Guo-Dong Zhan, Mamoru Mitomo, and Rong-Jun Xie: National Institute for Research in Inorganic Materials, Japan; Yuichi Ikuhara and Taketo Sakuma: Department of Materials Science, University of Tokyo, Japan

ABSTRACT In this study, high-resolution electron microscopy (HREM) was used to measure the thickness distribution of films in a beta-silicon nitride material before and after superplastic deformation with respect to the level of strain. The thickness of the grain boundary films showed a dependence on the angle between the grain boundary and the direction of the applied compressive stress. At a strain of –0.7, film thickness tended to decrease with increases in the compressive stress on the grain boundary. The most important finding is that, although parts of the grain boundaries were free of film, no boundaries were completely film-free under such a deformation, leading to a unimodal distribution in the thickness of the grain-boundary film. However, at a large strain of –1.1, although the thickness distribution was unimodal in boundaries perpendicular to the direction of the applied stress, in parallel boundaries it was bimodal. Moreover, it was found that many of the film-free boundaries observed were parallel to the direction of applied stress in the extremely deformed sample. This finding might be related to the processes involved in extreme deformation.

INTRODUCTION

Although a number of reports have addressed the issues of creep and superplastic behavior, very few have examined the nature of grain boundaries, using high-resolution electron microscopy (HREM).[1-3] Recently, several researchers have reported changes in the thickness of intergranular films and intergranular film chemistry after creep in studies employing high-resolution and analytical microscopy.[4-6] These techniques provide direct evidences of redistributions of grain-boundary glass during creep. However, Burger et al[7] investigated the

To the extent authorized under the laws of the United States of America, all copyright interests in this publication are the property of The American Ceramic Society. Any duplication, reproduction, or republication of this publication or any part thereof, without the express written consent of The American Ceramic Society or fee paid to the Copyright Clearance Center, is prohibited.

microstructure of a superplastic silicon nitride by TEM, and reported finding a large number of strain whorls in the superplastic sample. Their observation of a shear-thickening phenomena and a transition from mild to strong strain hardening were attributed to rigid contacts between grains, leading to the strain whorls observed. In our previous work[8], excellent superplasticity was demonstrated in fine-grained β-silicon nitride ceramics. In this study, HREM was used to measure the thickness distribution of films in β-silicon nitride materials before and after superplastic deformation. Special attention was given to the measurement of film thickness at the grain-boundaries after deformation with respect to the parallel and perpendicular orientations relative to the direction of the applied stress, which involve the maximum tensile and the maximum compressive stresses, respectively.

EXPERIMENTAL PROCEDURES

The material used for this study was prepared by hot-pressing a fine β-Si_3N_4 powder with 5wt% Y_2O_3 and 2wt% MgO as sintering additives at 1700°C and 20 MPa in a N_2 atmosphere. The superplastic deformation of the material was assessed by the uniaxial compression test under a constant displacement rate. The experiments were performed under a nitrogen atmosphere in a resistance furnace at 1600°C, with a tungsten-heating element. A rectangular specimen was deformed to a true strain of −1.1 (the true strain is defined as ln (h/h_0), with h_0 and h being the height of the sample before and after deformation) under a constant strain rate of 5 x 10^{-4}/s. Once superplastic deformation was completed, the force was immediately removed, and the samples were left to cool in the furnace. A detailed description of the deformation testing procedure can be found elsewhere.[8] For comparison, a disk-shaped sample was subjected to a true strain of − 0.7 under a constant load of 40 MPa in a vacuum furnace between 1400°C and 1500°C.

Thin foils for transmission electron microscopy of the sample both before and after deformation were prepared from longitudinal slices parallel to the compression axis, using the standard procedures of grinding, dimpling, and argon-ion-beam thinning, followed by carbon coating to minimize charging during observation. The HREM technique was used to determine the thickness of grain boundary amorphous films using a microscope (Topcon 002B, Tokyo, Japan) with a point-point resolution of 0.17 nm at 200 kV. In order to measure the film thickness accurately, the specimen was tilted so as to align the grain boundary in the "edge-on" position, and to orient both adjacent grains under good diffraction conditions using procedures described by Cinibulk et al[6] and Kleebe et al[5]. Specifically, measurements were carried out to correlate the film thickness at the grain-boundaries with the direction of the applied stress. A statistical evaluation

Grain Boundary Engineering in Ceramics

for the distribution in film thickness before and after deformation was performed for 15 ~ 20 selected grain boundaries for each orientation (giving 30 to 50 data points for each boundary)[5].

RESULTS AND DISCUSSION

Several noticeable features in the present deformed materials were identified from the TEM observations. Firstly, the materials exhibited a high degree of grain-size stability against dynamic grain growth during superplastic deformation.[8] The microstructure of the initial material was rather homogeneous, consisting of fine, and mainly equiaxed, β-Si_3N_4 grains with an average grain size of 180 nm and an aspect ratio of 2.3. After deformation, the β-silicon nitride grains were found to be unchanged in terms of size and shape. Secondly, no other crystalline phases were observed. Thirdly, cavitations along the grain boundaries or pockets as a result of negative pressure induced by grain-boundary sliding were not observed, even in the extremely deformed sample. Finally, the so-called strain whorls, which are generally located at the center of the grain boundaries, and far away from the triple-point junctions, did not emerge because the deformed samples were not subjected to a load during cooling in the present study. This finding is consistent with the fact that although strain whorls were found in samples deformed and cooled under loads, they disappear when samples are annealed under zero stress.[7]

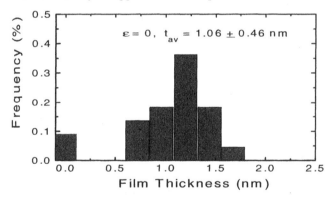

Fig 1 Histogram for the distribution in the thickness of the intergranular film in the experimental material before deformation.

HREM observation of the materials, both before and after deformation, reveals a number of interesting features depending on the level of strain. In the sample before deformation, an amorphous phase was found at most of the grain boundaries as a thin film, although some were free of this film. Statistical analysis showed that the mean thickness of the film was 1.06 nm with a standard deviation of 0.46 nm (Fig. 1).

In the superplastically deformed sample with a strain of – 0.7, the main feature is that the film thickness decreased with decreases in the angle between a grain boundary normal and the direction of applied force (Fig. 2). This behavior reflects the basic dependence of the film thickness on the level of local stress. As given by Chen and Huang[9], the average normal stress, σ, on any inclined grain boundary can be expressed as the following equation, in which grain boundaries are assumed to be randomly oriented, but with a locally equilibrated configuration containing segments intersecting at an angle of 120°:

$$\sigma = \sigma_a(\frac{1}{2} + \cos 2\theta) \qquad (1)$$

where σ_a is the applied compressive force and θ is the angle between a grain boundary normal and the direction of the applied force. Equation (1) indicates that for $60° < \theta < 90°$, a tensile stress prevails, and for $0° < \theta < 60°$, a compressive stress dominates at the grain boundaries and for $\theta = 0°$, the average compressive stress is at a maximum, i. e., $\sigma = 1.5 \, \sigma_a$.

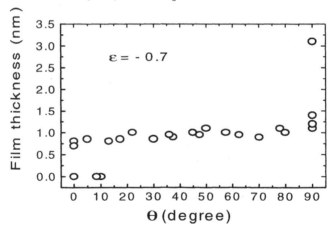

Fig 2 Intergranular film thickness as a function of the angle, θ, between the grain-boundary normal and the direction of applied stress after deformation with a strain of – 0.7

However, the most important finding is that, although parts of the grain boundaries were free of film, no boundaries were completely film-free under superplastic deformation. The distribution for the thickness of the films at the grain boundaries ($0\,° \leq \theta \leq 90\,°$) in the deformed samples is shown in Fig. 3.

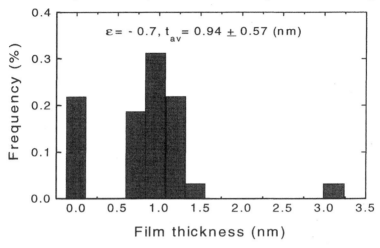

Fig 3 Histograms for the distribution in the thickness of the intergranular film in superplastically deformed Si_3N_4 with a strain of -0.7

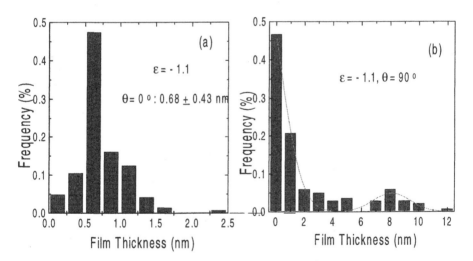

Fig 4 Histograms for the distribution in the thickness of the intergranular film in superplastically deformed Si_3N_4: (a) at grain boundaries oriented perpendicular to the direction of the applied stress and (b) for grain boundaries oriented parallel to the direction of the applied stress.

Compared to the boundaries before deformation, there is a wider distribution in the thickness of the film and an increase in the proportion of the boundaries that are film-free. Moreover, it is very interesting to note that there is a complete absence of film with a thickness between 0 and 0.5 nm. This is probably linked to both the discrete structural nature of the glassy phase (a network of SiO_4^{4-} units), and the structure of the interface between the glassy phase and the silicon nitride grain surfaces, where the structural unites of the glassy phase, SiO_4^{4-}, can be either attached epitaxially to the surface of silicon nitride or be nonepitaxial.[3]

In the extremely deformed sample with a strain of − 1.1, interestingly, the thickness of the film at boundaries perpendicular to the compressive direction of superplastic deformation ($\theta = 0$ °) displayed a normal distribution, with an average film thickness of 0.68 nm and a standard deviation of 0.43 nm (Fig. 4a), while films at boundaries parallel to the compressive direction ($\theta = 90$ °) exhibited a bimodal distribution in thickness (Fig. 4b). It should be noted, however, that at least three features of the extremely deformed sample were unexpected. Firstly, although the thickness of films along boundaries perpendicular to the compressive stress axis became thinner after superplastic deformation, no completely film-free boundaries were observed. Fig 5a shows a representative HREM image of a thinner boundary perpendicular to the direction of the compressive stress. Secondly, there was a wider distribution in the thickness of the films at grain boundaries parallel to the compressive stress axis, ranging form 0 to 12.5 nm (Fig. 4b). A HREM image of a thicker film at a grain boundary after superplastic deformation is shown in Fig. 5b. The most surprising observation, however, is that fact that the majority of grain boundaries parallel to the direction of compressive stress were free of films, although, it should be pointed out that none of the boundaries ($\theta = 90$ °) that were found to be film-free were low-angle boundaries.

According to the Clarke model[10], the net normal force acting per unit of area in an intergranular phase between two opposingly oriented grains can be given in terms of the van der Waals dispersion force and the repulsive steric force. Upon application of a load, the glass phase must be redistributed in order to establish a new equilibrium between the normal local stresses on the facets of each grain boundary. Although the response at the grain boundary to an applied force depends in theory on the nature of the force, namely whether it is a tension or compression force, it has been predicted that a stable grain-boundary glass phase of finite thickness will remain, even under compressive stress. With increases in compressive stress, the equilibrium thickness decreases. The compressive stress required to remove a liquid film entirely is extraordinarily high and greatly dependent on the value of an ordering parameter ($0 \leq \eta_0 \leq 1$), whereas the tensile stress required to separate two grains is relatively small. The dependency of the equilibrium film thickness on the application of a stress at the intersection angle is plotted in Fig. 6 as a function of parameter η_0.

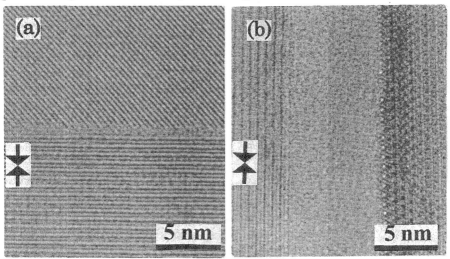

Fig 5 HREM images of grain-boundary films in superplastically deformed Si_3N_4, showing a dependency between film thickness and the local stress conditions: (a) a thinner grain boundary oriented perpendicular to the direction of the applied stress and (b) a thicker grain boundary oriented parallel to the direction of the applied stress. Arrows indicate the applied stress direction during deformation.

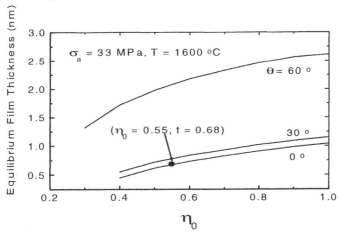

Fig 6 The equilibrium film thickness calculated as a function of the parameter η_0 at 1600 °C and at an applied stress of 33 MPa for θ equals 60 °, 30 °, and 0 °, respectively

This indicates, as noted by Clarke,[10] that higher values of η_0 and higher intersection angle both lead to thicker films. The latter reflects the effect of local stress levels on the equilibrium thickness, as indicated in Eq. (1). In the case of η_0 = 0.55, the equilibrium film thickness at $\theta = 0\,°$ is calculated to be 0.68 nm, which is consistent with our HREM observations.

The fact that most of the film-free boundaries were parallel to the direction of the applied stress, where maximum tensile stresses are involved, might be due to the following reasons. Local stress can vary greatly depending on how the grains actually support the applied stress. For granular materials with a distribution in grain size, the load is carried principally by a small number of chains of grains spanning from one side of the sample to another. On the other hand, it might also be related to the deformation processes involved with large strains. At intermediate strains, grain-boundary sliding in the direction of the compressive force might dominate the deformation process, whereas under extreme deformation the process might proceed mainly in a direction perpendicular to the direction of the applied compressive force. It was noted that some of the boundaries parallel to the direction of the applied stress became very wide. This may have allowed grains surrounding the boundaries under applied compressive stress to enter these extremely wide boundaries, leading to some boundaries parallel to the direction of applied stress being subjected to very high compressive stress. Thus, with very large deformations, some boundaries parallel to the direction of applied stress may experience high compressive stresses, while some boundaries may widen under maximum tensile stress.

CONCLUSIONS

By means of high-resolution electron microscopy, the thickness of films at grain boundaries was measured before and after superplastic deformation with respect to the level of strains. It was found that a change in the grain boundary phases took place after superplastic deformation. This redistribution depended on the orientation of the grain boundary with respect to the direction of the applied force. At a strain of –0.7, the thickness of the film decreased with decreases of the angle between a grain boundary normal and the direction of applied force. Although parts of the grain boundaries were free of film, no boundaries were completely film-free under such a deformation, leading to a unimodal distribution in thickness of the grain boundary film. At a large strain of –1.1, boundaries that were perpendicular to the direction of applied stress had thin films, and although the majority of the parallel boundaries were film-free, some had thick films. This finding is probably related to the processes involved in extreme deformation.

Grain Boundary Engineering in Ceramics

REFERENCES

1. Q. Jin, X.-G. Ning, D. S. Wilkinson, and G. C. Weatherly, "Redistribution of a Grain-Boundary Glass Phase during Creep of Silicon Nitride Ceramics," *J. Am. Ceram. Soc.*, **80** [2] 685-91 (1997).
2. Q. Jin, D. S. Wilkinson, and G. C. Weatherly, "High-Resolution Electron Microscopy Investigation of Viscous Flow Creep in a High-Purity Silicon Nitride," *J. Am. Ceram. Soc.*, **82** [6] 1492-96 (1999).
3. C.-M. Wang, M. Mitomo, T. Nishimura, Y. Bando, "Grain Boundary Film Thickness in Superplastically Deformed Silicon Nitride," J. Am. Ceram. Soc., 80[5] 1213-21 (1997). *J. Am. Ceram. Soc.*, **80** [5] 1213-21 (1997).
4. I. Tanaka, H.-J. Kleebe, M. K. Cinibulk, J. Bruley, D. R. Clarke, and M. Rühle, "Calcium Concentration Dependence of the Intergranular Film Thickness in Silicon Nitride," *J. Am. Ceram. Soc.*, **77,** 911-14 (1994).
5. H.-J. Kleebe, M. K. Cinibulk, R. M. Cannon, and M. Rühle, "Statistical Analysis of the Intergranular Film Thickness in Silicon Nitride Ceramics," *J. Am. Ceram. Soc.*, **76** [8] 1969-77 (1993).
6. M. K. Cinibulk, H.-J. Kleebe, and M. Rühle, "Quantitative Comparison of TEM Techniques for Determining Amorphous Intergranular Film Thickness," *J. Am. Ceram. Soc.*, **76** [2] 426-32 (1993).
7. P. Burger, R. Duclos, and J. Crampon, "Microstructural Characterization in Superplastically Deformed Silicon Nitride," *J. Am. Ceram. Soc.*, **80** [4] 879-85 (1997).
8. G.-D Zhan, M. Mitomo, T. Nishimura, R.-J. Xie, T. Sakuma, and Y. Ikuhara, "Superplastic Behavior of a Fine-Grained β-Silicon Nitride Material Under Compression," *J. Am. Ceram. Soc.*, **83** [4] 841-47 (2000).
9. I.-W. Chen and S.-L. Hwang, "Shear Thickening Creep in Superplastic Silicon Nitride," *J. Am. Ceram. Soc.*, **75** [5] 1073-79 (1992).
10. D. R. Clarke, "On the Equilibrium Thickness of Intergranular Glass Phases in Ceramic Materials," *J. Am. Ceram. Soc.*, **70** [1] 15-22 (1987).

Hereto Interface and Others

PREPARATION AND STRUCTURE OF EPITAXIAL CeO$_2$/YSZ/Si BUFFER LAYER

Naoki Wakiya, Makoto Yoshida, Tomoaki Yamada, Takanori Kiguchi, Kazuo Shinozaki and Nobuyasu Mizutani
Department of Metallurgy and Ceramics Science, Graduate School of Science and Engineering,
Tokyo Institute of Technology, 2-12-1, O-okayama, Meguro-ku, Tokyo 152-8552, Japan

ABSTRACT

Growth of an epitaxial insulating ceramic layer on silicon substrate is a key technology to realize memory devices such as DRAM and FRAM. For these devices, preparation of epitaxial buffer layer between ferroelectric layer and Si substrates is essential to prevent inter-diffusion and to form electrically clean interface between ferroelectric layers and Si substrates. This study describes preparation and interface structure of CeO$_2$/YSZ buffer layer on Si(001) substrates. The buffer layer was prepared by either sputtering or pulsed laser deposition (PLD). For sputtering, (Zr+Y) metal film was deposited on SiO$_2$/Si(001) substrate. The metals were reacted with SiO$_2$ layer having a few nm thick on the surface of Si substrate to form epitaxial YSZ thin layer (reactive sputtering). Then additional YSZ thin films and CeO$_2$ films were deposited using YSZ and CeO$_2$ ceramic target, respectively. For PLD, a KrF excimer laser was focussed on the YSZ sintered target and formed a very thin YSZ film on a Si(001). After YSZ layer formation, CeO$_2$ was also deposited on YSZ layer and epitaxial CeO$_2$ layer was prepared. The critical thickness of YSZ layer in order to prepare epitaxial CeO$_2$ layer was only 0.5 nm, which corresponds to only one atomic layer of YSZ. The mechanism of epitaxial layer formation, crystal structure and interface structure were discussed for the buffer layers prepared by both sputtering and PLD.

INTRODUCTION

Recently, preparation of epitaxial ceramic thin films have been attracting much interest from the point of potential applications to improve the performance of several devises such as ferroelectric random access memory (FRAM). However, it is well known that ferroelectric materials, such as Pb(Zr,Ti)O$_3$ (PZT) are difficult to grow epitaxially on Si(001) for constructing electronic devices due to the formation of SiO$_2$ or inter-diffusion at the interface between ferroelectric layer and Si substrate. To achieve such epitaxy on Si(001), a buffer layer is required. YSZ (Y$_2$O$_3$ stabilized ZrO$_2$) is one oxide that permits such direct epitaxial growth [1,2]. However, the lattice mismatch between YSZ and Si is considerably large (-5.4%). On the other hand, in the case of CeO$_2$, the lattice mismatch between CeO$_2$ and Si is fairly small (-0.4%). Along with the small mismatch, CeO$_2$ has comparatively high dielectric constant (εr=26 [3]), which is similar to that of YSZ (εr=30 [4]). Therefore, it is expected that CeO$_2$ thin films can be grown epitaxially on Si(001) with (001) orientation. However, it has been reported that CeO$_2$ thin films have (110) and/or (111) orientation but no (001) orientation on Si(001) substrate [3, 5-7]. The orientation of CeO$_2$ changes with oxygen pressure during deposition [8]. Under high oxygen pressure (> 1x10^{-5} Torr at 900

To the extent authorized under the laws of the United States of America, all copyright interests in this publication are the property of The American Ceramic Society. Any duplication, reproduction, or republication of this publication or any part thereof, without the express written consent of The American Ceramic Society or fee paid to the Copyright Clearance Center, is prohibited.

°C), CeO$_2$ shows (111) orientation. Under low oxygen pressure (< 5x10^{-6} Torr at 900 °C), CeO$_2$ shows (110) orientation [5, 8]. These reports indicate that it is impossible to obtain heteroepitaxially grown CeO$_2$ thin films on Si(001) with a cube-on-cube relation. The CeO$_2$/YSZ double buffer structure is widely used to achieve heteroepitaxial growth of several oxides having perovskite related structure, such as YBa$_2$Cu$_3$O$_{7-\delta}$ (YBCO) [9-12] on Si(001). However, the relationship between preparation conditions and crystal structure as well as electrical properties has not been clarified yet. In this work, CeO$_2$/YSZ heteroepitaxial double buffer layers were prepared by both pulsed laser deposition (PLD) and rf-magnetron sputtering to clarify the relationship.

EXPERIMENTAL

Substrate treatment

In both PLD and sputtering, 10 x 10 mm^2 sized Si(001) were used as substrates. In this work, two types of surface treatments were carried out prior to deposition, and the results were compared each other. One was to degrease using only 2-propanol. In this case, the surface of Si substrate was covered with natural SiO$_2$. The thickness of SiO$_2$ was determined to be around 2nm by XPS [13]. The other surface treatment method used was RCA cleaning [14], followed by dilute HF (HF:H$_2$O=1:10) dipping. In this case, the surface of the Si substrate was oxide-free (ascertained by XPS [13]) and hydrogen-terminated [15]. These substrates were immediately put into the processing chamber for both PLD and sputtering.

PLD

PLD was carried out using a KrF excimer laser (λ=248 nm). The laser beam was focused by a quartz lens up to an energy density of about 1.0 J/cm^2 and at an angle of 45 ° on targets of YSZ and CeO$_2$, which were rotated during the deposition. These targets are located on the multiple target holders, therefore, CeO$_2$/YSZ thin films were prepared in a single run process. The detailed deposition conditions of YSZ and CeO$_2$ are shown in Table I. The YSZ and CeO$_2$ targets were synthesized using YSZ powder (8 mol% Y$_2$O$_3$, Tosoh Corporation, Japan) and CeO$_2$ powder (3 N, Soekawa Chemicals, Japan), respectively, and sintered at 1500 °C for 2 h. The substrates were heated up to 800°C at a heating rate of 20°C/min and base pressure (5 x 10^{-6} Torr at 800°C). After the temperature reached 800°C, YSZ deposition was started at the base pressure without the introduction of O$_2$ gas. O$_2$ gas was introduced after laser pulsing began up to 5.5 x 10^{-4} Torr. After the prescribed deposition time, the YSZ target was changed to CeO$_2$ target.

Sputtering

Sputtering was carried out by a rf-magnetron sputtering. Three targets (all 2 inches in diameter) were used in this work. The first target was metal Zr. On the surface of this target, four small metal Y chips were placed. The second target was YSZ sintered between 1200 and 1400°C for 2 h. The third target was hot pressed CeO$_2$. The starting powders used for the preparation of targets were same as that of PLD. The detailed deposition conditions of metal (Zr+Y), YSZ and CeO$_2$ are shown in Table II. The substrates were heated up to 770°C at a heating rate of 10°C/min at a base pressure (5 x 10^{-6} Torr at 770°C). After the temperature reached 770°C, the substrates were heated at this pressure. Then Ar gas was introduced and pre-sputtering was carried out. For metal (Zr+Y), the pre-sputtering time was 40 min, and for YSZ and CeO$_2$, the pre-sputtering time was 20 min. O$_2$ gas was introduced after metal (Zr+Y) deposition was completed.

Grain Boundary Engineering in Ceramics

Analysis

The constituent phase was identified by a powder X-ray diffraction (X'Pert-MPD (θ-θ), Phillips) and texture and rocking curve (omega-scan) were obtained using an X-ray pole figure measurement apparatus (X'Pert-MPD (Open Eulerian Cradle), Phillips), respectively, using CuKα radiation operated at 40 kV-40 mA. The chemical composition of the films was analyzed by energy dispersive spectroscopy (EDS)(DX-95T, EDAX). The thickness of the thin films was measured with a surface profile meter (Dektak[3], Sloan). To measure current-voltage (C-V) characteristics, metal-insulator-semiconductor (MIS) structure should be formed. To form the top electrodes (0.2 mm in diameter), a metal mask was placed on the surface of the thin film, and Al was vacuum evaporated. To form the bottom electrode, SiO$_2$ layer at the back side of the substrates was removed using dilute HF, and Al was also vacuum evaporated. After the formation of both top and bottom electrodes, current-voltage (C-V) characteristics were measured by an impedance analyzer (4192A, Hewlett Packard).

Table I. Deposition conditions for YSZ and CeO$_2$ thin films prepared by PLD.

Laser	KrF excimer laser (248 nm)
Target	YSZ(8Y) and CeO$_2$ ceramics
Target-substrate distance	55 mm
Repetition frequency	7 Hz
Thickness / nm	YSZ:0.5-127, CeO$_2$: 80
Laser energy density	1.0 J/cm^2
Substrate temperature	800 $^\circ$C
O$_2$ gas pressure	5.5x10^{-4} Torr

Table II. Deposition conditions for YSZ and CeO$_2$ thin film prepared by sputtering

	Metallic film	YSZ film	CeO$_2$ film
Target	Metal Zr+Y	YZS ceramics	CeO$_2$ ceramics
Substrate temperature / $^\circ$C	770	770	650
Sputtering power / W	50	70	70
Gas pressure / mTorr	8	8	8
Sputtering gas	O$_2$/(Ar+O$_2$)=0/10	O$_2$/(Ar+O$_2$)=1/10	O$_2$/(Ar+O$_2$)=1/10
Deposition rate / nm·s^{-1}	0.2	1.5	3.0

RESULTS AND DISCUSSION

Crystal Structure of PLD CeO$_2$/YSZ Thin Film

Figures 1(a)-(d) show XRD spectra of the CeO$_2$/YSZ thin film on H-terminated Si(001) substrates prepared by PLD. When the CeO$_2$ thin film was deposited directly on Si(001), (111)-oriented CeO$_2$ was obtained but no trace of CeO$_2$(002) was detected (Fig. 1(d)). When YSZ was deposited prior to CeO$_2$ deposition, c-axis oriented CeO$_2$ thin films were obtained, as shown in Figs. 1(a)-(c). This means that YSZ thin film serves as a buffer layer to achieve c-axis oriented CeO$_2$ thin film even if the thickness of YSZ is as small as 0.5 nm. However, for 0.5 nm-thick YSZ buffered CeO$_2$ thin films, a trace CeO$_2$(111) peak was also detected. The CeO$_2$(111) peak was not detected at all for 1.5 nm-thick YSZ buffered CeO$_2$. This suggests that 0.5 nm-thick YSZ is not necessarily enough to completely cover the surface of the Si substrate. Figures 2(a) and (b) indicate X-ray pole figures for (a) CeO$_2$ and (b)

CeO₂/YSZ thin films deposited on H-terminated Si(001). These figures indicate that the YSZ buffered CeO₂ thin films not only have a c-axis orientation but also show heteroepitaxial growth on Si(001). It should be noted that the Si substrate before deposition was H-terminated and no SiO₂ was detected by XPS [15]; however, around 3 nm-thick SiO₂ layer was clearly observed by HRTEM for both CeO₂-Si and CeO₂/YSZ-Si interfaces [16], as schematically shown in Fig. 2. This fact indicates that a SiO₂ layer grows during the deposition of CeO₂ and YSZ. This also implies that the H-termination treatment is not necessary to achieve heteroepitaxial growth of YSZ. Therefore, a Si(001) substrate having natural SiO₂ was used during deposition of YSZ. The FWHM (omega-rocking curve) measurements were carried out and the effect of the initial state of Si substrates was compared. Table III(a) shows the result. For Si(001) having natural SiO₂, the value of FWHM was 2.33°. On the other hand, for H-terminated Si(001) , the value of FWHM was 2.50°. Comparing these data, the FWHM for Si having SiO₂ is relatively small; however, the difference is not so large. Actually RHEED patterns of YSZ and CeO₂/YSZ thin films deposited on Si(001) having natural SiO₂ show streaks, which indicate that both YSZ and CeO₂/YSZ are epitaxially grown with a smooth surface.

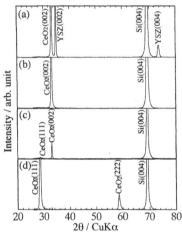

Figure 1. XRD spectra of CeO₂/YSZ thin films on H-terminated Si(001) by PLD. Thickness of YSZ is (a) 127, (b) 1.5, (c) 0.5 and (d) 0 nm, respectively. Thickness of CeO₂ is 80 nm.

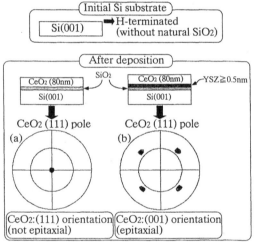

Figure 2. Change of the texture of CeO₂ thin films on H-terminated Si(001) with the introduction of YSZ buffer layer prepared by PLD.

486   Grain Boundary Engineering in Ceramics

Table III. Change of FWHM (omega) of YSZ(002) with the initial state of Si(001) substrates prepared by (a) PLD and (b) sputtering.

Preparation method	(a) PLD	(b) Sputtering
Thickness / nm	YSZ: 20	Metal (Zr+Y): 3 YSZ: 90
FWHM (omega) of YSZ(002) on Si(001) having natural SiO_2 / degree	2.33	1.48
FWHM (omega) of YSZ(002) on H-terminated Si(001) / degree	2.50	4.59

Crystal Structure of Sputtered Prepared CeO_2/YSZ Thin Film

For rf-magnetron sputtering, Si(001)substrates having natural SiO_2 was used. Figures 3(a)-(c) show schematic drawing of the X-ray pole figure results. In the case that CeO_2 was directly deposited on Si(001), c-axis oriented CeO_2 was obtained. By the X-ray pole figure measurement, it was clarified that CeO_2 was not epitaxial but in-plane random (Fig. 3(a)). It is interesting that the orientation of the CeO_2 thin film prepared by sputtering was completely different from that prepared by PLD as mentioned above. In the literature, the orientation of CeO_2 thin film was reported to be (111) or (011) (epitaxial) [3, 5-7] and no (001) orientation was reported. The reason why CeO_2(001) was formed is under consideration. In the case of PLD, the introduction of the YSZ buffer layer helped to achieve the heteroepitaxial CeO_2 thin film on Si(001). Therefore, for sputtering, a YSZ buffer layer was also introduced. However, as shown in Fig. 3(b), the X-ray pole figure indicated that CeO_2 thin film was in-plane random. Horita et al. succeeded in the preparation of heteroepitaxial YSZ on Si(001) by dc-reactive sputtering [17]. They deposited a metal (Zr+Y) thin film at the initial stage of deposition without O_2 flow. After the metallic film deposition they started to flow oxygen during deposition using a metal target. In this work, a metal (Zr+Y) film was also deposited at the initial stage of deposition, then the target was changed to YSZ or CeO_2 and O_2 was introduced. Figure 3(c) shows that heteroepitaxial CeO_2 thin film was achieved for both CeO_2/YSZ/(metal (Zr+Y)) and CeO_2/metal (Zr+Y)) thin films, even if the thickness of metal (Zr+Y) is as small as 1 nm. This indicates that metal (Zr+Y) is essential for epitaxial film formation by sputtering, which suggests that a redox reaction between metal (Zr+Y) and SiO_2 would be the key to achieve heteroepitaxial growth. Figure 4 shows the change of FWHM (omega) with the thickness of metal layer. The thickness of YSZ and CeO_2 was 90 nm. For both YSZ(002) and CeO_2(002) peaks, minimum FWHM is obtained for a 2 nm-thick metal layer. This suggests that a 1 nm-thick metal layer is insufficient to achieve this redox reaction, while a > 3 nm-thick metal layer is excessive. A metallic or low valent Zr and Y cations remains in the structure. This consideration was confirmed by XPS. For sputtering, the effect of the initial state of Si substrates was also compared. Table III(b) shows FWHM (omega) of YSZ(002) for Si(001) initially having SiO_2 and H-terminated. For these samples, the thickness of metal (Zr+Y) and YSZ is 3 and 90 nm, respectively. As shown in Table III(b), in the case of Si(001) having natural SiO_2, the value of FWHM was 1.48°. On the other hand, in the case of H-terminated Si(001), the value of FWHM was 4.59°. It is surprising that so large a FWHM was obtained on H-terminated Si(001). This indicates that SiO_2 on the Si substrate is essential to achieve heteroepitaxial growth.

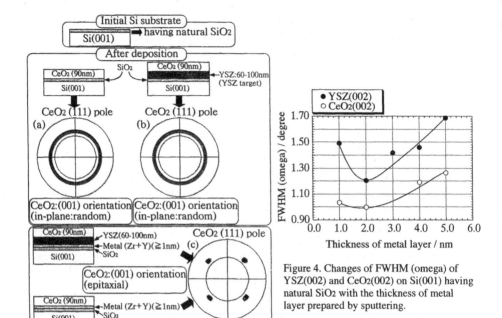

Figure 3. Change of the texture of CeO2 thin films on Si(001) having natural SiO2 with the introduction of metal (Zr+Y) buffer layer prepared by sputtering.

Figure 4. Changes of FWHM (omega) of YSZ(002) and CeO2(002) on Si(001) having natural SiO2 with the thickness of metal layer prepared by sputtering.

C-V Characteristics of PLD and Sputtered Prepared CeO₂/YSZ Thin Films

Figure 5 shows the C-V curve for CeO₂/YSZ thin films deposited on Si(001) having natural SiO₂, prepared by PLD. The thickness of CeO₂ and YSZ was 32 and 1 nm, respectively. This curve does not show any hysteresis caused by "ion drift" or "charge injection", which indicates a clear interface is achieved. Such an electrically-clean interface is required for FRAM applications. The overall dielectric constant εr for the thin film was calculated to be around 10. This value is smaller than the reported εr of CeO₂ (εr=26) [3]; however, this value is reasonable because the overall dielectric constant was calculated for Al/CeO₂/YSZ/SiO₂/Si/Al structure. In this case, low dielectric layer (SiO₂, εr=3.6) are in series connection.

On the other hand, Figure 6 shows C-V curve for CeO₂/YSZ thin film deposited on Si(001) having natural SiO₂ prepared by sputtering. The thickness of metal (Zr+Y) and CeO₂ was 2 and 90 nm, respectively. Fig. 6 shows that a slight hysteresis caused by "ion-drift" is observed. This "ion-drift" type hysteresis is commonly observed for YSZ/Si thin film [18], therefore, the hysteresis would be derived from the YSZ layer. Actually, the thickness of metal (Zr+Y) is twice larger than that of YSZ, as shown in Fig. 5. The overall dielectric constant εr for Fig. 6 is calculated to be 26, which value is close to bulk YSZ (εr=30) [4].

Grain Boundary Engineering in Ceramics

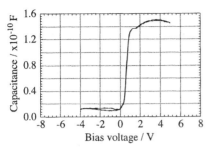

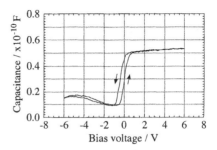

Figure 5. C-V curve for CeO2/YSZ thin film on Si(001) having natural SiO2 prepared by PLD. The thickness of CeO2 and YSZ was 32 and 1 nm, respectively.

Figure 6. C-V curve for CeO2/YSZ thin film on Si(001) having natural SiO2 prepared by sputtering. The thickness of metal (Zr+Y) and CeO2 was 90 and 2 nm, respectively.

General Discussion

Heteroepitaxial YSZ and CeO$_2$/YSZ thin films were successfully prepared on Si(001) substrate by both PLD and sputtering. By comparing the films made by these two methods, some general conclusions can be made. It was impossible to obtain heteroepitaxial CeO$_2$ directly on Si(001) for both methods. The orientation of CeO$_2$ was completely different in the two methods, i.e., (111) orientation by PLD and (001) orientation by sputtering. For PLD, heteroepitaxial CeO$_2$ was achieved by the introduction of a YSZ buffer layer; on the other hand, for sputtering, heteroepitaxial CeO$_2$ was not achieved by the deposition of YSZ but was achieved by the deposition of a metal (Zr+Y) layer. These facts imply that the redox reaction accompanying oxygen diffusion at the SiO$_2$/Si surface should be analyzed. For PLD, the YSZ target is ablated by laser beam to form plasma plume and the ablated species reaches the surface of the substrate. The fact that epitaxial YSZ is obtained by the use of YSZ target suggests that partially reduced zirconia (ZrO$_X$) or yttria (YO$_X$) are formed in the plasma plume. It is believed that the ablated species would have suitable valence or energy to achieve the redox reaction. On the other hand, for sputtering, almost no reduced species would be generated from the YSZ target; therefore no oxygen is released from the natural SiO$_2$ layer in Si substrates. Therefore, metal (Zr+Y) is needed to produce the redox reaction; however, the species sputtered from the target is believed to be metal. Therefore, SiO$_2$ is needed for the redox reaction. If the SiO$_2$ was removed by the H-termination, Si and metal (Zr+Y) cannot form an oxide but form a silicide. Comparing Table III(a) and III(b), the FWHM (omega) of PLD YSZ is much larger than sputtered YSZ. This means that the quality of YSZ thin film prepared by sputtering is much higher from the point of mosaicity. Whereas as shown in Figs. 5 and 6, the quality of PLD YSZ is higher than sputtered YSZ from the point of electrical applications. Optimizing processing conditions would improve these properties.

CONCLUSIONS

Heteroepitaxial YSZ and CeO$_2$/YSZ thin films were successfully achieved on Si(001) substrates by PLD and sputtering. For PLD, heteroepitaxial CeO$_2$ was achieved by the introduction of an ultra thin (0.5 nm-thick) YSZ buffer layer. The FWHM (omega) of YSZ(002) was similar irrespective of the surface treatment of the Si substrates. For sputtering, heteroepitaxial CeO$_2$ was achieved by the introduction of metal (Zr+Y) layer. The optimum thickness of the metal layer was 2 nm. Comparing PLD heteroepitaxial

CeO_2/YSZ with the sputtered one, sputtering is superior to PLD for preparing low mosaicity thin films, on the other hand, PLD is superior to sputtering for preparing electrically clean interface.

REFERENCES

[1] E. J. Tarsa, K. L. McCormick and J. S. Speck, "Common Themes in the Epitaxial Growth of Oxides on Semiconductors", *Mat. Res. Soc. Symp.Proc.* **341** 73 (1994).

[2] M. Suzuki and T. Ami, "A proposal of epitaxial oxide thin film structures for future oxide electronics", *Mater. Sci. Eng.* **B41** 166 (1996).

[3] T. Inoue, Y. Yamamoto, S. Koyama, S. Suzuki and Y. Ueda, "Epitaxial growth of CeO_2 layers on silicon", *Appl. Phys. Lett.* **56** 1332 (1990).

[4] M. T. Lanagan, J. K. Yamamoto, A. Bhalla and S. G. Sankar, "Dielectric properties of yttria-stabilized zirconia", *Mater. Lett.* **7** 437 (1989).

[5] M. Yoshimoto, H. Nagata, T. Tsukahara and H. Koinuma, "*In Situ* RHEED Observation of CeO_2 Film Growth on Si by Laser Ablation Deposition in Ultrahigh-Vacuum", *Jpn. J. Appl. Phys.* **29** L1199 (1990).

[6] C. Pellet, C. Schwebel and P. Hest, "Physical and electrical properties of yttria-stabilized zirconia epitaxial thin films deposited by ion beam sputtering on silicon", Thin Solid Films, **175** 23 (1989).

[7] T. Inoue and T. Ohsuna, "Growth of (110)-oriented CeO_2 layers on (100) silicon substrates", *Appl. Phys. Lett.* **59** 3604 (1991).

[8] T. Hirai, K. Teramoto, H. Koike, K. Nagashima and Y. Tarui, "Initial Stage and Growth Process of Ceria, Yttria-Stabilized-Zirconia and Ceria-Zirconia Mixture Thin Films on Si(100) Surfaces", *Jpn. J. Appl. Phys.* **36** 5253 (1997).

[9] C. A. Copetti, H. Soltner, J. Schubert, W. Zander, O. Hollricher, Ch. Buchal, H. Schulz, N. Tellmann and N. Klein, "High quality epitaxy of $YBa_2Cu_3O_{7-x}$ on silicon-on-sapphire with the multiple buffer layer YSZ/CeO_2", *Appl. Phys. Lett.* **63** 1429 (1993).

[10] R. Haakenaasen, D. K. Fork and J. A. Golovchenko, "High quality crystalline $YBa_2Cu_3O_{7-\delta}$ films on thin silicon substrates", *Appl. Phys. Lett.* **64** 1573 (1994).

[11] Y. Li, S. Linzen, P. Seidel, F. Machalett, F. Schmidl, H. Schneidewind, T. Schmauder, R. Cihar and S. Schaller, "Epitaxial growth of $YBa_2Cu_3O_{7-x}$ thin films on $CoSi_2/Si$ substrates with combined CeO_2/YSZ buffer layers", *Inst. Phys. Conf. Ser.* **148** 911 (1995).

[12] L. Mechin, J. C.Villegier, G. Rolland and F. Laugier, "Double CeO2/YSZ buffer layer for the epitaxial growth of $YBa_2Cu_3O_{7-\delta}$ films on Si(001) substrates", Physica C **269** 124 (1996).

[13] N. Wakiya, T. Yamada, K. Shinozaki and N. Mizutani, "Heteroepitaxial growth of CeO_2 thin film on Si(001) with ultra thin YSZ buffer layer", *This Solid Films* (in printing).

[14] W. A. Kern and D. A. Poutinen, "Cleaning Solutions Based on Hydrogen Peroxide for use in Silicon Semiconductor Technology", *RCA rev.* **31** 187 (1970).

[15] D. B. Fenner, A. M. Viano, D. K. Fork, G. A. N. Connel, J. B. Boyce, F. A. Ponce and J. C. Tramontana, "Reactions at the interfaces of thin films of Y-Ba-Cu- and Zr-oxides with Si substrates", *J. Appl. Phys.* **69** 2176 (1991).

[16] T. Kiguchi, N. Wakiya, K. Shinozaki and N. Mizutani, "Structure Analysis of $CeO_2/YSZ/Si$ Multilayer Thin Films by HRTEM", *Mat. Res. Soc. Symp. Proc.*, (in printing).

[17] S. Horita, Y. Abe, T. Kawada, "Heteroepitaxial growth of yttria-stabilized zirconia film on oxidized silicon by reactive sputtering", *Thin Solid Films* **281-282** 28 (1996).

[18] M. Watanabe, T. Naruse, A. Masuda and M. Horita, "Heteroepitaxial growth of YSZ films with controlled Y content on Si by reactive sputtering", *Technical Report of IEICE* **CPM96-98** 19 (1996).

REDUCTION OF NITROGEN OXIDE BY AN ELECTROCHEMICAL METHOD USING PROTON AND OXIDE-ION CONDUCTING CERAMICS

Tetsuro Kobayashi, Katsushi Abe, Yoshio Ukyo
Toyota Central R&D Labs., Inc., Nagakute, Aichi, 480-1192 Japan
Hiroyasu Iwahara
Center for Integrated Research in Science and Engineering, Nagoya University,
Furo-cho, Chikusa-ku, Nagoya, 464-8603 Japan

ABSTRACT

A steam electrolysis cell was constructed with a proton conductor $SrZr_{0.9}Yb_{0.1}O_{3-\alpha}$ as an electrolyte. The steam electrolysis cell efficiently reduced nitrogen oxide (NO) on the cathode, using hydrogen produced by steam electrolysis as a reducing agent at 460 °C. When a $Pt/Ba/Al_2O_3$ catalyst was placed on the cathode, NO was reduced even under an O_2-rich atmosphere. An electrolysis cell was also constructed with an oxide-ion conductor $8Y-ZrO_2$ as an electrolyte. The cell could reduce NO on the cathode, not only electrochemically electrolyzing NO directly but also chemically using hydrogen produced by steam electrolysis. However, the cell with the oxide-ion conductor could not reduce NO in an atmosphere containing O_2.

INTRODUCTION

The reduction of NO in the exhaust gas of automobiles is very important for the environment and the restriction of the exhaust gas. Chemical catalysts have been mainly used in order to reduce NO into N_2, using hydrocarbon (HC) and carbon monoxide (CO) as reducing agents in the exhaust gas. For improving fuel efficiency, lean-burn engines have become popular. The amount of O_2 in the exhaust gas of lean-burn engines is more than ten times that of NO, whereas the amount of HC and CO as reducing agents for NO are nearly equal to that of NO. Because almost all reducing agents react with O_2, the reduction of NO is difficult

To the extent authorized under the laws of the United States of America, all copyright interests in this publication are the property of The American Ceramic Society. Any duplication, reproduction, or republication of this publication or any part thereof, without the express written consent of The American Ceramic Society or fee paid to the Copyright Clearance Center, is prohibited.

under the lean-burn condition. Therefore, the selective reduction of NO will be necessary and many studies on this problem have been conducted [1]. Some perovskite-type oxides based on $SrCeO_3$, $BaCeO_3$, $CaZrO_3$ and $SrZrO_3$, such as $SrCe_{0.95}Yb_{0.05}O_{3-\alpha}$, exhibit good protonic conduction (10^{-3} S cm^{-1}) around 700 °C [2], where α is the mole fraction of oxide-ion vacancies. These high-temperature proton conductors have been studied widely for their applications to hydrogen sensors [3-5], SOFCs [6-8] and gas reactors [9,10].

EXPERIMENTAL PROCEDURE

We investigated the electrochemical reduction of NO using a steam electrolysis cell constructed with this proton conductor as the electrolyte [11]. Figure 1 shows a schematic diagram of a steam electrolysis cell for reducing NO. Water vapor is electrochemically oxidized on the anode and the produced protons are transported through the electrolyte to the cathode. On the cathode, protons are first reduced to atomic hydrogen, and then hydrogen gases are produced. Therefore, in this cell, it is expected that NO will be reduced not only electrochemically but also chemically. Water vapor in the exhaust gas can be used as a water source for the anode in this system, therefore, the proposed cell seems to be a simple and effective device for NO reduction in exhaust gases. An electrolysis cell with an oxide-ion conductor as an electrolyte was also constructed for reducing NO, as shown in Fig. 2. Figure 2 (a) shows the method of directly electrolyzing NO into N_2 and O_2. Figure 2 (b) shows the method of chemically reducing NO by H_2 produced by steam electrolysis when H_2O is supplied to the cathode with NO.

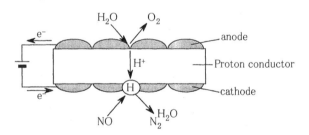

Figure 1. Schematic diagram of a steam electrolysis cell with a proton conductor for the reduction of nitrogen oxide.

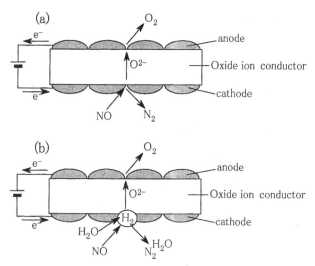

Figure 2. Schematic diagram of an electrolysis cell with an oxide-ion conductor for the reduction of nitrogen oxide.

Figure 3 shows a cross section of the experimental apparatus. Closed one-end tubes of sintered $SrZr_{0.9}Yb_{0.1}O_{3-\alpha}$ [12] (TYK Co., Ltd.) and 8 mol% Y_2O_3-doped ZrO_2 (NIKKATO Corp.) were used as proton and oxide-ion conductors, respectively. Platinum for electrodes was plated on both sides of the sintered tubes by an electroless plating method. The areas of the anode and the cathode were 6 cm^2 and 5 cm^2 in the cell with the proton conductor, and 8.3 cm^2 and 7.4 cm^2 in the cell with the oxide-ion conductor. The thickness of all electrodes was about 2 μm. A Pt/Ba/Al_2O_3 catalyst [13,14] was placed on the Pt plated cathode for facilitating the reduction of NO. The Pt/Ba/Al_2O_3 catalyst was prepared by the following process: γ-Al_2O_3 powder was added into platinum dinitro diammine solution and stirred for 3 hrs. After evaporating the water, the resulting powder was calcined in air at 300 °C for 3 hrs. This powder was put into barium acetate solution and treated by the same way to obtain Pt/Ba/Al_2O_3 catalyst. The composition of this catalyst was 1.2 wt% Pt, 25.1 wt% Ba and 73.7 wt% Al_2O_3. Because the specific surface area of γ-Al_2O_3 powder was about 170 m^2 g^{-1}, Pt and Ba were thought to be dispersed well on γ-Al_2O_3. Pt/Ba/Al_2O_3 catalyst and Pt paste (U-3820, N.E. Chemcat corp.) were mixed well, and then spread on the Pt plated cathode and calcined at 800 °C. The weight fractions of Pt/Ba/Al_2O_3 catalyst and Pt of Pt paste were 17 wt% and 83 wt%, respectively. The total amount of Pt/Ba/Al_2O_3 catalyst was 60 mg.

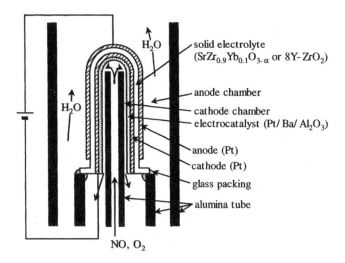

H₂O

solid electrolyte
(SrZr$_{0.9}$Yb$_{0.1}$O$_{3-\alpha}$ or 8Y-ZrO$_2$)

anode chamber

cathode chamber
electrocatalyst (Pt/ Ba/ Al$_2$O$_3$)

anode (Pt)
cathode (Pt)

glass packing

alumina tube

H₂O

NO, O₂

Figure 3. Cross-section of the experimental apparatus

The anode chamber and cathode chamber were constructed as shown in Fig. 3. In the cell with the proton conductor, 100 % H$_2$O was introduced into the anode chamber at a rate of 200 ml min^{-1}. On the other hand, in the cell with the oxide-ion conductor, the atmosphere of the anode was air. Inert gases with or without NO, O$_2$ and H$_2$O were introduced to the cathode at a rate of 30 ml min^{-1}. Direct current was applied galvanostatically to the cell. Experiments were carried out in the temperature range from 400 to 600 °C under atmospheric pressure. An hour after the desired temperature was obtained and the constant current was applied, the concentration of NO in the outlet gas from the cathode chamber was determined by Saltzman method. Gas chromatography with porapak Q and molecular sieve 5A columns was used to analyze the concentration of N$_2$, O$_2$ and N$_2$O. The concentration of NH$_3$ was determined by an absorbing-type gas detector (GASTEC Co. No. 3L, 3M). N$_2$, N$_2$O and NH$_3$ were the products of the reduction of NO.

RESULTS AND DISCUSSIONS

The results of the cell with the proton conductor without Pt/Ba/Al$_2$O$_3$ catalyst, of which the cathode was pure Pt plated electrode, is described first. Figure 4 shows the dependence of the NO-removal efficiency on current density when He gas containing 1000 ppm NO was introduced to the cathode at 450 °C. The NO-

Grain Boundary Engineering in Ceramics

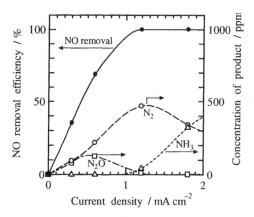

Figure 4. Change in NO-removal efficiency and concentrations of products with current density for the cell of Pt I Sr(Zr,Yb)O$_3$ I Pt. (Feed gas: 1000 ppm NO in He, Temp: 450 °C)

removal efficiency is defined as the ratio of the amount of removed NO to that of supplied NO according to equation (1).

NO-removal efficiency = removed NO/ supplied NO   (1)

The removal efficiency increased with increasing current density, and reached 100 % at current densities higher than 1.2 mA cm^{-2}. The main products were N$_2$O and N$_2$ at low current densities, and N$_2$ and NH$_3$ at high current densities. From this result, it was verified that the steam electrolysis cell constructed with the proton conductor could reduce NO effectively. Figure 5 shows O$_2$- and NO-removal efficiencies when He gas containing 1000 ppm O$_2$ and 1000 ppm NO was introduced to the cathode at 450 °C. At lower current densities, the reduction of O$_2$ was predominant. The reduction of NO started at current densities above 1.6 mA cm^{-2} at which most of O$_2$ was reduced. The product distribution of the NO reduction was similar to that of Fig. 4. It was found that O$_2$ was reduced easier than NO in this cell.

Figure 6 shows the temperature dependence of the NO-removal efficiency when He gas containing 8 % O$_2$ and 1000 ppm NO (O$_2$/NO=80) was introduced to the cathode. The concentrations of O$_2$ and NO correspond to those of exhaust gases from combustion engines operating under lean-burn condition. The current density was 2.4 mA cm^{-2}. If H$_2$ gas will be produced in an atmosphere

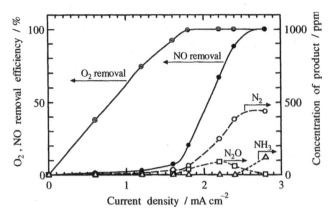

Figure 5. Change in O_2- and NO-removal efficiencies and concentrations of products with current density for the cell of Pt I Sr(Zr,Yb)O_3 I Pt. (Feed gas: 1000 ppm O_2 and 1000 ppm NO in He, Temp: 450 °C)

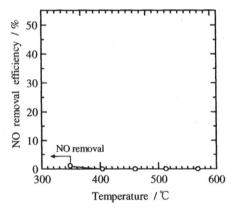

Figure 6. Temperature dependence of NO-removal efficiency for the cell of Pt I Sr(Zr,Yb)O_3 I Pt.   (Feed gas: 1000 ppm NO and 8 % O_2 in He, Current density: 2.4 mA cm^{-2})

containing O_2 and NO, the amount of H_2 gas produced at this current density is three times of the amount of NO and only one-twenty seventh of the amount of O_2 in the cathode gas.   NO was not removed at all on the pure Pt plated cathode with excess of O_2 as expected.

The NO-removal efficiency for the cell with Pt/Ba/Al$_2$O$_3$ catalyst on the Pt plated cathode under the same condition as Fig. 6 is shown in Fig. 7.   The cell with the catalyst could remove NO even under the O$_2$-rich condition, and the removal efficiency reached 20 % at 400 °C.   N$_2$ and N$_2$O were detected as products of reduction of NO.   Pt/Ba/Al$_2$O$_3$ catalysts are known to oxidize NO on Pt and absorb NO forming nitrate such as Ba(NO$_3$)$_2$ when O$_2$ coexists [13,14]. Therefore, the reaction cycle was considered as follows: (i) NO might be absorbed into Ba, (ii) absorbed NO might be reduced to N$_2$ and N$_2$O by hydrogen, and (iii) Ba formed by this reduction would absorb NO again.

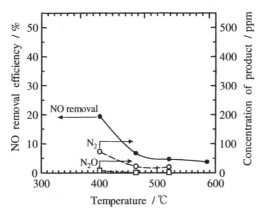

Figure 7. Temperature dependence of NO-removal efficiency and concentrations of products for the cell of Pt/Ba/Al$_2$O$_3$ I Pt I Sr(Zr,Yb)O$_3$ I Pt.   (Feed gas: 1000 ppm NO and 8 % O$_2$ in He, Current density: 2.4 mA cm^{-2})

Next, the reduction of NO by the cells with oxide-ion conductors was examined at 460 °C.   The feed gas to the cathode was Ar gas containing 1000 ppm NO without O$_2$.   Figure 8 shows the dependence of the NO-removal efficiency and the concentrations of the products on current density for the cell with only Pt plated cathode (without Pt/Ba/Al$_2$O$_3$ catalyst).   The removal efficiency was about 10 % and the reaction products were N$_2$ and N$_2$O.   Figure 9 shows the NO-removal efficiency for the cell with Pt/Ba/Al$_2$O$_3$ catalyst under the same condition as Fig. 8.   The removal efficiency increased with increasing current density and reached 80 % when the current density was above 0.5 mA cm^{-2}.   The reaction products were also N$_2$ and N$_2$O.   The reduction of NO by the direct electrolysis, as shown in Fig. 2 (a), was found to be possible from these results.

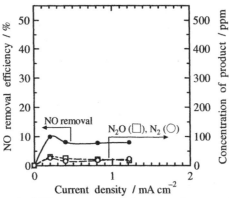

Figure 8. Change in NO-removal efficiency and concentrations of products with current density for the cell of Pt I 8Y-ZrO$_2$ I Pt. (Feed gas: 1000 ppm NO in Ar, Temp: 460 °C)

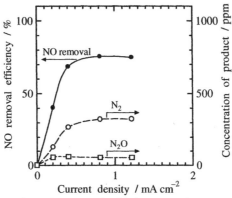

Figure 9. Change in NO-removal efficiency and concentrations of products with current density for the cell of Pt/Ba/Al$_2$O$_3$ I Pt I 8Y-ZrO$_2$ I Pt. (Feed gas: 1000 ppm NO in Ar, Temp: 460 °C)

Figure 10 shows the NO-removal efficiency for the cell without Pt/Ba/Al$_2$O$_3$ catalyst when Ar gas containing 1000 ppm NO, 1600 ppm O$_2$ and 1.6 % H$_2$O was fed to the cathode. At lower current densities, only the reduction of O$_2$ was observed. After complete reduction of O$_2$ at current densities above 2 mA cm^{-2}, the production of H$_2$ by steam electrolysis was started. The reduction of NO and the production of H$_2$ started at the same current. However, the NO-removal efficiency was only 10 % at 4 mA cm^{-2}. The reduction product was only NH$_3$.

Grain Boundary Engineering in Ceramics

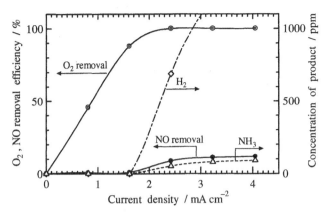

Figure 10. Change in O_2- and NO-removal efficiency and concentrations of products with current density for the cell of Pt I 8Y-ZrO_2 I Pt.   (Feed gas: 1000 ppm NO, 1600 ppm O_2 and 1.6 % H_2O in Ar, Temp: 460 °C)

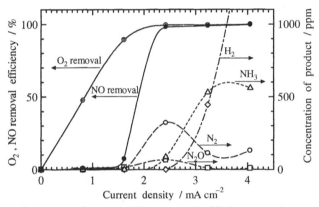

Figure 11. Change in O_2- and NO-removal efficiency and concentrations of products with current density for the cell of Pt/Ba/Al_2O_3 I Pt I 8Y-ZrO_2 I Pt. (Feed gas: 1000 ppm NO, 1600 ppm O_2 and 1.6 % H_2O in Ar, Temp: 460 °C)

The NO-removal efficiency for the cell with Pt/Ba/Al_2O_3 catalyst under the same condition as Fig. 10 is shown in Fig. 11.   The reduction of O_2 was only observed at lower current densities even though the cell with Pt/Ba/Al_2O_3 catalyst was used. The reduction of NO proceeded at current densities above 1.6 mA cm^{-2} where O_2 was reduced almost completely.   The NO-removal efficiency reached 100 % and H_2 was produced at current densities above 2.4 mA cm^{-2}.   When the removal of

NO was observed, the productions of N_2O and N_2, and that of N_2 and NH_3 proceeded at lower and higher current densities, respectively. Because the production of NH_3 was observed, as shown in Figs. 10 and 11, it was revealed that the removal of NO proceeded through the reduction of NO by hydrogen produced by steam electrolysis, as shown in Fig. 2 (b). The NO-removal efficiency for the cell with $Pt/Ba/Al_2O_3$ catalyst was much higher than that for the cell without the catalyst. The reduction of NO partially oxidized and absorbed into Ba (such as $Ba-NO_x^*$) is thought to be easier than the direct reduction of NO.

Then, the reduction of NO for the cell without the catalyst was tried when the reducing (positive) and oxidizing (negative) pulse currents, of which pulse interval was 0.25 or 0.5 second, were applied alternately. When a negative pulse current is applied to the cell, NO will be partially oxidized by oxygen ions on the anode, as shown in Fig. 12. Then the partially oxidized NO will be reduced when a positive pulse current is applied. Because the partially oxidized NO was expected to be easier to reduce than NO, it was considered that the NO-removal efficiency would be improved by this method. Figure 13 shows the NO-removal efficiency and the concentrations of the products when pulse currents (positive and negative current densities: +5.8 and −1.1 mA cm^{-2}, pulse interval: 0.25 or 0.50 second) and a constant current (only positive current density: +3.4 mA cm^{-2}) were applied. The feed gas to the cell was Ar gas containing 1000 ppm NO, 1900 ppm O_2 and 1.6 % H_2O. Oxygen produced by a negative pulse current must be reduced and the net charge for reducing NO by pulse currents is half of that by a constant current. Therefore, the reducing ability by electrolysis with the pulse currents, of which experimental conditions are described above, is estimated to correspond to that with the constant current of +2.35 mA cm^{-2}. The NO-removal efficiency when the pulse currents were applied alternately was higher than that when the constant current was applied, as shown in Fig. 13. This result also suggested that the partial oxidation of NO enhanced the reduction of NO.

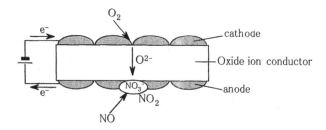

Figure 12. Schematic diagram of oxidation of nitrogen oxide by a cell with an oxide-ion conductor.

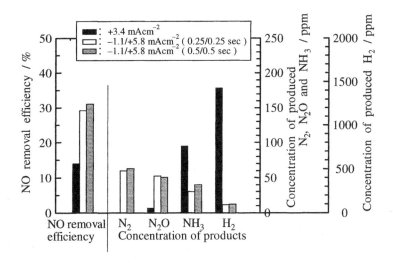

Figure 13. NO-removal efficiency and concentrations of products for the cell of Pt I 8Y-ZrO$_2$ I Pt when pulse current (negative and positive current densities: -1.1 and +5.8 mA cm^{-2}, pulse interval: 0.25 or 0.50 second) and constant current (positive current density: +3.4 mA cm^{-2}) were applied. (Feed gas: 1000 ppm NO, 1900 ppm O$_2$ and 1.6 % H$_2$O in Ar, Temp: 460 °C)

When the Pt/Ba/Al$_2$O$_3$ catalyst was placed on the cathode, O$_2$ was preferably reduced by the cell with the oxide-ion conductor, while NO was preferably reduced by the cell with the proton conductor even in the existence of excess O$_2$. This difference is thought to result from the strong reducing ability of H$^+$ or atomic hydrogen produced at the cathode of the cell with a proton conductor. This study has verified three ways to reduce NO electrochemically by the cell with high-temperature type ionic conductors, and also showed that the catalyst placed on the cathode had large effect on the reduction of NO. Therefore, an extensive study of catalysts will be necessary for improving the removal efficiency of NO in these electrolysis cells.

CONCLUSIONS

A steam electrolysis cell was constructed using a high-temperature type proton conductor, $SrZr_{0.9}Yb_{0.1}O_{3-\alpha}$. As a result of the examination of the reduction of NO by hydrogen produced by the steam electrolysis cell, NO was effectively reduced. When a $Pt/Ba/Al_2O_3$ catalyst was placed on the cathode of the cell, NO was preferably reduced even in the existence of excess O_2. An electrolysis cell with 8 mol% Y_2O_3-doped ZrO_2 as oxide-ion conductor was also constructed. It was revealed that the reduction of NO was possible by direct electrolysis of NO and by hydrogen produced by steam electrolysis. However, the reduction of NO was impossible in the cell with the oxide-ion conductor when O_2 coexisted.

REFERENCES

[1] *SAE SP-938*, "Automotive Emissions and Catalyst Technology," ISBN 1-56091-302-9 (1992).

[2] H.Iwahara, "Proton Conducting Ceramics and Their Applications," *Solid State Ionics*, **86-88** [1], 9-15 (1996).

[3] T.Yajima, K.Koide, H.Takai, N.Fukatsu and H.Iwahara, "Application of Hydrogen Sensor Using Proton Conductive Ceramics as a Solid Electrolyte to Aluminum Casting Industries," *Solid State Ionics*, **79**, 333-337 (1995).

[4] N.Kurita, N.Fukatsu, S.Miyamoto, F.Sato, H.Nakai, K.Irie and T.Ohashi, "The Measurement of Hydrogen Activities in Molten Copper Using an Oxide Protonic Conductor," *Metall. Mater. Trans. B*, **27** [6], 929-935 (1996).

[5] N.Fukatsu, N.Kurita, T.Ohashi and K.Koide, "Hydrogen Sensor for Molten Metals Usable up to 1500 K," *Solid State Ionics*, **113-115**, 219-227 (1998).

[6] H.Iwahara, T.Yajima, T.Hibino and H.Ushida, "Performance of Solid Oxide Fuel Cell Using Proton and Oxide Ion Mixed Conductors Based on $BaCe_{1-x}Sm_xO_{3-\alpha}$," *J. Electrochem. Soc.*, **140** [6], 1687-1691 (1993).

[7] H.Iwahara, "Oxide-ionic and Protonic Conductors Based on Perovskite-type Oxides and Their Possible Applications," *Solid State Ionics*, **52** [1-3], 99-104 (1992).

[8] N.Taniguchi, E.Yasumoto and T.Gamo, "Operating Properties of Solid Oxide Fuel Cells Using $BaCe_{0.8}Gd_{0.2}O_{3-\alpha}$ Electrolyte," *J. Electrochem. Soc.*, **143** [6], 1886-1890 (1996).

[9] S.Hamakawa, T.Hibino and H.Iwahara, "Electrochemical Hydrogen Permeation in a Proton-Hole Mixed Conductor and Its Application to a Membrane Reactor," *J. Electrochem. Soc.*, **141** [7], 1720-1725 (1994).

[10] T.Hibino, S.Hamakawa, T.Suzuki and H.Iwahara, "Recycling of Carbon Dioxide Using a Proton Conductor as a Solid Electrolyte," *J. Appl. Electrochem.*, **24** [2], 126-130 (1994).

[11] T.Kobayashi, S.Morishita, K.Abe and H.Iwahara, "Reduction of Nitrogen Oxide by a Steam Electrolysis Cell Using a Proton Conducting Electrolyte," *Solid State Ionics*, **86-88** [1], 603-607 (1996).

[12] T. Yajima, H. Suzuki, T. Yogo, and H. Iwahara, "Protonic Conduction in $SrZrO_3$-based Oxides," *Solid State Ionics*, **51** [1-2], 101-107 (1992).

[13] N. Miyoshi, S. Matsumoto, K. Katoh, T. Tanaka, J. Harada, N. Takahashi, K. Yokota, M. Sugiura, and K. Kasahara, "Development of New Concept Three-way Catalyst for Automotive Lean-burn Engines," *SAE paper* 950809 (1995).

[14] N. Takahashi, H. Shinjoh, T. Iijima, T. Suzuki, K. Yamazaki, K. Yokota, H. Suzuki, N. Miyoshi, S. Matsumoto, T. Tanizawa, T. Tanaka, S. Tateishi, and K.Kasahara, "The New Concept 3-way Catalyst for Automotive Lean-burn Engine: NO_x Storage and Reduction Catalyst," *Catalysis Today*, **27**, 63-69 (1996).

SURFACE DECOMPOSITION MECHANISMS ON SIC (0001) AND (000-1) FACES

M. Kusunoki, T. Suzuki, T. Hirayama and N. Shibata
Japan Fine Ceramics Center FCT Central Research Department, Atsuta, Nagoya, Japan

ABSTRACT

A remarkable difference of decomposed structures will be shown between the Si(0001)-face and the C(000-1)-face after heating at 1700°C for a half hour in a vacuum. On the C-face, an aligned carbon nanotube film are self-organized perpendicular to the surface. On the contrary, a layer of very thin graphite sheets parallel to the surface are formed on the Si-face. Decomposition mechanisms on both faces are proposed from high-resolution electron microscopy (HREM) results.

INTRODUCTION

Silicon carbide (SiC) is one of the most promising materials for engineering applications because of its attractive properties. It has wide band gaps, excellent mechanical properties and chemical stability, and exhibits poly typism [1]. Therefore, it is expected to be used for high-power and high-temperature semiconductor devices. The surface structures or the surface reconstruction have been investigated intensively by scanning tunneling microscopy (STM) [2-4] and low-energy electron diffraction (LEED) [2,5]. These experiments have shown that by increasing the heating temperature above 1000°C, surfaces of SiC are covered with carbon-rich structures such as $6\sqrt{3} \times 6\sqrt{3}$, 30° [2,3,5] or 6 x 6 [4,5]. This phenomenon has been confirmed by the selective desorption of Si atoms by annealing SiC in a vacuum.

Hexagonal α-SiC is a zinc blende type material with a polar [0001] axis, in which the ideal (0001) and (000-1) surfaces are terminated by a layer of Si and that of C atoms respectively. Van Bommel et al. [2] investigated the difference of surface decomposed

To the extent authorized under the laws of the United States of America, all copyright interests in this publication are the property of The American Ceramic Society. Any duplication, reproduction, or republication of this publication or any part thereof, without the express written consent of The American Ceramic Society or fee paid to the Copyright Clearance Center, is prohibited.

structures on both Si- and C-faces by heating up to 1500°C. They examined LEED patterns and Auger electron spectra (AES) obtained from each face, then concluded that a monocrystalline graphite layer and a polycrystalline graphite layer formed on the Si-face and C-face, respectively.

Recently, the authors discovered an aligned carbon nanotube film being self-organized by surface decomposition of SiC by TEM in-situ observations [6]. Additionally, it was shown by using HREM and electron energy-loss spectroscopy (EELS) that the surface decomposition progressed as residual oxygen gas produced an oxidation reaction[7,8].

In this paper, it will be clarified that the decomposed structures on the Si-face and the C-face are different, and they will be explained by the different oxidation mechanisms on the respective faces from cross-sectional HREM results.

EXPERIMENTAL PROCEDURE

Commercial single crystal wafers of 6H-SiC with the polished Si-face and those with the polished C-face (CREE Research, Inc.) were cut to the size of 1.0 x 4.0 x 0.2 mm. The wafers with the C-face were heated at 1200, 1250, 1300 and 1700°C for 0.5 h, and those with the Si-face at 1350 and 1700°C for 0.5 h in a vacuum furnace (1×10^{-4} Torr) with an electric resistance carbon heater. These specimens were carefully thinned by both dimpling and ion bombardment for HREM observation of the cross section using a Topcon 002B TEM operated at 200 kV.

RESULTS AND DISCUSSION

Figures 1(a) and (b) show images and the diffraction patterns obtained from the Si-face and the C-face of SiC single-crystal wafers heated to 1700°C for 0.5 h. On the C-face, aligned carbon nanotubes 3-5 nm in diameter and 0.25μm in length were formed perpendicular to the surface, as shown in Fig. 1(a) [7,8]. On the contrary, in the case of the Si-face, seventeen graphite sheets, 5 nm in thickness, were formed parallel to the Si-face as shown in Fig. 1(b). The 0002 lattice distances of the graphite were 0.353 nm and 0.343 nm on the C-face and on the Si-face respectively. The carbon layer formed on the C-face is almost 50 times thicker than the one on the Si-face. The differences in the respective faces regarding thickness and in the structures of carbon layers are remarkable, nevertheless both bulk structures are common. The difference in the thickness of decomposed layers on the respective faces has already been reported from the results of ellipsometric measurements [9] and AES results [2]. A number of researchers dealing with SiC single crystals have shown that heating SiC gave a rougher surface on the C-

Grain Boundary Engineering in Ceramics

face than on the Si-face. However, the cause of it has not been explained.

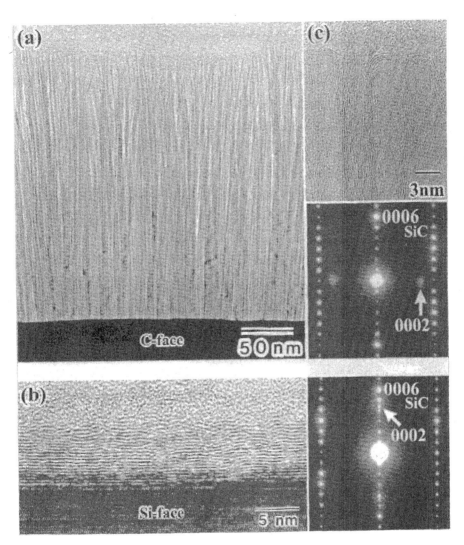

Figure 1. Micrographs and the diffraction patterns of the surface of SiC single-crystal wafers heated to 1700℃ for 0.5 h, on (a) the C(0001)-face and (b)the Si(0001)-face.

Two- and three- layer collapse mechanisms [2,10] were proposed to explain the orientation and the carbon density of the graphite layers formed by the surface decomposition of SiC. These mechanisms can explain the formation of the monocrystalline graphite layer parallel to the surface as seen on the Si-face. However, there are some difficulties to explain the differences between the C-face and the Si-face, as shown in Figs. 1(a) and (b) by these mechanisms.

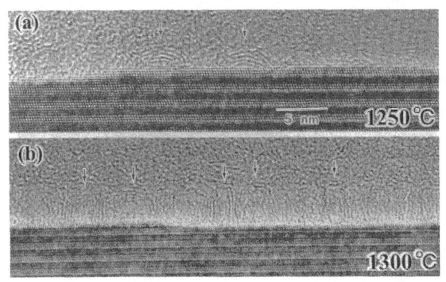

Figure 2. HREM images showing surfaces of the C-face heated at (a) 1250°C and (b)1300°C for 0.5 h.

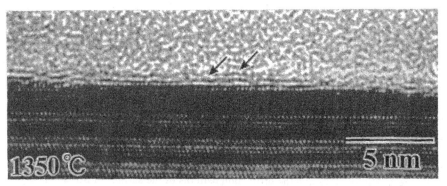

Figure 3. A HREM image showing the surface of the Si-face heated at 1350°C for 0.5 h. Two graphite sheets parallel to the surface are indicated by arrows.

Grain Boundary Engineering in Ceramics

Cross-sectional HREM of decomposed surfaces on the initial stage was carried out to investigate the mechanism of the surface decomposition on both the Si-face and the C-face. Figures 2(a) and (b) show micrographs of surfaces of the C-face heated at 1250°C and 1300°C for 0.5 h, respectively. The amorphous contrasts on the upper side of each micrograph in Fig. 2 and in Fig. 3 are due to the glue used during the specimen preparation. The glue played an important role in protecting the carbon structures from the ion bombardment, though the amorphous contrast strongly influences the weak contrasts of carbon structures at the initial stage.

At 1250°C, 4 layers of slightly arced graphite sheets of 5 nm in diameter and 1-2 nm in height could be found dispersed on the C-face as shown by arrows in Fig. 2(a). This corresponds to the initial stage of the decomposition of the C-face. Furthermore, when it was heated to 1300°C, 2-3 layered carbon nanocaps 3-5nm in diameter and 3-5 nm in height were generated densely over the surface as shown in Fig. 2(b). Here, cross-sectional HREM was used to clarify an atomically uneven-surface structure of the C-face, which has not been easy to detect by STM.

When the Si-face was heated to 1350°C for 0.5 h, only two graphite sheets were observed on the surface, as shown by arrows in Fig. 3. The graphite sheets were observed flat and parallel to the surface for every location.

From these experimental results, the following mechanisms are proposed to explain the decomposition process of the Si-face and of the C-face as shown in figure 4. In the case of the Si-face, Si atoms are removed from the surface of the Si-face due to the oxidation mechanism (Fig. 4(a)). This oxidation of Si results in the formation of SiO gas molecules, which eventually evaporate from the surface. The C atoms released from the Si-C bonding with the Si atom in the 1st layer, are individually supported by the vertical bonding with Si in the 3rd layer, as shown in Fig. 4(b). Therefore, each carbon atom remains at the initial position A (noted in Fig. 4(a) and (b)) without falling into disorder by the desorption of Si atoms. As soon as the removal of Si atoms in the 3rd layer (position B, noted in Fig. 4(b)) takes place by the ongoing oxidation process, the C atoms located at position A would drop to B to maintain the surface structures. As the results, a net of only carbon atoms are formed transitionally, as shown in Fig. 4(c). The net of carbon is also supported by C-Si bonding with Si in the next 5th layer. Furthermore, repeating the selective oxidation of Si atoms on odd layers, one graphite sheet would become completed dense at every collapse of three carbon layers [2].

STM and LEED experiments have shown that the Si-face exhibits a graphite surface with $(6\sqrt{3} \times 6\sqrt{3})$ or (6×6) reconstruction after annealing at about 1000°C, which was

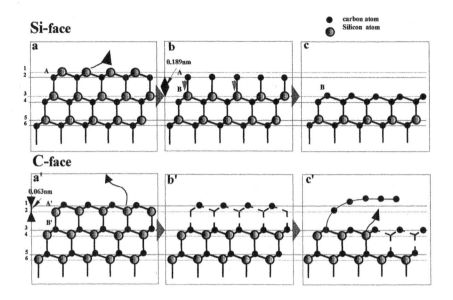

Figure 4. Mechanisms of the decomposition process of (a)-(c) the Si-face and (a')-(c')
the C-face due to the selective oxidation of Si atoms.

rotated $30°$ to have a lattice continuity with respect to the SiC lattice [2]. In the present
model, each graphite net is always drawn back to the Si atoms in the next layer of the
SiC crystal with the vertical bonding as shown in Fig. 4(c). This mechanism explains the
specific orientation of the monolayer or monocrystalline graphite layers, with respect to
SiC crystal, generated on the Si-face.

On the other hand, in the case of the C-face, Si atoms at position A' on the 2nd layer
(noted in Fig. 4(a')), are easily desorped by oxidation by breaking the Si-C bonding with
the C atoms on the 1st and the 3rd layers. Accordingly, the C atoms on the 1st layer would
become free as they lose any bonding with the surface (Fig. 4(b')). These free carbon
atoms would accumulate and form island carbon clusters on the surface or graphite sheet
fragments of size about 5 nm parallel to the SiC surface, as seen in Fig. 2(a). The edge of
the nano-scale graphite sheets was expected to be active from the calculation of the π-
electronic state [11]. Accordingly, the edge of the graphite tends to be curved to connect
with carbon atoms on the surface, as shown in Fig. 4(c'). Furthermore, SiO gas
molecules are continuously generated at the boundary between SiC and the graphite
clusters. Finally, as a result, carbon nanocaps [8] are formed as seen in Fig. 2(b).

These mechanisms could result in the difference in thickness of the carbon layers between the Si-face and the C-face. This is because, on the Si-face, the dense graphite sheet grown one-by-one parallel to the surface, would protect the inner side of the Si layer from oxidation process. In the case of the C-face, carbon nanocaps are formed from the initial stage of the decomposition and the aligned straight carbon nanotubes grow perpendicular to the surface. This makes it easy for the oxygen atoms to migrate to the SiC surface through the decomposed carbon layer.

As mentioned above, the difference of the decomposition on both faces was explained simply due to the atomic geometrical arrangement of the SiC crystal. However, the atomic oxidation mechanism of SiC is thought to be very complicated similar to that of the one of Si. In the case of a Si single crystal, the complicated oxidation process has been vigorously investigated, and formation of a Si-O-Si bond has been confirmed mainly by the first principle calculation [12,13]. It was also indicated that the direction of oxidation depended on the surface structures [14] and the surface orientation of Si. On the contrary, the oxidation or the decomposition of SiC has been rarely studied theoretically. Recently, Ventra et al.[15] indicated the formation of Si-O-C bonds during the oxidation process of SiC by the first principle theory. The difference of the complicated decomposition phenomena on Si- and C-face of SiC should be explained theoretically on the atomic level.

This mechanism indicates the possibility that the surface decomposition of some metal carbide crystals may play some interesting roles in the field of "new-carbon engineering" in the near future.

ACKNOWLEDGEMENT

The authors thank Dr. K. Kaneko for valuable suggestions. This work was supported by the FCT project, which was cosigned to JFCC by NEDO.

REFERENCES

[1] W. F. Knippenberg, "Growth Phenomena in Silicon Carbide," *Philips Research Reports,* **18,** 161-274 (1963).

[2] A. J. Van Bommel, J.E. Crombeen, and A. Van Tooren, "LEED and Auger Electron Observations of the SiC(0001) Surface," *Surf. Sci.,* **48,** 463-72 (1975).

[3] F. Owman and P. Martensson, "The SiC(0001)6$\sqrt{3}$x6$\sqrt{3}$ reconstruction studied with STM and LEED," *Surf. Sci.,* **369,**126-36 (1996).

[4] M. A. Kulakov, P. Heuell, V. F. Tsvetkov, and B. Bullemer, "Scanning Tunnelling Microscopy on the 6H SiC(0001) Surface," *Surf. Sci.*, **351**, 248-54 (1994).

[5] C. S. Chang and I.S.T. Tsong, " Scanning Tunneling Microscopy and Spectroscopy of Cubic β-SiC(111) Surfaces," *Surf. Sci.*, **256**, 354-60 (1991).

[6] M. Kusunoki, M. Rokkaku, and T. Suzuki, "Epitaxial Carbon Nanotube Film Self-Organized by Sublimation Decomposition of Silicon Carbide," *Appl. Phys. Lett.*, **71**, 2620-22 (1997).

[7] M. Kusunoki, M. Rokkaku, and T. Hirayama, Aligned Carbon Nanotube Film Self-Organized on a SiC Wafer," *Jpn. J. Appl. Phys.*, **37**, L605-L606 (1998).

[8] M. Kusunoki, T. Suzuki, K. Kaneko, and M. Ito, "Formation of Self-Aligned Carbon Nanotube Films by Surface Decomposition of Silicon Carbide," *Phil. Mag. Lett.*, **79**, 153-61 (1999).

[9] F. Meyer and G. J. Loyen, " Ellipsometry Applied to Surface Problems," *Acta Electronica,* **18**, 33-38 (1975).

[10] D.V. Badami, "X-ray Studies of Graphite Formed by Decomposing Silicon Carbide," *Carbon,* **3**, 53-57 (1965).

[11] M. Fujita, K. Wakabayashi, K. Nakada and K. Kusakabe, " Peculiar Localized State at Zigzag Graphite Edge," *J. Phys. Soc. Jpn.*, **65**, 1920-23 (1996).

[12] K. Kato, T. Uda and K. Terakura, "Backbond Oxidation of the Si(001) Surface: Narrow Channel of Barrierless Oxidation," *Phys. Rev. Lett.*, **80**, 2000-03 (1998).

[13] T. Uchiyama and M. Tsukada, "Scanning-Tunneling-Microscopy Images of oxygen Adsorption on the Si(0001) Surface," *Phys. Rev. B,* **55**, 9356-59 (1997).

[14] H. Kageshima and K. Shiraishi, "First-Principles Study of Oxide Growth on Si(100) Surfaces and at SiO2/Si(100) Interfaces," Phys. *Rev. Lett.*, **81**, 5936-39 (1998).

[15] M. D. Ventra and S. T. Pantelides, "Atomic-Scale Mechanisms 0f Oxygen Precipitation and Thin-Film Oxidation of SiC," *Phys. Rev. Lett.*, **83**, 1624-27 (1999).

SELECTIVE REACTION OF TITANIUM WITH CARBON AND NITRIDE IN Zr-Ti-O OXIDES AND THE RELATED MICROSTRUCTURES

Jianbao LI, and Yong Huang
The State-Lab of New Ceramics and Fine Processing,
Department of Materials Science and Engineering, Tsinghua University,
Beijing 100084, P.R.China
Tel:86-10-62782753, Fax:86-10-62771160,
E-mail:ljb-dms@mail.tsinghua.edu.cn

Hideaki Matsubara and Hiroaki Yanagida
Japan Fine Ceramics Center, 2-4-1 Mutsuno, Atsuta-ku,Nagoya 456-8587,Japan
Tel:81-52-871-3500.

ABSTRACT

Complex oxides of the Zr-Ti-O system are one of the most important groups of ceramic materials for functional and structural applications. The differences in behavior of the elements-titanium and zircon at high temperature and under a changing atmosphere affect not only the stability but also the properties. This study experimentally investigated the difference of these two kinds of the cations on the ionic mobility and their combination with other anions. Zirconium oxide (ZrO_2) was mixed with TiO_2 in solid solution by the co-precipitation technique. After mixing with carbon or explored in nitrogen gas at elevated temperatures, titanium cations preemptively moved out from the oxide compound against zirconium cations and formed carbide or nitride, depending on the existing species of reductive anions, leaving zirconium in oxide form. The reaction where the reductive anions of carbon and nitrogen selectively react with titanium in the complex oxides is called selective reaction and occurs at temperatures between 1350-1450°C. Both the titanium carbide and nitride formed around the grains but the nitride formed thin layers that closely contacted the grains. The carbide formed a small grain unattached to with oxide grains. The newly-formed nitride strengthened the grain boundaries and the carbide weakened the boundaries The diffusion process of titanium cation induced a weak bonding inside the matrix grains and formed nanolayered structures that divided the grains into subgrains with nanometer layeres and demonstrated stepped fractures. The microstructures of nitride and carbide examined under SEM are reported in detail.

INTRODUCTION

Ceramics have a modulus much higher than metals but the apparent strength and toughness are not significant. The reason is due to its less tolerance to the defect size and deformation rate. The effective deformation of ceramics mainly occurs due to grain boundary sliding, in contrast to the complex lattice sliding occurring in metallic materials at ambient temperature. An increase in

To the extent authorized under the laws of the United States of America, all copyright interests in this publication are the property of The American Ceramic Society. Any duplication, reproduction, or republication of this publication or any part thereof, without the express written consent of The American Ceramic Society or fee paid to the Copyright Clearance Center, is prohibited.

weak bonding in ceramics is expected to improve the toughness and strength. One approach to create the weak interface in the ceramic products is to make the grains or phases in nanometer sizes and another one is to introduce multilayered structures. The nanomaterials and nanocomposite approach have shown the great improvement results on the fracture strength of ceramic materials. Niihara[1] et al reported that the addition of 5% nanometer sized SiC particles increased the strength of alumina ceramics from 350 MPa to 1500 MPa, approximately four time of that in the monolithic ceramic materials. This type of nanometer materials also exhibits excellent super-plastic deformation which is important for forming the complicated mechanical parts of zirconia[2]. Multilayered structural ceramics probably break deflectively along the weak interlayer planes and exhibit a reasonable increase in apparent toughness[3][4]. Biological materials such as nacreous shells combine the micrometered layeres of calcium carbonate with organic nanophases to form a strong body of inorganic/organic laminated-composites,. Some of the prospective methods to combine the nanophase with layered structures in ceramic system are eutectic crystallization, spinodal descomposition[5], and annealing dissolution of the supersaturated solid solution compounds. The differences in reaction behaviors; for example, diffusion rates, reduction/oxidation ability and high temperature stability, between the composing elements can also be used to control the formation of fine structures.

ZrO_2, one of the toughest ceramics, has been called as ceramic steel. Its high toughness resulted from the phase transformation of the tetragonal phase (t-ZrO_2) to monoclinic phase (m-O_2) that accompanies the approximately 5vol% volumetric shrinkage, producing a constriction force to the crack tip opening. The oxide of MgO, Y_2O_3 and Ce_2O_3 are usually used to stabilize the tetragonal phase (t-ZrO_2) at ambient temperature. Titanium oxide also shows the stabilizing effect to t-ZrO_2 by forming the solid solution such as (Zr, Ti)O_2[6]. The highest solubility of TiO_2 in ZrO_2 is approximately 18 mol%[6] [7]. The excessive TiO_2 content reulted in the formation of TiZrO_2 and then TiO_2 solid solution, under the condition of dynamic equilibrium. However, the co-precipitation technique can produce a (Zr,Ti)O_2 solid solution with much higher TiO_2 content[6] [7]. In order to prepare the zirconia based Ti-nonoxide/Zr-oxide nanocomposite for wear-resistant applications, we have attempted to synthesize the nanoparticles of titanium carbide/nitride in situ in the zirconium-titanium complex oxides[8] .

This paper reports the reaction of carbonization and nitridation of co-precipitated (Zr,Ti)O_2 and their composite microstructures.

II. EXPERIMENTAL

The co-precipitated (Zr,Ti)O_2 solid solution was employed to show the difference in the carbonization/nitridation reaction. Chemicals of $ZrOCl_2 \bullet 8H_2O$ and $Ti(SO_4)_2$ were mixed in deioned water with ratio of TiO_2/ZrO_2=28 wt% and then NH_4OH was dropped into the solution to initiate the co-precipitation reaction at 9pH. Three mol% Y_2O_3 was dropped into the solution with chemical form of $Y(NO_3)_3$ to stabilize the t-ZrO_2. After drying and calcinning at 800°C in air for 1 h, the precipitates were identified as a single phase of tetragonal (Zr,Ti)O_2 solid solution using X-ray powder diffraction.

The carbonization and nitridation of the oxide: were accomplished as follows. Sucrose was mixed into the dried powders as the carbon source. These solid solution powders were pressed into pellets and then heated in a graphite chamber at the given reaction temperatures. The reaction atmosphere was flowing argon gas or nitrogen gas.

SEM and EDAX were used to observe the microstructure and the chemical composition ratios, respectively.

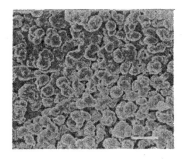

Figure 1. (Zr,Ti)O$_2$ solid solution particles (white bar=0.2 μm)

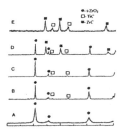

Figure 2. X-ray patterns showing the phase changes after annealing at the temperatures: A-as prepared, B-1350℃, C-1450℃,D-1550℃,E-1650℃

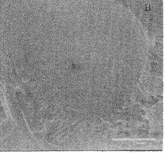

Figure 3. The morphology of TiC crystal formed around the oxide grain (white bar=0.1 μm). A-TiC fine particles on the surface of the grain; B-TiC surrounding the grain boundary

III. RESULTS AND DISCUSSIONS

Carbonization of (Zr,Ti)O$_2$ oxides powder

Figure 1 shows the solid solution powder of (Zr,Ti)O$_2$ as calcined. Particles were round shape and approximately 3 um in diameter. The X-ray powder diffraction peaks in Fig.2 proved that a pure t-ZrO$_2$ phase existed but the crystalline lattice was slightly less than the normal t-ZrO$_2$ due to the smaller ionic radii of Ti^{4+} than Zr^{4+}.

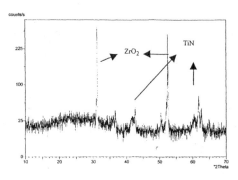

Figure 4. X-ray diffraction pattern of nitridated (Zr,Ti)O$_2$ powders

The mixture of the complex oxide with carbon source (sucrose) was heated at 1350℃, 1450℃, 1550℃ and 1650℃ in a flowing argon atmosphere. X-ray diffraction showed that temperatures between 1300-1450℃ was moderate to cause the selective reduction reaction of titanium ion, as shown in Fig.2. Temperature above 1500℃ caused the carbonization of zirconium oxide.

Fig.3a is the surface of reacted particles and Fig.3b is the fracture surface of the particles. The protuberance of TiC fine crystallites formed in the amorphous carbon surrounding the oxide grains. The EDAX data indicated that approximately 3-5 mol% of Zr was contained in the TiC phase.

Nitridation of the oxides (Zr,Ti)O$_2$

The co-precipitated oxide powder was also heated in flowing nitrogen atmosphere without addition of carbon source. The reaction temperature was 1450℃. After 2 hours of reaction in nitrogen gas, the titanium nitride (TiN) was detected in the X-ray diffraction pattern as shown in Fig.4. However, most of the peaks of the ZrO$_2$ phases were obviously shifted.

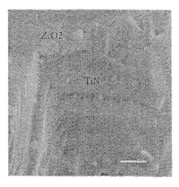

Figure 5. A- plate-like TiN crystallites formed along the ZrO$_2$ grains (the central lines is the chemical analysis position); B-lamellar structure of ZrO$_2$/TiN (white bar=0.4 µm)

Grain Boundary Engineering in Ceramics

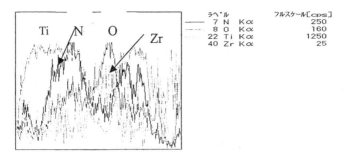

ラベル			フルスケール[cps]
7	N	Kα	250
8	O	Kα	160
22	Ti	Kα	1250
40	Zr	Kα	25

Figure 6. Line analysis of the elements: N,O, Ti and Zr.

The SEM observation found that the TiN phases in dense materials formed plate-like crystallites and arranged in lamellar structure with ZrO_2 phases. Figures 5 (A and B) show the arranged crystallites of TiN in ZrO_2ss matrix. The width of TiN plates was 0.1-0.3 um. Fig.6 is the line analysis profiles of the related four elements (Ti, N, Zr,O) along the line AB in Figure 5a. The heights of the lines of elements Ti and Zr changed in opposite directions. The lines of anion elements nitrogen N and oxygen O changed corresponding with Ti and Zr, respectively.

These profiles indicated the separation of the element Ti from $(Zr,Ti)O_2$ complex oxide to selectively form TiN in nitrogen atmosphere. The composition analysis with EDAX revealed that the TiN phase contained 3-18 atom% of zirconium. Titanium was also detected in the ZrO_2 grains. The separation was not complete, which is probably due to the short reaction time.

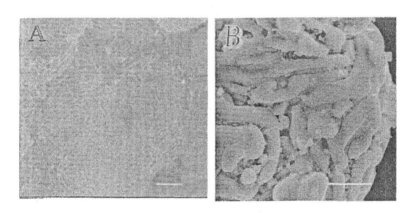

Figure 7. A-Fiber-like TiN crystallites grown on the surface of the oxide matrix grains; B-the close up of the TiN crystallites in A actually show plate-like morphologies (white bar=0.4μm).

Grain Boundary Engineering in Ceramics   517

Microstructures of the reaction boundary

Figures 7 (7A and 7B) show the reacted surfaces. The plate-like and fiber-like shaped crystallites protubed from the surface of the Ti-rich particles. These morphologies are quite deferent from one of TiC as shown in Fig.3.

The photographs in Fig.8 are the SEM images. The grain boundaries between $(Zr,Ti)O_2$ oxide grains were very clean (Fig.8(A)). The boundary contained titanium nitride that formed extended grain boundaries of the oxide. The extended thickness of the TiN boundary depended on the orientation relationship with the matrix grains, as shown in Fig.8(B). The expended boundary phase surrounded most of the matrix oxide. The composition of the expended boundary phase was TiN-based Ti-Zr-O-N solid solution, similar to that of Si_3N_4 based Si-Al-O-N solution (Sialon ceramics[9]). This expended phase was in close contact with the matrix grains on both sides and was expected to play a strengthening role for the boundary bonding on both chemically and mechanically. Fig.8(C) is the fracture section that reveals strong bonding between the oxide grains. This strengthening effect was in contrast with weakening of TiC boundary phase, as shown in Fig.8(D).

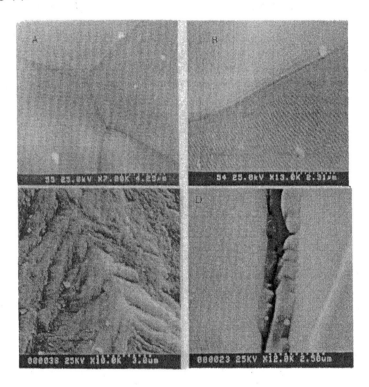

Figure 8. TiN boundary phase formed around the matrix grain (A-B). (C) shows the fractured interface with strong bonding with TiN phase; (D) shows TiC formed around the matrix grains with an interface defect.

Selective reduction of the complex oxides show the promising advantages: growing of nano-crystallites on the matrix grains, epitaxial growth of the crystallites with desirable shapes and orientation, preparation of hollow nanoparticles, creating nanometer structure inside the micrometer grain materials, improving the bonding and contact of boundaries. Figure 9 gives a demonstration of the fracture behavior of the fine structure of nanometer layers formed in the ZrO_2 grain. The $(Zr,Ti)O_2$ grains that partially suffered from the dissolution of titanium were fractured step by step. The heights of the fracture step was approximately 30-300 nm. This stepped fracture occurred due to the nanometer layer existing in the grain that can be also seen in Fig.8(A-B). The formation of this structure attributed to owe to the migration and dissolution of Ti ions in the matrix grain at the elevated temperature. The weak interfaces inside the grain divided the micrometer-sized grains into nanometer-sized subgrains with nanolayered structures. In this way, the controlled dissolution and selective reduction of the complex compound is expected to become a new approach to prepare nanocomposites.

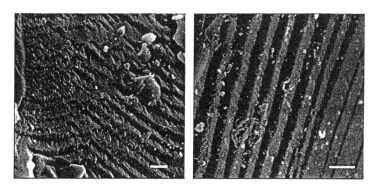

Figure 9. The step fractures observed on the oxide matrix grain (white bar=0.1um). (B) is the close-up of (A)

CONCLUSION

Co-precipitated $(Zr,Ti)O_2$ oxides were partially reduced by carbon and nitrogen. The results include:
1. Ti element was preemptively migrated out from the co-precipitated complex $(Zr,Ti)O_2$ grain to form its carbide or nitride compound (TiC or TiN). Nitridation had precedence over the carbonization if the nitrogen source existed.
2. Both the carbide and nitride of titanium were solid solutions with zirconium.
3. TiN_{ss} or TiC_{ss} phases surrounded the matrix oxide grains. TiN formed an extended boundary phase of the matrix oxide grains and strengthened the interfacial bonding, in contrast to the TiC

boundary phase which weakened the interfacial bonding of the matrix grains.

4. The partial nitridation and selective reaction of titanium with nitrogen also caused the formation of the nanolayered structure in the matrix grains. These partially nitridated oxide composites showed a step-like fracture with a step height of 30-300 nanometer in the matrix oxide grains.

ACKNOWLEDGEMENTS

This research was partially supported by Japan Science and Technology Agency (STA Fellowship). The work on the synthesis process was supported by NFS of China and The China Education Ministry for the Outstanding Youths Scientists Foundation Programs.

REFERENCES.

(1). K. Niihara, A. Nakahira and T. Sekino, "New Nanocomposite Structural Ceramics", pp.405-412 in *MRS Proceedings of the Symposium on Nanophase and Nanocomposite Materials*, Vol.286 (1993), Published by Materials Research Society of USA

(2). J. C. Parker, " Chapter 27 Commercialization Opportunities for Nanophase Ceramics" pp577 in *Nanomaterials Synthesis, Properties and Applications*, Ed by A. S. Edelstein and R. C. Cammarata, 1996, Institute of Physics Publish.

(3). W. J. Clegg, K. Kendall, N. M. Alford, T. W.Button and J. D. Birchall, "A Simple Way to Make Tough Ceramics", *Nature* (London), 34, 455-7(1990).

(4). D. B. Marshall, J. J. Ratto and F. F. Lange, "Enhanced Fracture Toughness in Layered Microcomposites of Ce-ZrO$_2$ and Al$_2$O$_3$", *J. Am. Ceram. Soc.*, 74 (12) 2979-87 (1991).

(5). J. B. Li and Y.Huang, "Microwave Interaction with Ceramic Materials and Its Application on Spinodal Discomposition" pp256-268 in *Mass Transport in Ceramics, Ceramic Transactions*, The American Ceramic Society, Vol71 (1996). Ed. by K. Koumoto et al.

(6). K. Tsukuma, "Transparent TiO$_2$-Y$_2$O$_3$-ZrO$_2$ Ceramics" pp11-20 in *Zirconia Ceramics* (Japanese) Vol.8.(1992), Japan Ceramic Society.

(7). C. Bateman, M. Notis and C. Lyman, "Phase Equilibria and Phase Transformation in ZrO$_2$-TiO$_2$ and ZrO$_2$-MgO-TiO$_2$ System" pp31-43 in *Proceedings of The Zirconia Vol.2*, (1986), (Japanese) Japan Ceramic Society.

(8). J. F. Liu, J. B. Li, H. L. Wang and Y. Huang, "In Situ Synthesis of Y-stabilized Tetragonal Zirconia Polycrystal Powder Containing Dispersed TiC by Selective Carbonization", *J. Am. Ceram. Soc.*, 82 (6) 1161-13(1999).

(9). K. H. Jack, "Sialons and Related Nitrogen Ceramics", *J. Mater. Sci.*, 11(8),1135-58(1976).

Key-Words: zirconium oxide, nanocomposite, titanium, carbide, nitride, selective reaction.

Observation of magnetic interfaces by electron holography

Tsukasa Hirayama

Japan Fine Ceramics Center
2-4-1 Mutsuno, Atsuta-ku, Nagoya 456-8587, Japan

TEL:+81-52-871-3401
FAX:+81-52-871-3599
E-mail:hirayama@jfcc.or.jp

Key words : electron holography, magnetic interfaces, observation, real-time

Electron holography [1] is a useful two-step imaging method to visualize electromagnetic microfields. In the first step, hologram is recorded by electron holographic microscope. Then, in the second step, the image is reconstructed by an optical system, Mach-Zehnder interferometer, or by a computer having a software equivalent to the optical interferometer. However, since electron holography is a two-step imaging method described above, electromagnetic fields could not be observed in real time. Although a few new methods of electron holography such as phase shifting method [1] and computer-assisted systems [2] succeeded in shortening the processing time of image-reconstruction, they did not achieve real-time observation. Nevertheless, dynamic observation is very important in materials research, because electromagnetic properties are often related to dynamic behavior of electromagnetic fields. Therefore, on-line real-time observation is needed for further study.

Recently, liquid-crystal spatial light modulators (LC-SLM) have been used in optical image processing systems or in optical holography. We have developed an on-line real-time electron holography system by using the LC-SLM to combine an electron holographic microscope and a Mach-Zehnder interferometer, and suceeded in observing magnetic domain walls in real time[3]. We would like to show this system and results of dynamic observatiuon of magnetic domain walls of a thin permalloy film.

References:
1. Q. Ru, J. Endo, T. Tanji and A. Tonomura, Appl. Phys. Lett. 59 (1991) 2372.
2. W. D. Rau, H. Lichte, E. Volkl and U. Weierstall, J. Compu. Assis. Micros. 3 (1991) 51.
3. T. Hirayama, J. Chen, T. Tanji and A. Tonomura, Ultramicroscopy 54 (1994) 9.

PAST, PRESENT AND FUTURE OF GRAIN BOUNDARY ENGINEERING

Tadao Watanabe
Laboratory of Materials Design and Interface Engineering,
Department of Machine Intelligence and Systems Engineering
Graduate School of Engineering, Tohoku University, Sendai, Japan

ABSTRACT

Grain boundary engineering based on the relationship between grain boundary structure and properties has been established and launched a new era of development of high performance structural and functional materials. It has been well proved that the grain boundary character distribution (GBCD) which can quantitatively describe the distribution of different types of existing grain boundaries,and the grain boundary connectivity are important microstructural factors controlling bulk properties in polycrystalline materials. This paper introduces the historical background of grain boundary engineering and discusses basic knowledge of designing and controlling grain boundaries in order to confer high performance on polycrystalline aggregates. Recent achievements of grain boundary engineering are briefly mentioned. Finally future prospect of grain boundary engineering is discussed.

INTRODUCTION

Grain boundaries and interphase boundaries are important microstructural elements which can exist and control bulk properties in polycrystals, but not in single crystals. These grain boundaries (hereafter including interphase boundaries) are special sites of the structural discontinuity and the heterogeneity being different from the grain interior having regular and periodic atomic structure. They are two-dimensional lattice defects and can play important roles in structure-sensitive metallurgical phenomena associated with bulk properties of polycrystalline aggregates. The importance of grain boundaries to the bulk properties of polycrystalline materials was recognized and seriously discussed by several pioneers like Cyril Stanley Smith [1],Karl Aust and Bruce Chalmers [2]. Donald Mclean [3], since 1940's almost half century after the first observation by Sorby on grain structure in blister steel with an optical microscope at the magnification of only 9 times [4]. More recently Hondros revisited the nature of the grain boundary from the earliest days of metallographic observations on grain boundaries in polycrystalline materials [5]. In my opinion, there have been three types of grain boundary research: the first one is concerned with the effect of grain size (grain boundary density) on bulk properties, particularly mechanical properties studied on engineering materials such as iron and steels. The second is a topological approach to geometrical configurations of grain boundaries existing in real polycrystals, as made by C.S.Smith [1]. The third type is a systematic study of the effects of grain boundary structure on boundary properties which was first discussed by Aust and Chalmers in detail [2] and gave us the most important and strongest impetus to recent progress in our understanding of the relationship between grain boundary structure and properties [6-8]. In fact, the concept of grain boundary design and

To the extent authorized under the laws of the United States of America, all copyright interests in this publication are the property of The American Ceramic Society. Any duplication, reproduction, or republication of this publication or any part thereof, without the express written consent of The American Ceramic Society or fee paid to the Copyright Clearance Center, is prohibited.

control which was first enunciated by the present author [9], is based on the fact that grain boundary phenomena and properties strongly depend on the boundary type and structure in both structural and functional materials. So let us firstly look at historical background of grain boundary engineering, revisiting the basis of "the concept of grain boundary design and control". Then we briefly look how much the concept has been effectively utilized as a powerful tool for development of high performance structural or functional materials, as recently demonstrated by Aust, Palumbo and coworkers [10-13] and by the present author's group [14-16].

HISTORICAL BACKGROUND OF GRAIN BOUNDARY ENGINEERING

(1) Early Stage of Grain Boundary Engineering

The influence of grain boundaries was found very early on plastic deformation in crystals; polycrystal specimens showed much higher flow stress than single crystal specimens [17]. This was an important recognition of the effect of the presence of grain boundaries on bulk properties and was explained by the effect of grain boundary as obstacle to crystal slip or dislocation motion. It is reasonable to expect that the presence of more grain boundaries may cause more significant effect on mechanical behaviour of polycrystals, as actually recognized from the observation that fine grained polycrystals show much higher yield and flow stresses than coarse grained ones. The effect of grain size (the effect of the grain boundary density) is of engineering importance demonstrating the influence of grain boundaries on bulk properties of polycrystalline materials [3]. The Hall-Petch type dependence on the grain size has been often observed and effectively used as a powerful tool in controlling the influence of grain boundaries. Grain size effect has been studied on mechanical properties [18] and also on electrical and magnetic properties of polycrystalline materials[ ]. Quantitative study of the influence of grain boundaries focussing on the grain size effect has helped our understanding of boundary-related bulk properties. However, quite recently, it has been demonstrated that the effect of grain size cannot be explained simply as previously understood because the opposite Hall-Petch slope was observed when the grain size becomes extremely small in the nanometer scale as observed in nanocrystalline materials [19]. Unfortunately the negative Hall-Petch slope has not been reasonably explained yet, needs to be clarified. On the other hand, a topological approach of grain boundary configuration in which Syril Stanley Smith [1] was deeply involved and made much effort at very early stage of the history of grain boundary study, has not been fully developed later, to understand, control and to utilize the influence of grain boundaries on bulk properties in two-dimensional (2D) thin or three-dimensional (3D) bulky polycrystalline materials.

(2) Relationship between Grain Boundary Character and Properties

Grain boundaries can play as preferential sites for metallurgical phenomena occurring in polycrystalline materials. When carefully observed, it is evident that grain boundary phenomena normally do not occur uniformly at every boundary,but very heterogeneously from boundary to boundary. We can see a large difference between grain boundaries of the propensity to grain boundary phenomena, for example, intergranular,segregation, migration, corrosion, fracture and so on. In fact this is due to the effect of grain boundary character (grain boundary type and structure). Figure 1 shows the propensity to intergranular fracture in Bi-doped copper polycrystal. We know very well that high purity OFHC copper never shows intergranular fracture without bismuth which is the most severe embrittling element to copper. Again it is worthwhile noting that all boundaries were not equally embrittled and that some boundaries easily broke but the portions of the weak boundary at which annealing twin meet,became extremely resistant to fracture (indicated by the letters A and B). Thus there is a significant difference of the propensity to corrosion or fracture among grain boundaries. Probably Aust and Chalmers [2] were the first who pointed out the importance of grain boundary structure to intergranular phenomena. So we must fully recognize the importance of grain boundary structure and take it into account in discussion of boundary-related bulk properties in polycrystalline materials. The relationship between grain boundary structure and properties (mechanical,physical and chemical) has been extensively studied since 1960's until recently [19-20] by using metal bicrystal specimens in which existing grain boundaries were well defined. More recently the influence of grain boundary structure on electric properties such as superconductivity and PTCR characteristics were carefully studied by using bicrystal specimens of YBCO [21] and BaTiO3 [22,23]. It has been revealed that special low-energy boundaries like low-angle boundary and low-Σ coincidence boundaries in these materials have quite different properties from those of high-energy random boundary, showing higher critical current Jc and much lower resistivity associated with PTCR characteristic,respectively.

MICROSTRUCTURAL FACTORS CONTROLLING BULK PROPERTIES

(1) Grain Size (GB Density) and Grain Shape (GB Inclination)
As easily understood from the difference of bulk property between a single crystal and a polycrystal, possible effects of grain boundaries on bulk properties in polycrystals is evident. In order to understand these effects of grain boundaries, we must take into account two kinds of factors associated with grain boundaries: one is geometrical factor, the second is structural factor. Firstly the geometrical factors associated with grain boundaries are the grain boundary density and grain boundary inclination. As far as the grain size is concerned, it is possible to change the average grain size (the grain boundary density) over a wide range covering 3 to 4 order of magnitudes of grain size, for example from 10nm to 100μm in polycrystalline specimens of the same material. As observed in polycrystalline nickels [24], the plasticity, hardness, corrosion and electric resistivity can drastically change with decreasing the grain size (increasing the grain boundary density) in the submicron range. On the other hand, when the grain size is extremely large as mm scale, the inclination of grain boundary becomes important to boundary-related metallurgical phenomena. We can easily imagine the extreme case from unidirectionally solidified or grown and lamellar-coarse grained polycrystals.

(2) Grain Boundary Character Distribution (GBCD).
The characterization of grain boundary is essential to establish the relationship between grain boundary structure and properties in bicrystals and polycrystals. In general, the grain boundary is characterized by determining the relative orientation relationship between adjoining grains, for example applying the SEM-ECP technique. Quite recently a computer-aided automated analyzing equipment, named "Orientation Imaging Microscope"(OIM) has been developed for orientation analysis and boundary characterization by Dingley, Adams and their coworkers [26,27]. Figures 2(a) and 2(b) show an example of the characterization by the OIM technique for grain boundaries in a molybdenum polycrystal produced by thermomechanical processing. The advantage and usefulness of the OIM is that for example the grain structure which shows similar image contrast in an optical micrograph (Fig.2(a)) can be clearly resolved into several smaller grains. Existing grain boundaries are characterized as high angle random boundary (R), low-angle boundary (L) and low-energy/low-Σ coincidence boundaries with Σ value smaller than 29. From the result of OIM analysis,we can easily know the distribution of different types of grain boundaries (the grain boundary character distribution,GBCD) and the grain boundary connectivity. Thus the development of OIM has enabled us to characterize a large number of grain boundaries in polycrystalline materials having the grain sizes ranging from mm to μm order without much difficulty and within a short time. We expect that the OIM technique will help to establish the standard method for the determination of GBCD in polycrystalline materials, even nanocrystalline material in the future.

Grain Boundary Engineering in Ceramics

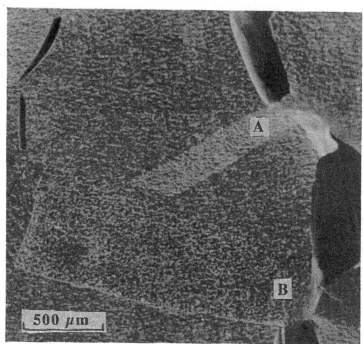

Fig. 1. Intergranular fracture occurring in a Bi-doped high purity copper polycrystal. Note that intergranular fracture will not proceed at the portion A and B where twins interat gran boundary.

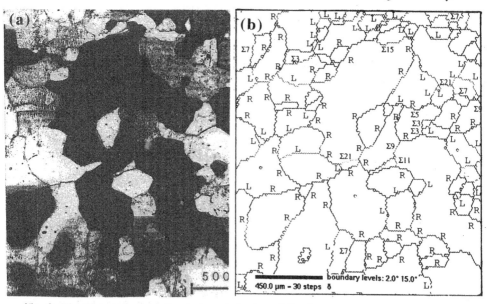

Fig. 2. The Characterization of grain boundaries by a SEM-EBSP-OIM system for molybdenum polycrystal produced by thermomechanically from single crystal. (a) SEM image, (b) GBCD.

Grain Boundary Engineering in Ceramics

(3) Grain Boundary Connectivity
In order to fully understand important roles of grain boundaries in bulk properties, we need to know the statistical distribution of different types of grain boundaries and also the grain boundary connectivity, because the magnitude of the influence of individual grain boundaries differs and strongly depends on the boundary type and structure. Figure 3 shows structure-dependent intergranular creep fracture observed in alpha iron-0.8at.% tin alloy. Intergranular creep fracture preferentially took place by sliding-assisted fracture mechanism at random boundaries which could easily slide during creep deformation [27], while low-Σ coincidence boundary and low-angle boundary are very difficult to slide and will not break. It has been found that the grain boundary fracture strength varies depending on the boundary type and misorientation in metal bicrystals [20]. Figure 4 schematically shows fracture processes occurring in a polycrystal in connection with the observed effects of the boundary character distribution and the grain boundary connectivity [9]. Since cracks tend to nucleate and propagate preferentially at random boundaries, a typical intergranular fracture can occur when a higher fraction (more than 2/3) of random boundaries are connecting to each other. On the other hand the mixture of tansgranular and intergranular fracture occurs when low-energy fracture-resistant low-Σ coincidence boundaries are connected with high-energy random boundaries in polycrystals kept in various environments (temperature, gas/liquid, external fields). Thus the importance of the grain boundary connectivity is rather easily understood to grain boundary-related percolative processes like corrosion, fracture, diffusion in a polycrystal.

Intergranular fracture becomes more serious when the material appears intrinsically brittle like refractory metals, intermetallics and ceramics and also when the material is used in environments and under the condition which promote intergranular fracture. Quite recently we have found that the grain boundary character distribution and the grain boundary connectivity are key factors controlling electric field-induced intergranular fracture in piezoelectric polycrystalline PZT ceramic in which grain boundaries had been characterized before electric-field induced fracture test.[28]. It was observed that the main crack propagated preferentially along random boundaries, avoiding low-angle boundary and low-Σ coincidence boundaries in the specimen. This is because the stress concentration is generated at surrounding grain boundaries due to the extension of individual domains and their interaction with the surrounding boundaries when an electric field is applied to a piezoelectric polycrystalline specimen during poling or service,. Again intergranular fracture takes place preferentially at weak random boundary when the stress concentration exceeds the cohesion of grain boundary, or the grain boundary fracture strength. It can be said generally that the grain boundary character distribution (GBCD) and the grain boundary connectivity are the most important microstructural factors controlling structure-dependent intergranular and percolative phenomena like intergranular corrosion and diffusion as well as intergranular fracture in polycrystalline materials.

HOW TO ENGINEER GRAIN BOUNDARY MICROSTRUCTURES

(1) Combined Effect of GBCD and GB Connectivity
The introduction of the grain boundary character distribution (GBCD) and the grain boundary connectivity has enabled us to quantitatively predict, design and control grain boundary-controlled phenomena and bulk properties in polycrystalline materials. Let us look at the prediction of GBCD-controlled fracture behaviour and characteristics made previously.

Lim and Watanabe predicted the fracture toughness as a function of GBCD and the brittle-ductile transition in two-dimensional [29] and three-dimensional polycrystals [30]. The effects of random or non-random GBCD on the fracture toughness were predicted for polycrystals having equiaxed grain structure and elongated grain structures with different grain aspect ratios. Figure 5 shows the

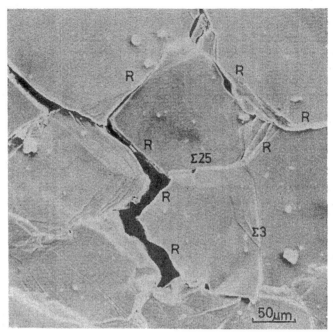

Fig. 3. Structure-dependent intergranular creep fracturein iron-o.8at.%tin allo crept at 973K under tensile stress of 29.4 M Pa. Random boundaries (denoted by R) could easily slide and fracture, while low-Σ coincidence boundareies like Σ3 will not break (the tensile stress axis is horizontal).

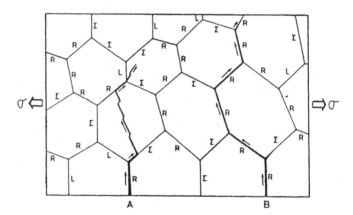

Fig. 4. Schematic representation of structure-dependent intergranular fracture processes in polycrystal. Path A: combined process of intergranular and transgranular fractrure, Path B: typical intergranular fracture. The grain boundary character distribution (GBCD) and the grain boundary connectivity determine possible fracture path and fracture mode [9].

Grain Boundary Engineering in Ceramics

fracture toughness as a function of the overall fraction, f, of fracture-resistant low-energy boundaries (low-Σ coincidence boundaries) for materials with different levels of the toughness given by the toughness ratio gt/gi for tansgranular fracture to intergranular fracture. We see that the fracture toughness monotonously increases with increasing the fraction of fracture-resistant low-energy boundaries when gt/gi is large, or the material is more brittle. This suggests that an increase of the fraction of fracture-resistant low-energy boundaries can improve more effectively the fracture toughness by restricting intergranular fracture without any change of fracture mode in 'intrinsically brittle' material with higher value of gt/gi. On the other hand, when the value of gt/gi is as low as 2, or the material is more ductile, the fracture toughness can increase with increasing the fraction of low-energy boundaries, changing the fracture mode from intergranular to transgranular fracture at the critical level of f=0.5, beyond which the fracture toughness controlled by transgranular fracture, does not depend on the fraction of low-energy boundaries anymore. These predictions have provided a useful guideline for improvement of fracture toughness of brittle materials by grain boundary engineering. It is worth remembering that grain boundary engineering is more effective as the brittleness of material becomes more severe. It has been found the grain boundary character distribution and the grain boundary connectivity can be effectively designed and controlled by the introduction of a sharp texture which is characterized by the localization of grain orientations around a specific crystal orientation,often low index orientations [31]. In fact this was achieved for B-free polycrystalline Ni$_3$Al which is normally brittle because of high propensity to intergranular fracture. As shown in Fig.6, the reduction of the fraction of weak random boundaries (R) up to about 30% by unidirectional solidification through floating zone melting resulted in high ductility (60%) at room temperature [32]. However such high ductility disappeared when a higher fraction of random boundaries exist again after rolling and annealing. It is evident that GBCD is a key factor controlling the brittleness.

(2) Combined Effect of GB Character Distribution (GBCD) and Grain Size

So far the control of the grain size has been extensively attempted in materials development by applying different processing methods. We have already found that the GBCD is closely related to the grain size for thermomechanically produced polycrystalline metals and alloys as shown in Fig.7; the frequency of low-Σ coincidence boundaries increases up to almost 100% with decreasing the grain size to about 1μm [32]. However the opposite grain size dependence of the frequency of low-Σ coincidence was found in iron-6.5mass% silicon ribbons produced by rapid solidification and subsequent annealing [24]. This demonstrates clearly that the relationship between GBCD and the grain size depends on processing method. In the rapidly solidified ribbons the driving force for grain growth was orientation-dependent surface energy being different from bulky polycrystals in which grain growth is driven by the grain boundary energy and strain energy associated lattice defects. It very likely that polycrystals which have the same grain size but not the same GBCD are produced in the same material by using different processing methods. Therefore we must remember that a large variety of combinations between GBCD and the grain size are possible. We can change the grain size over a wide range from the level of μm for conventional polycrystalline materials to nanocrystalline materials. Since the grain size is a measure of the grain boundary density so that grain refinement simply results in the introduction of more grain boundaries. When processing method used is specified, we can predict the GBCD in connection with the grain size and also optimal grain size and GBCD for desirable bulk property. The detail of the optimization of grain boundary microstructure given by at least the GBCD and grain size (grain boundary density) may be related to boundary-related property and service environment, as well as the material itself.

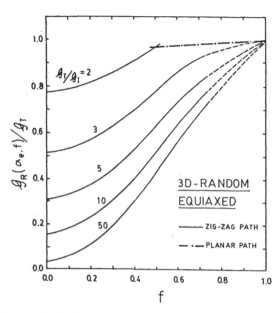

Fig. 5. Effect of the overall fraction of low-energy fracture resistant boundaries **f** on the fracture toughness of a three-dimensional tetrakaidekahedron-shaped polycrystal having random GBCD [30].

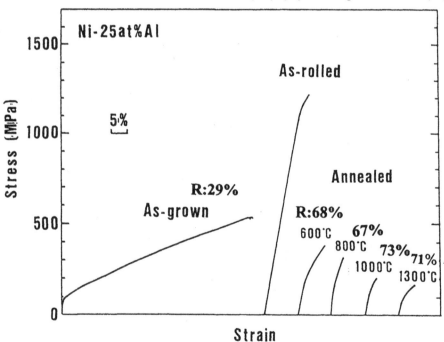

Fig. 6. Effect of the frequency of random boundaries (R) on room temperature ductility in B-free Ni₃Al specimens, as-grown or 25%rolled and annealed at different temperatures for 20min.

Grain Boundary Engineering in Ceramics

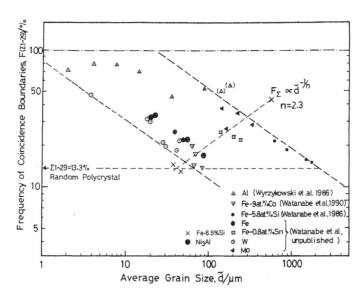

Fig. 7. The frequency of low-Σ (up to 29) coincidence boundaries as a function of the average grain size for thermomechanically processed polycrystalline metals and alloys. The data from rapidly solidified and annealed Fe-6.5mass%Si alloy ribbons are plotted indicating the inserve relationship.

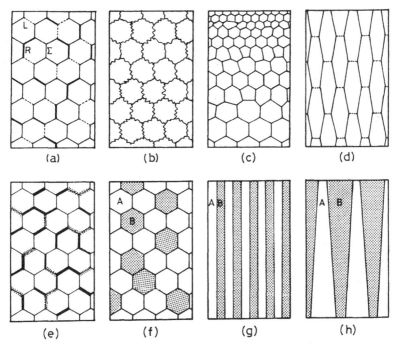

Fig. 8. Schematic representation of possible approaches to designing and controlling grain boundary microstructures in single-phase and two-phase polycrystalline materials [15].

Grain Boundary Engineering in Ceramics   531

(3) Combined Effect of GB Character Distribution (GBCD) and Grain Shape

It seems possible to produce a grain boundary microstructure which has a given GBCD and grain shape predicted for desirable property by applying an appropriate processing method. It is rather widely known that unidirectional solidification can produce lamellar or elongated grain structure, as similarly as in the case of eutectic alloy composite unidirectionally solidified. In cubic metals and alloys unidirectionally solidified polycrystalline specimens have a {100} texture, the sharpness of which normally depends on the condition of solidification. It was previously found that specific low-Σ coincidence boundaries could preferentially occur. The frequency of those low-Σ coincidence boundaries which could occur was higher in the order of Σ value as $\Sigma1,\Sigma5,\Sigma13$, for the {100} textured polycrystal, as predicted for coincidence orientation relationship for the <100> rotation axis. The presence of a close relationship between the GBCD and the type and sharpness of texture has been found experimentally by the present author and coworkers in iron-silicon alloys [24] and iron-chromium alloy [31], and theoretically by several research group [34-36]. Thus combined effect of GBCD and grain shape is very likely predicted and controlled to produce a desirable grain boundary microstructure or to optimize grain boundary-related bulk properties in polycrystalline material. However, unfortunately such combined effect of GBCD and grain shape has been little realized and studied so far. The combination of the control of structural configuration and of geometrical configuration of grain boundary microstructure will bring about much more benefit to development of high performance material by grain boundary engineering.

RECENT ACHIEVEMENTS OF GRAIN BOUNDARY ENGINEERING

Until presently, successful achievements of grain boundary engineering on the basis the control of GBCD have been reported for different types of materials by several groups as by Aust and Palumbo's group [11-13],Was's group [37], and Watanabe's group [16,32,38], although the present situation of the grain boundary engineering for high performance structural or functional materials is still on the half way to the goal. Aust and Palumbo's group has successfully achieved a drastic improvement of intergranular stress corrosion-resistance in nickel base nuclear-steam-generator tubing alloy (alloy 600) by increasing the fraction of low-Σ coincidence boundaries up to the level of 70% almost double of that for conventional alloy [39]. More recently they have also developed high performance lead-acid battery electrode material [13], on the basis of fundamental research on intergranular corrosion and stress corrosion cracking performed since 1980's [40].

It was found that high temperature creep strength could be improved by increasing the fraction of low-Σ coincidence boundaries in nickel base alloy [37] ,as shown that the presence of a single low-Σ coincidence boundary can increase creep strength of a bicrystal at high temperature [41]. On the contrary, it was found that an increase of the fraction of random boundaries which can slide more easily than low-Σ coincidence boundaries, can produce superplasticity by boundary sliding. However,it should be mentioned that random boundaries which slide easily become preferential sites for intergranular fracture so that there must be an optimal GBCD and grain boundary connectivity, as recently revealed in Al-Li alloy.[42].Moreover, quite recently it has been revealed that the control of grain boundary chemistry can drastically improve creep resistance of alumina [43] and SiC [44].

An example of successful achievement of grain boundary engineering for functional material is recent development of giant magnetostriction in ferromagnetic Fe-29.6at.%Pd alloy [16,38]. The magnetostriction of rapidly solidified and annealed ribbon specimens of the alloy was increased up to 10^{-4} which is almost two orders of magnitude higher than that of ordinary ferromagnetic alloys, by using only specific condition of rapid solidification and subsequent annealing. Thus an optimal condition of processing must be for high magnetostriction and good shape-memory performance.

FUTURE PROSPECT OF GRAIN BOUNDARY ARCHITECTURE

As we have discussed in the preceding sections, the grain boundary engineering based on the control of the grain boundary character distribution (GBCD) and the grain boundary connectivity have already been established to develop several high performance structural and functional materials. Now it is strongly pointed out that possible combined effects between GBCD, grain size and grain shape can be utilized for the control of grain boundary microstructure in order to produce high performance in single phase and two phases polycrystalline structural or functional materials [16], as schematically shown in Fig.8. So far the most simplest case of the grain boundary engineering which has taken into account only the control of the GBCD and the grain boundary connectivity (as is Case (a) single phase polycrystal without combined effects with grain size and shape), could still make considerable success of development of high performance materials. There are so many other possibilities of grain boundary design and control as, (b) boundary serration, (c) grain size grading, (d) grain elongation (boundary inclination), (e) boundary segregation and decoration, (f) two-phase polycrystal with grain and interphase boundaries, (g) bilayered, (h) volume grading. The author believes that there exist possible material performance and function which have not yet enlightened and await our calling, to be invoked by the grain boundary engineering of the second, the third stage in the future. In particular nanocrystalline materials are those which have a extremely high density of grain boundaries and need to have grain boundaries designed and controlled effectively and beneficially by interface engineering [45].

CONCLUSION

Grain boundary engineering based on the relationship between grain boundary structure and properties, has been established for development of high performance materials through the introduction of new microstructural factors termed the grain boundary character distribution (GBCD) and the grain boundary connectivity. The intergranular brittleness was found to be well controlled by increasing the frequency of fracture-resistant low-energy boundaries even for "intrinsically brittle materials". The combined effect of the grain boundary character distribution (GBCD) and the grain size is expected to play key role in the generation of optimal grain boundary microstructure and desirable bulk properties in structural and functional materials.

ACKNOWLEDGEMENTS

The author would like to thank his friends and coworkers who joined him and shared the realization of the vision and concept of grain boundary design and control for high performance materials in the past three decades since 1970. This work was supported by through Grant-in-Aid of Basic Research and Grant-in-Aid for COE Research, from the Ministry of Education,Science,Sports and Culture.

REFERENCES

[1] C.S.Smith, Trans.AIME.,175,15-51,(1948).
[2] K.T.Aust and B.Chalmers, "Metal Interfaces",ASM,153-178,(1952).
[3] D.McLean, "Grain Boundaries in Metals",Oxford University Press, (1957)
[4] H.C.Sorby, J.Iron and Steel Inst.,1,255,(1887).
[5] E.D.Hondros,"Structural Materials: Engineering Application through Scientific Insight", E.D.Hondros and M.Mclean (eds.),The Inst.Mater.,1-12, (1996).

[6] G.A.Chadwick and D.A.Smith (ed),"Grain Boundary Structure and Properties" Academic Press,London,(1976).

[7] M.Rühle,R.W.Balluffi,H.Fishmeister and S.L.Sass (ed), "Structure and Properties of Internal Interfaces",J.de Physique,46,Colloque,C4,No.4,(1985)

[8] Y.Ishida (ed.), "Grain Boundary Structure and Related Phenomena", Japan Inst.Metals, (1986).

[9] T.Watanabe, Res Mechanica,11[1],47-84,(1984).

[10] K.T.Aust and G.Palumbo,Proc.Intern.Symp.on Advanced Structural Materials, ed by D.S.Wilkinson, Pergamon Press, 215-226,(1989).

[11] K.T.Aust,U.Erb and G.Palumbo, Mater.Sci.Eng.,A176,329-334,(1994).

[12] G.Palumbo,E.M.Lehockey and P.Lin, J.Metals,50[2],40-43,(1998).

[13] E.M.Lehockey,G.Palumbo,P.Lin and A.Brennenstuhl,Metall.Mater.Trans.A29[1], 387-396,(1998).

[14] T.Watanabe, Mater. Sci. and Eng.,A166,11-28,(1993).

[15] T.Watanabe, Proc.K.T.Aust Intern.Symp.on "Grain Boundary Engineering", Can.Inst.Min.Met.Pet.,57-87,(1993).

[16] T.Watanabe and S.Tsurekawa, Acta Mater.,47[15],4171-85,(1999).

[17] E.Schmid and W.Boas, Plasticity of Crystals,F.A.Hughes,(1950).

[18] R.W.Armstrong, Advance in Materials Research,4,101-146,(1970).

[19] M.Biscondi and C.Goux ,Mem.Sci.Rev.Metallurg.,65,167-179,(1968).

[20] H.Kurishita and H.Yoshinaga, Materials Forum, 13,161-173,(1989).

[21] D.Dimos,P.Chaudhari,J.Mannhart and F.K.Legoues, Phys.Rev.Let.,61[2], 219-222,(1988).

[22] M.Kuwabara,K.Morimo and T.Matsunaga,J.Am.Ceram.Soc.,79[4],997-1001, (1996).

[23] K.Hayashi,T.Yamamoto and T.Sakuma, J.Am.Ceram.Soc.,79[6],1669-72,(1996).

[24] T.Watanabe, H.Fujii,H.Oikawa and K.I.Arai, Acta Met.,37,941-952,(1989).

[25] D.Dingley and D.Field,Structural Materials: Engineering Application through Scientific Insight,ed.by E.D.Hondros and M.Mclean,The Institute of Materials, 23-41,(1996).

[26] B.L.Adams,S.I.Wright and K.Kunze,Met.Trans.,23A[4],819-831,(1993).

[27] T.Watanabe, Met.Trans.,14A[4],531-545,(1983).

[28] T.Watanabe,S.Ikeda,T.Matsuzaki and X.Zhou, Proc.Japan-France Seminar on Intelligent Materials and Structures, 159-166,(1997).

[29] L.C.Lim and T.Watanabe,Scripta Met.,23[4], 489-495,(1989).

[30] L.C.Lim and T.Watanabe, Acta Metall.Mater.,38[12],2507-2516,(1990).

[31] T.Watanabe, Textures and Microstructures, 20,195-216,(1993).

[32] T.Watanabe and T.Hirano,Advanced Materials, Trans.Mater.Res.Soc.,16B,1457-1460,(1994).

[33] T.Watanabe, D.A.Smith Symp.on Boundaries and Interfaces in Materials, Min.Met.Mater.Soc.,19-28,(1998).

[34] A.Garbacz and M.W.Grabski, Acta.Metall.Mater., 41[2], 475-483,(1993).

[35] V.Y.Gertsman,A.P.Zhilyaev,A.I.Pshenichnyuk,R.Z.Valiev,Acta.Metall.Mater. 40[6],1433-1441,(1992).

[36] L.Zuo,T.Watanabe and C.Esling, Z.Metallkde, 85[8],554-558,(1994).

[37] G.S.Was,V.Thaveeprungsriporn and D.C.Crawford,J.Metals,50,44,(1999).

[38] Y.Furuya,N.Hagood,H.Kimura and T.Watanabe,Mater.Trans.JIM.,39[12]1248-1254, (1998).

[39] P.Lin,G.Palumbo,U.Erb and K.T.Aust,Sripta Metall.Mater.,33[9],1387-1392, (1995).

[40] G.Palumbo and K.T.Aust, Acta Metall.Mater.,38[11],2343-2352,(1990).

[41] T.Watanabe,M.Yamada and S.Karashima, Phil.Mag.,63A[5],1013-1022(1991).

[42] S.Kobayashi,T.Yoshimura,S.Tsurekawa and T.Watanabe, Mater.Sci.Forum, 304-306,591-596,(1999).

[43] H.Yoshida,Y.Ikuhara and T.Sakuma, Phil.Mag.Letters, 79[5],249-256,(1999).

[44] S.Kobayashi, Thesis of Dr.Engineering, Tohoku University, March,2000.

[45] H.Gleiter, Acta Materialia, 48[1],1-29,(2000).

Characteristics of Silicon Carbide Thin Films Prepared by Using Pulsed Nd:YAG Laser Deposition Method

Yoshiaki SUDA, Hiroharu KAWASAKI, Kazuya DOI and Satoshi HIRAISHI

Department of Electrical Engineering, Sasebo National College of Technology, Okishin-machi 1-1, Sasebo-city, Nagasaki 857-1193, Japan

ABSTRACT

Silicon carbide (SiC) thin films are synthesized on Si(100) substrates by using a pulsed Nd:YAG laser deposition method at different substrate temperatures(Ts) and methane gas pressure(P_{CH4}). The compositions of the SiC films are examined using an energy-dispersive X-ray analyzer (EDX). The C/(Si+C) composition ratio increases with increasing P_{CH4}. The bonding structures of the SiC films are identified by using a Fourier transform infrared spectroscope. A broad absorption band that corresponds to Si-C bond stretching (800 cm^{-1}) becomes sharper with the increasing Ts. At Ts = 800°C, the wave number of the absorption peak shifts toward 740~780 cm^{-1} due to the crystalline structure change of the film. The crystalline structure of the film is characterized by glancing-angle X-ray diffraction (GXRD). The crystallinity grade of the prepared films increases with increasing Ts, and with decreasing P_{CH4}. Those experimental results suggest that both the gas phase reaction in the plasma plume and the surface reaction on the Si substrate are important in the fabrication of crystalline SiC thin films.

INTRODUCTION

Silicon carbide (SiC) is arousing interest due to its promising properties as a wide-band-gap semiconductor[1-3]. SiC grows in many different crystalline structures called polytypes. The most common are the cubic polytype, called 3C-SiH or β-SiC, and a hexagonal polytype, called 6H-SiC or α-SiC. The difference between the structures of the polytypes is the sequence of stacking of double layers of Si and C atoms. In particular, β-SiC is a promising material for devices because of its high thermal stability, excellent resistance to chemical attack, high thermal conductivity, high electron mobility, and high breakdown electric field. Therefore, β-SiC films have been fabricated on silicon substrates by using several methods, including chemi-

To the extent authorized under the laws of the United States of America, all copyright interests in this publication are the property of The American Ceramic Society. Any duplication, reproduction, or republication of this publication or any part thereof, without the express written consent of The American Ceramic Society or fee paid to the Copyright Clearance Center, is prohibited.

cal vapor deposition (CVD) [4 - 8], sputtering [9], evaporation [10], molecular beam epitaxy [11, 12], and carbon-ion implantation [13, 14]. However, it has been difficult to fabricate stoichiometric and crystalline β-SiC thin films because of its high thermal and chemical stabilities.

Pulsed laser deposition (PLD) is a versatile method for depositing ceramic thin films that are superconducting, semiconducting or ferroelectric. We have deposited several kinds of functional thin films, such as tungsten carbide[15,16], chromium carbide[17], cubic boron nitride[18], carbon nitride[19] and silicon nitride[20] films by using PLD. Recently, we have also prepared β-SiC thin films using PLD at the substrate temperature (Ts) of 800°C[21]. However, detail deposition conditions and the growth mechanism of β-SiC thin films on the Si(100) substrate using PLD have not been studied[22]. In this paper, growth mechanism of the β-SiC film on the Si (100) substrate, and conditions required for preparing β-SiC thin films have been studied.

EXPERIMENTAL

The schematic of the experimental apparatus is shown in Fig. 1. A deposition chamber was made of stainless steel with a diameter of 400 mm and a length of 370 mm. The chamber was evacuated to a base pressure (below 4.0×10^{-4} Pa) by using a

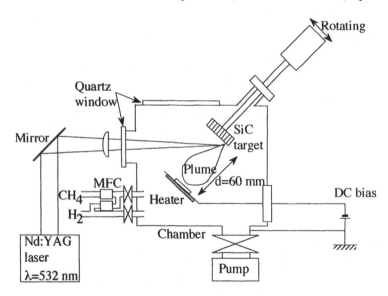

Figure 1   The schematic diagram of the experimental apparatus.

turbo-molecular pump and a rotary pump. The methane gas pressure(P_{CH4}) was varied from the base pressure to 10 Pa by feeding pure methane gas into the chamber. A pulsed YAG laser (Lumonics YM600; wavelength of 532 nm, pulse duration of 6.5 ns, maximum output energy of 340 mJ) was used to irradiate the SiC targets. The radiated area was kept at 2.8 mm^2. The laser energy density (Ed) was fixed at 3.8 J/cm^2. The targets were rotated at 20 rpm to avoid pitting of the target during the deposition. An ultrasonic agitator was used to clean the Si (100) substrates in consecutive baths of ethanol and rinses in high-purity deionized water prior to loading into the deposition chamber. The substrates were located at a distance of 60 mm from the facing target and could be heated up to Ts = 950 °C by a heater. After 18,000 laser pulses at a 10Hz repetition rate, the deposition process was completed.

An optical multichannel analyzer (OMA) with a 1024 photodiode array detected the optical emission from the plasma plume generated by the pulsed Nd:YAG laser radiation. The compositions of the SiC films were examined using an energy-dispersive X-ray analyzer (EDX; PHILIPS PV9900). The bonding structure of the SiC films was identified by using a Fourier transform infrared spectroscope (FT-IR:Shimadzu FT-IR 8300). The crystalline structure of the films was characterized by glancing-angle X-ray diffraction (GXRD; PHILIPS PW1350) using CuKα radiation, where the angle of incidence was kept at 1.0°.

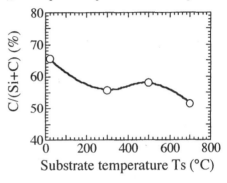

Figure 2 Dependence of C/(Si+C) composition ratio on Ts estimated from the EDX results (P_{CH4}=10Pa).

RESULTS AND DISCUSSIONS

The composition of the SiC films prepared by PLD were examined using EDX as the parameters of Ts and P_{CH4}. Figure 2 shows the dependence of the C/(Si+C) composition ratio on Ts estimated from the EDX results. The value of C/(Si+C) decreases at Ts =room temperature (R.T.) ~ 300°C, and after that it is almost constant to Ts = 700 °C. Figure 3 shows the dependence of the C/(Si+C) composition ratio on P_{CH4}. The value of C/(Si+C) increases

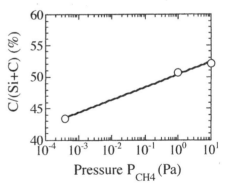

Figure 3 Dependence of C/(Si+C) composition ratio on P_{CH4} estimated from the EDX results (Ts = 700 °C).

with increasing P_{CH4}.

Figure 4 shows the FT-IR absorption spectrum of the SiC films at various substrate temperatures. There are no absorption peaks of Si-C at Ts = R.T. (Fig. 4(a)). A broad absorption band that corresponds to Si-C bond stretching is found near 800 cm^{-1} at Ts = 500 °C (Fig. 4(b)) and the absorption peak becomes sharper at Ts = 700 °C (Fig. 4(c)). At Ts = 800 °C, the wave number of the absorption peak shifts toward 740~780 cm^{-1} (Fig.4(d)). In our previous experiments, all of the films prepared at Ts ≤ 750°C were amorphous, and the film prepared at Ts = 800°C was crystalline SiC thin film[21]. Therefore, this peak shift may be due to the crystalline structure change of the film[22].

Detailed GXRD measurements were carried out to study the crystalline properties of the SiC films. Figure 5(a) shows the GXRD pattern of a reference SiC target pellet, in which the crystalline peaks of SiC can be observed. Figures 5(b) and 5(c) show the pattern of the film deposited at P_{CH4} =

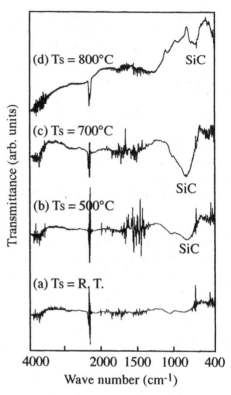

Figure 4   FT-IR absorption spectrum of the SiC films deposited at (a) Ts= R.T., (b) Ts = 500 °C, (c) Ts = 700 °C and (d) Ts = 800°C. (P_{CH4} = 10 Pa)

10 Pa and P = 4.0 × 10^{-4} Pa, respectively. In the result of the film prepared at P_{CH4} = 10 Pa, there are few peaks of SiC(102), (110), (116) (Fig. 5(b)). On the other hand, many peaks of SiC(101), (102), (103), (104), (110), (109), (116), (0012) can be observed in the result of the film prepared at P = 4.0 × 10^{-4} Pa (Fig. 5(c)).

To investigate the crystallinity grade of the prepared films, the crystal grain size was determined from the Full Width Half Maximum (FWHM) of the X-ray peaks, for example as shown in Fig. 5, throughout the equation:

$$\text{grain size} = \frac{0.9 \cdot \lambda}{\cos\theta \cdot \text{FWHM}},$$

where λ is the wavelength of the incident radiation and θ is the Bragg's angle. Figure 6 shows the dependence of the grain size on Ts. The grain size increases with increasing Ts. In general, the surface reaction on the Si substrates increases with increasing Ts. Therefore, this surface reaction is important when preparing crystalline SiC thin films.

Figure 7 shows the dependence of the grain size on P_{CH4}. The grain size increases with decreasing P_{CH4}. At P = 4.0×10^{-4} Pa, as the mean free pass of the species ablated from the target surface is much larger than the distance from the target to the substrate, the collisions between ablated species and CH_4 gas molecules can not occur. Therefore, gas phase reactions of its ablated species can not occur, and energy loss by the collisions of the species from the target is very small at P = 4.0×10^{-4} Pa. On the other hand, the mean free pass of the ablated species is ~1 mm at P_{CH4} = 10 Pa. Therefore, the ablated species can react with CH_4 gas.

On the basis of these experimental results, discussion will be given about the growth mechanism of the crystalline SiC thin film using PLD, qualitatively. At first, high energy light from Nd: YAG laser irradiates the crystalline SiC target surface. The surface of the target was ablated and several kinds of species, such as SiC molecules, Si and C atoms fly out in the direction of the substrate. When the pressure of methane gas is very low, these species can not react with CH_4 gas in the plasma plume. After that they react on the Si substrate by surface reaction, and SiC thin films are produced. In the case of high methane gas pressure, some part

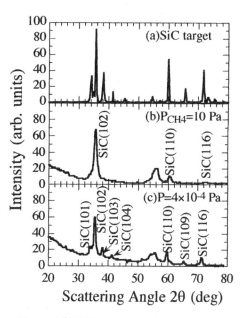

Figure 5 XRD patterns of (a)a reference SiC target pellet and the SiC films deposited at (b) P_{CH4} = 10 Pa, and (c) P= 4.0×10^{-4} Pa.

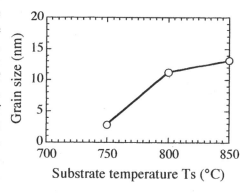

Figure 6 Dependence of the grain size on Ts. (P_{CH4}=10Pa).

of them may react with the CH_4 gas. These differences suggest that the crystalline structure of the SiC films can be controlled with the CH_4 gas pressure.

CONCLUSIONS

Silicon carbide (SiC) films were synthesized on Si(100) substrates by using a pulsed Nd:YAG laser deposition method at different P_{CH4} and Ts. The following conclusions can be drawn from the experimental results.

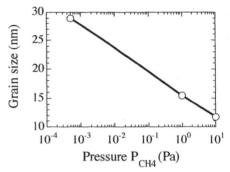

Figure 7   Dependence of the grain size on P_{CH4} (Ts = 700 °C).

1. The C/(Si+C) composition ratio increases with increasing P_{CH4}.

2. The broad absorption band near 800 cm^{-1}, that corresponds to Si-C bond stretching, becomes sharper with increasing Ts. At Ts = 800°C, the wave number of the absorption peak shifts toward 740~780 cm^{-1}.

3. The degree of crystallinity of the prepared films, characterized by GXRD, increases with increasing Ts, and decreasing P_{CH4}.

These experimental results suggest that both the gas phase reaction in the plasma plume and the surface reaction on the Si substrate are important for the growth of the crystalline SiC thin films.

ACKNOWLEDGMENTS

This work was supported in part by Grant-in Aid for Scientific Research (B) and the Regional Science Promoter Program and a Research Fund from the Nagasaki Super Technology Development Association.
The authors wish to thank Drs. Y. Watanabe and M. Shiratani of Kyushu University, Drs. K. Ebihara, T. Ikegami and Y. Yamagata of Kumamoto University, and Drs. H. Fujiyama and Y. Matsuda of Nagasaki University for their helpful discussions. The authors also wish to thank Dr. H. Abe and Mr. H. Yoshida of the Ceramic Research Center of Nagasaki for their technical assistance with the experimental data.

REFERENCES

[1] T. Kaneko, N. Miyakawa, H. Sone and M. Iijima,"Thin Film Growth of Silicon Carbide from Methyl-Trichloro-Silane by RF Plasma-Enhanced CVD", *Journal of Crystal Growth*, **174**[1] 658 (1997).

[2] J-W. Lee and K. S. Lim,"Dramatic Improvement of Performance of Visible Hydrogenated Amorphous Silicon Carbide Based p-i-n Thin-Film Light-Emitting Diodes by Two-Step Hydrogenation", *App. Phys. Lett.*, **69**[4] 547 (1996).

[3] M. Uchida, M. Deguchi, K. Takahashi, M. Kitabatake, M Kitagawa, "Heteroepitaxial Growth of 3C-SiC on Surface-Structure-Controlled MBE Layerby Low-Pressure CVD", *Mayerials Science Forum.*, **31** 343(1998) .

[4] S. Nishino, H. Suhara, H. Ono and H. Matsunami,"Reduction of Double Positioning Twinning in 3C-SiC Grown on a-SiC Substrates", *J. Appl. Phys.*, **61** 4889 (1987).

[5] J. A. Powell, L. G. Matus and M. A. Kuczmarski,"Effect of Tilt Angle on The Morphology of SiC Epitaxial Films Grown on Vicinal (0001)SiC Substrates", *J. Electrochem. Soc.*, **134** 1558 (1987).

[6] A. J. Steckl and J. P. Li,"Growth of SiC by CVD from Silacyclobutane", *IEEE Trans. Electron Devices*, **ED-39** 64 (1992).

[7] S. J. Toal, H. S. Reehal, S. J. Webb, N. P. Barradas, C. Jeynes, "Structural Analysis of Nanocrystalline SiC Thin Films Grown on Silicon by ECR Plasma CVD", *Thin Solid Films,* **343** 290 (1999).

[8] A. J. Steckl, C. Yuan. J. P. Li and M. J. Loboda,"Processed Silicon and Related Materials - Effect of The Temperature Ramp Rate During Carbonization of Si (111) on The Crystalline Quality of SiC Produced", *Appl. Phys. Lett.*, **63** 3347 (1993).

[9] H. Shimizu, M. Shiga,"Effect of Substrate Bias on 3C-SiC Deposition on Si by AC Plasma-Assisted CVD", Materials Science Forum, **264** 211 (1998).

[10] A. Fissel, B. Schroter, J. Krausslich, W. Richter, "Preparation of SiC Films by Solid State Source Evaporation", *Thin Solid Films,* **258** 64 (1995).

[11] T. Yoshinobu, H. Mitsui, Y. Tarui and H. Matsunami,"STM Observation of Wurtzite GaN(0 0 0 1) Surface Grown by MBE on 6H-SiC Substrates", *J. Appl. Phys.*, **72** 2006 (1992).

[12] E. Theodossiu, H. Baumann, E. K. Polychroniadis, K. Bethge, "Ion Beam Synthesis and Characterization of Thin SiC Surface Layers", *Nuclear Instruments and Methods in Phys Res-Section B Only-Beam Interact Mater Atoms,* **161**[1] 941 (2000).

[13] D. Chen, W. Y. Cheung, S. P. Wong, "Ion Beam Induced Crystallization Effect and Growth Kinetics of Buried SiC Layers Formed by Carbon Implantation into Silicon", *Nuclear Instruments and Methods in Phys Res-Section B*

Only-Beam Interact Mater Atoms, **148**[1] 589 (1999).

[14] A. R. Romano, C. Serre, L. Calvo-Barrio, A. Perez-Rodriguez, J. R. Morante, R. Kogler, W. Skorupa, " Detailed Analysis of β-SiC Formation by High Dose Carbon Ion Implantation in Silicon", *Materials Science and Engineering - B - Solid State Materials for Advanced Technology,* **36**[1] 282 (1996).

[15] Y. Suda, T. Nakazono, K. Ebihara and K. Baba, "Pulsed Laser Deposition of Tungsten Carbide Thin Films on Si(100) Substrate", *Nuclear Instruments and Methods in Phys Res-Section B Only-Beam Interact Mater Atoms,* **121**[1] 396 (1997) .

[16] Y. Suda, T. Nakazono, K. Ebihara, K. Baba and H. Hatada, "Properties of WC Films Synthesized by Pulsed Laser Deposition", *Materials Chemistry and Physics,* **54**[1] 177 (1998).

[17] Y. Suda, H. Kawasaki, R. Terajima and M. Emura, "Silicon Carbide Thin Films Synthesized by Pulsed Laser Deposition", *Jpn. J. Appl. Phys.,* **38**[6] 3619 (1999).

[18] Y. Suda, T. Nakazono, K. Ebihara and K. Baba, "Effect of RF Bias on the Cubic BN Film Synthesis by Pulsed YAG Laser Deposition", *Thin Solid Films,* **281-282** 324 (1996) .

[19] Y. Suda, T. Nakazono, K. Ebihara, K. Baba and S. Aoqui,"Pulsed Laser Deposition of Carbon Nitride Thin Films from Graphite Targets", *Carbon,* **36**[6] 771 (1998).

[20] Y. Suda, K. Ebihara, K. Baba, H. Abe and A. M. Grishin, "Crystalline Silcon Nitride Thin Films Grown by Pulsed Laser Deposition", *Nano Structured Materials,* **12** 291 (1999).

[21] Y. Suda, H. Kawasaki, R. Terajima, M. Emura, K. Baba, H. Abe, H. Yoshida, K. Ebihara and S. Aoqui,"Silicon Carbide Thin Films Synthesized by Pulsed Laser Deposition", *J. Korea. Phys. Soc.,* **35** S88 (1999).

[22] Z. Kántor, E. Fogarassy, A. Grob, J. J. Grob. D. Muller, B. Prévot and R. Stuck,"Evolution of Implanted Carbon in Silicon upon Pulsed Excimer Laser Annealing", *Appl. Phys. Lett.,* **69**[7] 969 (1996).

EFFECTS OF CRYSTALLOGRAPHIC ORIENTATION OF SILVER SUBSTRATE ON CRYSTALLINITY OF YBCO(Y123) FILM

Yoshiyuki Yasutomi and Yorinobu Takigawa
Japan Fine Ceramics Center, 2-4-1, Mutsuno, Atsuta-ku, Nagoya, 456-8587 Japan
Youji Yamada, Yoshihiro Koike and Izumi Hirabayashi
Superconductivity Research Laboratory, ISTEC, 2-4-1, Mutsuno, Atsuta-ku, Nagoya, 456-8587 Japan

ABSTRACT

The electron back scattering pattern (EBSP) method was applied for the first time to study the effect of the crystallographic orientation of the rolled Ag substrate on the crystal growth of the YBCO (Y123) film. Crystallizing of YBCO was observed on the Ag grain, which tilted within 10 degree from the crystal orientation of <110>. However, the YBCO film was strongly influenced by moisture in the atmosphere and the crystallinity was lost by degradation. When Ag grain tilted more than 20 degrees from <110>, no crystallization of YBCO film was observed.

INTRODUCTION

For practical application of high-Tc superconductors, the control of the crystal orientation and the current direction are important factors to achieve a proper system with desired properties. For the application of superconducting wires, the superconductor/metal system has been studied[1]. It is well known that the critical current (Jc) of YBCO film decreases, when the orientation distribution of the grains composing YBCO film becomes broad, for YBCO films on described a polycrystalline metallic substrate. The lower Jc is attributed to the depressed transport super-current due to the orientation mismatch with the high angled grain boundary[2]. Toward the better design of the crystal orientation, more attempts should be needed, especially, for identifying the

To the extent authorized under the laws of the United States of America, all copyright interests in this publication are the property of The American Ceramic Society. Any duplication, reproduction, or republication of this publication or any part thereof, without the express written consent of The American Ceramic Society or fee paid to the Copyright Clearance Center, is prohibited.

relevance between YBCO crystallization and the polycrystalline metallic substrate.

In general, transmission electron microscopy (TEM) is used for analyzing microstructures of ceramics, including high-Tc superconducting materials. TEM can determine the grain boundary structure of a limited part but not of the entire sintered body. On the other hand, EBSP[3] has been used to analyze the grain boundary structure of the entire sintered body by observing the crystallographic orientation of all particles, using field emission scanning electron microscopy (FE-SEM). We have successfully applied the EBSP method to examine the crystallographic orientation of in situ β -Si$_3$N$_4$ composites containing elongated β -Si$_3$N$_4$ particles and found that elongated Si$_3$N$_4$ particles grew in the alignment direction of seed grains[4]. The EBSP method, however, has never been applied to study the crystallographic orientation of thin film systems until now.

Ag is one of the materials available for YBCO coated conductors because YBCO film can be grown on a Ag substrate directly without any buffer layer. The orientation relationship between YBCO, deposited by a pulsed laser deposition method, and single crystalline silver has been reported[2]. However, the base substrate for the coated conductor should be polycrystalline and there are few studies focusing on the relationship between the YBCO film and polycrystalline silver substrate[5].

In this study, we apply the EBSP method to study the influence of the crystallographic orientation of the rolled (polycrystalline) Ag substrate on the crystal growth of the YBCO (Y123) film, formed by the pulse laser deposition method. The effect of moisture on the crystallinity of the oriented YBCO film is also reported.

EXPERIMENTAL PROCEDURE

Figure 1 shows the experimental procedure. First, a polycrystalline Ag substrate was heat treated at 900°C after a rolled processing. The Ag substrate consists of crystals in the range of millimeters, and twined crystals exist in Ag grains. The crystal orientation of polycrystalline Ag substrate was analyzed by

the EBSP method. 200 nm epitaxial YBCO film was then grown on the Ag
substrate by pulsed laser deposition at 730°C[6]. The laser source was a KrF excimer laser (248 nm) and was operated at 10 Hz for 600 seconds. The target composition was Y:Ba:Cu=1:2:3.3. During deposition, the ambient oxygen pressure was kept at 26.7 Pa. The crystal orientation of the YBCO film was analyzed again by EBSP to investigate the crystallographic relevance between the Ag substrate and YBCO film.

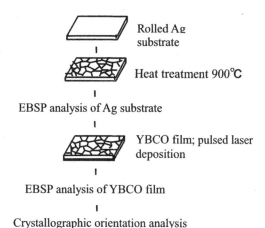

Rolled Ag substrate

Heat treatment 900°C

EBSP analysis of Ag substrate

YBCO film; pulsed laser deposition

EBSP analysis of YBCO film

Crystallographic orientation analysis

Figure 1 Process of EBSP analysis

The sample was installed at a 70-degree angle in the FE-SEM sample chamber with crystallization orientation imaging microscopy analyzer (EBSP analysis tool). When an electron beam hits a grain on the sample surface, back-scattered electrons were generated. The back-scattered electrons hit the phosphor screen and were detected by a high sensitive TV camera. The resulting image was called an electron back-scattered pattern (EBSP). The pattern was made up of several bands that were produced by diffracting planes in the crystal lattice. The crystallographic orientation can be determined from these bands.

An example of EBSP analysis for YBCO film is shown in Fig. 2. The parallel lines indicate a crystal plane, the interval between them is the lattice spacing and the intersection shows the orientation of the crystal (zone axis). The Euler angles can be determined from the EBSP pattern, and then the orientation of crystallization can be determined.

RESULTS AND DISCUSSION

Figures 3 and 4 show the SEM image and the crystal orientation of Ag substrate analyzed by EBSP, respectively. In the Ag grain, the twinned crystal

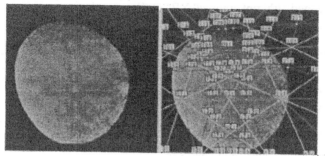

$\phi 1 = 105.55$ $\Phi = 1.01$ $\phi 2 = 348.25$

Figure 2 EBSP of a YBCO grain

Sample:70 deg. tilt

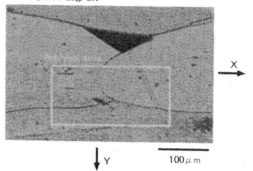

X

Y   100 μ m

Figure 3 SEM image of Ag substrate

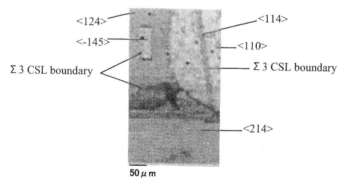

<124>
<-145>
Σ 3 CSL boundary

<114>
<110>
Σ 3 CSL boundary

<214>

50 μ m

Figure 4 Results of crystal orientation of Ag grains by EBSP

has the coincident site lattice (CSL) boundary of the sigma value 3. The crystal orientation of the Ag substrate has <114>, <214>, <124> and <-145>, rather than the priority orientation of <110>. Sequentially, the crystal orientation of YBCO film on this Ag crystal area was analyzed by EBSP method. Figure 5 shows the crystal orientation of YBCO film on the Ag crystal area. The mapped color corresponds to inverse pole figure showing the crystal orientation distribution of the direction of Z-axis of Ag substrate. In the <110> plane of Ag crystal, the YBCO film has grown to be a <001> direction. However, in case of the crystal orientation of Ag tilted above 20 degrees from <110>, the crystal of YBCO film was not crystallized. And, in case of the crystal orientation of Ag tilted about 10 degree from <110>, YBCO film deposit <001> crystal plane, but, the crystallinity of YBCO film was lost by degradation, after remaining several days in atmosphere (25°C/RH40%).

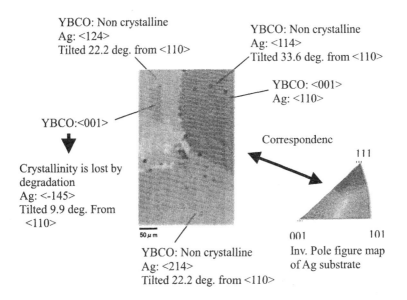

Figure 5 EBSP analysis of crystallographic orientation of Ag substrate and YBCO film

As mentioned above, the crystallinity of the YBCO film was strongly influenced by that of the Ag subtrate. In order to increase the alignment of YBCO crystal orientation, one should control the alignment of <110> crystal orientation of polycrystalline Ag substrate. On the crystal orientation of Ag <110> or <111>, the lattice of YBCO is almost crystallographically in fitness[2], and this is in agreement with the results of our EBSP analysis. The crystal orientation of Ag <110> is more suitable rather than Ag <111> for YBCO crystarization, because the lattice side is restrained at 90 degrees. We have also found that the degradation of YBCO crystallinity in atmosphere. This degradation can be attributed to the influence of moisture[7]-[8]. The detailed examination of this topic is left for subsequent study.

CONCLUSIONS

EBSP method was successfully applied to examine the crystallographic orientation of the rolled Ag substrate and the crystal growth of the YBCO (Y123) film.

We clarified that it is important to control the alignment of <110> crystal direction of polycrystalline Ag substrate. In the case of the grain of Ag tilted above 20 degrees from <110>, Crystallizing of YBCO did not occur. In the case of the grain of Ag tilted within 10 degree from <110>, YBCO <001> crystal plane was grown. The YBCO crystal plane, however, was strongly influenced by moisture in the atmosphere and the crystallinity was lost by degradation.

Acknowledgement

This work was supported by the New Energy and Industrial Technology Development Organization for R&D of Industrial Science and Technology Frontier Program.

REFERENCES

1) A. Goyal, D. Norton, D. Christen, E. Specht, M. Paranthaman, D. Kroeger, J. Budai, Q. He, F. List, R. Feenstra, H. Kerchner, D. Lee, E. Hatfield, P.

Martin, J. Mathis and C. Park, "Epitaxial Superconductors on Rolling -assisted Biaxially-textured Substrates (RABiTS): A Route Towards High Critical Current Density Wire," *Apllied Superconductivity*, **4**, [10-11], 403-27 (1996)

2) J. D. Budai, R. T. Young and B. S. Chao, "In-plane Epitaxial Alignment of $YBa_2Cu_3O_{7-x}$ Films Grown on Silver Crystals and Buffer Layer," *Appl.Phys. Lett.*, **62**, [15], 1836-38 (1993).

3) D. J. Dingray and V. Randle, "Review Microstructure Determination by Electron Back-scatter Diffration," *J. Mater. Sci.*, **27**, 4545-66 (1992).

4) Y. Yasutomi, Y. Sakaida, N. Hirosaki and Y. ikuhara, "Analysis of Crystallographic Orientation of Elongated β-Si_3N_4 Particles in In Situ Si_3N_4 Composite by Electron Back Scattered Diffraction Method," *J. Ceram. Soc. Japan*, **106**, [10], 980-83 (1988).

5) Y. Niiori, Y. Yamada, I. Hirabayashi, T. Fujiwara, K. Higashiyama, "Low Temperature LPE Growth of $YB_2Cu_3O_{6+x}$ Thick Film on Silver Substrate Using Silver Saturated Ba-Cu-O-F Flux," *Physica C*, **301**, 104 (1998).

6) H. Izumi, K. Ohata, T. Hase, K. Suzuki and T. Morishita, "Effects of Oxygen Pressure During Laser Deposition on Crystal Orientation in $YBa_2Cu_3O_{7-x}$ Films," Advances in Superconductivity II, ed. T. Ishiguro and K. Kajimura, p.861 (Springer- Verlag), Tokyo (1989).

7) S. Edo and T. Tanaka, "Effects of H_2O Absorption on Electrical Property and Crystal Structure in YBa_2CuO_x," *Jpn. J. Appl. Phys.*, **37**, 3956-60 (1988).

8) S. L. Qui, M. W. Ruckman, N. B. Brookes, P. D. Johnson, J. Chen, C. L. Lin, M. Strongin, B. Sinkovic, J. E. Crow and C. Jee, "Interaction of H_2O with a High-temperature Superconductor," *Phys. Rev.*, **B37**, 3747 (1988).

INTERFACIAL STRUCTURES OF YbBa$_2$Cu$_3$O$_{7-\delta}$ SUPERCONDUCTING FILMS DEPOSITED ON SrTiO$_3$(001) SUBSTRATES BY THE DIPPING-PYROLYSIS PROCESS

Junko Shibata and Tsukasa Hirayama
Japan Fine Ceramics Center, 2-4-1 Mutsuno, Atsuta-ku, Nagoya 456-8587, Japan

Katsuya Yamagiwa and Izumi Hirabayashi
Superconductivity Research Laboratory, ISTEC, 2-4-1 Mutsuno, Atsuta-ku, Nagoya 456-8587, Japan

Yuichi Ikuhara
Engineering Research Institute, School of Engineering, The University of Tokyo, 2-11-16 Yayoi, Bunkyo-ku, Tokyo 113-8656, Japan

ABSTRACT

YbBa$_2$Cu$_3$O$_{7-\delta}$ precursor films and final films formed on SrTiO$_3$(001) substrates by the dipping-pyrolysis process were characterized by transmission electron microscopy. The YbBa$_2$Cu$_3$O$_{7-\delta}$ precursor films were fabricated by two different heat treatments with changing the heating rate. As a result of this, we found that an amorphous precursor film was produced by rapid heating up to 698 K; in contrast, a polycrystalline precursor film was produced by slow heating. By heating at 998 K, the former film became a c-axis oriented YbBa$_2$Cu$_3$O$_{7-\delta}$ film with a thickness of 100nm and showed a sharp resistive transition around the critical temperature. However, the latter one was a polycrystalline film containing randomly oriented Yb123 and other crystals even after heating it at 998 K. This polycrystalline film exhibited a broad transition. In this paper, the effects of the heating rate were investigated from the interfacial structures for achieving the epitaxial growth of YbBa$_2$Cu$_3$O$_{7-\delta}$ film and the good superconducting properties.

INTRODUCTION

To the extent authorized under the laws of the United States of America, all copyright interests in this publication are the property of The American Ceramic Society. Any duplication, reproduction, or republication of this publication or any part thereof, without the express written consent of The American Ceramic Society or fee paid to the Copyright Clearance Center, is prohibited.

Many studies have been devoted to producing superconductors with high critical temperature(Tc) and high critical current density(Jc), since the La-Ba-Cu-O system and Y-Ba-Cu-O system were discovered in the latter half of the 1980s. Dipping pyrolysis is a promising method for producing the superconducting films of REBa$_2$Cu$_3$O$_{7-\delta}$(RE=rare-earth element) single crystal[1,2] at a low cost, compared with physical deposition techniques such as sputtering or laser-beam deposition which need a high-vacuum apparatus. This method was developed for preparing oxide films, such as BaTiO$_3$, PbTiO$_3$, SnO$_2$, In$_2$O$_3$ and PbO in the 1970s. The dipping-pyrolysis method has the advantage of producing thin films on substrates of any size and shape as well as the advantage of controlling the chemical composition of the films. Using this method, McIntyre et al.[3] and Manabe et al.[4] reported that they succeeded in fabricating YBa$_2$Cu$_3$O$_{7-\delta}$ films on SrTiO$_3$(100) substrates with higher Jc than those of films made by physical deposition techniques.

Generally, the dipping-pyrolysis process contains two heat treatment steps. The first step is for thermal decomposition of the metal organic compounds on the substrate at a relatively low temperature. In this step, a precursor film is prepared. The second step is for crystallization of this precursor film to form the superconducting final film. During these two heat treatments, it is important to prevent the random nucleation of crystals in the film and to realize the epitaxial growth of the single crystal film on the substrate.

In the present study, we prepared two precursor films of YbBa$_2$Cu$_3$O$_{7-\delta}$ under different conditions for the initial heat treatment at 698 K in air, and finally annealed these precursor films at 998 K under the low oxygen partial pressure. Interfacial microstructures of the precursor films and the final films were investigated by transmission electron microscopy. The effect of the initial heat treatment conditions on the microstructures of the precursor films and the final superconducting films is reported.

EXPERIMENTAL PROCEDURE

The YbBa$_2$Cu$_3$O$_{7-\delta}$(Yb123) precursor films and final films were deposited on SrTiO$_3$(STO) substrates as reported in detail in our previous papers.[5-7] First, the coating solution was prepared by dissolving constituent metal naphthenates of Yb, Ba and Cu in toluene. Secondly, this solution was deposited onto STO(001) substrates. Finally, each sample was heat treated by the process as shown schematically in Fig. 1(a) or 1(b). In the case of Fig. 1(a), the sample was treated as follows: (i)the substrate coated with the solution was inserted into a furnace

(a) the fast process

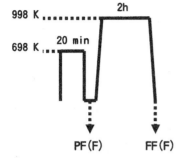

(b) the slow process

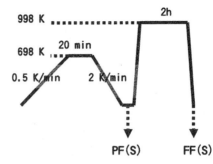

Figure 1. Schematic diagram of heating schedule for preparing Yb123 precursor films and final films: (a) the fast process. PF(F) was prepared by rapid heating up to 698 K. Then, FF(F) was produced by annealing PF(F) at 998 K in an Ar gas flow with the heating and cooling rate of 3 K/min; (b) the slow process. PF(S) was prepared by heating the sample up to 698 K at the heating rate of 0.5 K/min and cooling to room temperature at the cooling rate of 2 K/min. FF(S) was obtained by annealing PF(S) at 998 K.

kept at 698 K; (ii)after keeping for 20 min in air, the sample was taken out from the furnace; (iii) finally, the sample was heat treated at 998 K for 2 h in an Ar gas flow under the oxygen partial pressure of $p(O_2)=10^{-4}$ atm. From now on, we call this process the "fast process". The precursor film and the final film prepared by this fast process are referred to as precursor film(fast) and final film(fast), respectively. The former is abbreviated as PF(F) and the latter as FF(F). In the case of Fig. 1(b), the sample was placed in the furnace at room temperature, and then the temperature of the furnace was elevated to 698 K at the heating rate of 0.5 K/min. After holding this temperature for 20 min, the furnace was cooled to room temperature at the cooling rate of 2 K/min. Finally, the sample was annealed at 998 K. We refer to this process as the "slow process". The precursor film and the final film prepared by this process are named PF(S) and FF(S) respectively.

X–ray diffraction patterns of these films were obtained using Cu-Kα radiation, and their microstructures were observed by transmission electron microscopy (TEM).

Cross-sectional specimens for the TEM observation were made by standard processes used for TEM specimen preparation. The samples were cut in the proper orientation (electron beam was parallel to [100]STO.) Next, two of these pieces were glued together with the films surfaces each other. These 'sandwiches' were polished and dimpled to the thickness about $20\,\mu$ m. Finally, all the samples were thinned by Ar ion beam sputtering. During this process of ion thinning, the specimen was rotated fast when the ion beam was parallel to the interface between the substrate and the film, and slowly when the beam was perpendicular to the interface; otherwise, the film was removed and only the substrate remained. JEOL JEM-2010 transmission electron microscope operating at 200 kV was used.

The dependence of the electric resistance on the temperature for the final superconducting films after annealing at 998 K was measured by the standard DC four-probe method.[7]

RESULTS AND DISCUSSION

Precursor films

The X-ray diffraction patterns of both precursor films, PF(F) and PF(S), showed broad amorphous-like humps. However, further studies by TEM revealed differences between the two films. Figures 2 and 3 show cross-sectional electron micrographs of PF(F) and PF(S) respectively, taken along the direction of [100]STO. Electron diffraction pattern at the top left in each micrograph was

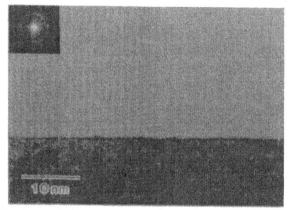

Figure 2. Cross-sectional transmission electron micrograph and electron diffraction pattern of PF(F), observed from the [100]STO direction. Note that the film is amorphous.

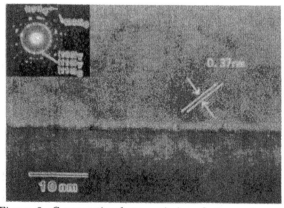

Figure 3. Cross-sectional transmission electron micrograph and electron diffraction pattern of PF(S), observed from the [100]STO direction. The film is polycrystalline. In the diffraction pattern of the film, "C" and "Y" indicate CuO and Yb123, respectively.

obtained from the selected area of the Yb123 precursor film. As shown in Fig. 2, PF(F) is amorphous. In contrast, PF(S) is polycrystalline. The crystals grown in PF(S) are about 10nm in size. This explains why strong sharp peaks did not appear in the X-ray diffraction pattern of PF(S). The lattice spacing that is remarkably seen in the vicinity of the interface is 0.37 nm. This almost corresponds to (101)Yb123, (110)Yb123 or (220)Yb$_2$O$_3$. In addition, this coincides approximately with the spacing of (100)STO. The crystals generated in the film seem to be grown along the lattice spacing of the substrate. From the electron diffraction pattern of this film, we calculated the lattice spacings in these crystals. Most of them agree with those for Yb123. The spacings of 0.125 nm and 0.106 nm, which do not agree with those for Yb123, correspond to (222) and (131) of CuO.

Yb123 final films

There was no noticeable difference between XRD profiles of FF(F) and FF(S). The peaks from (00l) reflection of Yb123 for both samples were anomalously strong compared with other peaks in these profiles. This implies that most Yb123 crystals in the films are highly oriented with the c-axes perpendicular to (001)STO. However, these two final films showed a different dependence of the electric resistance on the temperature. FF(F) showed a sharp resistive transition around T$_C$; the onset temperature for transition (T$_C$ onset) was 89 K and the temperature at which the resistance dropped to zero (T$_C$ zero) was 86 K. On the other hand, FF(S) showed the broad transition; T$_C$ onset was 90 K and T$_C$ zero was 81 K. These results indicate that the microstructures of the two final films are different.

Figure 4 shows a low-magnification micrograph of FF(F) observed from the [100]Yb123//[100]STO direction. This micrograph shows that a c-axis oriented Yb123 film with a thickness about 100nm is epitaxially grown on the substrate, and that a polycrystalline film of Yb123 and other grains is grown on the epitaxial film. Figure 5 shows a high-resolution micrograph of FF(F) in the vicinity of the interface between the film and the substrate. As shown in this micrograph, the c-axis of the Yb123 crystal is perpendicular to (001)STO. The film is defective with many stacking faults. This is consistent with the streaks observed along the c*-axis in the selected-area diffraction pattern shown at the top right of Fig.5.

Figure 6(a) shows a typical electron micrograph of FF(S) taken along the [100]STO direction. In this micrograph, Yb123 grains are seen to have many orientations, although strong (00l) peaks of Yb123 were visible in the XRD pattern. These crystals grown in the film are over 200 nm in size. Figure 6(b) shows a diffraction pattern obtained from region A in Fig. 6(a). In this pattern, spots indicated by arrows B are consistent with the lattice spacing (101) of $YbBa_2Cu_4O_8$ (Yb124) whose T_C is 80K. This implies the possibility that FF(S) contains Yb124 phase, although the spots of Yb124 were observed only in region A. In this case, Yb123 crystals in FF(S) also have many stacking faults, as shown in the grain "C" of Fig.7. Therefore, the spots (101) of Yb124 overlap with the streaks along the c^*-axis, and it is difficult to detect these spots in the diffraction patterns. As a result, other crystals of Yb123 besides the region A can include the Yb124 phase, even if the spots of the Yb124 are not seen in the pattern.

Figure 4. Low-magnification electron micrograph of FF(F), observed from the [100]Yb123//[100]STO direction. The film is a c-axis oriented Yb123 film with a thickness of 100 nm.

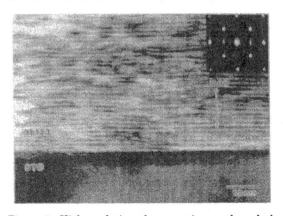

Figure 5. High-resolution electron micrograph and the electron diffraction pattern of FF(F), observed from the [100]Yb123//[100]STO direction. The film is defective with stacking faults.

(a)   **(b)**

Figure 6. Cross-sectional transmission electron micrograph and electron diffraction pattern of FF(S): (a) low-magnification micrograph of FF(S), observed from the [100]STO direction. Grain sizes of crystals grown in the film are over 200 nm in size; (b) electron diffraction pattern obtained from region A in Fig.6(a). Spots indicated by arrows B agree with (101) lattice spacing of Yb124. This suggests that the film contains Yb124 phase.

Grain Boundary Engineering in Ceramics

Figure 7 shows a high-resolution micrograph of the interface between the film and the substrate in FF(S). In this micrograph, grain "C" is a c-axis oriented Yb123 crystal. Nonsuperconducting phase, Yb$_2$O$_3$ is also seen at the top left. The remaining grains are Yb123 crystals tilted from the direction perpendicular to (001)STO.

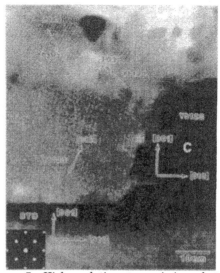

Figure 7. High-resolution transmission electron micrograph and electron diffraction pattern of FF(S), taken along the [100]STO direction. Grain "C" is a c-axis oriented Yb123 crystal. Other grains of Yb123 in the film are seen to have many orientations. Nonsuperconducting Yb$_2$O$_3$ grain is also visible at the top left.

From these results of the TEM observation, we suggest the relationship between the initial heat treatment conditions and the growth of Yb123 films prepared by the dipping-pyrolysis process as follows: (i)when the precursor film is prepared by the fast process, the metal organic compounds in the solution on the substrate are decomposed at the moment they are installed into the furnace kept at 698 K. Then, the solution is solidified to form an amorphous precursor film by cooling directly to room temperature. Finally, during heating this amorphous film at 998 K, random growth of crystals is restrained and the epitaxial growth of the c-axis oriented Yb123 film is achieved; (ii)on the other hand, in the case of the slow process, the metal organic compounds on the substrate are decomposed at different temperatures during heating gradually up to 698 K. This leads to the random nucleation of nonsuperconducting crystals such as CuO and Yb$_2$O$_3$ as well as Yb123 crystals. When this polycrystalline precursor film is heated at 998 K, the growth of those random nuclei in the film dominates the epitaxial growth of the Yb123 film.

We suggest that FF(S) showed the broad transition of electric resistance around T$_C$ as a result of the mixture of crystals with different T$_C$, such as Yb124, the grain boundaries among the Yb123 crystals, and the grain boundaries among the Yb123 crystals and nonsuperconducting crystals.

CONCLUSIONS

Cross sections of Yb123 precursor films and final films were investigated by transmission electron microscopy. These films were prepared on STO(001) substrates under two different conditions at the initial heat treatment by the dipping-pyrolysis method. It was found that an amorphous film was formed by rapid heating up to 698 K, and that this film became a c-axis oriented Yb123 film by annealing at 998 K. On the other hand, a polycrystalline precursor film was found to be produced by slow heating. In this case, when the film was finally annealed at 998 K, the random growth of the crystals dominates the epitaxial growth of Yb123. In conclusion, the rapid heating rate at the initial heat treatment is important and necessary for achieving the epitaxial growth and a sharp resistive transition of the Yb123 film.

Acknowledgement
This work was supported by the New Energy and Industrial Technology Development Organization (NEDO) .

References

1) T. Manabe, I. Yamaguchi, S. Nakamura, W. Kondo, T. Kumagai and S. Mizuta, "Crystallization and in-plane alignment behavior of YBa2Cu3O7-y films on MgO(001) prepared by the dipping-pyrolysis process", *J. Mater. Res.* **10**, 1635(1995).
2) T. Manabe, T. Tsunoda, W. Kondo, Y. Shindo, S. Mizuta and T. Kumagai, "Preparation and superconducting properties of Bi-Pb-Sr-Ca-Cu-O films (Tc =106K) by the dipping-pyrolysis process", *Jpn. J. Appl. Phys.* **31**, 1020(1992).
3) P. C. McIntyre, M. J. Cima and Man Fai Ng, "Metalorganic deposition of high-Jc Ba2YCu3O7-x thin films from trifluoroacetate precursors onto (100)SrTiO3", *J. Appl. Phys.* **68** [8], 4183(1990).
4) T. Manabe, W. Kondo, S. Mizuta and T. Kumagai, "Preparation of high-Jc Ba2YCu3O7-y films on SrTiO3(100) substrates by the dipping-pyrolysis process at 750°C", *Jpn. J. Appl. Phys.* **30**, L1641(1991).
5) K. Yamagiwa, I. Hirabayashi, "Structual and superconducting properties of biaxially aligned Yb123 films prepared by chemical solution deposition", *Physica C* **304**, 12(1998).
6) J.Shibata, K.Yamagiwa, I.Hirabayashi and T.Hirayama, "Transmission electron microscopic studies of YbBa2Cu3O7-δ -precursor films formed on SrTiO3 by dipping-pyrolysis process", *Jpn. J. Appl. Phys.* **37**, L1141(1998).
7) K. Yamagiwa, J. Shibata, T. Hirayama and I. Hirabayashi, "The influence of calcination processes on the superconducting properties of Yb123 films prepared by chemical solution deposition", *Physica C* **309**, 231(1998).

TRANSMISSION ELECTRON MICROSCOPIC STUDIES OF AlN FILMS FORMED ON OFF-ORIENTED R-PLANE OF SAPPHIRE SUBSTRATES BY MOCVD

Junko Shibata, Tomohiko Shibata, Yukinori Nakamura, Keiichiro Asai and Hiroaki Sakai
NGK Insulators, Ltd., 2-56 Suda-cho, Mizuho-ku, Nagoya 467-8530, Japan

Tsukasa Hirayama
Japan Fine Ceramics Center, 2-4-1 Mutsuno, Atsuta-ku, Nagoya 456-8587, Japan

Yuichi Ikuhara
Engineering Research Institute, School of Engineering, The university of Tokyo, 2-11-16 Yayoi, Bunkyo-ku, Tokyo 113-8656, Japan

ABSTRACT

AlN films were formed on just-oriented and off-oriented R-plane of sapphire substrates by metal-organic chemical vapor deposition (MOCVD). Cross-sectional specimens were prepared, and AlN/Al_2O_3 interfaces were investigated from two different directions by high resolution electron microscopy (HREM). It was found that the orientation relationship between the film and the substrate depends on the off-angle of R-plane of the sapphire substrates. In the AlN film formed on just-oriented R-plane, $(1\bar{2}10)AlN$ plane was parallel to $(1\bar{1}02)Al_2O_3$. In contrast, in the AlN film formed on $-4°$ off-oriented R-plane, $(1\bar{2}10)AlN$ was tilted by about $2°$ with respect to $(1\bar{1}02)Al_2O_3$. In addition, in the AlN film formed on $+4°$ off-oriented R-plane, moiré fringes were remarkably seen in the vicinity of the interface between the film and the substrate. This suggests that AlN crystals with different orientations were generated in this film. Based on the growth orientation relationship and HREM images, atomic structure models for the AlN/Al_2O_3 interfaces were proposed.

INTRODUCTION

AlN is a wide band gap (6.2eV) III-V compound (wurzite-type crystal structure)

To the extent authorized under the laws of the United States of America, all copyright interests in this publication are the property of The American Ceramic Society. Any duplication, reproduction, or republication of this publication or any part thereof, without the express written consent of The American Ceramic Society or fee paid to the Copyright Clearance Center, is prohibited.

with the space group P63mc, a=0.31114nm and c=0.49792nm. It has desirable thermal conductivity, electric resistivity and acoustic properties. AlN films are, therefore, promising materials for applications in microelectronic and optoelectronic devices, such as short wavelength emitters, dielectric layers in integrated circuits and surface acoustic wave (SAW) devices. In particular, AlN/Al₂O₃ heteroepitaxial films have been recently investigated for GHz SAW devices for applications in wireless communication systems. Tsubouchi et al. have reported that they succeeded in fabricating zero-temperature coefficient SAW delay lines operating over 1 GHz on an AlN/Al₂O₃ system.[1]

For such applications, it is essential to achieve a good crystal quality of AlN epitaxial films. The AlN/Al₂O₃ system has a large lattice mismatch, and different thermal expansion coefficients between the substrate and the film. In addition, inversion domain boundaries (IDB) in the AlN films are easily generated during the film deposition. IDB reduces the velocity of the surface acoustic wave. From these reasons, many researchers have studied the AlN epitaxial growth technology for improving the quality of the films.[2,3] On the other hand, the growth mechanism of the AlN films is not well understood. In the present work, we investigated the interfacial structure of the AlN films by transmission electron microscopy (TEM) to understand the growth mechanism for the films grown by MOCVD on just-oriented and off-oriented R-plane of sapphire substrates.

EXPERIMENTAL PROCEDURE

AlN films were formed by MOCVD on just-oriented and off-oriented R-plane of sapphire substrates. Before growth, the sapphire substrates were cleaned with organic solvents. These substrates were then heated up to 1000°C in a NH₃ gas flow. Subsequently, AlN films were grown to 2μ m in thickness at 1000°C in trimethyl aluminium (TMA) and NH₃ mix-gas flow in a vacuum of about 20Torr. We used two different off-oriented R-plane of sapphire substrates, $-4°$ off-oriented and +4° off-oriented. In the former substrate, the R-plane was rotated clockwise around the [11$\bar{2}$0]Al₂O₃ axis with an angle of 4° from the surface of the substrate. In the latter, the R-plane was rotated counter-clockwise with 4° around the [11$\bar{2}$0]Al₂O₃ axis as shown in Fig. 1.

Cross-sectional TEM specimens were prepared by the following procedure. First, AlN/Al₂O₃ samples were cut in two proper orientations so as to observe the interface parallel to [11$\bar{2}$0] Al₂O₃ or [1$\bar{1}$01]Al₂O₃. Secondly, two such pieces were glued together by G1 epoxy (Gatan Inc.) with the film surfaces facing each other. These 'sandwiches' were cut in strips with lengths of 3 mm, polished to about 0.08 mm in thickness, and attached to Mo reinforcement rings. Subsequently,

Grain Boundary Engineering in Ceramics

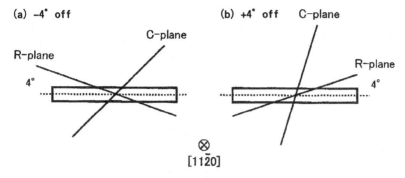

Figure 1. Relationship between (1$\bar{1}$02)Al$_2$O$_3$(R-plane) and the surface of the sapphire substrate: (a) -4° off-oriented R-plane; (b) +4° off-oriented R-plane, projective along the [11$\bar{2}$0]Al$_2$O$_3$ direction.

they were dimpled to a thickness of about 15 μ m, and finally all samples were thinned by Ar ion beam sputtering. The TEM system used in this work was a JEM-2010 with a point-to-point resolution of 0.194 nm. Image simulations were carried out (objective aperture: 0.67Å^{-1}, spherical aberration: 0.5 mm) using the "Tempas" software developed by Kilaas.[4]

RESULTS AND DISCUSSION

Orientation relationship and misfit dislocation

HREM images taken at the interfaces confirmed that the AlN film was epitaxially grown on the Al$_2$O$_3$ substrate. Figure 2 shows a typical cross-sectional transmission electron micrograph of the AlN film formed on just-oriented R-plane of the sapphire substrate, observed from the direction of [10$\bar{1}$0]AlN//[11$\bar{2}$0]Al$_2$O$_3$. As shown in this figure, the orientation relationship between the AlN film and the Al$_2$O$_3$

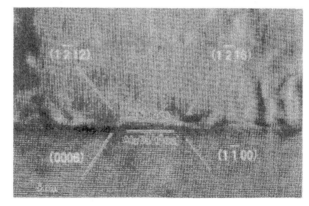

Figure 2. Cross-sectional electron micrograph of the AlN film formed on just-oriented R-plane of the sapphire substrate, observed from the direction of [10$\bar{1}$0]AlN//[11$\bar{2}$0]Al$_2$O$_3$.

substrate is as follows:
$(1\bar{2}10)\text{AlN}//(1\bar{1}02)\text{Al}_2\text{O}_3$,
$[0001]\text{AlN}//[1\bar{1}01]\text{Al}_2\text{O}_3$.
In this relationship,
$(1\bar{2}12)\text{AlN}$ is also
parallel to $(1\bar{1}00)\text{Al}_2\text{O}_3$
and $(1\bar{2}16)\text{AlN}$ is
almost parallel to
$(0006)\text{Al}_2\text{O}_3$ (the angle
is 176°). The growth of
the AlN film is
considered to be
dominated by this
three-dimensional
relationship.

Figure 3 shows the
cross-sectional
micrograph of the AlN
film formed on −4°
off-oriented R-plane of
the sapphire. In this
film, $(1\bar{2}10)\text{AlN}$ is tilted
by about 2° with
respect to R-plane of
the sapphire. This
misoriented growth of
the AlN film seems to
be attributed to the
interaction between
$(1\bar{2}16)\text{AlN}$ and
$(0006)\text{Al}_2\text{O}_3$, because
$(0006)\text{AlN}$ plane is
exposed at the terrace

Figure 3. Cross-sectional electron micrograph of the AlN film formed on −4° off-oriented R-plane of the sapphire substrate, observed from the direction of $[10\bar{1}0]\text{AlN}//[11\bar{2}0]\text{Al}_2\text{O}_3$. Note that $(1\bar{2}10)\text{AlN}$ plane is grown with a tilted angle about 2° with respect to the R-plane of the sapphire substrate. Misfit dislocations are indicated by arrows.

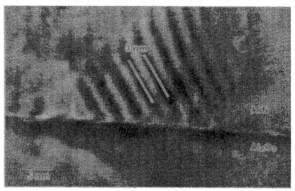

Figure 4. Cross-sectional electron micrograph of the AlN film formed on +4° off-oriented R-plane of the sapphire substrate, observed from the direction of $[10\bar{1}0]\text{AlN}//[11\bar{2}0]\text{Al}_2\text{O}_3$. Moiré fringes are remarkably seen in the vicinity of the interface.

on the surface of −4° off-oriented substrate. Such a misoriented growth was also reported for the metallic film on the R-plane of the sapphire substrate.[5] Figure 4 shows the cross-sectional micrograph of the AlN film formed on +4° off-oriented R-plane of the sapphire. As shown in this figure, moiré fringes with the spacing about 3 nm are remarkably seen in the vicinity of the interface between the film and the substrate. Generally, if the interplanar spacings of the two sets of overlapping planes are respectively d_1 and d_2, the moiré fringe spacing D is given as follows: $D = d_1 d_2 \cdot \cos\theta / |d_1 - d_2|$. These overlapping planes meet at an angle of

θ with each other. In the present case, the spacing of (0002)AlN is 0.249 nm, and θ is $4°$ from diffraction patterns. Assuming that AlN crystals with different orientations were formed, we approximately calculated the spacing d_2 by substituting 0.249 nm for d_1, 3nm for D and $4°$ for θ. Consequently, we got the spacing as 0.272 nm, which almost corresponds to the spacing of (1010)AlN.

Figures 5(a) and 5(b) show selected-area electron diffraction patterns of the AlN films formed on $-4°$ and $+4°$ off-oriented substrates, respectively, obtained in the vicinity of the interface. In the pattern of the film formed on $+4°$ off-oriented substrate, extra spots are visible. The spacing of the extra spot indicated by a circle in Fig. 5(b) agrees with that of the moiré fringes (3 nm).

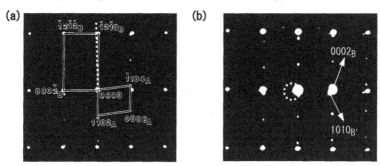

Figure 5. Selected-area electron diffraction patterns of AlN films formed on : (a) $-4°$ off-oriented R-plane; (b)$+4°$ off-oriented R-plane of the sapphire substrate. These patterns were obtained in the vicinity of the interface of each film, from the [10$\bar{1}$0]AlN // [11$\bar{2}$0]Al$_2$O$_3$ direction. In the diffraction patterns, 'A' and 'B' indicate Al$_2$O$_3$ and AlN, respectively. 'B' also shows AlN, which forms moiré fringes.

Figures 6(a) and 6(b) show cross-sectional electron micrographs of AlN films formed on $-4°$ off-oriented and $+4°$ off-oriented R-plane of the sapphire substrates, respectively, observed along the [0001]AlN // [1 1 01]Al$_2$O$_3$. Selected-area electron diffraction patterns at the top left of both micrographs were obtained in the vicinity of the interface between the substrate and the film. Misfit dislocations are seen at the interface as indicated by arrows in Fig. 6(b). The misfit parameter along [10$\bar{1}$0]AlN//[11$\bar{2}$0]Al$_2$O$_3$ is 12.5%, and the average distance between two misfit dislocations is 1.9 nm. This value corresponds to 8 planes of (11$\bar{2}$0)Al$_2$O$_3$ or 7 planes of (10$\bar{1}$0)AlN. Furthermore, in both films many stacking faults are observed along (01$\bar{1}$0)AlN plane. It was reported that GaN films were grown with "house-shaped" domains, which have boundaries planes (10$\bar{1}$0),(01$\bar{1}$0),(10$\bar{1}$1) and (10$\bar{1}$2).[6] SEM and TEM observations from the direction perpendicular to the surface of the films revealed that AlN films were also grown with "house-shaped" domains. The stacking faults appear to be formed during coalescence of these domains that have different stacking orders across the boundary. Actually, in the plan-view observation by TEM, many stacking faults along (0001)AlN were observed. High-resolution electron micrographs implied that these stacking faults were due to different hexagonal stacking sequences, such as ACAC or BCBC instead ABAB for the adjacent domain. (A,B and C were used for cation sites in the basal plane of the hexagonal crystal structure.) In addition, as shown in Fig.6(b) spots of AlN crystals have streaks. This indicates that misoriented AlN crystals exist rotating around [0001]AlN axis with an angle about 12°.

Atomic structure of the AlN/Al₂O₃ interface

From the results of the TEM observation, it was found that the orientation of AlN films is related to the off-angle of the R-plane of the sapphire substrates. In order to consider the

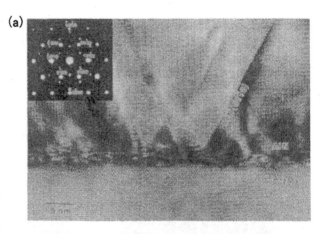
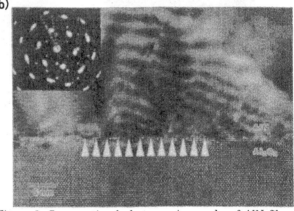

Figure 6. Cross-sectional electron micrographs of AlN films formed on : (a)-4° off-oriented R-plane; (b)+4° off-oriented R-plane of the sapphire substrate respectively, observed from the direction of [0001]AlN//[1$\bar{1}$01]Al₂O₃. Electron diffraction patterns at the top left of both micrographs were obtained in the vicinity of the interface.

reason for this, we constructed interfacial structure models and compared simulation images with the experimental images.

The computer simulations were carried out on the structure models shown in Figs. 7(a) − 7(c); these figures show the [11$\bar{2}$0]Al₂O₃ // [10$\bar{1}$0]AlN projection of the three-dimensional models. Separated by the misfit dislocations, a coherent region is thought of as a fully coherent interface. On the assumption that the O atoms from the terminating surface of the Al₂O₃ substrate bond to the Al atoms of the film at the interface, three different interfacial structures can be considered corresponding to models I (Fig. 7(a)), II (Fig. 7(b)) and III (Fig. 7(c)). The shortest interatomic distances across the interfaces are taken to be 0.185-0.323 nm (model I), 0.166-0.194 nm (model II) and 0.186-0.260 nm (model III), respectively. These distances coincide approximately with the Al-O interatomic spacing in bulk Al_2O_3, which is 0.185 nm. The parameters of the super-cells in the image simulation were 0.95 nm×1.229 nm×2.889 nm for model I, 0.95 nm×1.229 nm ×2.596 nm for model II and 0.95 nm×1.229 nm×3.006 nm for model III. Interface images for the different specimen thickness (from 2 nm to 20 nm) and various defocus values (from −20 nm to −50 nm) were simulated on the basis of these models. The simulated images of model I and model III agreed with the experimental micrographs of the AlN film formed on the just-oriented R-plane of the sapphire, and these two models were not able to be distinguished. The inset in Fig. 8 is the simulated image based on the model I at the specimen thickness of 4nm and the defocus value of −38 nm. As far as the interplanar spacing and the interface image characteristics are concerned, the simulated images give a good matching with the experimental results. It is noted that the model III can be obtained by removing O atoms from the first epitaxial layer in the model I. It seems of less importance to distinguish model III from model I. However, the formation of IDB seems to be related to these interfacial structures. More detailed calculations are necessary for definitely explaining the effect of the R-plane off-angle on the orientation and the growth of the AlN film, such as the simulation of the image projective from the other direction.

CONCLUSIONS

We investigated the interfacial structure of AlN/Al_2O_3 films by transmission electron microscopy. These films were formed on the just-oriented and off-oriented R-planes of sapphire substrates by MOCVD. In the AlN film formed on the just-oriented R-plane,

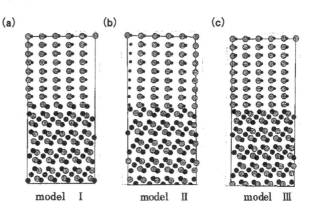

(a) (b) (c)

model I   model II   model III

Figure 7. Interfacial structure models showing the types of bonding of Al-O at the interface. ⬣ Al ● O •N

($1\bar{2}10$)AlN plane was epitaxially grown in parallel with ($1\bar{1}02$)Al_2O_3. However, in the film formed on −4° off-oriented R-plane of the sapphire substrate, the

(1̄210)AlN plane was tilted with an angle about 2° with respect to the (11̄02)Al₂O₃ plane. In the AlN film formed on +4° off-oriented R-plane, moiré fringes were seen in the vicinity of the interface, which indicated that AlN crystals with different crystallographic properties were generated. Furthermore, we proposed the interfacial structure models and rationalized them by image

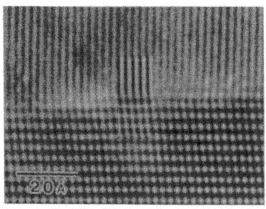

Figure 8 High-resolution transmission electron micrograph of the AlN film formed on the just-oriented R-plane of the sapphire substrate taken along the [101̄0]AlN//[112̄0]Al₂O₃ direction. The simulated image of the proposed structure model I at the specimen thickness of 4 nm and defocus value of −38 nm is shown as an inset.

simulation. Experimental images agreed with the simulated images calculated on the models, in which Al atoms in the first layer of the film connected to O atoms in the R-plane of the sapphire substrate.

References

1) K. Tsubouchi and N. Mikoshiba, "Zero-temperature-coefficient SAW devices on AlN epitaxial films", *IEEE Trans. Sonics. Ultrason.* **32**[5], 634(1985).
2) H. Kawakami, K. Sakurai, K. Tsubouchi and N. Mikoshiba, "Epitaxial growth of AlN film with an initial-nitriding layer on α-Al₂O₃ substrate", *Jpn.J.Apl.Phys.* **27**, L161(1988).
3) K. Masu, Y. Nakamura, T. Yamazaki, T. Shibata, M. Takahashi and K. Tsubouchi, "Transmission electron microscopic observation of AlN/ α-Al₂O₃ heteroepitaxial interface with initial-nitriding AlN layer", *Jpn.J.Apl.Phys.* **34**, L760 (1995).
4) R. Kilaas, Proceedings of the 49th Annual Meeting of the Electron Microscopy Society of America, edited by G.W.Bailey (San Francisco Pres), 528(1991).
5) Y. Ikuhara, P. Pirouz, A. H. Heuer, S. Yadavalli and C. P. Flynn, "Structure of V-Al₂O₃ interfaces grown by molecular beam epitaxy", *Phil.Mag.A* **70**, 75(1994).
6) X. H. Wu, L. M. Brown, D. Kapolnek, S. Keller, B. Keller, S. P .DenBaars and J.S.Speck, "Defect structure of metal-organic chemical vapor deposition-grown epitaxial (0001) GaN/Al₂O₃", *J.Appl.Phys.* **80**[6], 3228(1996).

ALIGNED CARBON NANOTUBE FILMS SELF-ORGANIZED BY SURFACE DECOMPOSITION OF SiC

T. Suzuki, M. Kusunoki, T. Hirayama and N. Shibata
Japan Fine Ceramics Center FCT Central Research Department
2-4-1 Mutsuno, Atsuta-ku, Nagoya, 456-8587 Japan

ABSTRACT

Aligned carbon nanotube (CNT) films were produced by a simple method based on the self-organization by surface decomposition of a SiC wafer in a vacuum furnace. From the results of cross-sectional observations by high-resolution transmission electron microscopy (HRTEM) and analytical electron microscopy, it was clarified that the CNT film grew by the active oxidation of SiC. The thickness of the film was controlled by the heating temperature and time. The formation mechanism of the CNT film was also proposed.

INTRODUCTION

Applications of CNTs are strongly expected for nanometre-scale engineering and electronics industries. For these purposes, well-aligned and size-controlled CNT films are in great demand. CNTs have been produced by such methods as carbon-arc discharge [1] and laser-ablation [2]. However, these methods usually produce randomly oriented CNTs along with other forms of carbon.

Recently, the present authors discovered that aligned CNTs were formed on the surface of SiC particles heated at 1700°C, using a laser system attached to a transmission electron microscope (TEM) [3]. Moreover, the CNT films were produced on an α-SiC (0001) wafer in a vacuum electric furnace [4,5]. The growth of CNTs was due to decomposition of SiC by selective desorption of Si atoms.

EXPERIMENTAL

Commercial α-SiC (0001) single-crystal wafers (Cree Research, Inc.) were heated at 1300, 1500 and 1700°C in a vacuum furnace (1×10^{-4} Torr) with an electric resistance carbon heater. The heating rate was 15°C min^{-1}. These specimens were prepared by ion bombardment and observed along the plan-

To the extent authorized under the laws of the United States of America, all copyright interests in this publication are the property of The American Ceramic Society. Any duplication, reproduction, or republication of this publication or any part thereof, without the express written consent of The American Ceramic Society or fee paid to the Copyright Clearance Center, is prohibited.

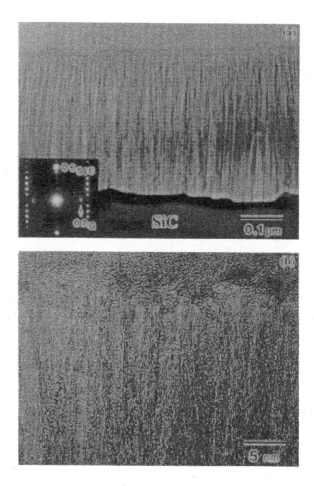

Figure 1 CNT film on a SiC(0001) wafer. (a) The TEM micrograph of the CNT film self-organized on the surface of an α-SiC (0001) wafer heated at 1700°C for 0.5 h. (b) The HRTEM micrograph of the top of the CNT shown in (a).

view and the cross-sectional directions using HRTEM (TOPCON 002B and JEOL 2010) accelerated at 200 kV.

Quantitative chemical microanalysis was done with a dedicated scanning transmission electron microscope (STEM) (Thermo, Vacuum-Generators HB 601 UX) operated at 100 kV, equipped with a high-resolution pole piece and facilities for electron energy loss spectroscopy (EELS). The energy resolution was better than 0.6 eV and the mean beam diameter was almost 0.3 nm.

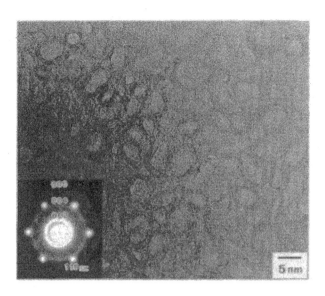

Figure 2 TEM micrograph of the CNT film observed along the CNT axis plan-view direction.

RESULTS AND DISCUSSION

Figure 1(a) shows the surface of the SiC wafer heated at 1700°C for 0.5 h. It can be seen that a film of well-aligned CNTs 0.3 μm in length was formed perpendicular to the (0001) plane of the SiC substrate. Figure 1(b) is a HRTEM image of the top of the CNTs shown in Fig. 1(a). The CNTs 3-5 nm in width were two to five layered, each with a cap. The amorphous contrast on the upper side of Fig. 1(b) is due to the glue, which was used in the preparation of a cross-sectional TEM specimen.

Figure 2 shows a HRTEM micrograph of the CNT film along the CNT axis plan-view direction. Two to five layered concentric circles 3-5 nm in diameter, which correspond to each CNT cross-section, are adjacent to each other. From this micrograph, the density of the CNTs in this film was estimated to be roughly $3 \times 10^4 \, \mu m^{-2}$. The insetted electron diffraction pattern was obtained also along the plan-view direction from the CNT film on an unremoved thin SiC substrate near the area shown in Fig. 2. The electron beam was parallel to the $[0001]_{SiC}$ direction. Only 000l (l = 2n) graphite reflections were excited around the direct spot as ring patterns. This means that the CNTs are well-oriented perpendicularly to the (0001) plane of SiC. The ring around each $11\bar{2}0$ spot of SiC is a 0002 graphite reflection ring excited by double diffraction.

To investigate the formation mechanism of the CNT film, SiC wafers were heated under several heating conditions, and then observed by TEM. Figure 3(a) is a TEM micrograph of the SiC wafers heated at 1300°C for 0.5 h. The small caps 3-5 nm in size consisting of a couple of layered graphite sheets were generated all over the surface of the SiC crystal. This corresponds to the initial stage of the

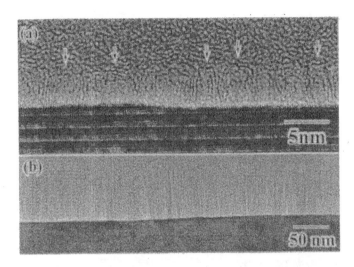

Figure 3 TEM micrograph of the surface of a $(0001)_{SiC}$ plane (a) heated at 1300°C for 0.5 h, (b) heated at 1500°C for 0.5 h.

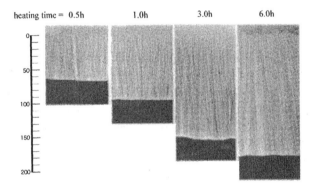

Figure 4 Thickness change of CNT films with increasing heating time at 1500°C.

self-organization of the CNT film. This result indicates that the cap generated at an early stage determines the diameter of the fully-grown CNT. When the SiC wafer was heated at 1500°C for 0.5 h, the length of CNTs increased to 65 nm, as shown in Fig. 3(b). When heated at 1700°C for 0.5 h, the length of CNTs increased to 0.3 μm, as shown in Fig. 1(a). Besides, Fig. 4 shows the change of CNT films as a function of heating time. Only the length of CNTs increased with an increase in heating time. These results show that the length of CNTs - that is, the thickness of the CNT film - can be controlled by varying the heating temperature and time.

Grain Boundary Engineering in Ceramics

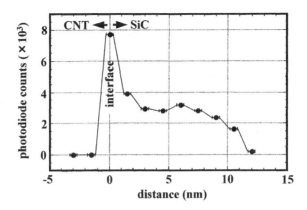

Figure 5 EELS line-scan profile of the integrated O K-edge intensity across the SiC/CNT

The EELS measurement was carried out to determine the real density of oxygen atoms near the interface. The electron beam was electronically moved along a direction perpendicular to the interface between the CNT film and the SiC substrate. The oxygen segregation was determined in terms of the intensity variations of the O K-edge relative to the intensities of both the C K-edge and Si L_{23}-edges. As shown in Fig. 5, a high-resolution line profile of the integrated O K-edge intensity across the SiC/CNT interface was obtained in a 1.5 nm step. The segregation of oxygen atoms can be seen at the interface between the CNT film and SiC substrate, that is, only the surface of the SiC was oxidized. This result suggests that residual oxygen in the vacuum chamber plays the most important role for the self-organization of CNTs. In the present experiment, since the SiC wafers were heated with the electric resistance carbon heater, most of the oxygen transforms to carbon monoxide (CO) over 1000°C [6].

From the above results, the model for the formation mechanism of the CNT film was proposed as shown in Fig. 6. When a wafer was heated at around 1300°C, Si atoms are preferentially oxidized to produce SiO gas molecules, and then the left carbon atoms form nanocaps of graphite (Fig. 6(a)). These reactions were explained from the 'active oxidation' process [7]. In the reaction, solid SiC is continuously oxidized to form a SiO gas without forming a passive SiO$_2$ solid film. The carbon atoms in each graphite sheet of nanocaps have σ-bonding with the carbon atoms in the ($10\overline{1}2$) plane of the SiC crystal, but not π-bonding. Once σ-bonds are generated, this bonding is preferentially kept and cylindrical graphite sheets with the diameter of the nanocaps grow perpendicularly on the SiC because σ-bonding is more stable than π-bonding. Therefore, CNTs grow as the SiC single crystal erodes by the continuous active oxidation (Fig. 6(b)).

CONCLUSIONS

The structure of the CNT film formed by surface decomposition of SiC was clarified by electron microscopy. From the results, a model for the formation mechanism has been proposed. This simple

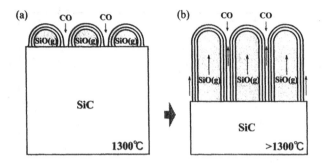

Figure 6 Schematic diagram for the formation mechanism of the CNT film on a SiC crystal by the surface decomposition method. (a) At around 1300°C, carbon nanocaps are formed on the SiC crystal. (b) The CNTs grow towards the interior of the SiC crystal, eroding the SiC single crystal.

method can achieve to prepare pure and aligned CNT films. Moreover, the length of CNTs is easily adjusted on inch-size SiC wafers. These characteristics will be of great advantage for industrial applications.

ACKNOWLEDGMENTS

The authors thank Dr. K. Kaneko, International Joint Research Program, Ceramics Superplasticity, JST, Japan, for EELS analyses.
A part of this work was supported by FCT project, which was cosigned to JFCC by NEDO.

REFERENCES

[1] T. W. Ebbesen, and P. M. Ajayan, "Large-scale synthesis of carbon nanotubes", *Nature*, **358**, [6383], 220 (1992).
[2] A. Thess, R. Lee, P. Nikolaev, H. Dai, P. Petit, J. Robert, C. Xu, Y. H. Lee, S. G. Kim, A. G. Rinzler, D. T. Colbert, G. E. Scuseria, D. Tománek, J. E. Fischer, and R. E. Smalley, "Crystalline Ropes of Metallic Carbon Nanotube", *Science*, **273**, 483 (1996).
[3] M. Kusunoki, M. Rokkaku, and T. Suzuki, "Epitaxial carbon nanotube film self-organized by sublimation decomposition of silicon carbide" *Appl. Phys. Lett.*, **71**, [18], 2620 (1997)
[4] M. Kusunoki, J. Shibata, M. Rokkaku, and T. Hirayama, "Aligned Carbon Nanotube Film Self-Organized on a SiC Wafer" *Jpn. J. Appl. Phys.*, **37**, [5B], L605 (1998).
[5] M. Kusunoki, T. Suzuki, K. Kaneko, and M. Ito, "Formation of self-aligned carbon nanotube films by surface decomposition of silicon carbide", *Phil. Mag. Lett.*, **79**, [4], 153 (1999).
[6] T. Yamaguchi, "New development of composite refractories", *Ceramics*, **23**, [11] 1072 (1988) in

Japanese.

[7] R. J. Fordham, "Graphical displays of the thermodynamics of high-temperature gas-solid reactions and their applications", pp.33-52 in *High Temperature Corrosion of Technical Ceramics* (Amsterdam: Elsevier Applied Science) (1990).

STRUCTURE OF FCC-Ti/6H-SiC INTERFACE GROWN BY ELECTRON BEAM EVAPORATION

Y. Sugawara and N. Shibata

Research and Development Laboratory, Japan Fine Ceramics Center, 2-4-1 Mutsuno, Atsuta-ku, Nagoya 456-8587, Japan

S. Hara

Materials Science Division, Electrotechnical Laboratory, 1-1-4 Umezono, Tsukuba, Ibaraki 305-8568, Japan

Y. Ikuhara

Engineering Research Institute, University of Tokyo, 2-11-16 Yayoi, Bunkyo-ku, Tokyo 113-8656, Japan

ABSTRACT

A titanium thin film was grown on the (0001) plane of 6H-SiC single crystal by an electron beam evaporation technique. High-resolution electron microscopy (HREM) was applied to characterize the interface structures between the Ti film and 6H-SiC substrate. The structure of the Ti film was confirmed to be face-centered cubic (fcc), which suggested that the formation of the unnatural fcc is due to the high coherency between Ti and SiC at the interface. The orientation relationship of fcc-Ti/6H-SiC interface was $(111)_{fcc-Ti}//(0001)_{6H-SiC}$ and $[\bar{1}10]_{fcc-Ti}//[11\bar{2}0]_{6H-SiC}$. This orientation relationship is considered to occur so that the lattice coherency across the interface is maximized, which was evaluated by the coincidence of reciprocal lattice points (CRLP). The fcc-Ti crystal is tilted by $\sim 4°$ with respect to the (0001) planes of the 6H-SiC crystal. This crystal tilt is due to the introduction of misfit dislocations to accommodate the lattice mismatch between Ti and SiC.

To the extent authorized under the laws of the United States of America, all copyright interests in this publication are the property of The American Ceramic Society. Any duplication, reproduction, or republication of this publication or any part thereof, without the express written consent of The American Ceramic Society or fee paid to the Copyright Clearance Center, is prohibited.

INTRODUCTION

It is important to design the electronic devices for an understanding the structure of metal/semiconductor interfaces.[1, 2] Since the crystal structure and electronic properties of metal thin film are influenced by the film/substrate interface structure, characterizing the interface structure is a key to understand the mechanism of thin film formation and electric properties of the system.

In the system using the SiC single crystal as a substrate, aluminum, titanium, molybdenum, nickel and gold have been used as a metallic thin film to develop new electronic devices.[3] In order to obtain an ideal metal/semiconductor interface, the schottky barrier height should be controlled to decrease the energy level of the interfaces. For this purpose, the surface treatment technique of semiconductor is now one of the most important topics in this field.[3]

In this study, Ti film was deposited on the ideal clean surface at the silicon layer of $(0001)_{6H-SiC}$ plane by electron beam evaporation method, and the Ti/SiC interface was mainly observed by high-resolution electron microscopy (HREM) and the lattice coherency across the interface was evaluated by the coincidence of reciprocal lattice points (CRLP).[4-6]

EXPERIMENTAL PROCEDURES

A single crystal of n-type 6H-SiC wafer (Cree Research, Inc., Durham, North Carolina, USA) was used as the substrate, and Ti was deposited on its (0001) Si-face. Firstly, the substrate was etched in 5% hydrofluoric acid (HF) for 10 min to remove the natural oxide layer, and then thermally oxidized at 1100°C for 150 min in dry oxygen gas. Next, it was etched again in 5% HF for 10 min to remove the oxide layer, and then immersed in boiling high purity water for 10 min. After these pre-treatments, the sample was transferred in a vacuum chamber, and Ti was deposited on it using electron beam evaporation at ~29°C in a vacuum of 5.1×10^{-8} Pa. TEM specimens were prepared by the following procedure. First, the sample was oriented by X-ray Laue back reflection from the SiC substrate side and cut by a diamond saw in the proper orientation. Then, two such pieces were glued together by G-1 epoxy (Gatan, Inc., Preasanton, California, USA) with the Ti film surfaces facing each other. This 'sandwich' was cut into strips with lengths of about 1 mm, polished to a thickness of about 0.1 mm, and they were attached to Mo rings by epoxy in order to reinforce the specimens. Subsequently, they were dimpled to a thickness of about 20 μm and finally thinned by ion beam sputtering at a voltage of 3-5 kV. TEM observations were performed using JEOL JEM-2010 (200 kV) and JEM-4000FX (400 kV) electron microscopes.

RESULTS AND DISCUSSION

Figure 1 shows selected-area diffraction patterns (SADP) obtained from (a) $[11\bar{2}0]_{6H\text{-SiC}}$ and (b) $[1\bar{1}00]_{6H\text{-SiC}}$ directions in which the two cross-sectional direction is orthogonal. The lattice spacings and the axial angles in the diffraction patterns of the present specimen do not coincide with those in bcc-Ti (β-Ti) and hcp-Ti (α-Ti) which are reported in the data of the Joint Committee on Powder Diffraction Standards (JCPDS). We observed SADPs at every 30° step to cover 180° around the growth direction of the Ti crystal,[1,2] and found that the crystal has 3-fold or 6-fold symmetry on the plane perpendicular to the primitive axis, and has screw operation of 1/3c along the primitive axis. Moreover, each interatomic distance between nearest neighbor atoms were the same in the tetrahedron in the crystal. Judging from these results, the crystal structure in the present Ti film can be identified to be face-centered cubic (fcc) with a $Fm\bar{3}m$ space group. The lattice constant and nearest neighbor atom spacing were determined to be 0.438 nm and 0.310 nm, respectively. The orientation relationship of fcc-Ti/6H-SiC interface is thus $(111)_{fcc\text{-Ti}}//(0001)_{6H\text{-SiC}}$, $[1\bar{1}0]_{fcc\text{-Ti}}//[11\bar{2}0]_{6H\text{-SiC}}$. The presence of fcc-Ti has also been reported in Ti thin films grown on Al,[7,8] in Ti/Al multilayers,[9] and in Ti/Ni multilayers.[10] The present result indicates that fcc-Ti can form even on covalent-bonded SiC crystal. It has been reported that Ti film on the 6H-SiC(0001) has the hcp structure.[11] In this case, interaction between the Ti overlayer and the SiC substrate is presumably weaker than that of the fcc-Ti/6H-SiC system because Ti has the most stable hcp structure with a large lattice mismatch of 4.3% with respect to the SiC substrate. The formation of

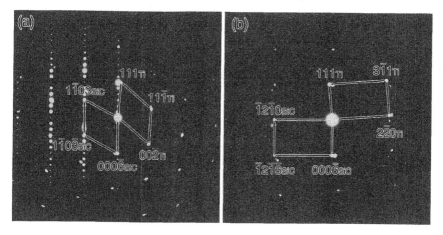

Figure 1. Selected-area diffraction patterns of the fcc-Ti/6H-SiC interface obtained along (a) $[110]_{fcc\text{-Ti}}//[11\bar{2}0]_{6H\text{-SiC}}$ and (b) $[112]_{fcc\text{-Ti}}//[1\bar{1}00]_{6H\text{-SiC}}$ cross-sectional directions. T and S denote fcc-Ti and 6H-SiC, respectively.

hcp-Ti or fcc-Ti is considered to depend on the condition and cleanness of the substrate. Figure 2 shows a cross-sectional high-resolution electron micrograph of the fcc-Ti/6H-SiC interface observed along the $[110]_{fcc-Ti}//[11\bar{2}0]_{6H-SiC}$. Lattice images of Ti and SiC are clearly seen in the micrograph. The fcc structure was observed throughout whole the Ti crystal from the interface to the surface (~80 nm thickness). The fcc-Ti/6H-SiC interface is almost atomically flat, as shown in Fig. 2. In the figure, the $(1\bar{1}1)_{fcc-Ti}$ and $(\bar{1}102)_{6H-SiC}$ planes, the $(\bar{1}13)_{fcc-Ti}$ and $(\bar{1}10\bar{2})_{6H-SiC}$ planes, and the $(002)_{fcc-Ti}$ and $(\bar{1}104)_{6H-SiC}$ planes are almost parallel as indicated in the figure. The lattice coherency across the interface can be quantitatively evaluated by the coincidence of reciprocal lattice points (CRLP) between the two crystals since the reciprocal lattice points contain information on both the interplanar spacings and the directions of the crystal lattice planes.[4] In order to formulate CRLP, each reciprocal lattice point hkl corresponding to the reciprocal lattice vector g, is represented by a sphere with radius r^* around it. The coincidence of reciprocal lattice spheres of the two crystals then corresponds to near-parallelness of various sets of planes with nearly equal interplanar spacings. It is then hypothesized that those orientation relationships between the two crystals are favored in which the sum of all intersection volumes is maximized.

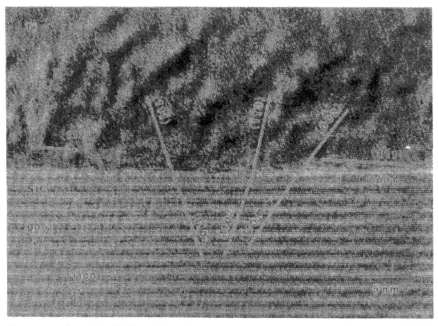

Figure 2. Cross-sectional high-resolution electron micrograph of the fcc-Ti/6H-SiC interface with the incident beam parallel to the $[110]_{fcc-Ti}//[11\bar{2}0]_{6H-SiC}$ axis. The periodic unit of the Ti crystal consists of 3 layers along the growth direction normal to the interface.

Grain Boundary Engineering in Ceramics

In order to apply this method three-dimensionally, one of the crystals is rotated about two orthogonal axes of the other crystal. The orientation between the two crystals is then specified by a set of rotation angles (ϕ, θ). At each value of (ϕ, θ), a number of reciprocal lattice spheres of the two crystals intersect, corresponding to those sets of planes in the two crystals that are nearby parallel and near equi-spaced. The volume v_{gG} at the intersection of reciprocal lattice spheres g and G of the two crystals is calculated for all the spheres in the reciprocal space and the sum $V(\phi, \theta) = \Sigma v_{gG}$ at this orientation relationship is computed. The method of calculating the sum total V was described in detail elsewhere.[4-6] Figure 3 shows the calculated V results between the fcc-Ti structure and the 6H-SiC structure as a function of ϕ and θ. In this calculation, the radius of r^* was taken as $0.2a^*$ (a^*: a-axis length of the 6H-SiC crystal in the reciprocal space), and the rotation angles were selected between $0°$ and $90°$ around the $[100]_{fcc-Ti}$ axis for ϕ and the $[010]_{fcc-Ti}$ axis for θ so as to cover the whole space between the two crystals. The size of r^* does not influence the locations of peaks in Fig. 3, but affects the peak resolution.[6] The initial orientation relationship ($\phi=0°$, $\theta=0°$) was set to be $(001)_{fcc-Ti}//(0001)_{6H-SiC}$, $[100]_{fcc-Ti}//[1\bar{1}20]_{6H-SiC}$ in Fig. 3. As seen in the figure, a maximum peak occurs at $\phi=35°$ and $\theta=45°$, which is consistent with the

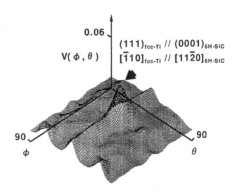

Figure 3. Three-dimensional plots of $V(\phi, \theta)$ for rotations ϕ and θ about the $[100]_{fcc-Ti}$ and $[010]_{fcc-Ti}$ axes, respectively. Note that the predominant peak at $\phi=35°$ and $\theta=45°$ corresponds to the orientation relationship experimentally observed.

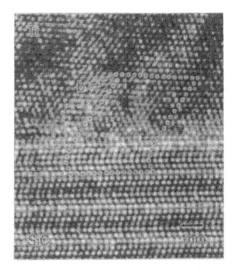

Figure 4. A Burgers circuit around a dislocation in the vicinity of the fcc-Ti/6H-SiC interface (incident electron beam parallel to $[110]_{fcc-Ti}//[1\bar{1}20]_{6H-SiC}$), indicating that the Burgers vector b of the misfit dislocation is 1/2<110> on the (111) plane.

observed orientation relationship in Fig. 2. Consequently, it is concluded that the interface with the highest coherency is formed between fcc-Ti and 6H-SiC. In fact, the misfit parameters between $(1\bar{1}1)_{fcc\text{-}Ti}$ and $(\bar{1}102)_{6H\text{-}SiC}$ planes, $(002)_{fcc\text{-}Ti}$ and $(\bar{1}104)_{6H\text{-}SiC}$ planes are 0.79% and 0.92%, respectively, as shown in Fig. 2. The high coherency of the interface is, thus, would be the origin of the formation of the fcc-Ti crystal.

As shown in Fig. 2, the (111) planes of fcc-Ti is tilted by ~4° to clockwise with respect to the (0001) planes of the 6H-SiC crystal. In this system, the Ti crystal is under a compressive stress because of the difference in the interplanar spacings of Ti and SiC. As shown in Figure 4, a magnification of Fig. 2, the misfit dislocations were introduced in the Ti crystal at the interface. These misfit dislocations are considered to be introduced to accommodate the lattice mismatch at the interface. Referring to the Burgers circuit, the Burgers vector of the misfit dislocation can be identified to be 1/2<110>, which corresponds to the Burgers vector of fcc crystal. According to Frank's formula, the tilt angle θ is expressed as $\theta = \tan^{-1}b/h$, where b is a magnitude of the Burgers vector of the misfit dislocation, h is distance between dislocations. Taking into account that the tilt angle θ is 4° and the magnitude of the Burgers vector is 0.310 nm for 1/2⟨110⟩ of the fcc crystal, the distance between misfit dislocations h is estimated to be 4.4 nm. It was considered that the fcc-Ti is almost relax the elastic strain along the interface by the crystal tilt because the observed distance between each dislocation is close to 4.4 nm.

CONCLUSION

A titanium film was deposited on the (0001) plane of 6H-SiC single crystal by an electron beam evaporation technique, and the interface structure was characterized mainly by HREM. It was found that fcc-Ti is epitaxially formed on the (0001) plane of the 6H-SiC crystal. The orientation relationship (OR) of the fcc-Ti/6H-SiC interface is $(111)_{fcc\text{-}Ti}//(0001)_{6H\text{-}SiC}$ and $[\bar{1}10]_{fcc\text{-}Ti}//[1\bar{1}20]_{6H\text{-}SiC}$, which corresponds to the OR with the highest coherency across the interface. The degree of coherency can be estimated by the coincidence of reciprocal lattice points. The Ti crystal is tilted by introducing misfit dislocations with the Burgers vector of 1/2<110> to accommodate the lattice strain along the interface.

ACKNOWLEDGMENT

This study was supported by the Special Coordination Funds of the Science and Technology Agency of the Japanese Government.

REFERENCES

1. Y. Sugawara, N. Shibata, S. Hara, and Y. Ikuhara, "Interface structure of fcc-Ti thin film grown on 6H-SiC substrate", *J. Mater. Res.* to be submitted.

2. Y. Sugawara, N. Shibata, S. Hara, and Y. Ikuhara, in preparation.

3. S. Hara, T. Teraji, H. Okushi, and K. Kajimura, "Control of schottky and ohmic interfaces by unpinning Fermi level", *Appl. Surf. Sci.*, **117/118**, 394 (1997).

4. Y. Ikuhara, and P. Pirouz, "Orientation relationship in large mismatched bicrystals and coincidence of reciprocal lattice points (CRLP)", *Materials Science Forum* **207-209**, 121 (1996).

5. Y. Ikuhara, and P. Pirouz, "High resolution transmission electron microscopy studies of metal/ceramics interfaces", *Microscopy Research and Technique* **40**, 206 (1998).

6. S. Stemmer, P. Pirouz, Y. Ikuhara, and R. F. Davis, "Film/substrate orientation relationship in the AlN/6H-SiC epitaxial system", *Phys. Rev. Lett.* **77** [9], 1797 (1996).

7. A. A. Saleh, V. Shutthanandan, R. Shivaparann, R. J. Smith, T. T. Tran, and S. A. Chambers, "Epitaxial growth of fcc Ti films on Al(001) surfaces", *J. Phys. Rev. B* **56** [15], 9841 (1997).

8. P. M. Marcus, and F. Jona, "Identification of metastable phases: face-centered cubic Ti", *J. Phys. Condens. Matter* **9**, 6241 (1997).

9. D. Shechtman, D. van Heerden, and D. Josell, "fcc titanium in Ti-Al multilayers", *Mater. Letters* **20**, 329 (1994).

10. D. Josell, D. Shechtman, and D. van Heerden, "fcc titanium in Ti/Ni multilayers", *Mater. Letters* **22**, 275 (1995).

11. L. M. Porter, R. F. Davis, J. S. Bow, M. J. Kim, R. W. Carpenter, and R. C. Glass, "Chemistry, microstructure, and electrical properties at interfaces between thin films of titanium and alpha (6H) silicon carbide (0001)", *J. Mater. Res.* **10** [3], 668 (1995).

MICROSTRUCTURES OF THE AlN/TiN/MgO(001) INTERFACES

X. L. Ma[1], Y. Ikuhara[1], and N. Shibata[2]
[1]*Department of Materials Science, University of Tokyo, 113 Tokyo, Japan*
[2]*Japan Fine Ceramics Center, 456 Nagoya, Japan*

ABSTRACT

AlN/TiN/MgO(001) interfaces, prepared by molecular beam epitaxy, have been characterized by high-resolution electron microscopy. It is found that the thin TiN buffer layer is epitaxially grown on the MgO(001) substrate and hexagonal AlN epitaxially on the as-received TiN(001). Based on the growth orientation relationship and high resolution images, atomistic structure models for the AlN/TiN interface are proposed, image-simulated, and compared with experimental images.

INTRODUCTION

AlN is a potential candidate for applications in microelectronics ranging from optoelectronic and high temperature devices to electronics packaging. Thin films of hexagonal AlN have been epitaxially prepared on several substrates such as Si,[1, 2, 3-6] SiC,[7,8] and sapphire.[4] In our recent study of the AlN film grown on MgO(001) with a TiN buffer layer,[9] we reported the orientation relationships between the as-grown films and MgO substrate, we also rationalized the experimental observation by theoretical calculations on the basis of three dimensional lattice continuity. The growth of hexagonal AlN on the substrate was dominated by the orientation relationship $(1\bar{2}10)_{AlN}//(110)_{TiN}//(110)_{MgO}$, and $(0001)_{AlN}//(001)_{TiN}//(001)_{MgO}$, although another oriented growth with less frequency was also observed. Interfacial structure contains two aspects from the viewpoint of crystallography: first, the orientation relationship between film and substrate; and second, the atomistic bonding between two materials across the interface.

Previous work on AlN films grown on various substrates only focused on the orientation relationships between the epitaxial film and employed substrate. Few papers have been published about atomistic structures at the interface between AlN films and substrate. In this paper, we propose atomistic structure models for the AlN/TiN interface on the basis of the observed orientation relationship and HREM images. We evaluate the HREM images by comparing

To the extent authorized under the laws of the United States of America, all copyright interests in this publication are the property of The American Ceramic Society. Any duplication, reproduction, or republication of this publication or any part thereof, without the express written consent of The American Ceramic Society or fee paid to the Copyright Clearance Center, is prohibited.

the micrographs with simulated images. The present study is also expected to shed some light on understanding of AlN films grown on other substrates.

EXPERIMENTAL

Details of the MBE process were reported in our previous paper [9]. Cross-sectional specimens for TEM observation were prepared by "sandwich-glueing", slicing, grinding, dimpling and finally ion-milling using a cold stage (liquid N2) in order to minimize damage to the specimen. A JEOL 2010 high-resolution electron microscope with a point resolution of 0.194 nm operated at 200 kV was used for the microstructural observation.

HREM OBSERVATION

Selected-area electron diffraction (SAED) pattern showed that the growth of the TiN buffer layer was governed by the parallel orientation relationship with the MgO substrate; i.e., $(001)_{TiN}$ // $(001)_{MgO}$, $(010)_{TiN}$ // $(010)_{MgO}$, and $(111)_{TiN}$ // $(111)_{MgO}$. Figure 1 shows a low magnification cross-section image of the as-grown AlN/TiN/MgO(001) layer-material observed with the electron beam along the MgO[110] zone axis. The growth direction is vertically upward.

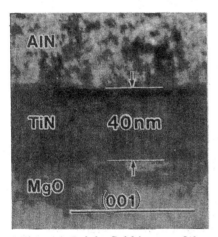

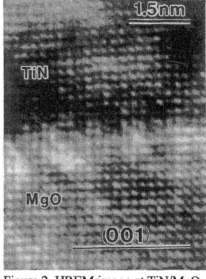

Figure 1. Bright-field image of the AlN/TiN/MgO(001) obtained with the electron beam parallel to the [110]TiN direction.

Figure 2. HREM image at TiN/MgO

Figure 2 shows the HREM image at the TiN/MgO interface obtained from the [100] direction. It is clearly seen that the (100) plane of TiN smoothly connects to the same-indexed planes of MgO since the lattice mismatch between them is as small as 0.7%. Due to the above orientation relationship between the epitaxial TiN and MgO(001), the TiN(001) sequentially acted as the substrate for the growth of AlN layer. The growth was dominated by the orientation relationship of $(1\bar{2}10)_{AlN}//(110)_{TiN}$, $(10\bar{1}0)_{AlN}//(1\bar{1}0)_{TiN}$, and $(0001)_{AlN}//(001)_{TiN}$.

Figure 3 shows a cross-sectional HREM image taken at an interface between the epitaxial AlN and the as-received TiN substrate. The orientation relationship between the hexagonal AlN and the TiN substrate is $[1\bar{2}10]_{AlN}//[110]_{TiN}$ and $(0001)_{AlN}//(001)_{TiN}$. In the following section, we shall discuss the atomistic structure models at the AlN/TiN interface based on this orientation relationship.

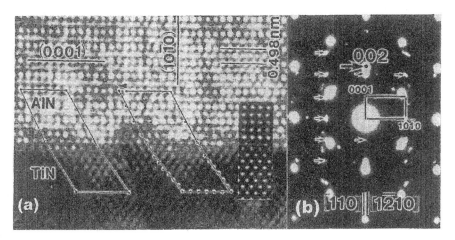

Figure 3. (a) HREM image at AlN/TiN, A Burgers circuit is marked. The inset is simulated image. (b) EDP corresponding to area in (a).

INTERFACE STRUCTURE

Structure of hexagonal AlN

The thermodynamically stable AlN has a hexagonal lattice with the space group of $P6_3mc$. The characteristics of this hexagonal structure are schematically illustrated in Fig. 4, exhibiting the three-dimensional tessellation of the Al and N atoms in one unit cell. Atoms at different layers, such as Z=0, 3/8, 4/8, and 7/8, are shown in Figs. 4(b), 4(c), 4(d), and 4(e), respectively. Fig. 4(f) is the atoms projection from the $[01\bar{1}0]$ direction onto the $(1\bar{2}10)$ plane. In the present study, the AlN/TiN interface was imaged along this direction and the film-growth direction is vertically upward.

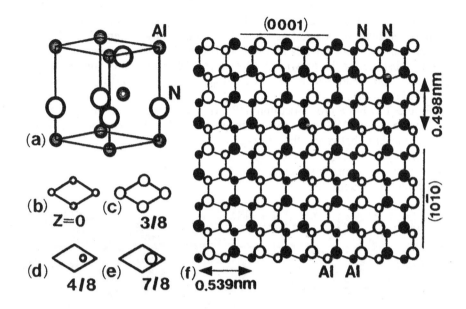

Figure 4. (a) Tessellation of Al and N atoms in hexagonal AlN. (b), (c), (d), and (e) show atoms at Z=0, 3/8, 4/8, and 7/8 respectively. (f) is the atoms projection from the [01̄10] direction.

Mismatch dislocation

The interface between a thin film and a substrate can be coherent, semi-coherent or incoherent depending on the lattice mismatch between corresponding planes of both constituents and bonding strength at the interface. A semi-coherent interface is characterized by coherent regions, which are separated by mismatch dislocations, which is applicable to the present AlN/TiN system. The atomistic structure of the coherent regions at the interface equals the structure of a fully coherent interface. The misfit dislocations accommodate the lattice mismatch partially or completely. A lattice mismatch can be calculated as:

$$f = \frac{2|d_1 - d_2|}{d_1 + d_2},$$

Where d corresponds to the interplanar spacing. In the present study, the HREM image was obtained along the $[01̄10]_{AlN}/[110]_{TiN}$ direction. The interplanar spacings of $(101̄0)_{AlN}$ and $(11̄0)_{TiN}$ are 0.27 nm and 0.30 nm respectively. Therefore, the misfit parameter can be calculated as

Grain Boundary Engineering in Ceramics

$f_{[2130]/[110]TiN}$=10.5%. The average distance D between two misfit dislocations can be obtained according to:

$$D = \frac{d_1 d_2}{d_1 - d_2},$$

which is as long as 5.4nm. This value corresponds to eighteen $(1\bar{1}0)_{TiN}$ planes or twenty $(10\bar{1}0)_{AlN}$ planes. A Burgers circuit around a misfit dislocation is indicated in the middle part of the HREM micrograph as shown in Fig. 3. The left circuit in Fig. 3 is a reference showing a coherent region.

Interface structural models

Separated by the misfit dislocations, a coherent region is thought of as a fully coherent interface. In the interface models, the lattice of film material is strained parallel to the interface plane according to the lattice mismatch f; under such a condition the models are then focused on atomistic bonding across the interface. In the SiC/AlN ceramic-ceramic system, Lambrcht and Segall [10] proposed that the cation-anion bonding across the interface, i.e. Si to N and Al to C, be favorable on the basis of energy consideration. In the present AlN/TiN system, we also propose that the cation-anion bonding, i.e, Al-N(TiN) and N(AlN)-Ti(TiN), be favorable energetically. However, taking into account of the four-layered structure of hexagonal AlN, the four sub-layers are successively considered to act as the first epitaxial layer connecting to the substrate surface and therefore forming four different bondings at the interface, as shown schematically in Fig. 5, which are cross-sectionally projected from the $[01\bar{1}0]_{AlN}/[110]_{TiN}$ direction. These models correspond to chemical bonding at coherent regions of the HREM image, so the lattice of AlN is strained by 10.5% along the $[21\bar{3}0]$ direction. Figs. 5(a) and 5(b) show the way of bonding of Al-N(TiN) at the interface, distinguished by the Al atoms of different layers (i.e., z=0 and z=4/8) being in turn as the first epitaxial layer (referred as model (I) and model (II), respectively); and also, Figs. 5(c) and 5(d) show the bonding of N-Ti(TiN), distinguished by N atoms of different layers (i.e., z=3/8 and z=7/8) in turn as the first epitaxial layer (referred as model (III) and model (IV), respectively). The shortest interatomic distance across the interface is taken to be 0.212 nm that is the same as the bonding distance in bulk TiN. Interface images for the different specimen thickness (from 3 nm to 15 nm) and various defocus values (from –20 nm to –50 nm) are simulated on the basis of these interface models. The simulated images of the four models and the experimental micrograph result in approximately the same agreement, thus the four models cannot be distinguished. However, they share a unique building principle at the interface, i.e., cation-anion bonding. The inset in

Fig. 3 is the simulated image based on the model (I) at the specimen thickness of 5 nm and the defocus value of –30 nm. As far as the interplanar spacing and the interface image characteristics are concerned, the simulated images match well with the experimental results.

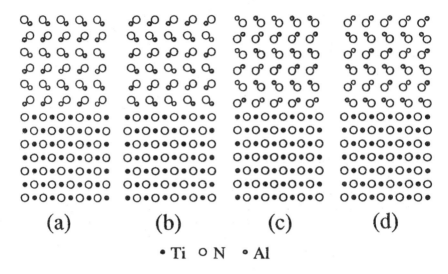

(a) (b) (c) (d)

• Ti ○ N ● Al

Figure 5. Four interface structural models showing the way of bonding of Al-N(TiN) at the interface.

It should be noted from Fig. 5 that, although model (I) and (II) are originally distinguished by the different Al layers being the first epitaxial atom layer, the model (II) can actually be obtained from (I) by a simple operation of being projected from $[1\bar{1}0]_{TiN}$ direction, or by a mirror-reflection across the $(1\bar{1}0)_{TiN}/(10\bar{1}0)_{AlN}$ plane. It seems of less importance to distinguish model (I) from model (II), since it depends on the imaged direction, $[110]_{TiN}$ or $[1\bar{1}0]_{TiN}$. However, it is of interest to note that the co-existence of model (I) and model (II) can result in a stacking fault, which was experimentally observed in AlN thin films by Ivanov et al [11] and in the present study. Similar to above, if model (III) and (IV) coexist, then a stacking fault also occurs; however, if model (I) and (III) or (II) and (IV) coexist, then an inverted domain boundary (IDB) forms, as observed in AlN thin film grown on sapphire[5] and on SiC[7]. Fig.6 shows HREM image taken at the AlN/TiN interface. It is clearly seen that a stacking fault occurs at the interface and remains in the entire AlN film. This experimentally supports the above suggestions that different atomic bonding locally controlled the epitaxial growth modes at the interface. The four interface models, as shown in Fig. 6,

Grain Boundary Engineering in Ceramics

have presented a way to understand the fact that a stacking fault and an inverted domain boundary are frequently observed in the AlN films grown on various substrates.

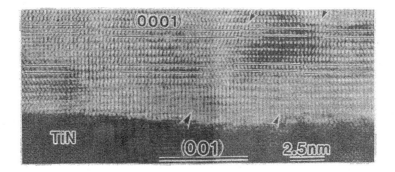

Figure 6. HREM image at AlN/TiN showing stacking faults in AlN film.

CONCLUSIONS

AlN/TiN/MgO(001) interfaces, prepared by molecular beam epitaxy, have been characterized by cross-sectional high-resolution electron microscopy. The growth of the hexagonal AlN film is dominated by the orientation relationship of $(1\bar{2}10)_{AlN}//(110)_{TiN}//(110)_{MgO}$, $(10\bar{1}0)_{AlN}//(1\bar{1}0)_{TiN}//(1\bar{1}0)_{MgO}$, and $(0001)_{AlN}//(001)_{TiN}//(001)_{MgO}$. It appears that both Al and N atoms occurring the first epitaxial layer connecting to the TiN surface layer with the understanding of cation-anion bonding. Taking into account the four-layered structure of the hexagonal AlN, four interface models are proposed and rationalized by image-simulation. These structural models can explain the formation of a stacking fault and an inverted domain boundary (IDB) observed in thin AlN films.

ACKNOWLEDGMENT

This study was performed through Special Coordination Funds of the Science and Technology Agency of Japan.

REFERENCES

1. M. Miyauchi, Y. Ishikawa, and N. Shibata, *"Growth of Aluminum Nitride films on silicon by electron-cyclotron-resonance-assisted molecular beam epitaxy"*, Jpn. J. Appl. Phys., **31**, L1714. (1992).

2. W. J. Meng, J. A. Sell, T. A. Perry, L. E. Rehn, and P. M. Baldo, *Growth of aluminum nitride thin films on Si(111) and Si(001): Structural characteristics and development of intrinsic stresses*, J. Appl. Phys., **75**, 3446 (1994).

3. R. D. Vispute, J. Narayan, H. Wu, and K. Jagannadham, *Epitaxial growth of AlN thin film on silicon substrates by pulsed laser deposition*, J. Appl. Phys., **77**, 4724 (1995).

4. J. Chaudhuri, R. Thokala, J. H. Edgar, and B. S. SywE, *X-ray double crystal characterization of single crystal epitaxial aluminum nitrid thin films on sapphire, silicon carbide and silicon substrates*, J. Appl. Phys., **77**, 6263 (1995).

5. K. Dovidenko, S. Oktyabrsky, J. Narayan, and M. Razeghi, *Aluminum nitride films on different orientations of sapphire and silicon*, J. Appl. Phys., **79**, 2439 (1996).

6. E. Calleja, M. A. Sánchez-Garcìa, E. Monroy, F. J. Sánchez, E. Muñoz, A. Sanz-Hervás, C. Villar, and M. J. Aguilar, *Growth kinetics and morphology of high quality AlN grown on Si(111) by plasma-asisted molecular beam epitaxy*, J. Appl. Phys., **82**, 4681 (1997).

7. S. Tanaka, R. S. Kern, and R. F. Davis, *Initial stage of aluminum nitride film growth on 6H-silicon carbide by plasma-assisted, gas-source molecular beam epitaxy*, Appl. Phys. Lett., **66**, 37 (1995).

8. S. Stemmer, P. Pirouz, Y. Ikuhara, and R. F. Davis, *Film/substrate orientation relationship in AlN/6H-SiC epitaxial system*, Phys. Rev. Lett., **77**, 1797 (1996).

9. X. L. Ma, N. Shibata, and Y. Ikuhara, *Interface characterization of AlN/TiM/MgO(001) prepared by molecular beam epitaxy*, J. Mater. Res., **14**, 1597-1603 (1999).

10. W. R. L. Lambreecht and B. Segall, *Electronic structure of a ceramic/ceramic interface*: SiC/AlN, Metal-Ceramics Interfaces (Pergamon press) M. Rühle, A. G. Evans, M. F. Ashby, and J. P. Hirth, (Eds.) P29-34 (1990).

11. I. Ivanov, L. Hultmn, K. Järandahl, P. Mårtensson, J. –E. Sundgren, B. Hjörvarsson, and J. E. Greene, *Growth of epitaxial AlN(0001) on Si(111) by reactive magnetron sputter deposition*, J. Appl. Phys., **78**, 5721 (1995).

EFFECT OF SINTERING ATMOSPHERES ON CREEP BEHAVIOR OF DENSE Al₂O₃ CERAMICS

Satoshi Kitaoka, Yorinobu Takigawa, Hideaki Matsubara,
Japan Fine Ceramics Center, 2-4-1 Mutsuno, Atsuta-ku, Nagoya 456-8587, Japan

Hiroki Hara, Hatsuhiko Usami, Junji Sugishita
Meijo Univ., Shiogamaguchi Tempaku-ku, Nagoya 468-8502, Japan

Mineaki Matsumoto, Motohiro Kanno
Univ. of Tokyo, 7-3-1 Hongo, Bunkyo-ku, Tokyo 113-0033, Japan

Nobuo Shinohara, Takashi Higuchi
Chubu Electric Power Co., 20-1 Kitasekiyama, Odaka, Midori-ku, Nagoya 459-8522, Japan

ABSTRACT
The effect of sintering atmosphere such as air, O_2 and vacuum on compressive and tensile creep behavior of dense Al_2O_3 ceramics was investigated. These materials contained gaseous elements, which were mainly attributed to ambient gases trapped during sintering. It was worth noting that the gases evolved from the sample sintered in air were mostly N_2. The total amount of the internal gases decreased as the atmosphere changed from air to O_2 to. Grain size dependence of the gas evolution suggests that the internal gases were predominantly segregated at the grain boundaries. The compressive creep deformation without a cavity formation, where the grain-boundary diffusion or grain-boundary sliding was rate-controlling, was not influenced by the internal gaseous elements. However, the tensile creep deformation obviously depended on the sintering atmospheres. The time to failure decreased in order of the samples sintered in O_2, vacuum and air. The shorter time to failure, the more apparent the tertiary creep region appears. This may be attributed to the promotion of the cavity nucleation by precipitation of the gas constituents into the cavity volume and to the acceleration of cavity growth because of decreasing the angle formed at the junction of the cavity and the grain boundary.

INTRODUCTION
An analytical technique for measuring a slight amount of internal gases in materials has been recently established. This technique detects emissions evolved during fracture and deformation of the materials in an ultra-high vacuum using a

To the extent authorized under the laws of the United States of America, all copyright interests in this publication are the property of The American Ceramic Society. Any duplication, reproduction, or republication of this publication or any part thereof, without the express written consent of The American Ceramic Society or fee paid to the Copyright Clearance Center, is prohibited.

quadrupole mass spectrometer.[1,2] This method has revealed that even fully dense ceramic polycrystals contain a large amount of various gases, which were mainly trapped and/or formed during.[3] The internal gases sometimes lead to serious changes in properties of the materials, such as hydrogen embrittlement in metallic materials,[4] and a reduction of lifetime and luminosity efficiencies in vitreous silica,[5] etc.

Diffusional creep, where stress-directed vacancy diffusion between grain boundaries gives rise to macroscopic deformation, appears to dominate the low-stress deformation of many ceramic polycrystals at high temperatures.[6] It is also well known that a reduction of ductility arises through formation, growth, and interlinkage of microcracks and cavities along the grain boundaries.[7] Therefore, the gas species that dissolve or segregate in the grain boundaries may affect such a high temperature deformation. The creep of fully dense alumina polycrystals, convenient structural materials, has been well-characterized an intrinsic deformation mechanisms have been proposed with great confidence. However, these studies have not been able to clearly identify an interaction of the internal gases.

In the present study, we have focused on the effect of the internal gases on the creep behavior of essentially dense alumina polycrystals.

EXPERIMENTAL PROCEDURES
Specimen Fabrication

The raw powders used in this study were commercial-grade Al_2O_3 (Taimei Chemical Co., TM-DAR). The powders were uniaxially pressed at 25 MPa in air, and then isostatically cold pressed at 300MPa. The green compacts were sintered at 1300° to 1500 °C for half an hour in air, O_2, and vacuum (approx. 1×10^2 Pa). The sintered ceramics were annealed under the same environments as sintering to control the grain size. The relative density of all the samples was over 99%.

Analysis of Internal Gases

Specimens were cut from the pressure-less sintered blocks and ground into 3 mm$\times$4 mm$\times$20 mm bars, where a V-notch (depth of 200 µm) was introduced on the center of the tensile surface to make a constant apparent fracture surface area. The specimens were fractured at room temperature under a background pressure of about 1×10^{-8} Pa by three-point bending to conform to JIS R 1601. The gaseous emissions evolved from the specimens were detected by a quadrupole mass spectrometer and total pressure gauge. Details of this evaluation technique are given elsewhere.[8]

Creep Properties

Compressive creep specimens with average grain sizes of 0.5 and 4 µm, with dimensions close to 5 mm$\times$5 mm$\times$7 mm, were tested in air under constant

Grain Boundary Engineering in Ceramics

stresses in the range of 25-100 MPa at 1200° and 1350 °C, respectively. Since the maximum total strain on each sample was < 4%, the tests could be considered to be constant stress. All of the compressive creep curves exhibited a normal primary creep period, during which the creep rate decreases with strain. This was followed by a steady-state period where the creep rate remained essentially constant with strain. Tensile creep specimens with the average grain size of 4 µm were ground into a gage length of 30 mm and its outside diameter of 3.5 mm. The tensile creep fracture testing was performed in air at 1350 °C under a constant stress of 50 MPa.

RESULTS AND DISCUSSION
Analysis of Internal Gases

Figure 1 shows typical output signals measured by the quadrupole mass spectrometer at masses from the specimens and the corresponding load-displacement curves. The masses are assigned to emissions related to the species of N_2+CO, O_2, Ar and CO_2. The signals detected before and after the fracture are related to background noise corresponding to the lower limit of the accuracy of the measurements. The emission of mass 28, overlapped the signals for N_2 and CO, seems to be attributed mainly to N_2 because the relative intensities of the species, formed by the ionization of N_2 during the measurements, were similar to the N_2 inherent pattern coefficients. The fracture events are accompanied by the gas emissions, which were related mainly to surrounding gases during sintering. For the samples sintered in air, mostly N_2 is evolved at the moment of the fracture. This suggests that oxygen is liberated from the bulk preferentially. For the samples sintered in O_2 and vacuum, N_2 and Ar, constituents of air, are barely detected at the moment of the fracture. Therefore, the gases trapped in the green bodies during molding in air and CIP, except components of the sintering atmosphere, were easily released from the samples during sintering in O_2 and vacuum.

Total pressure change during fracture as a function of average grain size of the samples is given in Fig. 2. The total pressure change corresponds to the total amounts of the gases released mainly from the grain boundaries, because the crack preferentially extended along the grain boundaries for all the samples. For the samples sintered in air, the amount of evolved N_2 increases with an increase in average grain size. Since nitrogen is too large to enter interstitial positions of Al_2O_3, nitrogen is thought to be segregated at defects of the grain boundaries. The grain growth leads to the decrease of the grain boundary density, resulting in the increment of the amount of the segregated nitrogen per unit grain boundary area. If the crack propagates along the grain boundaries of mono-sized and spherical grains with an average diameter d, the total quantity of the segregated nitrogen per

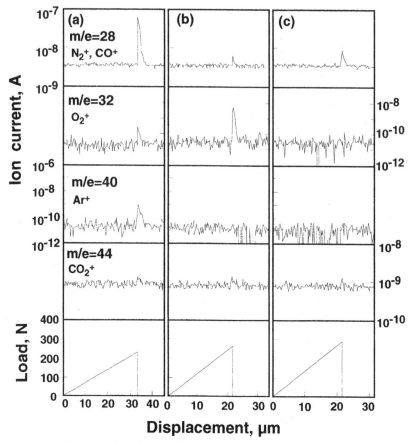

Figure 1. A series of typical output signals from the quadrupole mass spectrometer for the gas emissions and the corresponding load-displacement curves of the samples sintered in (a) air, (b) O_2, and (c) vacuum at 1500°C. The average grain size of these samples is 4 μm.

unit grain boundary area is approx. 1×10^{-7} mol/m². The samples sintered in vacuum contain a slight amount of the gases, not depending on the grain size. For the samples sintered in O_2, the amount of the internal gases increases with the grain size, and then decreases slightly. The solubility and diffusivity of oxygen are believed to be larger than those of nitrogen and Ar, etc.[9] Therefore, oxygen seems to be collected at the grain boundaries during the grain growth, simultaneously, diffusing oxygen outward.

Grain Boundary Engineering in Ceramics

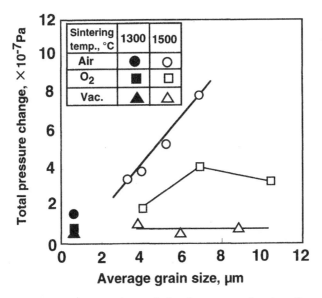

Figure 2. Total pressure change during fracture as a function of average grain size for samples sintered in air, O_2 and vacuum.

Compressive Creep

The cavitation was rarely observed in specimens after testing below the total strain of 4%. Figure 3 shows the steady strain rate as a function of log compressive stress for the samples with the grain size of 0.5 and 4 μm tested at 1200° and 1350°C, respectively. The dashed lines, L1 and L2 indicate the curves calculated by assuming the grain-boundary and lattice-diffusion control of Al^{3+}, respectively.[6] The stress exponents, n, for these creep mechanisms must be unity.[6] The lines L3 are determined by diffusional creep controlled by an interface reaction.[6] The internal gases do not affect the steady-state creep deformation for both grained samples. The n values determined from the strain rate at various applied loads are 2.5 and 1.6 for the finer and coarser-grained samples, respectively. The creep deformation at 1350°C for the larger-grained samples follows the grain boundary diffusion control as reported previously,[6] because the calculated line L1 relatively fits the data. The higher n value is caused by grain growth during the longer time tests under the lower stresses, when the grain size increases by a maximum of 30%. The creep deformation at 1200°C for the finer grained samples deviates largely from all the calculated lines. Therefore, the deformation at 1200°C does not follow any diffusional creep mechanisms, and may be described by grain-boundary sliding with the rapid grain growth during

Grain Boundary Engineering in Ceramics

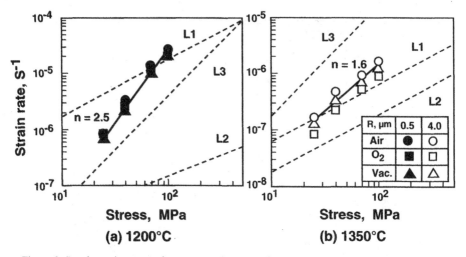

Figure 3. Steady strain rate vs. log compressive stress for samples with the average grain size of 0.5 and 4 µm tested at (a) 1200° and (b) 1350°C, respectively. The dashed lines in the figures indicate the creep controlled by grain-boundary diffusion, L1, lattice-diffusion, L2, or an interface reaction, L3.

the creep tests.

Tensile Creep

Creep curves for the larger-grained samples tested at 1350°C under the tensile stress of 50 MPa are shown in Fig. 4. The tensile creep deformation obviously depends on the sintering atmospheres, unlike the compressive creep behavior. The time to failure decreased in order of the samples sintered in O_2, vacuum and air. The shorter time to failure, the more apparent the tertiary creep region appears.

Figure 5 shows cavities near the fracture surfaces of the samples. The cavities appear only on the grain boundaries, as shown in Fig. 6. The total area fraction A_c of the cavities, which was determined using the SEM images, was 15% for the samples sintered in air and about 10% for those in O_2 and vacuum. The distance L_c between cavities was calculated by dividing the grain boundary length per unit area by the number of cavities per unit area, where the grains were assumed to be mono-sized and spherical with the average diameter d. L_c of the sample sintered in air was the smallest value of 0.7 µm, 1.3 µm for sintering in vacuum, and 1.5 µm for sintering in O_2. It is note that L_c becomes smaller for the specimens with a shorter time to failure. The smaller L_c suggests a larger number of cavity nuclei. Gas formation inside the cavity produces a change in Gibb's free energy, ΔG, associated with the formation of a cavity embryo. If the amount of gaseous elements available for each cavity embryo and the kinetics of gas permeation into the cavity are large enough to maintain an equilibrium gas pressure, p, in the

Grain Boundary Engineering in Ceramics

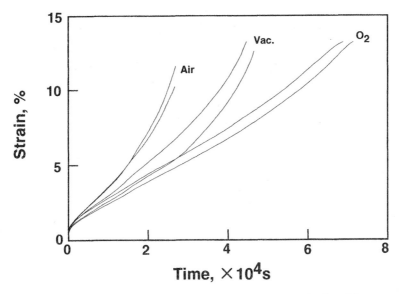

Figure 4. Tensile creep curves for samples with the average grain size of 4 μm tested at 1350°C under the stress of 50 MPa.

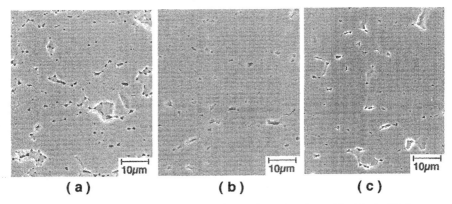

Figure 5. Cavities near the fracture surfaces of samples sintered in (a) air, (b) O_2, and (c) vacuum.

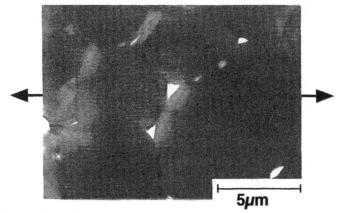

Figure 6. Typical microstructure of sample sintered in air at 1500°C after the creep test at 1350°C under the tensile stress of 50 MPa. The arrows show the tensile direction.

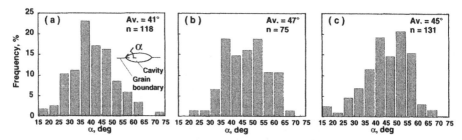

Figure 7. Distribution of angle α formed at the junction of the cavity and the grain boundary in samples sintered at 1500°C in (a) air, (b) O_2 and (c) vacuum.

cavity throughout the nucleation process, the maximum free energy, ΔG_c, for a cavity nucleation can be given as [7]

$$\Delta G_c = \frac{4 F_v \gamma_s^3}{\left(\sigma_n + p\right)^2} \tag{1}$$

where γ_s is the specific surface energy, σ_n is the normal boundary stress, F_v is the geometrical factor that depends on the ratios of the specific surface and interfacial energies. Because a large amount of nitrogen is segregated at the grain-boundaries for the samples sintered in air, the decrease of ΔG_c probably accelerates the nucleation of cavities. Figure 7 illustrates the distribution of angle α formed at the junction of the cavity and the grain boundary. The average angle

Grain Boundary Engineering in Ceramics

becomes larger for the specimens with longer time to failure. The variation of the angle is presumably caused by the difference of the gas species in the cavities affecting the specific surface energy and/or surface diffusion of the matter. Grain boundary diffusion of the matter is thought to be not very influenced by the internal gas species according to the compressive creep behavior shown in Fig. 3(b). Decreasing the angle leads to accelerate the cavity growth,[10] resulting in shorter time to failure.

Although the samples sintered in vacuum include a smaller amount of gaseous elements than those sintered in O_2, the cavity nucleation for sintering in vacuum is more readily because of the smaller L_c. Because the concentration of vacancies for the samples sintered in vacuum is believed to be larger than those for other samples, the supersaturated vacancies may promote the cavity nucleation compared with the samples sintered under the atmospheric pressure.

CONCLUSIONS

(1) The gaseous elements found in the materials were mainly attributed to ambient gases trapped during sintering and were predominantly segregated at the grain boundaries.

(2) The compressive creep deformation with a free cavity formation, where the grain-boundary diffusion or grain-boundary sliding was rate-controlling, was not affected by the internal gaseous elements.

(3) For tensile creep, the time to failure decreased as the sintering atmosphere changed from O_2 to vacuum to air. This may be related to the promotion of the cavity nucleation by precipitation of the gas constituents into the cavity volume and to the acceleration of cavity growth.

REFERENCES

1. J. T. Dickinson, L. C. Jensen and M. R. McKay, "The Emission of Atoms and Molecules accompanying Fracture of Single-crystal MgO," *J. Vac. Sci. Technol.*, **A4** [3], 1648-1652 (1986).
2. M. Kanno, H. Okada and G. Itoh, " Detection of Gasses Evolved from Metallic Materials during Deformation," *J. Jpn. Inst. Metals*, **59** [3] 296-302 (1995).
3. M. Matsumoto, H. Okada, M. Kanno, S. Kitaoka and H. Matsubara, "Gas Evolution Behavior during Fracture of Alumina," *Scripta Materialia*, **37** [12] 2047-2051 (1997).
4. R. E. Stoltz and A. J. West, "*Hydrogen Effects in Metals*," pp 541-553 Ed. by I. M. Bernstein and A. W. Thompson., The Metallurgical Society of AIME, Warrendale, (1981).

5. Y. Morimoto, T. Igarashi, H. Sugahara and S. Nasu, " Analysis of Gas Release from Vitreous Silica," *J.Non-Cryst.Sol.*, **139,** 35-46 (1992).

6. R. M. Cannon, W. H. Rhodes and A. H. Heuer, "Plastic Deformation of Fine-Grained Alumina (Al_2O_3): I, Interface-Controlled Diffusional Creep," *J. Am. Ceram. Soc.*, **63,** 46-53 (1980).

7. M. H. Yoo and H. Trinkaus, "Crack and Cavity Nucleation at Interface during Creep," *Metall. Trans.*, **14A,** 547-561 (1983).

8. S. Kitaoka, H. Matsubara, H. Kawamoto, H. Okada, M. Matsumoto and M. Kanno, "Detection of Gases Generated from Si_3N_4-Based Ceramics during the Fracture Test under Ultra-High Vacuum," *J. Ceram. Soc. Jpn.,* **105,** 915-917 (1997).

9. Y. -K. Paek, K. -Y. Eun and S. -J. L. Kang, "Effect of Sintering Atmosphere on Densification of MgO-Doped Al_2O_3," *J. Am. Ceram. Soc.*, **71,** C380-C382 (1988).

10. T. Chuang, K. I. Kagawa, J. R. Rice and L. B. Sills, "Non-Equilibrium Models for Diffusive Cavitation of Grain Interfaces," *Acta Met.,* **27,** 265-284 (1979).

SOLITONS IN THE VICINITY OF A MARTENSITIC PHASE TRANSITION

V.V. Kiseliev

Institute of Metal Physics, Urals Branch of the Academy of Sciences, Ekaterinburg 620219, Russian Federation

ABSTRACT

The low-amplitude two-dimensional solitons are predicted and analytically described, which can be interpreted as precursors of a martensitic transition. The conditions for formation of solitons are found and the experimental manifestations of solitons are discussed.

INTRODUCTION

In many types of crystals, the transformation from a phase with a cubic lattice (austenite) to a phase with a lattice of the orthorhombic or tetragonal symmetry (martensite) is observed with a decrease in temperature, as an outcome of a cooperative displacement of atoms. The martensitic transitions, related phenomena and processes have a considerable influence on physical properties of construction materials. The martensitic transitions are difficult to describe analytically. In this work, nonlinear excitations in the vicinity of a martensitic phase transition are considered in the context of the Ginsburg - Landau approach, based on the nonlinear theory of elasticity. We used a two-dimensional model [1] that is a good approximation for the $O_h - D_{4h}$ cubic-tetragonal transition in a number of real systems ($In_{0.76}Tl_{0.24}$, $Fe_{0.72}Pd_{0.28}$, Nb_3Sn, V_3Si). These systems have the common property that the atom displacements of quadrate/rectangle type ($4\,mm/2\,mm$) take place in a separate crystallographic plane.

Let's indicate the basic relations of a model. Assume that r_i is the position of a material point in a deformed medium, x_i is its position in the undeformed state, an $u_i = r_i(\mathbf{x},t) - x_i$ is the vector of a displacement. In the classical nonlinear theory of elasticity, the energy is taken as a function of the Lagrangian strain tensor:

$$\eta_{i,j} = \frac{1}{2}\left[u_{i,j} + u_{j,i} + u_{r,i}u_{r,j}\right], \quad u_{i,j} = \frac{\partial u_i}{\partial x_j} \ . \tag{1}$$

The energy for the $4\,mm/2\,mm$ phase transition has the following form [1]:

To the extent authorized under the laws of the United States of America, all copyright interests in this publication are the property of The American Ceramic Society. Any duplication, reproduction, or republication of this publication or any part thereof, without the express written consent of The American Ceramic Society or fee paid to the Copyright Clearance Center, is prohibited.

$$w = \frac{1}{2}A_1 e_1^2 + \frac{1}{2}A_2 e_2^2 + \frac{1}{2}A_3 e_3^2 + \frac{1}{4}B_2 e_2^4 + \frac{1}{6}C_2 e_2^6 + \frac{1}{2}d_1\left(e_{1,1}^2 + e_{1,2}^2\right) +$$

$$+ \frac{1}{2}d_2\left(e_{2,1}^2 + e_{2,2}^2\right) + \frac{1}{2}d_3\left(e_{3,1}^2 + e_{3,2}^2\right) + d_4\left(e_{1,1}e_{2,1} - e_{1,2}e_{2,2}\right) +$$

$$+ d_5\left(e_{1,1}e_{3,2} + e_{1,2}e_{3,1}\right) + d_6\left(e_{2,1}e_{3,2} - e_{2,2}e_{3,1}\right). \tag{2}$$

Here $A_i, B_2, C_2, d_j \; (i = 1,2,3; \; j = 1,2...6)$ are the elastic modules. The fields $e_1 = \left(\eta_{1,1} + \eta_{2,2}\right)/\sqrt{2}$ and $e_3 = \eta_{1,2}$ describe a dilatation and a shear strain, respectively. The field $e_2 = \left(\eta_{1,1} - \eta_{2,2}\right)/\sqrt{2}$ describes the extension of a square into a rectangle and, therefore, in Landau theory it represents the order parameter for the martensitic transition. In a context of the Landau theory the modulus A_2 is temperature dependent, so that A_2 is equal to zero in the transition point. As it usually is, the temperature dependence of other elastic modules is disregarded.

The dynamical equations can be derived by variation of the Lagrange function L (Eq. 3) over the fields u_i :

$$L = \int dx_1\, dx_2 \left[\frac{\rho}{2}\partial_t u_i \partial_t u_i - w\right]. \tag{3}$$

Here ρ is the mass density in an undeformed state. Further we assume $\rho = constant$.

The phase diagrams and hysteresis of the martensitic transitions were qualitatively described previously within the framework of such an approach. However, the inconsistency exists in the works [1,2] at the theoretical description of boundaries and centers of new phases. On the one hand, the degrees of the strain tensor up to the sixth order are included in expression for an energy of a system (the problem is essentially nonlinear), and on the other hand, the own nonlinearity of the strain tensor is disregarded for simplification of the problem. We restrict our attention to a simpler task, when there are the low-amplitude nonlinear excitations in the background of the equilibrium state (austenite). In this case it is possible to overcome correctly difficulties connected to nonlinearity. We show that neglecting nonlinearity of the strain tensor changes significantly the nature of the interaction even for the low-amplitude modes. The effective 2D+1 model describing the interaction of phonon modes has been constructed. Remarkably, the model permits a rich set of exact solutions, which demonstrate formation of precursors of a new phase in a vicinity of the martensitic transition.

NONLINEAR DYNAMICS OF THE PHONON MODES

The energy minimum conditions $\partial w / \partial u_{i,j}^{(0)} = 0$ result in the $4mm$ phase with the square lattice ($u_{i,j}^{(0)} = 0$) and two variants of the $2mm$ phase with the rectangular lattice ($u_{i,j}^{(0)} \neq 0$). In the present work we consider nonlinear excitations against the background of the $4mm$ phase (austenite). Nonlinear excitations on the background martensite have also been investigated [4].

Grain Boundary Engineering in Ceramics

The phase with the square lattice is metastable for $0 < \tau \equiv 16C_2A_2/3B_2^2 < 1$ and it is thermodynamically stable in the interval $1 < \tau < 4/3$ [1].

Let's consider propagation of the nonlinear elastic wave along a fixed direction in the (x_1, x_2) plane when its field change in the transverse direction is smaller. To describe such waves it is convenient to go over from x_1, x_2 to the new variables:

$$\xi = x_1 \cos\varphi + x_2 \sin\varphi, \quad \eta = -x_1 \sin\varphi + x_2 \cos\varphi$$

Let ξ be the coordinate along the direction of the wave propagation. Then, the η - dependence of the field variables is weak.

To develop the effective equations for low-amplitude nonlinear excitations against the background of the ground state (4mm phase), we write out the dynamical equations for the deviations $\bar{u}_i$ of the displacement fields from equilibrium state up to terms quadratic in amplitudes of the deviations:

$$\rho \partial_t \mathbf{V} + \mathbf{D}(\nabla) : \mathbf{V} + \mathbf{E}(\varphi)\partial_\xi : \partial_\xi \mathbf{V} \circ \partial_\xi \mathbf{V} = 0. \tag{4}$$

Here $\mathbf{V} = (\bar{u}_1, \bar{u}_2)$, the differential operators $\mathbf{D}(\nabla)$ and coefficients $\mathbf{E}(\varphi)$ have the cumbersome form, but can be easily calculated from (3). As the η-dependence of the field variables is weak, we have kept the η - derivatives in linear terms of the Eq. (4) only, and in linear terms even we have neglected the derivative of the fourth order with respect to η. We go over from the fields $\bar{u}_1$ and $\bar{u}_2$ to the normal modes R and Q:

$$\mathbf{S}(\nabla)\begin{pmatrix} \bar{u}_1 \\ \bar{u}_2 \end{pmatrix} = \begin{pmatrix} R \\ Q \end{pmatrix}; \quad \begin{pmatrix} \bar{u}_1 \\ \bar{u}_2 \end{pmatrix} = \mathbf{S}^{-1}(\nabla)\begin{pmatrix} R \\ Q \end{pmatrix}. \tag{5}$$

The operator $\mathbf{S}(\nabla)$ is chosen so that it provides diagonal form for Eq. (4) linearized near the equilibrium state ($\mathbf{V} = 0$):

$$\rho \partial_t^2 R + \rho \omega_1^2 (i\partial_\xi, i\partial_\eta)R + \ldots = 0, \quad \rho \partial_t^2 Q + \rho \omega_2^2 (i\partial_\xi, i\partial_\eta) + \ldots = 0. \tag{6}$$

The dispersion laws of linear modes $\left(u_i \sim \exp i[k_1\xi + k_2\eta - \omega_i(k_1, k_2)t], k_1 \gg k_2\right)$:

$$\rho \omega_i^2 (k_1, k_2) = \alpha_i^{(20)} k_1^2 - \alpha_i^{(11)} k_1 k_2 + \alpha_i^{(02)} k_2^2 + \alpha_i^{(40)} k_1^4, \quad i = 1, 2 \tag{7}$$

determine the explicit form of the differential operators $\rho \omega_i^2 (i\partial_\xi, i\partial_\eta)$ in Eq. (6). The explicit forms of the parameters $\alpha_i^{(mn)}$ depending on elastic modules and angle φ have been found in [3]. When one of the spectral branches is excited ($R \gg Q$), we can neglect the interactions with other branch modes. To transform the nonlinear terms in the dynamical equations in the long-wavelength limit it is sufficient to use the expression for $S(\nabla)$, which is independent of the derivatives. As a result, we constructed the effective wave equation for the dominant spectral branch

$$\rho \partial_t^2 R + \rho \omega_1^2 (i\partial_\xi, i\partial_\eta) R - g_1 \partial_\xi^2 R^2 = 0.$$
(8)

The parameter g_1 depends on elastic modules, angle φ and was found in [3,4]. The equation for Q has the similar form at condition $Q \gg R$. In the subsequent discussion we shall consider equation (8). The subscripts of the parameters g_1, $\alpha_1^{(ik)}$ and so on will be omitted.

If we consider waves propagating in one side along a direction (110), then $\alpha^{(11)} = 0$, $R = R(x, \eta, t)$, $x = \xi + st$, $\left[\rho \partial_t^2 - \alpha^{(20)} \partial_\xi^2 \right] R \cong 2\rho s \partial_t \partial_x R$ and equation (8) is reduced to the integrable Kadomtsev- Petviashvili (KP) model:

$$\partial_x \left[2\rho s \partial_t R + \alpha^{(40)} \partial_x^3 R - g \partial_x R^2 \right] = \alpha^{(02)} \partial_\eta^2 R,$$
(9)

where $s = \sqrt{\alpha^{(20)}/\rho}$ is the sound velocity.

For one-dimensional motion, when the η-dependence can be neglected, equation (8) is transformed into the Korteweg–de Vries (KdV) equation with a quadratic nonlinearity. It is important that there will be other results, when the geometric nonlinearity in the expression for the strain tensor η_{ij} is neglected as in Refs. [1,2]. Namely, such an approximation gives the modified KdV equation with a stronger cubic nonlinearity, which is of secondary importance for the low-amplitude excitations.

TWO-DIMENSIONAL SOLITONS AS PRECURSORS
OF THE PHASE TRANSITION

We found that the Eq. (8) has a Backlund transformation: If u_0 is some solution of Eq. (8) and the function φ satisfies the equation

$$\left[\rho D_t^2 + \rho \omega^2 (iD_\xi, iD_\eta) - 2gu_0 D_\xi^2 \right] \varphi \cdot \varphi = 0,$$
(10)

then $u_1 = -6\alpha^{(40)} g^{-1} \partial_\xi^2 \ln \varphi + u_0$ will be a solution of the equation (8) also. Here, $D_t f \cdot g = (\partial_t - \partial_{t'}) f(t) f(t') \big|_{t=t'}$ etc. The bilinear form (10) allows obtaining the soliton-like solutions of Eq. (8) by the Hirota method. In particular, we have found the multi-soliton exponential solutions [3,4], that describe the elastic pair collisions of the "planar" solitons of the following type:

Grain Boundary Engineering in Ceramics

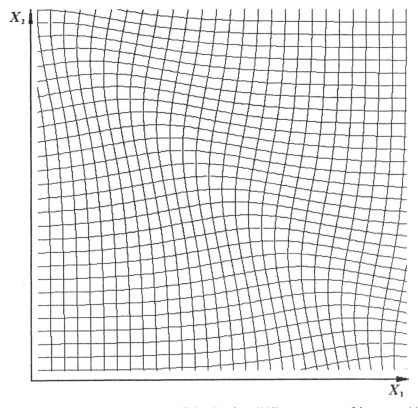

Figure 1. The "planar" soliton (11) parallel to the plane (110) – a precursor of the martensitic transition, $\varphi = \pi/4, \alpha^{(40)}g > 0$.

$$R^{(0)} = -\frac{6\alpha^{(40)}d^2}{g\,\text{ch}^2\theta}, \quad \theta = d(\xi + \text{v}t), \quad d^2 = \frac{\alpha^{(20)} - \rho\text{v}^2}{4\alpha^{(40)}} > 0. \quad (11)$$

Here v is the real-valued parameter (the soliton velocity).
It is interesting that as a principal approach we obtain

$$e_1 \approx 0, \quad e_3 \approx 0, \quad e_2 \neq 0. \quad (12)$$

for the "planar" solitons propagating in the directions [110] on the background of the 4 mm phase. Inasmuch as the field e_2 is a parameter of the order of the theory and e_2 has a nonzero value inside the soliton, it is evident that inside the soliton the square lattice is transformed into the rectangular lattice (Fig. 1). Although the rectangular lattice corresponds to a new phase, the phase

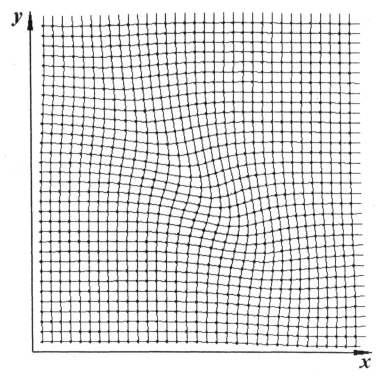

Figure 2. Deformations of the square lattice for the "cigar-shaped" soliton (13) parallel to the plane (110), $\varphi = \pi/4$, $\alpha^{(40)}g > 0$.

transition does not happen in the system yet, since the soliton amplitudes are small. However, a formation of the inhomogeneous interior structure of the ground state with deformations typical for other phase testify to emerging in a system of the tendencies to phase transition. It is the beginning of formation of centers of a new phase. These solitons can be interpreted as the low-amplitude twin patches parallel to the planes (110) or remaining tracks of the low-temperature 2 mm phase (martensite) against the background of the basic 4 mm phase (austenite). Quasi-periodic finite-gap solutions of the KP model (9) correspond to the low-amplitude twin bands. The features of a new phase are less bright for the interior structure of solitons formed along other directions. It is interesting that the twins and the tetragonal modulations parallel to planes (110) are observed experimentally on the background of austenite in the vicinity of the martensitic transition (see references in [1]).

We have found not only the exponential solutions of the equation (8) but also polynomial ones. The simple polynomial solution is the "cigar-shaped" soliton:

$$R = \frac{12\alpha^{(40)}\left[A^2(\xi + vt)^2 - (AB - 2C^2)\eta^2 + 2CA\eta(\xi + vt) - A\right]}{g\left[1 + A(\xi + vt)^2 + B\eta^2 + 2C\eta(\xi + vt)\right]^2},$$

$$A = \left[4\alpha^{(02)}(\alpha^{(20)} - \rho v^2) - (\alpha^{(11)})^2 \right] \left[12\alpha^{(40)}\alpha^{(02)} \right]^{-1},$$

$$B = (\alpha^{(20)} - \rho v^2)\left[\alpha^{(02)} \right]^{-1} A, \quad C = \alpha^{(11)} A \left[2\alpha^{(02)} \right]^{-1}, \quad A, B > 0, \ AB - C^2 > 0. \tag{13}$$

The two-dimensional soliton (13) (Fig. 2) is spatially localized ($R = O(r^{-2})$ at $r \to \infty$). It can be motionless ($v = 0$).

Excitation of the predicted solitons is possible because of the action of pulse like forces, which arise locally at deformation and on the grain boundaries undergoing a local phase transition. Besides, solitons may be created as a result of the energy interchange between different phonon branches and instability developing in the phonon system.

STABILITY OF SOLITONS

It is very important for the development of a possible soliton scenario to answer the question of the stability of the solitons. In the KP model (9) the stability or instability of solitons depends on the sign of the combination $\alpha^{(40)}\alpha^{(02)}$. It is established that for $\alpha^{(40)}\alpha^{(02)} < 0$ the "planar" solitons of type (11) are stable against two-dimensional perturbations. A nontrivial picture occurs for $\alpha^{(40)}\alpha^{(02)} > 0$. In this case the "planar" solitons of the KP model are unstable; however, the "cigar-shaped" solitons of type (13) exist, which are stable against two-dimensional perturbations. The nonlinear stage of instability developing for the "planar" solitons of the KP model was investigated in [5,6]. It was established [5] that depletion of the "planar" soliton arises from excitation on its front of small oscillations, which propagate faster then soliton. They carry away a part of the energy. The emitted energy is dispersed in space further. The alternative variant was found in the work [6]. As a result of a decay of the "planar" soliton, chains of the "cigar-shaped" solitons and a new "planar" soliton with the smaller amplitude can be formed. The energy does not disperse; the condensation of energy into new soliton-like structures takes place.

The stability conditions of solitons for the more general Eq. (8) must be modified. We obtain the instability criterion for two-dimensional perturbations of the "planar" soliton of type (11). We shall find the solution of the perturbed Eq. (8) in the form:

$$R = R^{(0)}\left[\theta + \alpha(Y_i, T_i) \right] + \sum_{n=1}^{\infty} \varepsilon^n R^{(n)}\left[\theta + \alpha(Y_i, T_i), Y_i, T_i \right]. \tag{14}$$

Here, ε is a small parameter characterizing the relation of diffraction and nonlinearity, $Y_i = \varepsilon^i \eta$, $T_i = \varepsilon^i t$ are the slow variables, i is natural numbers. Inserting (14) into (8) and grouping of the same powers ε we obtain a string of the linked equations. The solvability condition for these equations determines the evolution of α in an approximation of a geometric optics:

$$\gamma_1 \frac{\partial^2 \alpha}{\partial T_1^2} - \gamma_2 \frac{\partial^2 \alpha}{\partial Y_2^2} - \gamma_3 \frac{\partial^2 \alpha}{\partial Y_1 \partial T_1} = 0, \quad \gamma_1 = \rho(\alpha^{(20)} - 4\rho v^2),$$

$$\gamma_2 = \alpha^{(02)}(\alpha^{(20)} - \rho v^2) + \frac{3}{4}(\alpha^{(11)})^2, \quad \gamma_3 = 3v\rho\alpha^{(11)}. \tag{15}$$

The instability criterion is evident from Eq. (15):

$$\gamma_3^2 + 4\gamma_1\gamma_2 < 0. \tag{16}$$

In particular, with the conditions $\alpha^{(11)} = 0$, $\alpha^{(20)} > 0$, $\alpha^{(40)} > 0$ soliton (11) can be stable in the region of small velocities $0 < v^2 < \alpha^{(20)}/4\rho$ when $\alpha^{(02)} > 0$ and in the region of high velocities $\alpha^{(20)}/4\rho < v^2 < \alpha^{(20)}/\rho$ when $\alpha^{(02)} < 0$ only.

The stability threshold can be reached by a soliton because of the temperature nonhomogeneity or local stress. A stress field reduces the value of the elastic modules and influences the stability of solitons in a similar way as temperature does. As the instability evolves, solitons can emit phonons. It gives new insight into the problem of an anomalous acoustic emission in the vicinity of the martensitic phase transitions.

CONCLUSION

The effective 2D+1 nonlinear model for the martensitic ferro-elastic phase transition has been constructed. The model is a close approximation for a number of real crystals. Multidimensional solitons have been predicted and described analytically. Solitons can be interpreted as precursors of the phase transformation. As the amplitudes of solitons are small, the phase transition is not realized. However, the tendency for the transition arises in the system because the deformations inside the solitons are typical for a new phase. We have found that multi-solitons are formed along the preferred crystallographic directions. Twin patches and the tetragonal modulations of austenite along these directions were observed experimentally. The regions of existence and stability of the various types of solitons are pointed out. The soliton stability threshold can be reached due to temperature inhomogeneities and a local stress on the grain boundaries. The new soliton mechanism for anomalous sonic emission in the vicinity of the martensitic transitions has been suggested: as the soliton instability evolves, solitons can emit phonons.

REFERENCES

1. G.R. Barsh and J.A. Krumhansl, "Nonlinear and Nonlocal Continuum Model of Transformation Precursors in Martensites", *Metall. Trans. A*, **19** 761-775 (1988).
2. F. Falk, "Ginzburg-Landau Theory of Static Domain Walls in Shape-memory Alloys", *Z. Phys. B: Condensed Matter.* **51** [2] 177-187 (1983).
3. V.V. Kiseliev, "Weakly- nonlinear Dynamics and Solitons in the Vicinity of a Martensitic Phase Transition", *Phys. Lett. A*, **196** 97-100 (1994).
4. V.V. Kiseliev, "Weakly-nonlinear Soliton-like Excitations in a Two-dimensional Model of a Martensitic Transition", *Sov. Phys. Solid State*, **36** [11] 3321-3331 (1994).
5. V.E. Zakharov, "Instability and the Nonlinear Oscillations of Solitons", *Sov. Phys. JETP Lett,* **22** [7] 364-367 (1975).
6. D.E. Pelinovsky and Yu.A. Stepaniants, "The Self-focusing Instability of the Planar Solitons and Chains of Two-dimensional Solitons in the Media's with a Positive Dispersion", *Sov. Phys. JETP*, **104** [4] 3387-3400 (1993).

KEYWORD AND AUTHOR INDEX

Grain Boundary Engineering in Ceramics

Printed and bound by CPI Group (UK) Ltd, Croydon, CR0 4YY

16/04/2025

14658452-0005